Signal Processing

IEEE Press
445 Hoes Lane
Piscataway, NJ 08854

Signal Processing

An Applied Decomposition Approach

James Vincent Candy
Lawrence Livermore National Laboratory
University of California, Santa Barbara, USA

IEEE Press
WILEY

Published by John Wiley & Sons, Inc., Hoboken, New Jersey.
Published simultaneously in Canada.

For general information on our other products and services or for technical support, please contact our Customer Care Department within the United States at (800) 762-2974, outside the United States at (317) 572-3993 or fax (317) 572-4002.

Wiley also publishes its books in a variety of electronic formats. Some content that appears in print may not be available in electronic formats. For more information about Wiley products, visit our web site at www.wiley.com.

Library of Congress Cataloging-in-Publication Data applied for:

Hardback ISBN: 9781394207442

Cover Design: Wiley
Cover Image: © Patricia Candy

Set in 9.5/12.5pt STIXTwoText by Straive, Chennai, India

"In Christ and through his blood we have redemption and forgiveness for our sins" (Ephesians: 1:7).

Contents

About the Author

James Vincent Candy is the Chief Scientist for Engineering, a Distinguished Member of the Technical Staff and founder/former Director of the Center for Advanced Signal & Image Sciences (CASIS) at the Lawrence Livermore National Laboratory. Educationally, he received his BSEE degree from the University of Cincinnati and his MSE and PhD degrees in Electrical Engineering from the University of Florida. He is a registered professional Control System Engineer in the state of California. He is an Adjunct Full-Professor at the University of California, Santa Barbara. Dr. Candy is a Life-Fellow of the IEEE "for contributions to model-based ocean acoustic signal processing" and a 25Year-Fellow of the Acoustical Society of America (ASA) "for contributions to model-based acoustic signal processing." He was elected as a Life Member (Fellow) at the University of Cambridge (Clare Hall College). He is a member of Eta Kappa Nu and Phi Kappa Phi honorary societies. Dr. Candy received the IEEE Distinguished Technical Achievement Award for the "development of model-based signal processing in ocean acoustics." He is an IEEE Distinguished Lecturer for oceanic signal processing. Dr. Candy has been awarded the Interdisciplinary Helmholtz-Rayleigh Silver Medal in Signal Processing/Underwater Acoustics by the Acoustical Society of America. He received the R&D100 award for his innovative invention in radiation threat detection. He has published over 250 journal articles, book chapters, and technical reports as well as written six texts in signal processing, "Signal Processing: the Model-Based Approach," (McGraw-Hill, 1986) and "Signal Processing: the Modern Approach," (McGraw-Hill, 1988), "Model-Based Signal Processing," (Wiley/IEEE Press, 2006), "Bayesian Signal Processing: Classical, Modern and Particle Filtering" (Wiley/IEEE Press, 2009), "Bayesian Signal Processing: Classical, Modern and Particle Filtering, 2nd Ed." (Wiley/IEEE Press, 2016), and "Model-Based Processing: An Applied Subspace Identification Approach" (Wiley/IEEE Press, 2019). He was the IEEE Chair of the Technical Committee on "Signal and Image Processing" and was the Chair of the ASA Technical Committee on "Signal Processing in Acoustics." He was elected to the Administrative Committee of IEEE OES. His research interests include Bayesian learning, signal decomposition, estimation, identification, spatial estimation, signal and image processing, array signal processing, nonlinear signal processing, structural signal processing, tomography, sonar/radar processing and biomedical applications.

Dr. Candy is considered a subject matter expert in signal processing, estimation and detection with a career extending over 48 years. He is an IEEE Distinguished Lecturer and has presented over 50 short courses in statistical signal processing at universities, conferences, businesses, laboratories, etc.

Preface

This text is focused on the decomposition of complex signals into independent sub-signals or equivalently components enabling subsequent analysis and design. Decomposing a signal into its constituent components not only enables an effective mechanism for analysis, but also provides a means to process each individual component more effectively, while mitigating disturbances and noise [1–4]. Decomposition techniques have evolved historically as the Fourier decomposition of a signal and/or system into sinusoidal components for subsequent frequency domain analysis, to statistical decompositions in classifications, to modal decompositions in structural analysis, to wavelet decompositions providing a novel domain for subsequent analysis of variable signals [1, 5–7]. This text is focused on such a decomposition essentially decoupling signals and their underlying systems into more manageable independent sub-signals or subsystems enabling effective analysis and subsequent designs [8]. What motivates this approach are the large number of available numerical methods capable of performing such a decomposition in a robust and efficient manner [9, 10]. Perhaps the most prominent method is the singular value decomposition (SVD), well known for its robust methodology of detecting the rank of a matrix that has evolved as the "go-to" solution ranging from eigen-decompositions to provide reduced rank filtering of noisy data [9–15]. This method has been applied to many problems with great success ranging from the usual tracking problem in radar/sonar, to decomposing images into subimages in optical processing, to decomposing transient speech signals in voice recognition, to medical application in cardiac as well as neurological signals [16–18]. In any case this method is the premier workhorse in signal processing applications that will carefully be interwoven throughout each approach developed within this text.

For instance, acoustical response data encompasses a wide variety of signal classes ranging from sharp transients similar to seismic events or speech to muffled sinusoids in oceans or structural vibrations [6, 17]. Distinct resonant frequency peaks evolving from the measured material vibrations of resonant ultrasound, to step-frequency excitations of vibrating systems enable the evaluation of structural properties critical for the detection of potential failures [19, 20]. Control systems have often depended on somehow decomposing the underlying complex system into more manageable subsystems decoupled for both analysis and design of effective controllers [10, 21]. The response of a large bridge or building with complex and coupled substructures also depends on their decomposition into independent subsystems with their superposed responses producing the resulting measurement data. Biological signals also exhibit this complexity while they evolve from the well-understood heart beat (EKG) to the complex evoked responses of the neurological process (EEG) [18]. In all of these applications, the analysis, design, detection, and information extraction depend heavily on one significant connection–the decomposition of the signal or system into independent subsystems that can easily be evaluated and modified to achieve a specific objective (estimation, detection, control, modification, etc.) [22]. This decomposition is well known mathematically based on simple linear algebraic techniques such as diagonalizing a complex matrix by performing an eigen-decomposition [9, 11]. For mechanical systems producing complicated vibrational responses, this decomposition is called a modal decomposition where the independent modes of the system determined its measured response, while in controls, the overall plant or system is decomposed mathematically into a decoupled set of subsystems for both centralized or decentralized designs [23–25]. This decomposition usually evolves again in the form of an eigen-decomposition or the more robust SVD revealing the underlying modes dominating the composite system signal or frequency response enabling the extraction of prominent features for such applications as estimation, identification, detection, and classification [22, 26, 27]. In any case, the underlying thread that is woven throughout the "signal space" is that of the modal description of the measured data in one form or another [28]. Mechanical (control) engineers have long known about these decomposition approaches as well as signal processors in terms of modal decompositions in which a system is decomposed for analysis and design and then reconstructed through transformations (similarity) to return to the physical coordinates governing its operation [23, 25, 29].

Perhaps the mathematical notions of orthogonality and independence are two concepts that imply "decomposition" especially in the signal processing jargon. The application of techniques to decompose a signal in terms of its individual constituent components enables not only a better understanding both from a mathematical perspective, but also from a physical perspective for analysis and design [4, 5, 30–32]. It is this perspective that can lead to the extraction of crucial information directly from uncertain measured data sets–the primary objective of signal decomposition [33]. Applying this approach leads to immediate and effective as well as significantly robust techniques with a multitude of applications in tracking, estimation, detection, classification, and many other applications [26, 27, 29]. In fact, the classification of various signals depends heavily on the ability to rank the importance of features and particularly their inherent independence as in a multitude of pattern recognition applications [27]. The detection problem is quite challenging in a highly reverberant environment, but can be attacked more readily when signal decomposition techniques are applied leading to the design of decentralized rather than centralized detection solutions [34].

The essential engine that drives the solution to the signal decomposition problem is the SVD of the underlying "data matrix" that evolves directly from measurements or simulation [35, 36]. This matrix appears in a variety of forms relating the "data" directly to the underlying physical system creating it. For example, pure measurement data is produced by sensors through an acquisition system directly or can take the form of correlation estimates, frequency spectra, wavelet estimates, image pixels, and power spectra [14–18]. These data are typically placed into the data matrix that is processed by the algorithm designed to extract the desired information. The matrix can be arranged into various structures depending on the generating physical system or model. For instance, in system identification, that is, constructing a parametric model from noisy data, a Hankel or Toeplitz matrix or both are operated on by the SVD to extract the underlying model, while spectral data can be arranged into a frequency response matrix enabling the extraction of resonant frequencies [36, 37]. In any case, it is the application of the SVD or its counterparts to these matrices that enable the signal decomposition to be successfully performed [37].

Processing structural system data is characterized not just by the usual uncertainties created by environmental disturbances and noise, but also by their underlying complexities resulting from multichannel measurements and coupled responses [6]. State-space techniques have demonstrated the capability of unraveling these complexities due to their physics-based designs while mitigating these uncertainties primarily because of their robust performance all related to one primary linear algebraic methodology–the matrix SVD [17, 36]. This approach has successfully been applied to solving a wide variety of signal processing problems ranging from simple signal enhancement [4, 17, 36], to multichannel spectral estimation and array signal processing [22], anomaly detection and to even the more challenging problem of extracting a coupled structural response from noisy vibrational measurements [26]. The SVD provides the essential engine that drives these state-space processors especially when extracting the structural response of an uncertain vibrational system. In this text, the decomposition approach to signal processing is motivated by the decomposition of response data from a so-called "structural system model" [25]. Such a system is typically a civil engineered structure like a building, bridge, or machinery or even decomposing structured material all induced to vibrate revealing their internal structural properties in the form of modal responses [17, 20]. Similar to modal control systems of system theory, the modes are individual features of the structure [8, 21, 23, 25]. It is the task of the signal processor or the structural engineer or the analyst to extract this critical information from a vibrating structure to design a stable, efficient structure. Much effort has been accomplished in "operational modal analysis" (OMA) to provide a mechanical system decomposed (mathematically) into a decoupled set of subsystems for decentralized control [6, 17, 23–25]. A wealth of techniques has been developed to perform the extraction task for exciting structures, while the design task has not been explained as strongly, but evolves from the control systems perspective [25]. The processing perspective of this text is to focus on the extraction of modal information to analyze complex systems providing the necessary information to evaluate its performance. This approach can lead directly to the evaluation of other mechanical systems such as motors, automobiles, and self-driving vehicles that could potentially fail due to the unforeseen or unpredictable response to vibrations, disturbances, and noise [22, 26].

The objective of this text is aimed at simplifying the analysis of complex signals (systems) by performing a signal decomposition into constituent components that combine to capture the uncertain measured response(s). This approach, although well documented, has not been sufficiently provided to the signal processing and other technical communities because it is somewhat veiled by the complexity of the underlying mathematics along with complex applications. This material is strongly motivated, mathematically, by the modal decomposition methodology evolving from control systems and more specifically the structural and civil engineering technical communities incorporating signal processing applications produced within [6, 17, 36]. It has been this pigeon-holed effect that has confined this information and preventing a more general approach to be made available to the overall technical communities. This model-based decomposition approach is presented primarily as an advanced undergraduate or first-year graduate text with applications and data in both the time and frequency domain. These data and subsequent algorithms (in MATLAB®) have not been made readily available by the communities in a single text, while some attempts have been made but are unfortunately embedded within specialized application areas (structures, materials, etc.) [6, 16, 17, 20]. In this text, the model-based signal decomposition approach is presented that is directly motivated by deterministic as well as random signals leading to fundamental transformation techniques that are more easily comprehended within this framework.

The model-based perspective taken in this text is to introduce the concept of signal decomposition by demonstrating how each of the major approaches decomposes a complex (composite) signal or system into a set of its independent constituent components. Starting with the well-known Fourier decomposition followed by the SVD applied to both time series (signals) and frequency spectra and subsequently the statistical approach based on principal/independent component theory are incorporated to elucidate the concepts of dimensionality reduction as well as simplification and processing leading to a full exposure of this approach [27]. In the introduction, a simple sinusoidal system is simply developed to elucidate the approach initially through its Fourier decomposition followed by the SVD representation terminating with the underlying principal component decomposition. This system, although simple, is rich enough to motivate the entire text demonstrating not only how a signal can be decomposed effectively, but also a system into subsystems leading to eventual analysis and design.

This proposed text is designed primarily as a graduate-level text; however, it will prove useful to practicing signal processing professionals and scientists, since a wide variety of case studies are included to demonstrate the applicability of the decomposition approach to real-world problems. The prerequisite for such a text is a melding of undergraduate work in linear algebra (especially matrix decomposition methods), random processes, linear systems, and some basic digital signal processing. It is somewhat unique in the sense that many texts cover some of its topics in piecemeal fashion, but limit exposure to frequency domain approaches especially without the notion of signal/spectral decompositions. The underlying approach of this text is the thread that is embedded throughout in the algorithms, examples, applications, and case studies. It is an embedded model-based theme, together with the developed hierarchy of physics-based models, that contributes to its uniqueness coupled with SVD methods that even enable potential real-time methods to become a reality. This text has evolved from five previous texts and has been broadened by a wealth of practical applications to real-world problems [27, 34, 36–53]. The introduction of robust SVD methods for model-building and signal/system decomposition that have been available in the literature for quite a while, but require more of a systems theoretical background to comprehend. After introducing these fundamental concepts in the first chapter, random signals and systems are briefly discussed leading to the concept of spectral estimation. Once developed, the parametric estimators lead directly into the concept of subspaces and the decomposition of the data space into the signal and noise subspaces evolving to the well-known multiple signal classification (MUSIC) spectral estimator [28, 36].

The third chapter is focused on representing a signal in a nonparametric as well as parametric constructs. Nonparametrically, the usual time-series representation of a signal can take the structure of a set of coefficients or weights with specific properties (e.g. orthogonality) or simply a polynomial-in-time, while its frequency domain representation offers a variety of structures ranging from the usual Fourier decomposition to the statistical approach of component theory. Random signals can be converted or transformed to a deterministic form using correlation and power spectra replacing time and frequency domain characterizations.

Parametric representations offer an even larger variety of constructs ranging from the fundamental transfer function along with its statistical characterization to capture noise and uncertainties with the autoregressive family along with related exponential/sinusoidal models leading to the decomposed modal model. Parametric physics-based models evolve that are captured by state-space models that are easily transformed to any of the previous representations as well as enabling both single to multichannel characterizations. Signal decomposition approaches, the heart of this text, are developed in the fourth chapter. Here, starting with the so-called "eigen-signals" that are based on statistical projection theory and shown to easily be extracted from both deterministic and random data where the basics are developed. The usual Fourier modal and principal component decompositions are also briefly discussed to offer some viable alternatives to the fundamental approach completing the chapter. Chapter 5 provides the heart of signal/system decomposition methods starting with data-based methods followed by spectral-based techniques and concluded with model-based decomposition.

Chapter 6 develops the "subspace" approach for generic state-space systems with the underlying system theoretical properties that enable the application of many algorithms to capture physics-based data [36]. It is here that both the nonparametric and parametric techniques are married to show equivalence and compare performance. Modal examples are discussed along with actual application data.

The sixth chapter provides the technical breath of this text in the time domain, briefly developing the time domain and frequency domain approaches to processing data. It is here that signal decomposition techniques are carefully developed and applied to both simulation and application data sets. Starting with the usual preprocessing of data based on a priori knowledge, both noise and disturbances are mitigated to improve the overall signal-to-noise ratio (SNR). Two of the most popular and robust techniques, N4SID and MOESP, are carefully developed based on projection theory and shown how the SVD plays a significant role in their implementation. The well-known prediction error method (PEM) is also briefly discussed to illustrate the usual approach to decomposing the signal as well.

Next, in chapter seven not as well known, frequency domain techniques are developed starting with the simplest basic frequency domain (BFD) method of peak-picking the estimated frequency response function (FRF) data and evolving to the more sophisticated least squares-based methods completing the chapter. Both the least-squares frequency domain (LSFD) method as well as the polyreference (P-LSFD) approaches to resonant frequency estimation are developed leading to the sophisticated iterative maximum likelihood method (ML-LSFD). Here, the complex mode indicator function (CMIF) is developed providing a viable technique to extract modal resonant frequencies from measured spectra–again based on the SVD methodology to decompose the FRF matrix.

Chapter eight focuses on the performance analysis of the signal decomposition techniques based on statistical methods and many of the techniques inherited from modal analysis methodologies. This performance analysis metric coupled with the well-known modal assurance criterion (MAC) enables a complete extraction of a decoupled signal and underlying system.

Finally, the text is completed with a set of real-world applications and simulations briefly developed to reinforce the signal decomposition approach developed throughout the text. The applications include: the response evolving from rocket flight dynamics, a simple vibrating cylinder (pipe) experiment, followed by material property extraction using resonant ultrasound spectroscopy, to the analysis of an 8-story (mass) structure and its subsystem decomposition, concluding with the time-reversal decomposition of a spatiotemporal signal for flaw detection in nondestructive evaluation.

Appendices are included for critical review as well as problem sets and notes for the *MATLAB* software used in the signal processing/controls/identification areas at the end of each chapter.

12 August 2022
Danville, CA

James Vincent Candy

References

1 Oppenheim, A. and Schafer, R. (1989). *Discrete-Time Signal Processing*. New Jersey: Prentice Hall.

2 Hayes, M. (1996). *Statistical Signal Processing and Modeling*. New Jersey: Wiley.

3 Vaccaro, R. (1991). *SVD and Signal Processing II: Algorithms, Analysis and Applications*. New York: Elsevier.

4 Cadzow, J., Baseghi, B., and Hsu, T. (1983). Singular-value decomposition approach to time series modeling. *IEE Proc.* 130 (Pt. F, 3): 203–210.

5 Scharf, L.L. (1991). The SVD and reduced-rank signal processing. In: *SVD and Signal Processing, II: Algorithms, Analysis and Applications* (ed. R.J. Vaccaro). New York: Elsevier.

6 Rainieri, C. and Fabbrocino, G. (2014). *Operational Modal Analysis of Civil Engineering Structures*. New York: Springer.

7 Mallet, S. (2009). *A Wavelet Tour of Signal Processing*. New York: Elsevier.

8 Moore, B.C. (1979). Singular value analysis of linear systems. In: *Proceedings of IEEE Conference on Decision and Control*, 66–73.

9 Golub, G. and Van Loan, C.F. (1984). *Matrix Computations*. Baltimore, MD: Johns Hopkins University Press.

10 Klemma, V. and Laub, A. (1980). The singular value decomposition: its computation and some applications. *IEEE Trans. Autom. Control* AC-25: 164–176.
11 Eckart, C. and Young, G. (1936). The approximation one matrix by another of lower rank. *Psychometrika* 1: 211–218.
12 Kung, S. (1978). A new identification and model reduction algorithm via singular value decomposition. In: *12th Asilomar Conference on Circuits, Systems and Computers*, 305–314. Asilomar, CA.
13 Scharf, L. (1987). Rank reduction for modeling stationary signals. *IEEE Trans. Acoust. Speech Signal Process.* ASSP-35 (3): 350–355.
14 Tufts, D., Kumaresan, R., and Kirsteins, I. (1982). Data adaptive signal estimation by singular value decomposition of a data matrix. *Proc. IEEE* 70: 684–685.
15 Tufts, D. and Kumaresan, R. (1982). Singular value decomposition and frequency estimation by linear prediction. *IEEE Trans. Acoust., Speech, Signal Process.* ASSP-30 (4): 671–675.
16 Bienvenu, G. and Kopp, L. (1981). Adaptive high resolution spatial discrimination of passive sources. In: (ed. L. Bjorno), —Underwater Acoustics and Signal Processing-L. Dordrecht, The Netherlands: D. Reidel.
17 Reynders, E. (2012). System identification methods for (operational) modal analysis: review and comparison. *Arch. Comput. Methods Eng.* 19 (1): 51–124.
18 Vandewalle, J. and De Moor, B. (1988). A variety of applications of singular value decomposition in identification and signal processing. In: *SVD and Signal Processing Algorithms, Applications and Architectures* (ed. E. Deprettere), 43–91. North Holland: Elsevier.
19 Lebedev, A. (2002). Method of linear prediction in the ultrasonic spectroscopy of rock. *Acoust. Phys.* 48 (3): 339–346.
20 Migliori, A. and Sarrao, J. (1997). *Resonant Ultrasound Spectroscopy: Applications to Physics, Materials, Measurements, Nondestructive Evaluation.* New York: Wiley.
21 Moore, B. (1981). Principal component analysis in linear systems: controllability, observability and model reduction. *IEEE Trans. Autom. Control* AC-26: 17–22.
22 Candy, J., Franco, S., Ruggiero, E. et al. (2017). Anomaly detection for a vibrating structure: a subspace identification/tracking approach. *J. Acoust. Soc. Am.* 142 (2): 680–696.
23 Juang, J. and Phan, M. (2001). *Identification and Control of Mechanical Systems*. Cambridge: Cambridge University Press.
24 Meirovitch, L. (1990). *Dynamics and Control of Structures*. New York: Wiley.
25 Gawronski, W. (2004). *Advanced Structural Dynamics and Active Control of Structures*. London: Springer.

26 Candy, J., Case, J., Fisher, K. et al. (2020). Vibrational processing of a dynamic structural flight system: a multichannel spectral estimation approach. *J. Acoust. Soc. Am.* 147 (4): 2694–2712.

27 Duda, R., Hart, P., and Stork, D. (2001). *Pattern Classification*, 2e. New Jersey: Wiley.

28 Schmidt, R. (1986). Multiple emitter location and signal processing estimation. *IEEE Trans. Antennas Propag.* 34 (3): 276–280.

29 Johnson, D. and Dudgeon, D. (1993). *Array Signal Processing: Concepts and Techniques*. New Jersey: Prentice Hall.

30 Cadzow, J. (1987). Analysis of transient data in noise. *IEE Proc.* 134 (Pt. F, 1): 69–78.

31 Van Der Veen, A., Deprettere, E.D.F., and Swindlehurst, A. (1993). Subspace-based signal analysis using singular value decomposition. *Proc. IEEE* 81 (9): 1277–1308.

32 Rao, B. and Arun, K. (1992). Model-based processing of signals: a state-space approach. *Proc. IEEE* 80 (2): 283–309.

33 Roy, R., Paulraj, A., and Kailath, T. (1986). ESPIRIT-a subspace approach to estimation of parameters of cisoids in noise. *IEEE Acoust. Speech Signal Process.* ASSP-34 (5): 1340–1342.

34 Candy, J. (2016). *Bayesian Signal Processing: Classical, Modern and Particle Filtering Methods*, 2e. New Jersey: Wiley.

35 Cadzow, J. (1983). Singular-value decomposition approach to time series modelling. 130 (Pt. F, 3): IEE Proc., 202–210.

36 Candy, J. (2019). *Model-Based Processing: An Applied Subspace Identification Approach*. New Jersey: Wiley.

37 Xu, G. and Kailath, T. (1994). Fast subspace decomposition. *IEEE Trans. Signal Process.* 42 (3): 539–551.

38 Ljung, L. (1987). *System Identification -Theory for the User*. Englewood Cliffs, NJ: Prentice Hall.

39 Soderstrom, T. and Stoica, P. (1989). *System Identification*. Englewood Cliffs, NJ: Prentice Hall.

40 Ljung, L. and Soderstrom, T. (1983). *Theory and Practice of Recursive Identification*. Boston, MA: MIT Press.

41 Haykin, S. (1991). *Adaptive Filter Theory*. Englewood Cliffs, NJ: Prentice Hall.

42 Ho, B.L. and Kalman, R.E. (1966). Effective construction of linear, state-variable models from input/output functions. *Regelungstechnik* 14: 545–548.

43 H. P. Zeiger and A. J. McEwen, "Approximate linear realizations of given dimension via Ho's algorithm," *IEEE Trans. Automat. Control*, vol. AC-19, p. 153, Apr. 1974.

44 Kailath, T. (1980). *Linear Systems*. Englewood Cliffs, NJ: Prentice Hall.

45 Rugh, W.J. (1992). *Linear System Theory. A Graduate Course.* Englewood Cliffs, NJ: Prentice Hall.

46 Kalman, R.E., Falb, P.L., and Arbib, M.A. (1970). *Topics in Mathematical System Theory*, International Series in Pure and Applied Mathematics. New York: McGraw-Hill.

47 Lawson, C.L. and Hanson, R.J. (1974). *Solving Least-Squares Problems.* Englewood Cliffs, NJ: Prentice Hall.

48 Rao, C.R. (1965). The Use and Interpretation of Principal Component Analysis in Applied Research. Sankhya, Tech. Rep. *9*.

49 Pratt, W. (1978). *Digital Image Processing.* New York: Wiley-Interscience.

50 Owlsley, N. (1981). Modal decomposition of data adaptive spectral estimates. Proceedings of the Workshop on Applications of Adaptive Systems Theory Yale University, New Haven, CT.

51 Candy, J. (2006). *Model-Based Signal Processing.* New Jersey: Wiley/IEEE Press.

52 Candy, J. (1986). *Signal Processing: The Model-Based Approach.* New York: McGraw-Hill.

53 Candy, J. (1988). *Signal Processing: The Modern Approach.* New York: McGraw-Hill.

Acknowledgments

My wife, Patricia, is the major motivational element needed to undertake this endeavor through her support and encouragement. My family, extended family, and friends having endured the many regrets, but still offer encouragement in spite of all of my excuses. Of course, the constant support of my great colleagues and friends, especially Mr. B. Beauchamp, Dr. K. Fisher, and Dr. S. Lehman who carefully reviewed the manuscript and suggested many improvements, cannot go without a acknowledgment.

Glossary

AIC	Akaike information criterion
AR	autoregressive (model)
ARMA	autoregressive moving average (model)
ARMAX	autoregressive moving average exogenous input (model)
ARX	autoregressive exogenous input (model)
BFD	basic frequency domain (method)
BSP	Bayesian signal processing
BW	bandwidth
CC-T	correlation coefficient statistical test
CD	central difference
CDF	cumulative distribution function
CDM	common denominator model
CM	conditional mean
CMIF	complex mode indicator function
CRLB	Cramér–Rao lower bound
C-Sq	Chi-squared (distribution or test)
CT	continuous-time
DPSS	discrete prolate spheroid sequences
DExp	damped exponential decomposition
DFT	discrete Fourier transform
DOF	degrees of freedom
DtFT	discrete-time Fourier transform
EMA	experimental modal analysis
FDD	frequency domain decomposition (method)
FFT	fast Fourier transform
FIR	finite impulse response
FPE	final prediction error
FRF	frequency response function
GLRT	generalized likelihood-ratio test

G-M	Gaussian mixture
GM	Gauss Markov
G–S	Gram–Schmidt (orthogonal decomposition)
HD	Hellinger distance
ICA	independent component analysis
IDtFT	inverse discrete-time Fourier transform
IFFT	inverse fast Fourier transform
i.i.d.	independent-identically distributed (samples)
IIR	infinite impulse response
IRF	impulse response function
KD	Kullback divergence
KL	Kullback Leibler
KLD	Kullback–Leibler divergence
LD	lower-diagonal (matrix) decomposition
LE	Lyapunov equation
LKF	linear Kalman filter
LMFD	left matrix fraction description (system)
LMS	least mean square
LR	lower residual (terms)
LS	least-squares
LSCD	least-squares common denominator (method)
LSFD	least-squares frequency domain (method)
LSCF	least-squares complex frequency (method)
LTI	linear time-invariant (system)
MA	moving average (model)
MAC	modal assurance criterion
MAD	median absolute deviation (statistic)
MAICE	minimum Akaike information criterion
MAP	maximum a posteriori
MATLAB®	mathematical software package
MBID	model-based identification
MBP	model-based processor
MBSP	model-based signal processing
MC	Monte Carlo
MCK	mass damper spring (system)
MDL	minimum description length
MDOF	multiple degree of freedom (system)
MIMO	multiple input/multiple output (system)
MinE	minimum probability-of-error
MNORM	minimum norm (spectral estimator)
ML	maximum likelihood

MOC	modal observability criterion
ML-LSCF	maximum likelihood polyreference least-squares complex frequency (method)
MOESP	multivariable output error state-space algorithm
MMSE	minimum mean-squared error
MSE	mean-squared error
MSV	modal singular value criterion
MTM	multiple taper method (spectral estimator)
MV	minimum variance
MVDR	minimum variance distortionless response (spectral estimator)
MUSIC	multiple signal classification (spectral estimator)
N4SID	numerical algorithm for subspace state-space system identification
NMSE	normalized mean-squared error
OMA	operational modal analysis
PCA	principal component analysis
PHD	Pisarenko harmonic decomposition (spectral estimator)
pLSCF	polyreference least-squares complex frequency (method)
PDF	probability density function (continuous)
PEM	prediction error method
PID	proportional integral derivative (control)
PI-MOESP	past-input multivariable output error state-space algorithm
PMF	probability mass function (discrete)
PO-MOESP	past-output multivariable output error state-space algorithm
PSD	power spectral density
RC	resistor capacitor (circuit)
RLC	resistor–inductor–capacitor (circuit)
RLS	recursive least-squares
RMFD	right matrix fraction description (system)
RMS	root mean squared
RMSE	root minimum mean-squared error
RPE	recursive prediction error
RPEM	recursive prediction error method
RUS	resonant ultrasound spectroscopy
SDIAG	stabilization diagram
SDOF	single degree of freedom (system)
SID	subspace identification
SISO	single input/single output (system)
SNR	signal-to-noise ratio
SSA	singular spectral analysis
SSP	state-space processor decomposition

SSQE	sum-squared error
SVD	singular value (matrix) decomposition
UD	upper-diagonal matrix decomposition
UR	upper residual (terms)
WPM	Welch periodogram method
WSSR	weighted sum-squared residual statistical test
W-Test	whiteness statistical test
Z	Z-transform
Z-M	zero-mean statistical test

About the Companion Website

This book is accompanied by a companion website:

www.wiley.com/go/CandySignProc

The website includes:

- Instructor Solutions Manual

1 Introduction

1.1 Background

When we are confronted by a problem, simple or complex, we tend to decompose it in order to achieve a solution. For instance, if the lights go out in our room, we first ask ourselves, if the switch is still on; if on, we next search for a fuse box and if unsuccessful, we proceed to question whether the actual power source is available. In more complex scenario, if our car stops unexpectedly while driving, we first check the fuel gauge, then the battery, and then search various components of the engine based on telltale sounds that may have occurred. These are problems that occur in everyday life and exemplify the concept of decomposition. Engineers and scientists "decompose" complex signal/system problems into separate components in order to comprehend their superposition or overall performance. Thus, the understanding of signal components or subsystems is necessary in order to analyze or modify complex signals or systems. In this text, we develop techniques that enable us to perform this decomposition in order to simplify the signal/system into more comprehensible components. For example, in chemistry, when presented with an unknown substance to be analyzed, a chemist will perform the analysis by applying mass spectrometry to decompose it into constituent components. In criminal forensics, investigators decompose the crime scene to determine the cause, weapons, to piece together a scenario that will lead to the identification of the perpetrators. In energy science, scientists decompose nuclear reactions into constituent components to achieve higher energy yields. Control/structural engineers intensely study second-order system performance that can be used to evaluate complex systems like motors or bridges or building responses based on decomposing them into these simpler subsystem responses. Given noisy, uncertain measurement data, signal processors perform routine spectral analysis to reveal constituent components in order to separate signals from the noise. Applications like extracting the pregnant mother's heart beats from those of the fetus are achieved by decomposition techniques. Complex

Signal Processing: An Applied Decomposition Approach, First Edition. James Vincent Candy.

Companion website: www.wiley.com/go/CandySignProc

detection problems can also be solved by techniques such as decentralized detector design by decomposing the problem into local detectors reporting to a central processor. All-in-all, the decomposition approach to problem-solving from the simple to the complex is routine, often going unnoticed, in everyday life as well as in sophisticated scientific engineering applications. This text presents "decomposition" from the signal/system perspective.

We must first decide on the essence of the signal itself. Is it deterministic or random? A *deterministic signal* is characterized by its repeatability, that is, it is free from variations and repeatable from measurement to measurement, while if it varies extraneously, then it is no longer repeatable and it is defined simply as a *random signal*. Here we are concerned with the development of processing techniques to extract the pertinent information from random signals using any a priori information available. These techniques are called *signal estimation* with enhancement achieved by applying a particular algorithm called a *signal estimator* or just *estimator* that is annotated by a "∧." We start with a simple example.

Example 1.1 Suppose we are given sinusoidal data, $x(t_k)$, sampled at approximately 0.1 sec depicted in Figure 1.1a, and wish to decompose it by separating the signal, $s(t_k)$, from any disturbances, $d(t_k)$, and noise, $n(t_k)$.

$$x(t_k) = s(t_k) + d(t_k) + n(t_k) \tag{1.1}$$

That is, we would like to analyze its components that represent the desired signal to be extracted along with undesirable disturbances to be removed along with inherent measurement (numerical) noise. The first step a signal processor would pursue is to perform an analysis of the frequency spectrum as shown in Figure 1.1b. The signal is assumed to be deterministic (nonrandom) ignoring any uncertainties. Performing a discrete Fourier transform (*DFT*) on the data reveals spectral peaks corresponding to the sinusoids,

$$\mathcal{F}(\Omega_m) = \sum_{k=1}^{N} x(t_k) e^{-j\Omega_m t_k} \quad \text{for} \quad \Omega_m := \left(\frac{2\pi}{N}\right) \times m \tag{1.2}$$

It is clear from the spectrum that resonances exist at three temporal frequencies of 0.1225, 0.3228, and 0.496 Hz. If we were analytically aware (a priori) that the signal was composed of three sinusoids

$$s(t_k) = A_1 \cos(\omega_1 t_k + \phi_1) + A_2 \cos(\omega_2 t_k + \phi_2) + A_3 \cos(\omega_3 t_k + \phi_3) \quad \text{for} \quad \omega := 2\pi f \tag{1.3}$$

the sinusoidal transform relation assuming zero phases ($\phi_k \approx 0$) is given by

$$S(\omega) = \sum_{k=1}^{3} A_k \left[\delta(\omega - \omega_k) + \delta(\omega + \omega_k)\right] \quad \text{for} \quad \omega_k := 2\pi f_k \tag{1.4}$$

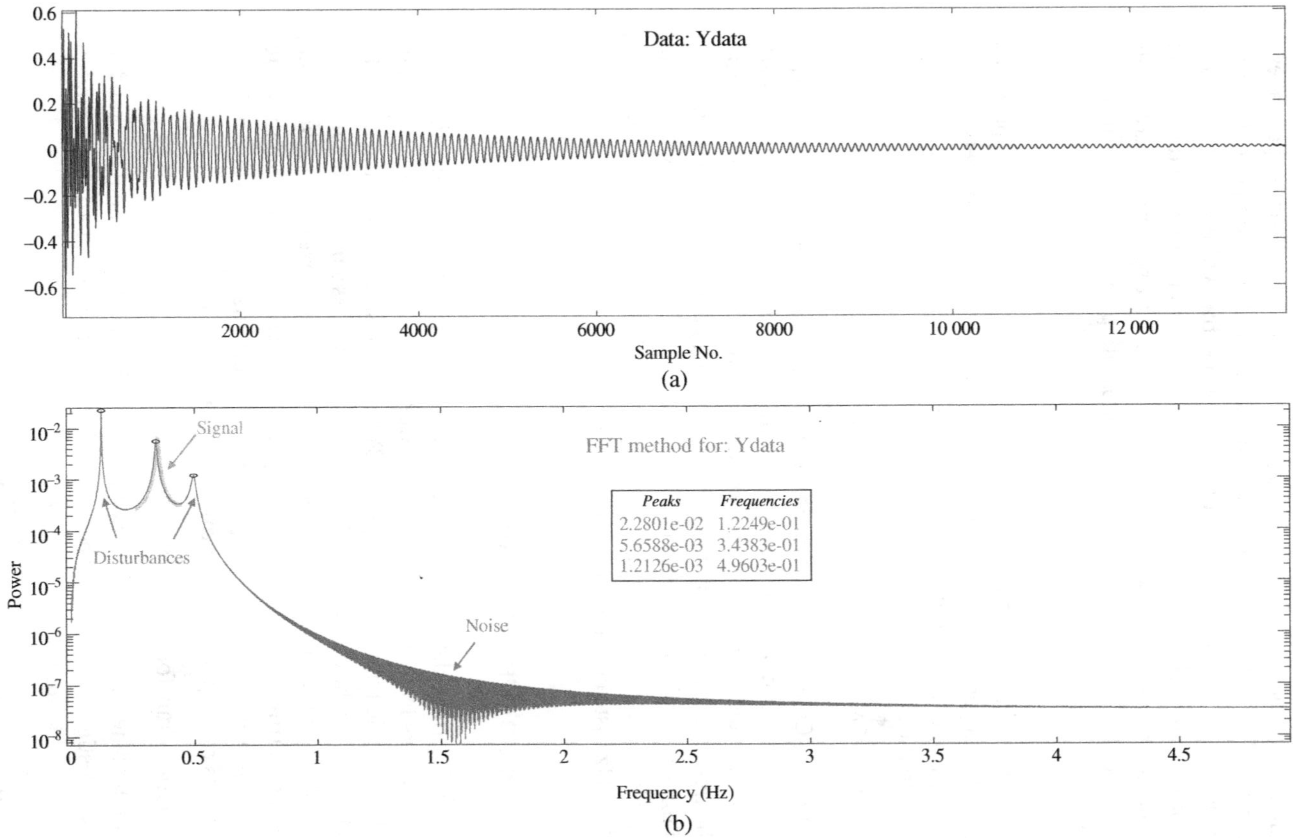

Figure 1.1 *Deterministic* sinusoidal data: (a) composite signal consisting of three sinusoids at frequencies of 0.1225, 0.3228, and 0.496 Hz, zero phase at a large *SNR* (200 dB). (b) Composite Fourier spectrum illustrating three sinusoidal peaks consisting of two disturbances and the actual signal (thick line) along with numerical noise.

and we expect to observe perfect spectral lines (impulses) located at these specified frequencies. However, since there is leakage in the discrete Fourier spectral estimation created by truncating the data abruptly, a small DC-component appears creating a smearing of the peaks that result in a spectral width yet still providing correct peak frequencies. So we see ideally that the Fourier transform decomposes the composite signal of superposed sinusoids into the three spectral peaks enabling us to analyze its composition and consider separating them into its constituent components.

△△△

Next, the signal processor may know (a priori) from the underlying physics of the problem that generated this data that the actual signal is located between 0.15 and 0.4 Hz, while a pair of disturbances, $d(t_k)$ caused by local signals lie outside that band, that is,

$$0.15\,\text{Hz} < s(t_k) < 0.40\,\text{Hz} \quad \text{and} \quad 0.15\,\text{Hz} > d(t_k) \quad \text{or} \quad d(t_k) > 0.4\,\text{Hz}$$

Example 1.2 Consider the same sinusoidal data of Example 1.1 contaminated by instrumentation noise and disturbances at a 0 dB signal-to-noise ratio (*SNR*). The noisy data and corresponding spectrum are shown in Figure 1.2a,b. Clearly from the figure, the corresponding spectral (Fourier) data still indicate the presence of three sinusoidal peaks as before; however, the measurement instrumentation noise has distorted their locations somewhat at frequencies of 0.1219, 0.3450, and 0.4960 Hz instead of the true frequencies.

△△△

1.2 Spectral Decomposition

Simply treating these data as stochastic requires an improved spectral estimator to extract the peaks leading to the basis of power spectral density (*PSD*) estimation rather than simple Fourier (*FFT*) spectra. Perhaps the simplest spectral estimator follows by treating the data as a random signal. Techniques similar to deterministic theory hold when the random signal is transformed to its *correlation sequence* and its Fourier spectrum is transformed to its *power spectrum*. We know that the correlation sequence and power spectrum form a *DFT* pair, analogous to a deterministic signal and its corresponding spectrum, that is, we have that

$$\text{Fourier Transform:} \qquad R_{xx}(k) \longleftrightarrow S_{xx}(\Omega)$$

and as in the deterministic case, we can analyze the spectral content of a random signal by investigating its power spectrum. The *PSD* function for a discrete random process, $x(t_k)$ is defined as:

$$S_{xx}(\Omega) := \lim_{N\to\infty} E\left\{\frac{X(\Omega)X^*(\Omega)}{2N+1}\right\} \tag{1.5}$$

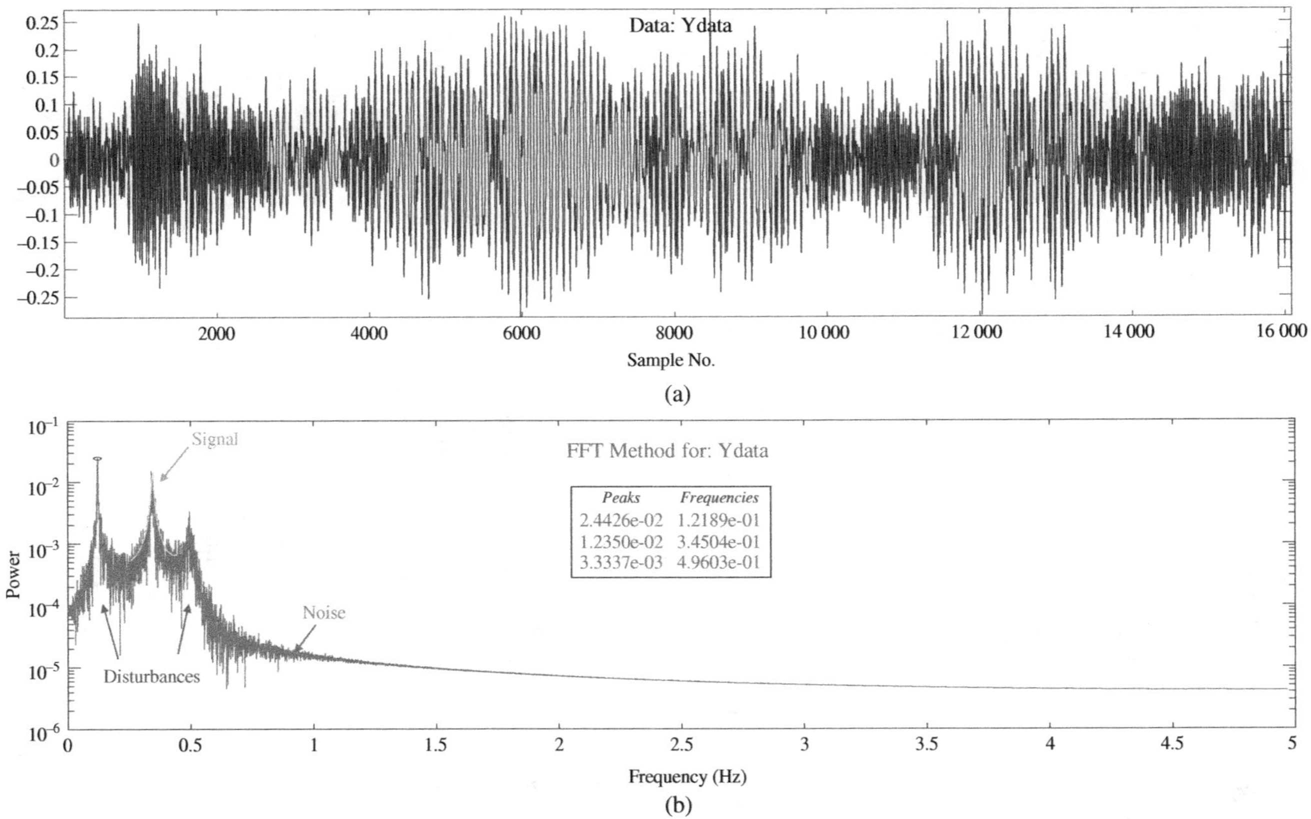

Figure 1.2 *Random* sinusoidal data with disturbances and noise: (a) Composite signal consisting of three sinusoids at frequencies of 0.1225, 0.3228, and 0.4960 Hz. (b) Composite Fourier spectrum illustrating three noisy sinusoidal peaks at frequencies of 0.1219, 0.3450, and 0.4960 Hz consisting of two disturbances and the actual signal (thick line) along with numerical noise.

where * is the complex conjugate. The expected value operation $E\{.\}$ can be thought of as "mitigating" the randomness. Similarly, the *correlation* (at lag k) is

$$R_{xx}(k) := E\{x(t)x(t+k)\} \tag{1.6}$$

Recall that the *DFT pair* is defined by

$$\begin{aligned} X(\Omega_m) &:= \text{DFT}\,[x(t)] = \sum_{t=0}^{M-1} x(t)e^{-j\Omega_m t} \\ x(t) &= \text{IDFT}\,\left[X(\Omega_m)\right] = \frac{1}{M}\sum_{t=0}^{M-1} X(\Omega_m)e^{j\Omega_m t} \end{aligned} \tag{1.7}$$

for $\Omega_m = \left(\frac{2\pi}{M}\right) m$.

Therefore, for $x(t_k)$, a random signal, we have that the correlation-power spectrum pair is (*Wiener–Khintchine theorem* [1])

$$\begin{aligned} \hat{S}_{xx}(\Omega_m) &= DFT\,\left[\hat{R}_{xx}(k)\right] \\ \hat{R}_{xx}(k) &= IDFT\,\left[\hat{S}_{xx}(\Omega_m)\right] \end{aligned} \tag{1.8}$$

With the a priori information available about the signal and disturbance spectra, it is possible for the signal processor to perform a simple extraction by applying bandpass filters illustrating how the uncertain composite signal can be decomposed into its constituent components as previously performed by the Fourier decomposition (*DFT*) of the deterministic sinusoidal data of Figure 1.1. We illustrate this application in the following example.

Example 1.3 Consider the noisy composite data of Example 1.2 with measurements acquired at a 0 dB *SNR* shown in Figure 1.2a. Bandpass filters are designed by the processor to extract the desired signal at 0.345 Hz rejecting disturbances at 0.1225 and 0.4960 Hz. The 256-weight finite impulse response (*FIR*) filters are designed to operate within the respective frequency bands of (0.025–0.15) Hz, (0.2–0.4) Hz, and (0.4–0.5) Hz. The results are shown in Figure 1.3 for each band.

△△△

This example illustrates the fundamental concepts of decomposing a composite noisy measurement into its constituent components using any a priori information to perform a simple decomposition and extraction of the desired signal. Next, we consider more sophisticated methods of achieving such a decomposition.

1.3 Data Decomposition

Data are acquired in a variety of domains depending on the particular application and hardware available. In this text, we will consider the usual time domain

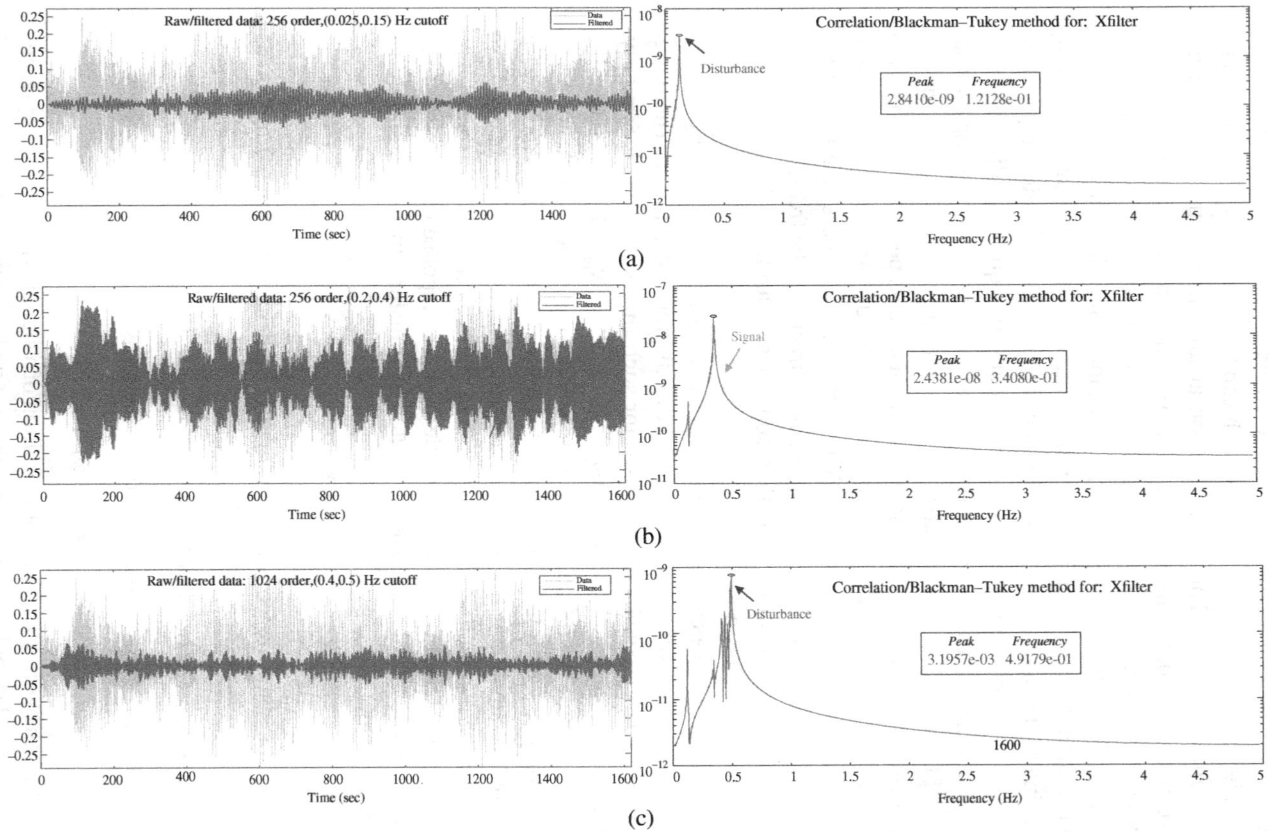

Figure 1.3 Filtering *random* sinusoidal data for decomposition/extraction: (a) Disturbance at 0.1225 Hz separation using (0.025, 0.15) Hz bandpass filter: signal and power spectrum. (b) Signal at 0.3450 Hz extraction using (0.2, 0.4) Hz bandpass filter: signal and power spectrum. (c) Disturbance at 0.4960 Hz separation using (0.4, 0.5) Hz bandpass filter: signal and power spectrum.

data (time series), frequency domain data (spectra) along with space–time (spatiotemporal) or space–frequency (frequency–wavenumber) data sets. Once data, either single or multichannel, are acquired, it can be arranged in a variety of data structures for processing. The most common arrangement is the vector-matrix structure, that is, the data can be placed into a so-called data matrix and processed accordingly. Data decomposition methods follow based on the underlying nature of the problem and the domain selected. For instance, vibration responses are usually measured in the time domain using multichannel tri-axial (vector) accelerometers leading to the data matrix, while in a nondestructive evaluation application such as resonant ultrasound, frequency domain data are acquired leading to a complex frequency data matrix for subsequent processing. It is the underlying internal structure of the data matrix that leads to a particular decomposition technique.

For our previous sinusoidal problem, we showed how spectral decomposition and processing (bandpass filtering) led to a simple decomposition and signal extraction technique that is easily applied to analyze and process the noisy time series data. These data can then be transformed directly to the frequency domain or have been measured directly in that domain and arranged in a particular structural matrix for decomposition.

With this in mind, we can define a sequence of time domain data by a set of N_t-samples, $\{x(t)\}; t = 0, 1, \ldots, N_t - 1$ for t the discrete-time index. Therefore, this sequence can be considered an N_t-vector such that $\mathbf{x}(t) \in \mathcal{R}^{N_t \times 1}$. Similarly, the set of frequency domain data $\{X(\omega_k)\}; k = 0, 1, \ldots, N_\omega - 1$ for ω_k, the discrete frequency can be defined by the complex N_ω-vector, with $\mathbf{X}(\omega_k) \in \mathcal{C}^{N_\omega \times 1}$. Spatiotemporal or spatio-frequency data also follow as $\mathbf{x}(t, z)$ and $\mathbf{X}(\omega, \kappa)$ with z, the spatial variable, and κ, the spatial frequency or wavenumber. In any case with the "data vector" defined, a variety of data matrices can be constructed. For instance, if $x(t)$ represents the deterministic impulse response of a system, then the corresponding frequency response vector or transfer function is simply its Fourier transform, that is,

$$\mathbf{x}(t) \Longrightarrow \mathcal{F}(x(t)) \Longrightarrow \mathbf{X}(\omega_k)$$

However, if the data are considered random, then its correlation function is estimated leading to the correlation matrix

$$\mathbf{x}(t) \Longrightarrow E\{\mathbf{x}(t)\mathbf{x}^\dagger(t)\} \Longrightarrow \mathbf{R}_{xx}$$

with corresponding power spectral matrix evolving as

$$\mathbf{X}(\omega) \Longrightarrow E\{\mathbf{X}(\omega)\mathbf{X}^\dagger(\omega)\} \Longrightarrow \mathbf{S}_{xx}(\omega)$$

So we see that the vector-matrix representation of a signal is a common representation in statistical signal processing leading to a wide expanse of techniques and decompositions based on the internal structure of the constructed data matrix.

For example, in realization theory impulse response data are constructed in a Hankel matrix that possesses critical system theoretical properties, while in linear prediction theory data are constructed in Toeplitz matrices dominating algorithm developments [1, 2]. In fact, we note in passing that the important data matrices constructed in signal processing are typically the impulse response matrix, $\mathbf{H}$, the correlation matrix, $\mathbf{R}_{xx}$, the frequency response or transfer function matrix, $\mathbf{F}$—all of which are easily decomposed using numerical factorization methods such as Cholesky decomposition, LU-decomposition, eigen-decompositions, and singular value decompositions [3]. For example,

$$\mathbf{H} = \mathbf{LU}; \quad \mathbf{R}_{xx} = \mathbf{U\Lambda U}^{\dagger}; \quad \mathbf{F} = \mathbf{U}''\mathbf{V}^{\dagger}; \quad \mathbf{S}_{xx} = \mathbf{LDL}^{\dagger}$$

It is these types of decomposition approaches that will be developed throughout this text. Let us return to our sinusoidal example and apply the decomposition approach to illustrate this concept [4, 5].

Example 1.4 Consider the same sinusoidal data of Example 1.1 contaminated by instrumentation noise and disturbances at a 0 dB *SNR*. The noisy data are shown in Figure 1.4a. First, the N_t-data vector $\mathbf{x}(t)$ is used to estimate the correlation matrix, $\mathbf{R}_{xx}$. Next, this matrix is processed by performing an eigen-decomposition such that

$$\mathbf{R}_{xx} = \mathbf{E} \times \mathbf{\Lambda} \times \mathbf{E}^{\dagger}$$

From this decomposition, a subset of the eigenvectors, $\{\mathbf{e}_i\}; i = 1, \ldots, N_s$, are selected and a corresponding power estimator created such that

$$P(\omega) = \frac{1}{\sum_{i=N_s+1}^{N_t} |\mathbf{v}^{\dagger}(\omega)\mathbf{e}_i|}$$

where $\mathbf{v}(\omega) = [1 \; e^{-j\omega} \; e^{-j2\omega} \; \cdots \; e^{-j(N_t-1)\omega}]'$ is a complex vector of discrete frequencies. Applying this decomposition processor to the reduced correlation data results in the spectrum shown in Figure 1.4b. The spectrum is very smooth and results in reasonable frequency (peak) estimates at frequencies of 0.1231, 0.3444, and 0.4973 Hz as annotated in the figure. The power estimator is the 24th-order multiple signal classification (*MUSIC*) harmonic estimation algorithm completing the example and illustrating a data decomposition approach [6].

△△△

Spatiotemporal (space–time) data or wave phenomena occur in a wide variety of applications ranging from the simple dropping of a pebble into a puddle generating spherical waves; to ocean acoustic wave propagation searching for unknown targets; to tracking aircraft/missiles during conflicts; to locating flaws in materials [1, 7–9]. Typically, the problem is to separate or decompose sources or scatterers for analysis, whether they are benign or malignant tumors or friendly/unfriendly

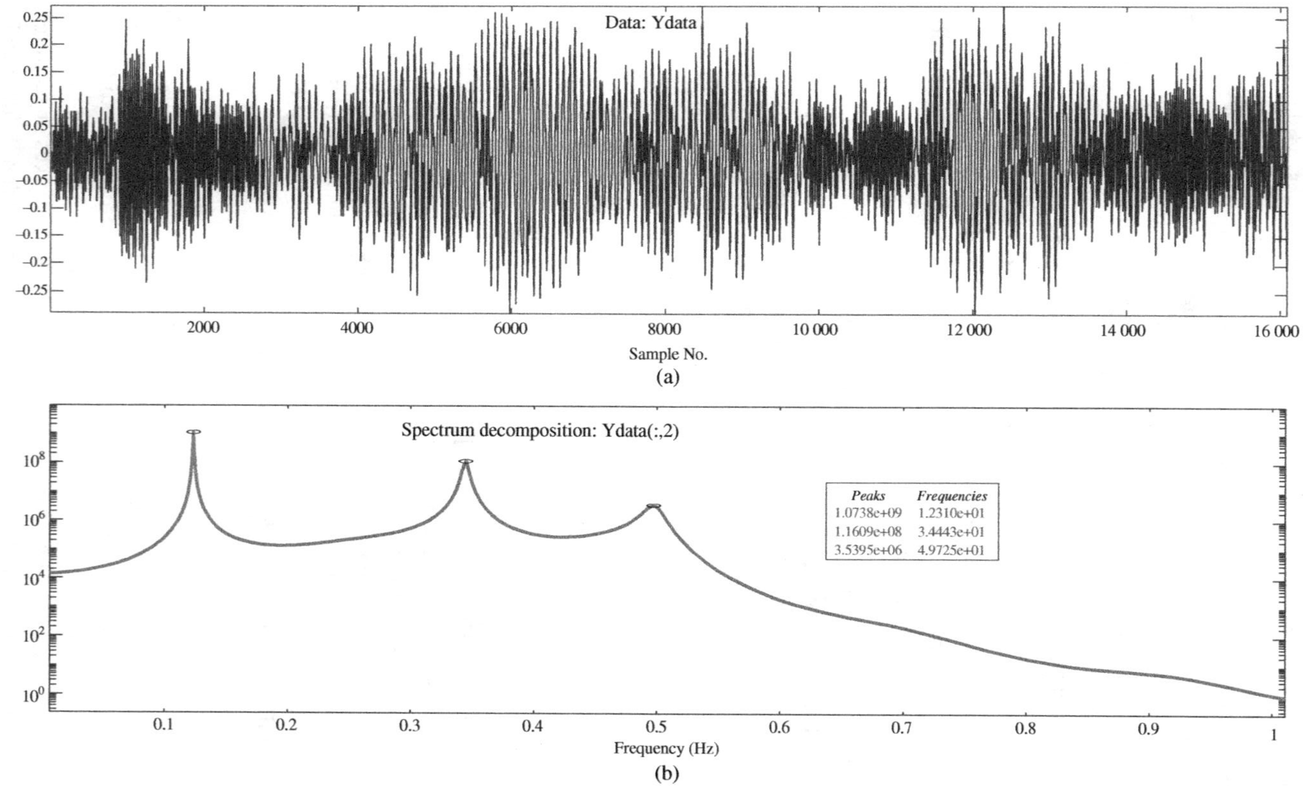

Figure 1.4 Spectral decomposition of noisy (0 dB) sinusoidal data: (a) Composite signal consisting of three sinusoids at frequencies of 0.1225, 0.3228, and 0.4960 Hz. (b) Decomposed power spectrum (*MUSIC*: Order = 24) with three noisy estimated peaks at frequencies of 0.1231, 0.3444, and 0.4973 Hz.

targets [8, 9]. In any case, the goal remains the same—the decomposition of the space-time waves into individual scattering components. The following example illustrates how spatiotemporal processing can achieve such a decomposition.

A narrowband, *spatiotemporal wave signal* can be defined by an L-element sensor array measurement model contaminated by random noise as [8]

$$\mathbf{y}(t) = D(\kappa)\mathbf{s}(t) + \eta(t) \tag{1.9}$$

where $\mathbf{y}, \eta \in C^{L\times 1}$ are the array measurement and noise vectors, $\mathbf{s} \in C^{N_s\times 1}$ is the signal vector and $D \in C^{L\times N_s}$ direction matrix that can be defined uniquely for planar or spherical wavefronts, and κ_m the corresponding spatial frequency or wavenumber ($\kappa_o = \frac{\omega_o}{c} = \frac{2\pi}{\lambda_o}$) for c the propagation speed.

The wavenumber is a vector specifying the direction-of-arrival (*DOA*) of a wavefront as it impinges on the multichannel measurement array at an angle of incidence, θ_o, which the direction vector makes with the array vertical reference, defined in terms of the wavenumber and sensor locations.

For instance, a harmonic plane wave signal vector can be represented by

$$\mathbf{s}(\mathbf{r};t) = \alpha_o e^{j\omega_o t} e^{-j\kappa_o\cdot\mathbf{r}} = \alpha_o e^{j\omega_o t} e^{-j(\kappa_x x+\kappa_y y)} = \alpha_o e^{j\left(\omega_o t-(|\kappa_o|\sin\theta_o x+|\kappa_o|\cos\theta_o y)\right)} \tag{1.10}$$

Therefore, the complex wavenumber vector has magnitude, $|\kappa| = \sqrt{\kappa_x^2 + \kappa_y^2}$, and the angle is $\angle\kappa_{\mathbf{o}} = [\sin\theta_o \quad \cos\theta_o]$. We see that in this case the plane wave is characterized by the parameter vector, $\Theta_o = \{\alpha_o, \kappa_o\}$ or equivalently $\Theta_o = \{\alpha_o, \omega_o, \theta_o\}$.

Consider the following example demonstrating the spatial–temporal method of decomposing noisy multichannel time series measurements in the spatial (Fourier) domain to extract independent source locations (bearing angles)—a common problem in sonar signal processing [9].

Example 1.5 Suppose a 16-element, uniformly spaced at 2.5 m, linear array is impinged upon by a planar wavefront generated from three sources ($N_s = 3$) emanating from 20°, 40°, and 55° incidence angles. The temporal frequency is $\omega_o = 300$ Hz with corresponding propagation speed of $c = 1500$ m/sec. The sensor data are generated at 0 dB *SNR* and shown in Figure 1.5a. The order or number of sources is estimated using the information-theoretic techniques giving an $\hat{N}_s = 3$ with the corresponding plots shown in Figure 1.5b. The results of applying the power estimators are shown in Figure 1.5a,b, where we observe the ensemble results of the 16 sensor channels and the corresponding median spectral estimates. The results demonstrate that the algorithms are capable of decomposing the spatial spectrum providing reasonable *DOA* estimates. This completes the example.

△△△

So we see that the power spectral estimation techniques developed for temporal harmonic model-based parametric estimation map over to the wave type problems

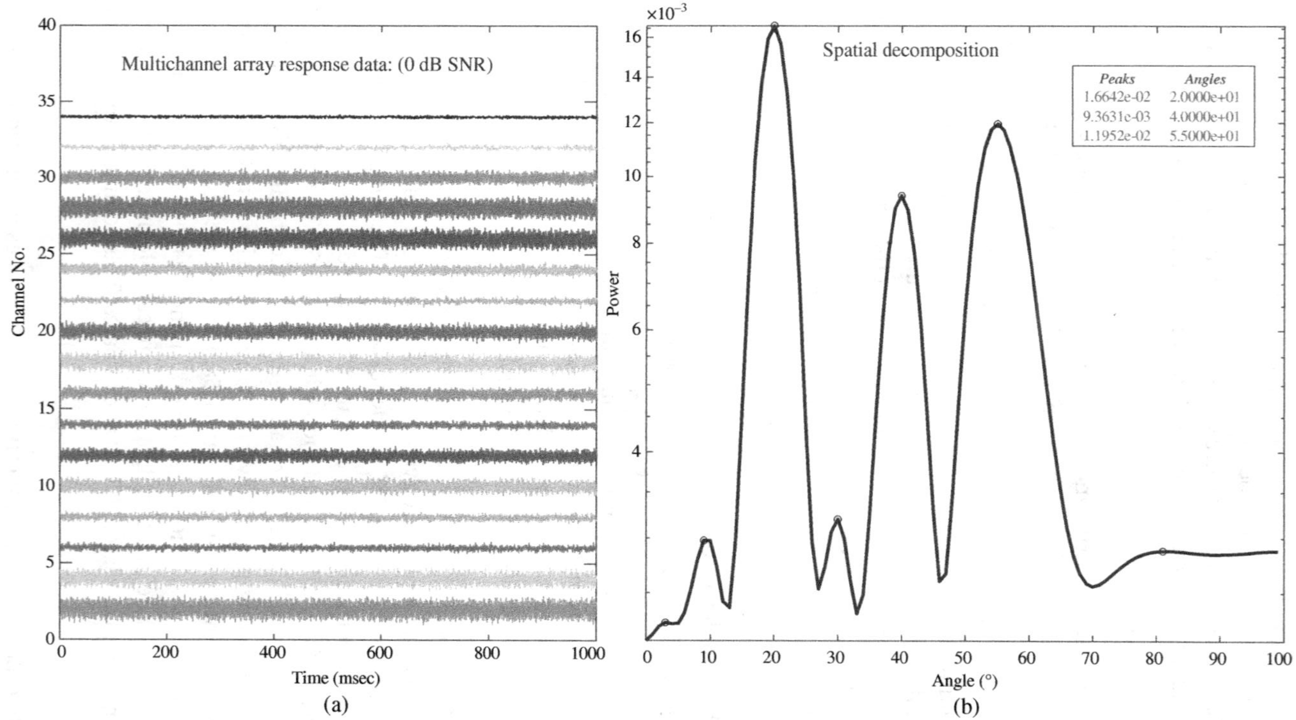

Figure 1.5 Spatial spectral decomposition of multiple source data: (a) Multichannel array measurements (0 dB *SNR*). (b) Spectral source decomposition (degrees).

directly with the temporal harmonic signal vectors replaced by the spatial signal direction vectors.

Thus, we see how various data matrices can be simply decomposed using linear algebraic techniques applied to vectors/matrices of a specific data matrix. Next, we consider the model-based approach to perform signal/system decompositions.

1.4 Model-based Decomposition

Signal processing relies on any a priori knowledge of the phenomenology generating the underlying measurements. Characterizing this phenomenology and propagation physics along with the accompanying measurement instrumentation and noise are the preliminaries that must be tackled to solve a complex processing problem. In many cases, this is much easier said than done. The first step is to determine what the desired information is. In our case, we assume that the investigation is to extract information stemming from signals either emanating from a source whether it be an autonomous unmanned vehicle (*AUV*) passively operating in the deep ocean or a vibrating structure responding to ground motion. Applications can be very complex especially in the case of ultrasound propagating through complex media such as tissue in biomedical applications or through heterogeneous materials of critical parts in nondestructive evaluation (*NDE*) investigations [1, 10]. In any case, the processing usually involves manipulating the measured data to extract the desired information, such as location and tracking of the *AUV*, to failure detection for the structure, or tumor/flaw detection and localization in both biomedical and *NDE* [1, 5, 10].

Another view of the same problem is to decompose it into a set of steps that capture the strategic essence of the processing scheme. Inherently, the more a priori knowledge about the measurement and its underlying phenomenology that can incorporated into the processor, the better we can expect it to perform as long as the information that is included is correct. One strategy, called the "model-based approach," provides the essence of model-based signal processing (*MBSP*) [1, 10, 11]. Some believe that all of the signal processing schemes can be cast into this generic framework. Simply, the model-based approach is "incorporating mathematical models of both physical phenomenology and the measurement process (including noise) into the processor to extract the desired information." This approach provides a mechanism to incorporate knowledge of the underlying physics or dynamics in the form of mathematical propagation models along with measurement system models and accompanying uncertainties such as disturbances, instrumentation noise, or ambient noise as well as model uncertainties directly into the resulting processor. In this way, the signal processor enables the interpretation of results directly in terms of the underlying problem physics.

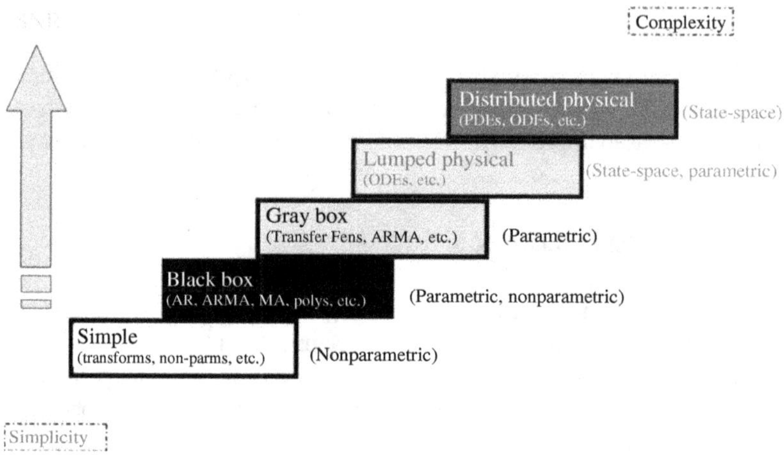

Figure 1.6 Signal processing approach: the model-based "staircase." [Step 1] "Simple" nonparametric implicit models. [Step 2] "Black-box" models (transfer function, autoregressive, moving average, polynomial, etc.). [Step 3] "Gray-box" models (transfer function, moving average, etc.). [Step 4] "Lumped" physical ordinary differential equation models (state-space, parametric, etc.). [Step 5] "Distributed" physical partial differential equation models (state-space, etc.).

The model-based processor is really a modeler's tool enabling the incorporation of any a priori information about the particular application problem to extract the desired information. As illustrated in Figure 1.6, the fidelity of the incorporated model determines the complexity of the processor. These models can range from simple implied nonphysical representations of the measurement data such as the Fourier or wavelet transforms to parametric black-box models used for data prediction, to lumped mathematical physical representations characterized by ordinary differential equations, and to full physical partial differential equation models from finite element models capturing the critical details of propagation physics in a complex medium. The dominating factor of which model is the most appropriate is usually determined by how severe the measurements are contaminated with noise and the underlying uncertainties encompassing the philosophy of "letting the problem dictate the approach." If the *SNR* of the measurements is high, then simple nonphysical techniques can be used to extract the desired information. This approach of selecting the appropriate model is illustrated in the signal processing staircase of Figure 1.6 where we note that as we progress up the "modeling" steps to increase the *SNR*, the complexity of the model increases to achieve the desired results. In the following, we will apply the model-based framework to develop decomposition techniques for signal data as well as system models. This is our roadmap [10].

Example 1.6 Suppose we have a noisy acoustical measurement of a single oscillation frequency contaminated by random noise at a 0 dB *SNR* and we would like to extract the desired information (the single oscillation frequency) as shown in Figure 1.7a. The "simple" approach to analyze the measurement data would be to take its Fourier transform and investigate the various frequency bands for resonant peaks. The result is shown in Figure 1.7b, where we observe a noisy spectrum and a set of potential resonances but nothing really conclusive. Next, we apply a broadband power spectral estimator with the resulting spectrum shown in Figure 1.7c. Here we note that the resonances have clearly been enhanced and appear in well-defined bands while the noise is attenuated by the processor, but there still remains a significant amount of uncertainty in the spectrum due to all of the resulting spectral peaks. Upon seeing these resonances in the power spectrum,

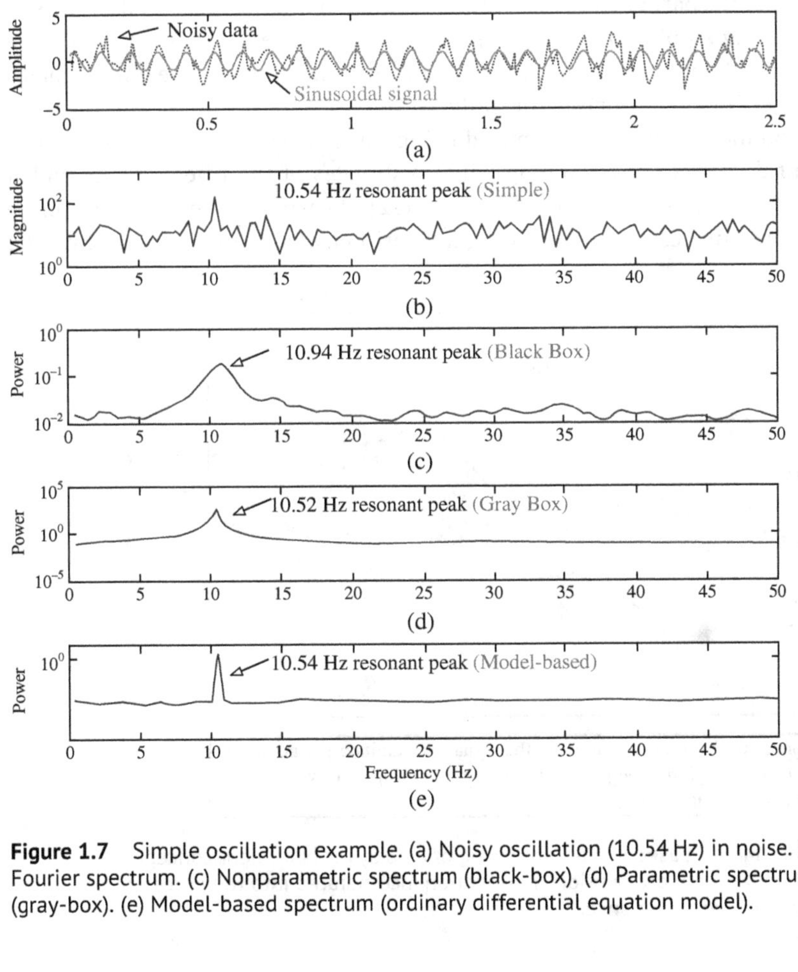

Figure 1.7 Simple oscillation example. (a) Noisy oscillation (10.54 Hz) in noise. (b) Raw Fourier spectrum. (c) Nonparametric spectrum (black-box). (d) Parametric spectrum (gray-box). (e) Model-based spectrum (ordinary differential equation model).

we might proceed next to a gray-box model to enhance the resonances even further by using our a priori knowledge that there is essentially one dominant resonance. The results of applying this processor are shown in Figure 1.7d. Finally, we use this extracted model to develop an explicit model-based processor (*MBP*) by developing a set of harmonic equations for a sinusoid in noise and construct the *MBP* based on these relations. The results are shown in Figure 1.7e. So we see that once we have defined the processing problem, assessed the a priori information including the underlying phenomenology, then we can proceed up the staircase and exit any time we are satisfied with the result [1, 10].

△△△

It is important to investigate what, if anything, can be gained by processing the data. The amount of information available in the data is related to the precision (variance) of the particular measurement instrumentation used as well as any signal processing employed to enhance the outputs. As we utilize more and more information about generating the given data, we expect to improve our estimates (decrease estimation error) significantly.

A typical measurement y is depicted in Figure 1.8 where we see the noisy measurement data and corresponding "final" estimates of the true signal bounded by its corresponding confidence intervals [12]. As we develop more and more sophisticated processors by including a priori knowledge in the algorithm, the

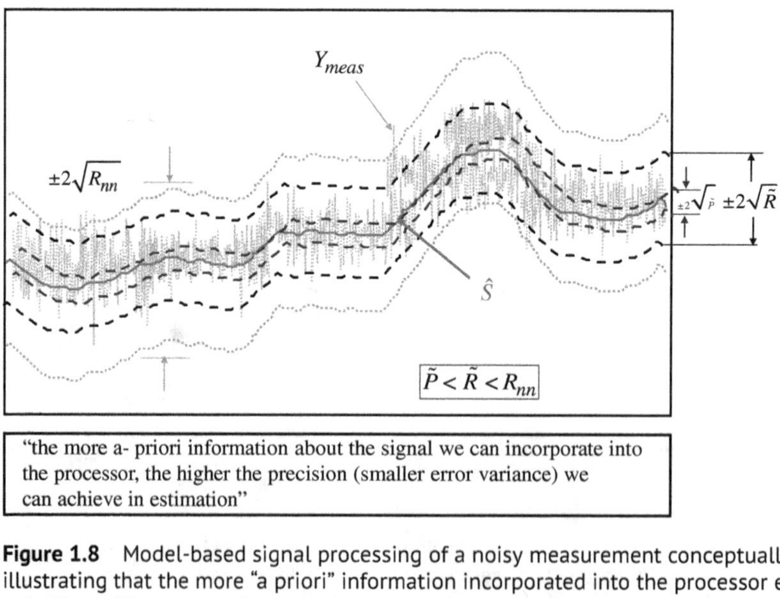

Figure 1.8 Model-based signal processing of a noisy measurement conceptually illustrating that the more "a priori" information incorporated into the processor enables a reduction of the error variance.

uncertainty in the estimate decreases further as shown by the dashed bounds. Mathematically, we illustrate these ideas by the following.

In Figure 1.9a, the ideal measurement instrument is given by

$$Y_{meas} = S_{true} + N_{noise} \quad \text{[Ideal Measurement]}$$

and along with and a more realistic model of the measurement process is depicted in Figure 1.9b.

If we were to use Y to estimate S_{true} (that is, $\hat{S}$), we have the noise lying within the $\pm 2\sqrt{\tilde{R}}$ confidence limits superimposed on the signal (see Figure 1.8). The best estimate of S_{true} we can hope for is only within the accuracy (bias) and precision (variance) of the instrument. If we include a model of the measurement instrument (see Figure 1.9b) as well as its associated uncertainties, then we can improve the SNR of the noisy measurements. This technique represents the processing based on our instrument and noise (statistical) models given by

$$Y_{meas} = C(S_{true}) + N_{noise}, \quad N \sim \mathcal{N}(0, R_{nn}) \quad \text{[Measurement and Noise]}$$

where $C(S_{true})$ is the measurement system model and $\mathcal{N}(0, R_{nn})$ is the noise statistics captured by a *Gaussian* or *normal distribution* of zero-mean and variance specified by R_{nn}.

Figure 1.9 Classes of model-based signal processors: (a) Simple measurement systems. (b) Model-based signal processing with measurement instrument modeling. (c) Model-based signal processing with measurement uncertainty and process modeling.

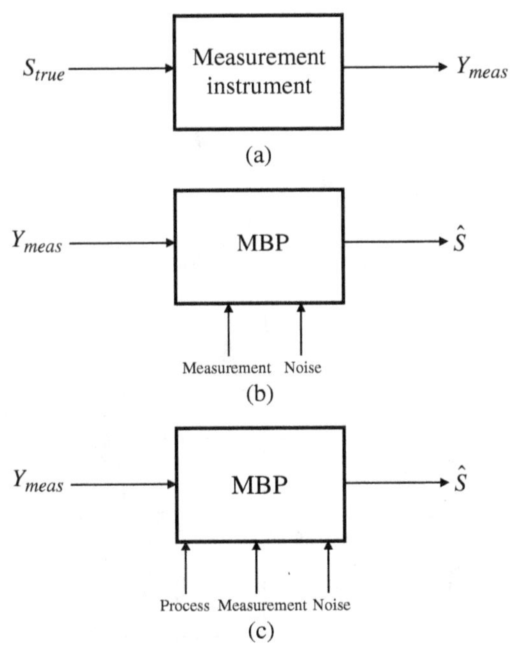

We can also specify the estimation error variance $\tilde{R}$ or equivalently the quality of this processor in estimating S. Finally, if we incorporate not only instrumentation knowledge, but also knowledge of the physical process (see Figure 1.9c), then we expect to do even better, that is, we expect the estimation error ($\tilde{S} = S_{true} - \hat{S}$) variance $\tilde{P}$ to be small (see Figure 1.8 for $\pm 2\sqrt{\tilde{R}}$ confidence limits or bounds) as we incorporate more and more knowledge into the processor, i.e.

$$\hat{S} = A(S_{true}) + W_{noise}, \quad W \sim \mathcal{N}(0, R_{ww}) \quad \text{[Process Model and Noise]}$$
$$Y_{meas} = C(S_{true}) + N_{noise}, \quad N \sim \mathcal{N}(0, R_{nn}) \quad \text{[Measurement Model and Noise]}$$

where $A(S_{true})$ is the process system model and $\mathcal{N}(0, R_{ww})$ is the process noise statistics captured by a zero-mean, Gaussian distribution with variance, R_{ww}.

In fact, this is the case because it can be shown that

$$\tilde{P} < \tilde{R} < \text{Instrument variance}$$

This is the basic idea in *MBSP*: "the more a priori information we can incorporate into the algorithm, the smaller the resulting error variance." Consider the following example to illustrate these ideas.

Example 1.7 The voltage at the output of an RC circuit is to be measured using a high-impedance voltmeter shown in Figure 1.10. The measurement is contaminated with random instrumentation noise, which can be modeled by

$$e_{out} = K_e e + v$$

where

$$\begin{aligned} e_{out} &= \text{measured voltage} \\ K_e &= \text{instrument amplification factor} \\ e &= \text{true voltage} \\ v &= \text{zero-mean random noise of variance, } R_{vv} \end{aligned}$$

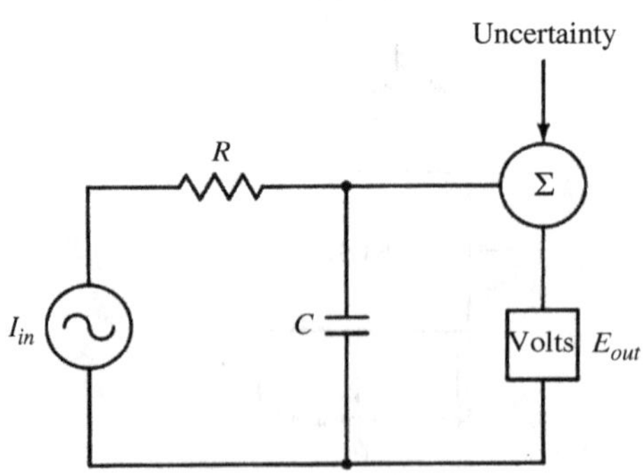

Figure 1.10 Model-based signal processing "model" for an RC circuit.

This model corresponds to that described in Figure 1.9b. A processor is to be designed to improve the precision of the instrument. Then, as in the figure we have

- Measurement:

$$\begin{aligned} S &\rightarrow e \\ Y &\rightarrow e_{out} \\ C(\cdot) &\rightarrow K_e \\ N &\rightarrow v \\ R_{nn} &\rightarrow R_{vv} \end{aligned}$$

- and for the filter,

$$\begin{aligned} \hat{s} &\rightarrow \hat{e} \\ \tilde{R}_{ss} &\rightarrow \tilde{R}_{ee} \end{aligned}$$

The precision of the instrument can be improved even further by including a model of the process (circuit). Writing the Kirchhoff node equations we have the differential equation

$$\dot{e} = \frac{1}{C} I_{in} - \frac{e}{RC} + q$$

where

$$\begin{aligned} R &= \text{resistance} \\ C &= \text{capacitance} \\ I_{in} &= \text{excitation current} \\ q &= \text{zero-mean random noise of variance } R_{qq} \end{aligned}$$

The improved model-based processor employs both measurement and process models as in Figure 1.9c. Thus, we have

- Process:

$$\begin{aligned} \dot{s} &\rightarrow \dot{e} \\ A(\cdot) &\rightarrow -\frac{e}{RC} + \frac{1}{C} I_{in} \\ W &\rightarrow q \\ R_{ww} &\rightarrow R_{qq} \end{aligned}$$

- Measurement:

$$\begin{aligned} S &\rightarrow e \\ Y &\rightarrow e_{out} \end{aligned}$$

$$C(\cdot) \rightarrow K_e$$
$$N \rightarrow \upsilon$$
$$R_{nn} \rightarrow R_{\upsilon\upsilon}$$

- Therefore, the filter becomes,

$$\hat{s} \rightarrow \hat{e}$$
$$\tilde{P}_{ss} \rightarrow \tilde{P}_{ee}$$

- such that

$$\tilde{P}_{ee} < \tilde{R}_{ee} < R_{\upsilon\upsilon}$$

This completes the example. △△△

With this approach in mind, let us proceed to investigate a mechanical system characterized by a multichannel representation and see how it can easily be decomposed to yield decoupled subsystems available for analysis. First, we briefly present the generic linear time-invariant, multivariable, mass-damper-spring (*MCK*) mechanical system representation and illustrate its decomposition into constituent subsystems [13, 14].

Linear, continuous time, time-invariant multiple input/multiple output (*MIMO*) mechanical systems are characterized by the vector-matrix differential equations creating the *vibrational process model* in continuous time, t given by

$$M\ddot{\mathbf{d}}(t) + C_d\dot{\mathbf{d}}(t) + K\mathbf{d}(t) = B_f\mathbf{f}(t) \tag{1.11}$$

where $\mathbf{d}$ is the $N_d \times 1$ displacement vector, $\mathbf{f}$ is the $N_f \times 1$ excitation force, and M, C_d, K are the $N_d \times N_d$ lumped mass, damping, and spring constant matrices with B_f the $N_d \times N_f$ input transmission matrix. The structure of these matrices, typically, takes the form as

$$M = \begin{bmatrix} M_1 & 0 & 0 & 0 & 0 \\ 0 & M_2 & 0 & 0 & 0 \\ \vdots & \vdots & \ddots & \vdots & \vdots \\ 0 & 0 & 0 & M_{N_d-1} & 0 \\ 0 & 0 & 0 & 0 & M_{N_d} \end{bmatrix}, \quad C_d = \left[C_{d_{ij}}\right], \quad \text{and}$$

$$K = \begin{bmatrix} (K_1+K_2) & -K_2 & 0 & 0 & 0 \\ -K_2 & (K_2+K_3) & \ddots & 0 & 0 \\ 0 & \ddots & \ddots & -K_{N_d-1} & 0 \\ 0 & 0 & -K_{N_d-1} & \left(K_{N_d-1}+K_{N_d}\right) & -K_{N_d} \\ 0 & 0 & 0 & -K_{N_d} & K_{N_d} \end{bmatrix}$$

The corresponding measurement or output vector relation can be characterized by

$$\mathbf{y}(t) = \mathbf{C_a}\ddot{\mathbf{d}}(t) + \mathbf{C_v}\dot{\mathbf{d}}(t) + \mathbf{C_{dp}}\mathbf{d}(t) \tag{1.12}$$

where the constant matrices: $\mathbf{C_a}$, $\mathbf{C_v}$, $\mathbf{C_{dp}}$ are the respective acceleration, velocity, and displacement weighting matrices of appropriate dimension.

By performing a so-called modal (coordinate) transformation ($\mathbf{T}_m$) such that $\mathbf{d} = \mathbf{T}_m \times \mathbf{q}_m$ with $\mathbf{q}_m$ the modal vector. The resulting *modal model* evolves as [14]

$$\begin{aligned}
&\ddot{\mathbf{q}}_m(t) + \underbrace{2\mathbf{Z}\boldsymbol{\Omega}}_{\mathcal{Z}}\,\dot{\mathbf{q}}_m(t) + \underbrace{\boldsymbol{\Omega}^2}_{\mathcal{O}}\,\mathbf{q}_m(t) = \mathcal{B}_m\mathbf{f}_m(t)\\
&\mathbf{y}_m(t) = \mathcal{C}_m\mathbf{q}_m(t)
\end{aligned} \tag{1.13}$$

where the *modal damping matrix* is $\mathcal{Z} = \text{diag}\left[\zeta_1, \ldots, \zeta_{N_d}\right]$ and the corresponding *modal frequency matrix* is $\boldsymbol{\Omega} = \text{diag}\left[\omega_1, \ldots, \omega_{N_d}\right]$. This representation is the "decoupled" or "decomposed" *MCK*-model that can be expressed as a set of uncoupled subsystems as

$$\begin{aligned}
&\ddot{\mathbf{q}}_{m,i}(t) + 2\zeta_i\omega_i\dot{\mathbf{q}}_{m,1}(t) + \omega_i^2\mathbf{q}_{m,i}(t) = \mathbf{b}_{m,i}^T\mathbf{f}_{m,i}(t)\\
&\mathbf{y}_{m,i}(t) = \mathbf{c}_{m,i}^T\mathbf{q}_{m,i}(t), \qquad i = 1, \ldots, N_d
\end{aligned} \tag{1.14}$$

Next, we investigate a simple mechanical system developed previously in [14].

Example 1.8 Consider the simple 3-mass (spring-damper) mechanical system[1] illustrated in Figure 1.11 from [14] where $m_i = 1; k_i = 3; i = 1, \ldots, 3$ and $C_d = 0.01 \times K$. These parameters along with the input/output transmission matrices lead to the so-called "nodal model" of structural dynamics with $N_d = 3$.

$$M\ddot{\mathbf{d}}(t) + C_d\dot{\mathbf{d}}(t) + K\mathbf{d}(t) = B_f\mathbf{f}(t)$$

$$\underbrace{\begin{bmatrix} 1 & 0 & 0 \\ 0 & 1 & 0 \\ 0 & 0 & 1 \end{bmatrix}}_{\mathbf{M}} \ddot{\mathbf{d}}(t) + \underbrace{\begin{bmatrix} -0.06 & 0.03 & 0 \\ 0.03 & -0.06 & 0.03 \\ 0 & 0.03 & -0.03 \end{bmatrix}}_{\mathbf{C}_d} \dot{\mathbf{d}}(t) + \underbrace{\begin{bmatrix} -6 & 3 & 0 \\ 3 & -6 & 3 \\ 0 & 3 & -3 \end{bmatrix}}_{\mathbf{K}} \mathbf{d}(t) = \underbrace{\begin{bmatrix} 0 \\ 0 \\ 1 \end{bmatrix}}_{\mathbf{B}_f} \delta(t)$$

with the corresponding accelerometer output given by

$$y(t) = \underbrace{\begin{bmatrix} 1 \end{bmatrix}}_{\mathbf{C}_a} \ddot{\mathbf{d}}(t)$$

Exciting this system with an impulsive hammer strike yields the 3-mass response shown in Figure 1.12a along with its accompanying Fourier spectra in

1 See [14] for a detailed analysis of this mechanical system.

1.12b and descretizing with a sampling interval of $\Delta t = 0.01$ sec. This coupled *MCK* system can be decomposed by performing an eigen-decomposition of the system matrices [14]. That is, performing the modal decomposition of the *MCK* system with coordinate transformation (eigenvector matrix) from

$$\mathbf{T}_m = \begin{bmatrix} 0.5910 & 0.7370 & 0.3280 \\ -0.7370 & 0.3280 & 0.03 \\ 0.3280 & -0.5910 & 0.7370 \end{bmatrix}$$

yields the resulting modal system of Eq. (1.13) given by

$$\ddot{\mathbf{q}}(t) + 2 \times \underbrace{\begin{bmatrix} 0.0156 & 0 & 0 \\ 0 & 0.0108 & 0 \\ 0 & 0 & 0.0039 \end{bmatrix}}_{\mathbf{Z}} \times \underbrace{\begin{bmatrix} 3.1210 & 0 & 0 \\ 0 & 2.1598 & 0 \\ 0 & 0 & 0.7708 \end{bmatrix}}_{\mathbf{\Omega}} \dot{\mathbf{q}}(t)$$

$$+ \underbrace{\begin{bmatrix} 3.1210 & 0 & 0 \\ 0 & 2.1598 & 0 \\ 0 & 0 & 0.7708 \end{bmatrix}^2}_{\mathbf{\Omega}^2} \mathbf{q}(t) = \underbrace{\begin{bmatrix} 0.3280 \\ -0.5910 \\ 0.7370 \end{bmatrix}}_{\mathbf{B}_m} \delta(t)$$

with the velocity output matrix given by

$$y_m(t) = \mathbf{c}_{mv,1}^T \mathbf{q}(t) + \mathbf{c}_{mv,2}^T \dot{\mathbf{q}}(t) = \underbrace{\begin{bmatrix} 0.7370 & 0 & 0 \end{bmatrix}}_{\mathbf{c}_{mv,1}^T} \mathbf{q}(t) + \underbrace{\begin{bmatrix} 0 & 0.7370 - 0.5910 \end{bmatrix}}_{\mathbf{c}_{mv,2}^T} \dot{\mathbf{q}}(t)$$

The decomposed modal model is now available. Each of the modal subsystem responses and spectra can be obtained as illustrated by the corresponding impulse responses and spectra shown in Figure 1.11. Comparing the individual spectra with the composite spectra consisting of the set of frequencies $\{f_i = 0.497, 0.343, 0.123\,\text{Hz}\}$ have been extracted by the modal decomposition. This completes the model-based decomposition example.

△△△

Before we depart from this chapter, let us define the notation to be used throughout the text and then summarize the chapter.

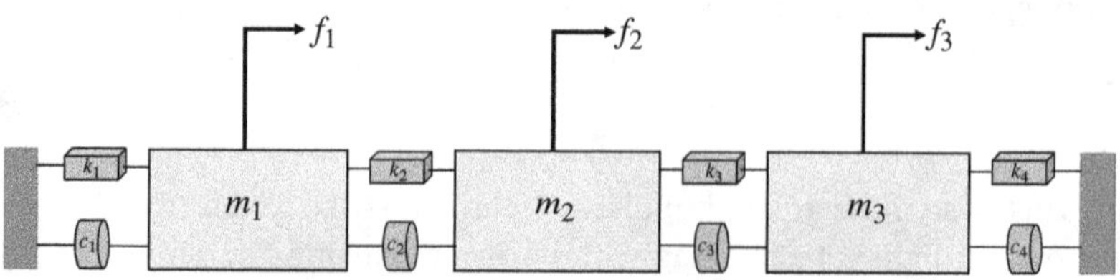

Figure 1.11 The 3-mass mechanical system under investigation.

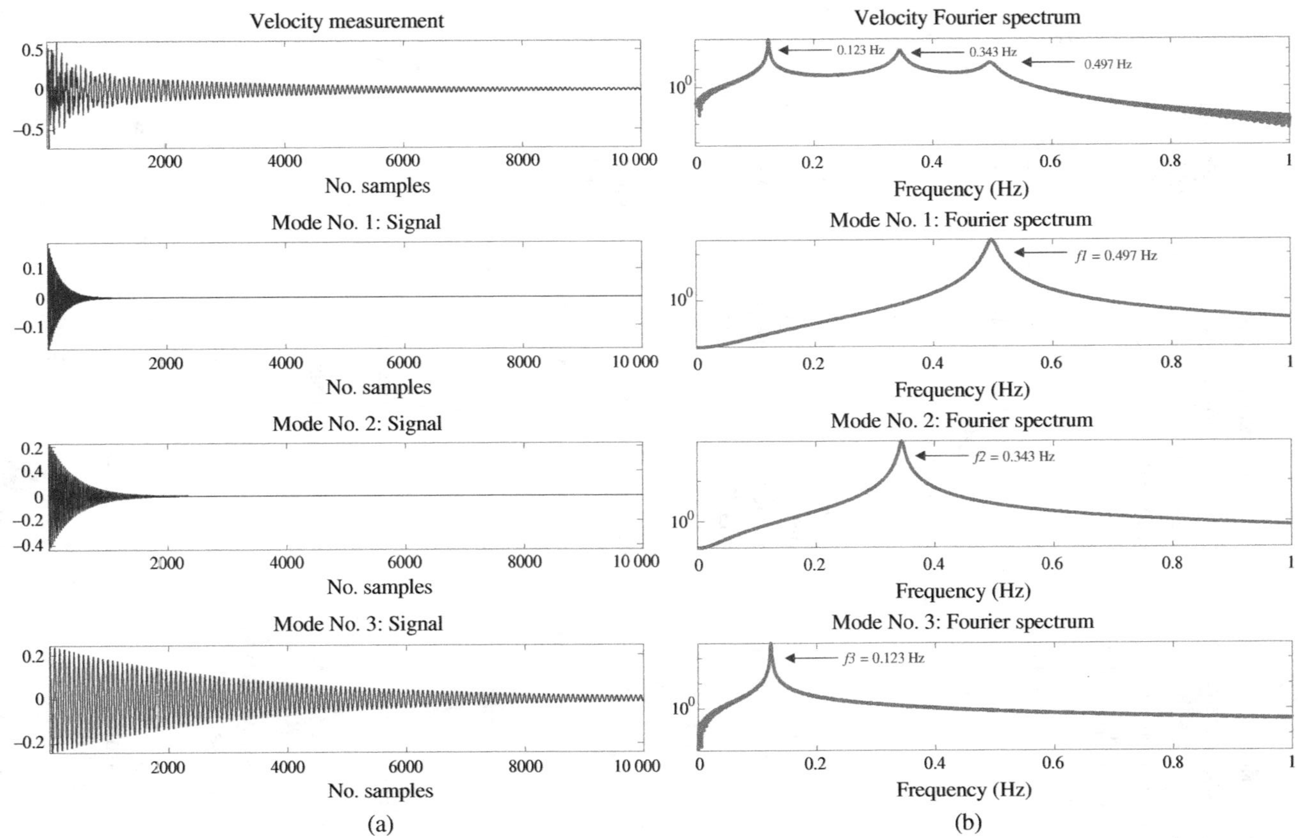

Figure 1.12 Decomposition of 3-mass mechanical system. (a) Composite channel impulse responses and individual modal responses. (b) Fourier transforms of composite channel responses with spectral peaks at 0.497, 0.343, 0.123 Hz and individual constituent Fourier spectra obtained from system modal decomposition.

1.5 Notation and Terminology

The notation used throughout this text is standard in the literature. Where necessary, vectors are represented by boldface, lowercase, $\mathbf{x}$, and matrices by boldface, uppercase, $\mathbf{A}$. The *adjoint* of the matrix $\mathbf{A}$ is annotated as $\text{adj}\,(\mathbf{A})$ and its corresponding *determinant* by $\det\,(\mathbf{A})$. We define the notation N to be a shorthand way of writing $1, 2, \ldots, N$. It will be used in matrices, $A(N)$ to mean there are N-columns of A. We also use the equivalent *MATLAB* notation $(1\,:\,N)$. We denote the real part of a signal by Re and its imaginary part by Im. As mentioned previously, estimators are annotated by the caret, such as $\hat{x}$. We also define partial derivatives at the component level by $\frac{\partial}{\partial\theta_i}$, the N_θ-gradient vector by ∇_θ and higher order partials by ∇^2_θ.

The most difficult notational problem will be with the "time" indices. Since this text is predominantly discrete-time, we will use the usual time symbol, t to mean a discrete-time index, i.e. $t \in \mathcal{I}$ for $\mathcal{I}$ the set of integers. However, and hopefully not too confusing, t will also be used for continuous-time, i.e. $t \in \mathcal{R}$ for $\mathcal{R}$ the set of real numbers denoting the continuum. When used as a continuous-time variable, $t \in \mathcal{R}$ it will be represented as a subscript to distinguish it, i.e. $x_{\mathbf{t}}$. This approach of choosing $t \in \mathcal{I}$ primarily follows the system identification literature for the ease of recognizing discrete-time variable in transform relations (e.g. *DFT*). The rule-of-thumb is therefore to "interpret t as a discrete-time index unless noted by a subscript as continuous in the text." With this in mind, we will define a variety of discrete estimator notations as $\hat{x}(t|t-1)$ to mean the estimate at time (discrete) t based upon all of the previous data up to $t-1$. We will define these symbols prior to their use with the text to assure no misunderstanding of their meaning.

With a slight abuse of notation, we will use the terminology *distribution* of X, $\Pr\,(X)$ in general, so as not to have to differentiate between density for continuous random variables or processes and mass for discrete variates. It will be obvious from the context which is meant. In some cases, we will be required to make the distinction between cumulative distribution function (*CDF*) and density (*PDF*) or mass (*PMF*) functions. Here we use the uppercase notation, $\mathrm{P}_X(x)$ for the *CDF* and lower case $\mathrm{p}_X(x)$ for the *PDF* or *PMF*.

Subsequently, we will also need to express a discrete *PMF* as a continuous *PDF* using impulse or delta functions as "samplers" much the same as in signal processing when we assume there exists an impulse sampler that leads to the well-known Nyquist sampling theorem [1]. Thus, corresponding to a discrete *PMF*, we can define a continuous *PDF* through the concept of an *impulse sampler*, that is, given a discrete *PMF* defined by

$$\mathrm{p}_X(x) \approx \mathrm{p}(X = x_i) = \sum_i p_i\, \delta(x - x_i) \tag{1.15}$$

then we define the *equivalent continuous PDF* as $p_X(x)$. Moments follow from the usual definitions associated with a continuous *PDF*, for instance, consider the definition of the expectation or mean. Substituting the equivalent *PDF* and utilizing the sifting property of the impulse function gives

$$E\{x\} = \int_{-\infty}^{\infty} x\, p_X(x)\, dx = \int_{-\infty}^{\infty} x \left(\sum_i p_i \delta(x - x_i) \right) dx = \sum_i x_i\, p_i \tag{1.16}$$

which is precisely the mean of the discrete *PMF* (see Appendix A for more details).

Also, as mentioned, we will use the symbol $\sim$ to mean "distributed according to" as in $x \sim \mathcal{N}(m, v)$ defining the random variable x as Gaussian distributed with mean m and variance v. We may also use the extended notation: $\mathcal{N}(x : m, v)$ to include the random variable x as well. When *sampling* we use the *nonconventional* right arrow "action" notation $\longrightarrow$ to mean "draw a sample from" a particular distribution such as $x_i \rightarrow \Pr(x)$–this again will be clear from the context. When *resampling*, that is, replacing samples with new ones we use the "block" right arrow such as $x_j \Rightarrow x_i$ meaning new sample x_j replaces current sample x_i. In a discrete (finite) probabilistic representation, we define a purely discrete variate as $x_k(t) := \Pr(x(t) = \mathcal{X}_k)$ meaning that x can only take on values (integers) k from a known set $\mathcal{X} = \{\mathcal{X}_1, \dots, \mathcal{X}_k, \dots, \mathcal{X}_N\}$ at time t. We also use the symbol $\triangle\triangle\triangle$ to mark the end of an example.

We define two projection operators: (i) orthogonal; and (ii) oblique or parallel. The *orthogonal projection operator* "projects" $\bullet$ onto $\mathcal{Y}$ as $\mathcal{P}_{\bullet|\mathcal{Y}}$ or its complement $\mathcal{P}^{\perp}_{\bullet|\mathcal{Y}}$, while the *oblique projection operator* "projects" $\bullet$ onto $\mathcal{Y}$ "along" $\mathcal{Z}$ as $\mathcal{P}^{\|}_{\bullet|\mathcal{Y}\circ\mathcal{Z}}$. For instance, orthogonally projecting the vector $\mathbf{x}$, we have $\mathcal{P}_{\mathbf{x}|\mathcal{Y}}$ or $\mathcal{P}^{\perp}_{\mathbf{x}|\mathcal{Y}}$, while obliquely projecting $\mathbf{x}$ along $\mathcal{Z}$ is $\mathcal{P}_{\mathbf{x}|\mathcal{Y}\circ\mathcal{Z}}$ (see Appendix B). The matrix projection operator evolves from decompositions such as the LQ-decomposition (see Appendix C) where the operator is given in terms of its orthonormal bases as $\mathcal{P} = Q \times Q'$.

1.6 Summary

In this chapter, we have introduced the concept of decomposition starting with the simple notion of transforming a signal into an alternate domain to analyze and extract the desired information. With this in mind, data-based decomposition was investigated with the primary idea that signal/measurement data could be used to construct a variety of formats leading to the usual vector-matrix form that dominates the signal processing techniques. Finally, *MBSP* was introduced based on the idea of incorporating more and more any available a priori information into the processing scheme leading to the model-based decomposition approach. A simple 3-mass mechanical system was analyzed and followed throughout the chapter demonstrating how the various approaches can be applied to achieve a decomposition of the composite signal into constituent components. Various signals were classified as deterministic or random. When the signal is random the resulting processors are called estimation filters or estimators. A procedure was defined (estimation procedure) leading to a formal processing approach to extract information based on prior knowledge of the composition which will be used throughout this text. After a couple of examples motivating the concept of *MBSP*, a more formal discussion followed with *RC* and *MCK* circuit examples completing the chapter. Finally, the notation utilized throughout the text was discussed completing this introductory chapter.

MATLAB® Notes

MATLAB® is command-oriented vector-matrix package with a simple yet effective command language featuring a wide variety of embedded *C* language constructs making it ideal for signal processing and graphics. All of the algorithms we have applied to the examples and problems in this text are *MATLAB*-based in solutions ranging from simple simulations to complex applications. We will develop these notes primarily as a summary to point out to the reader many of the existing commands that already perform the signal processing operations discussed in the presented chapter and throughout the text.

References

1 Candy, J. (2006). *Model-Based Signal Processing*. Hoboken, NJ: Wiley/IEEE Press.
2 Hayes, M. (1996). *Statistical Digital Signal Processing and Modeling*. New York: Wiley.
3 Golub, G. and Van Loan, C. (1984). *Matrix Computations*. Baltimore, MD: Johns Hopkins University Press.
4 Rainieri, C. and Fabbrocino, G. (2014). *Operational Modal Analysis of Civil Engineering Structures*. New York: Springer.
5 Reynders, E. (2012). System identification methods for (operational) modal analysis: review and comparison. *Arch. Comput. Meth. Eng.* 19 (1): 51–124.
6 Schmidt, R. (1986). Multiple emitter location and signal processing estimation. *IEEE Trans. Antennas Propg.* 34 (3): 276–280.
7 Fink, M. (1997). Time reversed acoustics. *Phys. Today* 50: 34–40.
8 Johnson, D. and Mersereau, R. (1993). *Array Signal Processing*. New Jersey: Prentice-Hall.

9 Nielson, R. (1995). *Sonar Signal Processing*. New Jersey: Prentice-Hall.
10 Candy, J. (2008). Signal processing in acoustics: science or science fiction? *Acoust. Today* 4 (3): 6–15.
11 Candy, J. (1986). *Signal Processing: The Model-Based Approach*. New York: McGraw-Hill.
12 Jazwinski, A. (1970). *Stochastic Processes and Filtering Theory*. New York: Academic Press.
13 Meirovitch, L. (1990). *Dynamics and Control of Structures*. New York: Wiley.
14 Gawronski, W. (2004). *Advanced Structural Dynamics and Active Control of Structures*. New York: Springer.

Problems

1.1 A one-story structure is monitored for potential damage after being subjected to an earthquake. Cars and trucks passing by provide the excitation modeled as random noise. The displacement of the structure is acquired by placing an accelerometer at a critical location sensitive to these vibrations. The accelerometer has inherent inaccuracies which is modeled as

$$y = K_b x + n$$

with y, x, n the measured and actual displacement and white measurement noise of variance R_{nn} and K_b the instrument gain. The dynamics of the vibrating structure is modeled by a simple mass-spring-damper (MCK) system.

(a) Construct and identify the measurement model of this system identifying each of its components.
(b) Construct and identify the process model and model-based estimator for this problem.
(c) What is the variance associated with each modeling phase: measurement, process, and noise?

1.2 We are asked to estimate the displacement of large vehicles (semi-trailers) when parked on the shoulder of a freeway and subjected to wind gusts created by passing vehicles. We measure the displacement of the vehicle by placing an accelerometer on the trailer. The accelerometer has inherent inaccuracies which is modeled as

$$y = K_a x + n$$

with y, x, n the measured and actual displacement and white measurement noise of variance R_{nn} and K_a the instrument gain. The dynamics of the vehicle can be modeled by a simple mass-spring-damper.

(a) Construct and identify the measurement model of this system.
(b) Construct and identify the process model and model-based estimator for this problem.

1.3 Think of measuring the temperature of a liquid in a breaker heated by a burner. Suppose we use a thermometer immersed in the liquid and periodically observe the temperature and record it.
(a) Construct a measurement model assuming that the thermometer is linearly related to the temperature, that is, $y(t) = k\triangle T(t)$. Also model the uncertainty of the visual measurement as a random sequence $v(t)$ with variance R_{vv}.
(b) Suppose we model the heat transferred to the liquid from the burner as

$$Q(t) = CA\triangle T(t)$$

where C is the coefficient of thermal conductivity, A is the cross-sectional area, and $\triangle T(t)$ is the temperature gradient with assumed random uncertainty $w(t)$ and variance R_{ww}. Using this process model and the models developed above, identify the model-based processor representation.

1.4 We are given an *RLC* series circuit driven by a noisy voltage source $V_{in}(t)$ and we use a measurement instrument that linearly amplifies by K and measures the corresponding output voltage. We know that the input voltage is contaminated by an additive noise source, $w(t)$ with covariance, R_{ww} and the measured output voltage is similarly contaminated with noise source, $v(t)$ with R_{vv}.
(a) Determine the model for the measured output voltage, $V_{out}(t)$ (measurement model).
(b) Determine a model for the circuit (process model).
(c) Identify the general model-based processor structures depicted in Figure 1.9. In each scheme, specify the models for the process, measurement, and noise.

1.5 A communications satellite is placed into orbit and must be maneuvered using thrusters to orientate its antennas. Restricting the problem to the single axis perpendicular to the page, the equations of motion are

$$J\frac{d^2\theta}{dt^2} = T_c + T_d$$

where J is the moment of inertia of the satellite about its center of mass, T_c is the thruster control torque, T_d is the disturbance torque, and θ is the angle of the satellite axis with respect to the inertial reference (no angular acceleration) A. Develop signal and noise models for this problem and identify each model-based processor component.

1.6 Consider a process described by a set of linear differential equations

$$\frac{d^2c}{dt^2} + \frac{dc}{dt} + c = Km$$

The process is to be controlled by a proportional-integral-derivative (PID) control law governed by the equation

$$m = K_p \left(e + \frac{1}{T_i} \int edt + T_d \frac{de}{dt} \right)$$

and the controller reference signal r is given by

$$r = e + c$$

Suppose the reference is subjected to a disturbance signal and the measurement sensor, which is contaminated with additive noise, measures the "square" of the output. Develop the model-based signal and noise models for this problem as in Figure 1.9.

1.7 The elevation of a tracking telescope is controlled by a DC motor. The telescope has a moment of inertia J and damping B due to friction, the equation of motion is given by

$$J\frac{d^2\theta}{dt^2} + B\frac{d\theta}{dt} = T_m + T_d$$

where T_m and T_d are the motor and disturbance torques and θ is the elevation angle. Assume a sensor transforms the telescope elevation into a proportional voltage that is contaminated with noise. Develop the signal and noise models for the telescope and identify all of the model-based processor components.

2

Random Signals and Systems

2.1 Introduction

We can apply a filter to a random signal, but since its output is still random, we must find a way to eliminate or reduce this randomness in order to employ the powerful techniques available from systems theory. We shall show that techniques from statistics combined with linear systems theory can be applied to extract the desired signal information and reject disturbances or noise. In this case, the filter is called an estimation filter or simply an *estimator* that is required to extract the useful information (signal) from noisy or random measurements. Techniques similar to linear deterministic systems theory hold when the random signal is transformed to its *covariance sequence* and its Fourier spectrum is transformed to its *power spectrum*. Once these transformations are accomplished, then the techniques of linear systems theory can be applied to obtain results similar to deterministic signal processing with $x(t) \rightarrow X(\Omega)$. In fact, we know that the covariance sequence and power spectrum form a *discrete* Fourier transform (*DFT*) pair, analogous to a deterministic signal and its corresponding spectrum, that is, we have that

$$\text{Fourier Transform:} \qquad R_{xx}(k) \longleftrightarrow S_{xx}(\Omega)$$

and as in the deterministic case, we can analyze the spectral content of a random signal by investigating its power spectrum. The *power spectral density* (*PSD*) function for a discrete random process is defined as:

$$S_{xx}(\Omega) = \lim_{N\rightarrow\infty} E\left\{ \frac{X(\Omega)X^*(\Omega)}{2N+1} \right\} \tag{2.1}$$

where * is the complex conjugate, $E\{\cdot\}$ is the expected value operation that can be thought of as "mitigating" the randomness. Similarly, the *covariance*[1] of the

1 It is also common to use the so-called *correlation* function, which is merely the mean-squared function and identical to the covariance function for a mean of zero.

Signal Processing: An Applied Decomposition Approach, First Edition. James Vincent Candy.

Companion website: www.wiley.com/go/CandySignProc

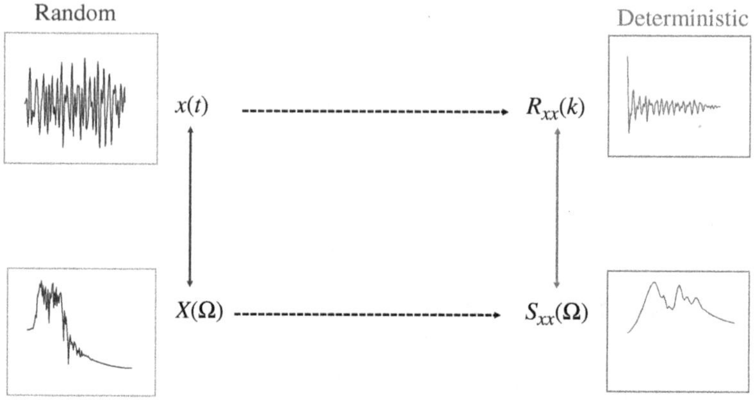

Figure 2.1 Random signal comparison: random signal and spectrum and the deterministic covariance and power spectrum.

process is defined by:

$$R_{xx}(k) := E\{x(t)x(t+k)\} - E^2\{x(t)\} \tag{2.2}$$

In a sense, these relations replace the random signals but play the same role as their deterministic counterparts. These concepts are illustrated in Figure 2.1 where we see the random signal and its random Fourier transform replaced by the deterministic covariance and its deterministic power spectrum.

Techniques of linear systems theory for random signals are valid, just as in the deterministic case where the covariance at the output of a system excited by a random signal $x(t)$, is given by the convolution relationship in the temporal or frequency domain characterized by its impulse response, $h(t)$ or transfer function, $H(\Omega)$, that is,

$$\begin{aligned} &\text{Convolution:} && R_{yy}(k) = h(t) * h(-t) * R_{xx}(k) \\ &\text{Multiplication:} && S_{yy}(\Omega) = H(\Omega)H^*(\Omega)S_{xx}(\Omega) = |H(\Omega)|^2 S_{xx}(\Omega) \end{aligned} \tag{2.3}$$

Analogously, the filtering operation is performed by an estimation filter, $\hat{H}_f$, designed to shape the output *PSD*, similar to the deterministic filtering operation

$$\text{Filtering:} \qquad S_{yy}(\Omega) = |\hat{H}_f(\Omega)|^2 S_{xx}(\Omega) \tag{2.4}$$

Consider the following example of analyzing a random signal using covariance and spectral relations.

Example 2.1 Suppose we have a measured signal given by

$$\underbrace{x(t)}_{\text{Measurement}} = \underbrace{s(t)}_{\text{Signal}} + \underbrace{n(t)}_{\text{Noise}}$$

and we would like to extract the signal and reject the noise, so we design an estimation filter,

$$\begin{aligned} \hat{y}(t) &= \hat{h}_f(t) * x(t) \\ \hat{Y}(\Omega) &= \hat{H}_f(\Omega)X(\Omega) \end{aligned} \tag{2.5}$$

Our signal in this case will be the sinusoids at 10 and 20 Hz contaminated with additive random noise. In Figure 2.2a, we see the random signal and raw (random) discrete Fourier spectrum. Note that the noise severely obscures the sinusoidal signals and there are many false peaks that could erroneously be selected

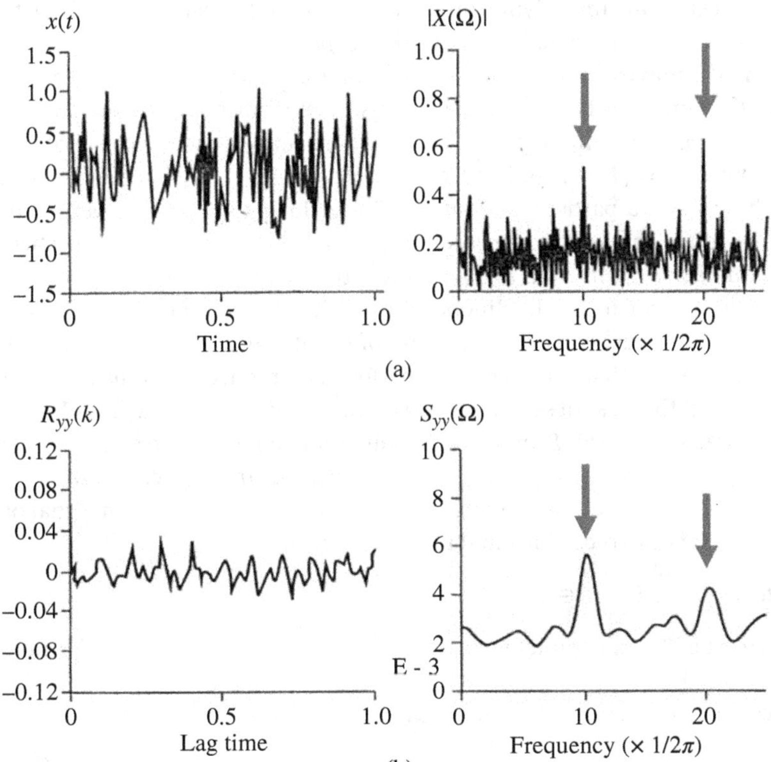

Figure 2.2 Random signal processing: (a) data and spectrum, (b) covariance and power spectrum.

as sinusoidal signals. Estimating the covariance of the random signal along with its corresponding power spectrum using statistical methods, we can now easily see the sinusoidal signals at the prescribed frequencies as depicted in Figure 2.2b. Therefore, analogous to deterministic signal processing techniques, the covariance and power spectra enable the powerful methods of linear systems theory to be applied to extract useful signal information for noisy data. △△△

In this chapter, we are concerned with the development of analysis methods for random signals that will prove useful in applications to follow.

2.2 Discrete Random Signals

A random signal or equivalently a stochastic process finds its roots in probability theory. The idea of an underlying probability or sample space is essential to the treatment of random signals. Recall, that a *probability/sample space*, Ξ, is a collection of samples or experimental outcomes, ξ, which are elements of that space. Certain subsets of Ξ or collections of outcomes are the *events*. For example, if we consider flipping a coin, then the sample space consists of the set of outcomes, $\Xi = \{H, T\}$ and the events, $\{0, \{H\}, \{T\}\}$. When we attempt to specify how likely a particular event will occur during a trial or experiment, then we can define the notion of *probability*. In terms of our probability space, we define a *probability function*, $\Pr(\cdot)$, as a function defined on a class of events that assigns a number to the likelihood of a particular event occurring during an experiment. We must constrain the class of events on which the probability function is defined so that set operations performed produced sets that are still events. Therefore, the class of sets satisfying this constraint is called a *field (Borel)*. If we have a space, Ξ, a field, B, and a probability function, $\Pr(\cdot)$, we can precisely define and *experiment* as the triple, $\{\Xi, B, \Pr\}$. Stochastic processes can be thought of as a collection of indexed random variables. More formally, a random signal or *stochastic process* is a two-dimensional function of t and ξ:

$$X(t, \xi) \qquad \xi \epsilon \Xi, \quad t \in T \tag{2.6}$$

where T is a set of index parameters (continuous or discrete) and Ξ is the sample space.

So we see that a *discrete random signal* can be precisely characterized by

$$x_i(t) := X(t, \xi_i) \tag{2.7}$$

A *realization* of a stochastic process and in fact each distinct value of t can be interpreted as a random variable. Thus, we can consider a stochastic process, simply a sequence of ordered random variables. A collection of realizations of a stochastic process is called an *ensemble*.

A common representation of a random process is a random walk given in the following example.

Example 2.2 (Random Walk) Suppose we toss a fair coin at each time instant, that is,

$$X(t, \xi_i), \qquad t \in T = \{0, 1, \dots, N-1\}, \qquad \xi \in \Xi = \{0, \{H\}, \{T\}\}$$

where the random variable is given by

$$X(t, \xi_i) = \begin{cases} K & \xi_1 = H \\ -K & \xi_2 = T \end{cases}$$

We define the corresponding probability mass function as

$$P_X(X(t, \xi_i) = \pm K) = 1/2$$

Here we see that for each $t \in T, X(t, \xi)$ is just a random variable. Now, if we define the function,

$$y(N, \xi_i) = \sum_{t=0}^{N-1} X(t, \xi_i) \qquad i = 1, 2, \dots$$

as the position at t for the ith realization, then for each t, y is a sum of discrete random variables, $X(., \xi_i)$, and for each i, $X(t, \xi_i)$ is the ith realization of the process (a time series). For a given trial, we may have to say $\xi_1 = \{HHTHTT\}$, and $\xi_2 = \{THHTTT\}$, then for these realizations we see the ensemble shown in Figure 2.3. Note that $y(5, \xi_1) = 0$, and $y(5, \xi_2) = -2K$, which can be predicted using the given random variables and realizations. This completes the example.

△△△

In summary, we see that a random signal can be represented precisely as a discrete stochastic process that is a two-dimensional function of an index set (e.g. time or frequency) and samples of the probability space [1–3]. Since a random signal can be interpreted as a sequence of random variables, we expect the basic statistics and properties to apply as well. In fact, this is the case, the resulting statistic merely becomes a function of the index parameter t (time) in our case. Basic statistics and properties hold for stochastic processes and are listed in Appendix A. Note that the expected value operation implies that, for stochastic processes, these basic statistics are calculated *across* the ensemble. For example, if we want to calculate the mean of a process, that is,

$$m_x(t) = E\{X(t, \xi_i) = x_i(t)\}$$

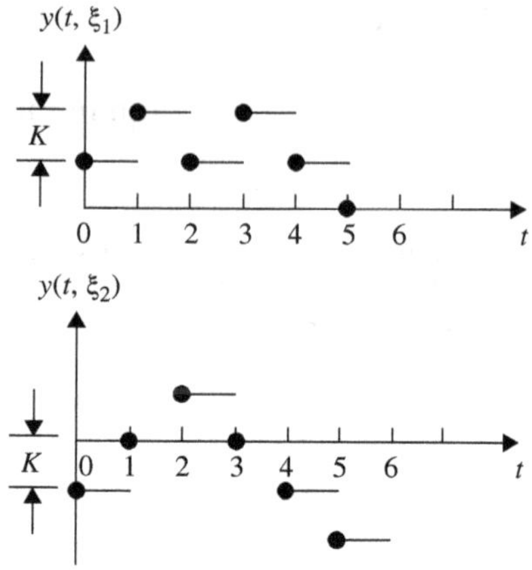

$$y(5,\xi_1) = \sum_{t=0}^{4} x(t,\xi_1) = 0$$

$$y(5,\xi_2) = \sum_{t=0}^{4} x(t,\xi_2) = -2K$$

Figure 2.3 Ensemble of random walk realizations.

we simply take the values of $t = 0, 1, \ldots$ and calculate the mean for each value of time across ($i = 1, 2, \ldots$) the ensemble. So we see that dealing with stochastic processes is similar to dealing with random variables except that we must account for the (time) indices. Knowledge that a random signal is characterized by a stochastic process is not sufficient to make this information useful. We must consider properties of stochastic processes themselves that make them useful, but as before we must first develop the ideas from probability theory and then define them in terms of stochastic processes. Suppose we have more than one random variable, then we define the *joint* probability mass and distribution functions of an N-dimensional random variable as

$$p_X(x_1, \ldots, x_N), \qquad P_X(x_1, \ldots, x_N)$$

All of the basic statistical definitions remain, except that we replace the single variate probability functions with the joint variate functions. Clearly, if we think of a stochastic process as a sequence of ordered random variables, then we are dealing with joint probability functions. Suppose we have two random variables, x_1 and x_2, and we know that the latter has already assumed a particular value, then we can define the *conditional* probability mass function of x_1 given that $X(\xi_2) = x_2$ has occurred by:

$$p(x_1 \mid x_2) := p_X(X(\xi_1) \mid X(\xi_2) = x_2) \tag{2.8}$$

and it can be shown from basic probabilistic axioms (see Papoulis [1]) that

$$p(x_1 \mid x_2) = \frac{p(x_1, x_2)}{p(x_2)} \tag{2.9}$$

Note that this equation can also be written as

$$p(x_1, x_2) = p(x_2 \mid x_1)p(x_1) \tag{2.10}$$

Substituting this relation above gives *Bayes' rule*, that is,

$$p(x_1 \mid x_2) = p(x_2 \mid x_1)\frac{p(x_1)}{p(x_2)} \tag{2.11}$$

If we use the definition of joint mass function, then we obtain the probabilistic *chain rule*,

$$p(x_1, \ldots, x_N) = p(x_1 \mid x_2, \ldots, x_N)p(x_2 \mid x_3, \ldots x_N) \cdots p(x_{N-1} \mid x_N)p(x_N) \tag{2.12}$$

Along with these definitions follows the idea of conditional expectation, that is,

$$E\{x_i \mid x_j\} = \sum_i X_i p(x_i \mid x_j)$$

Note that the conditional expectation is also a *linear* operator that possesses many useful properties, that is,

$$E\{ax_i + b \mid x_j\} = aE\{x_i \mid x_j\} + b$$

With these definitions in mind, we can now examine some further properties of stochastic processes that will prove useful in characterizing random signals. A common class of useful stochastic processes is called Markov processes. Markov processes are defined in terms of a sequence of random variables $\{X(t)\}$, we say that a process is *Markov*, if and only if

$$p(X(t+1) \mid X(t), X(t-1), \ldots, X(1)) = p(X(t+1) \mid X(t)) \tag{2.13}$$

that is, the future $X(t+1)$ depends only on the present $X(t)$ and not the past $\{X(t-1), \ldots, X(1)\}$. Thus, if a process is Markov, then the chain rule simplifies considerably, that is,

$$p(x_1, \ldots, x_N) = p(x_1 \mid x_2)p(x_2 \mid x_3) \cdots p(x_{N-1} \mid x_N)p(x_N)$$

Just as in systems theory, the concept of a time-invariant system is essential for analytical purposes, so is the equivalent concept of stationarity essential for the analysis of stochastic processes. We say that a stochastic process is *stationary*, if the joint mass function is time invariant, that is,

$$p(X(1) \ldots X(N)) = p(X(1+k) \ldots X(N+k)) \quad \forall k \tag{2.14}$$

If this relation is only true for values of $t \leq M < N$ then it is said to be *stationary of order* M. Special cases arising are:

- For $M = 1$, $p(X(t)) = p(X(t+k)) = p(X)\forall k \Rightarrow m_x(t) = m_x$a constant (*mean-stationary*).
- For $M = 2$, $p(X(t_1), X(t_2)) = p(X(t_1+k), X(t_2+k))$ or for $k = t_2 - t_1$, $p(X(t_2), X(t_2+k))$ (*covariance stationary*) which implies

$$R_{xx}(t, t+k) = R_{xx}(k) = E\{x(t)x(t+k)\} - m_x^2$$

- A process that is *both* mean and covariance stationary is called *wide-sense stationary*.
- A process is said to be *ergodic* if its time average, E_T, is identical to its ensemble average, E, that is,

$$E\{x(t, \xi_i)\} \Longleftrightarrow E_T\{x(t)\} = \lim_{N\to 0} \frac{1}{2N+1} \sum_{t=-N}^{N} x(t)$$

- An ergodic process *must* be wide-sense stationary, but the *inverse* does *not* necessarily hold.

This means that *all* of the statistical information in the ensemble can be obtained from *one realization* of the process. For instance, in order to calculate the ensemble mean of this process, we must average across the ensemble for each instant of time to determine, $m_x(t)$; however, if the process is ergodic, then we may select *any member* (realization) of the ensemble and calculate its time average, E_T to obtain the required mean, that is,

$$m_x(t) = E_T\{x_i(t)\} = m_x$$

This discussion completes the introductory concepts of relating random signals to the probabilistic notion of stochastic processes. Throughout this text, the ideas of probability spaces are implicitly defined; sometimes defined that $x(t)$ is a stochastic process and it is assumed that the reader will know that [1–3]

$$x(t) \longrightarrow X(t, \xi_i) \quad t \in T, \quad \xi \epsilon \Xi$$

We now have all of the probabilistic ingredients to begin to characterize useful random signals. In Section 2.3, we discuss the covariance and power spectrum "tools" and show how they can be used to specify the properties of random signals.

2.3 Spectral Representation of Random Signals

In many engineering problems before the design of processing algorithms can proceed, it is necessary to analyze the measured signal. As we have stated, analogous to the deterministic case, the covariance function transforms the random signal and the power spectrum (transforms) its Fourier transform for analysis as illustrated in Figure 2.1. In this section, we derive the relationship between the covariance function and power spectrum and show how they can be used to characterize fundamental random signals.[2]

First, we begin by defining the power spectrum of a discrete random signal. Recall that the *discrete-time Fourier transform* (*DtFT*) pair is given by [4]

$$X(e^{j\Omega}) := \text{DtFT}[x(t)] = \sum_{t=-\infty}^{\infty} x(t)e^{-j\Omega t}$$
$$x(t) = \text{IDtFT}[X(e^{j\Omega})] = \frac{1}{2\pi}\int_{2\pi} X(e^{j\Omega})e^{j\Omega t}d\Omega \tag{2.15}$$

If $x(t)$ is random, then $X(e^{j\Omega})$ is *also* random because

$$x(t, \xi_i) \Longleftrightarrow X(e^{j\Omega}, \xi_i) \quad \forall\, i$$

both are simply realizations of a random signal over the ensemble generated by i. Also, and more importantly, $X(e^{j\Omega})$ for stationary processes almost never exists because any nonzero realization $x(t, \xi_i)$ is not absolutely summable in the ordinary sense (these integrals can be modified (see [2] for details)).

Now let us develop the relationship between the *PSD* and its corresponding covariance. We have

$$S_{xx}(e^{j\Omega}) = \lim_{N\to\infty} \frac{1}{2N+1} E\left\{X_N(e^{j\Omega})X_N^*(e^{j\Omega})\right\} \tag{2.16}$$

Substituting for the *DtFT*, moving the expectation operation inside the summation and introducing a change of variable gives the relation

$$S_{xx}(e^{j\Omega}) = \sum_{k=-\infty}^{\infty} E_T\{R_{xx}(m+k, m)\}e^{-j\Omega k} \tag{2.17}$$

If we further assume that the process is wide-sense stationary, then $R_{xx}(m + k, m) \to R_{xx}(k)$ is no longer a function of time, but only of lag k and therefore,

$$S_{xx}(e^{j\Omega}) = \sum_{k=-\infty}^{\infty} R_{xx}(k)e^{-j\Omega k} \tag{2.18}$$

evolves as the well-known *Wiener–Khintchine* relation with the corresponding covariance given by *IDtFT*

$$R_{xx}(k) = \frac{1}{2\pi}\int_{2\pi} S_{xx}(e^{j\Omega})e^{j\Omega k}d\Omega \tag{2.19}$$

Recall that the *DtFT* is just the $\mathcal{Z}$-transform of $x(t)$ evaluated on the unit circle, that is,

$$S_{xx}(\Omega) = S_{xx}(z)|_{z=e^{j\Omega}}$$

2 We shall return to more mathematically precise notation for this development, that is, $X(\Omega) = X(e^{j\Omega})$.

and we obtain the equivalent pair in terms of the $\mathcal{Z}$-transform as

$$S_{xx}(z) = Z[R_{xx}(k)] = \sum_{k=-\infty}^{\infty} R_{xx}(k)z^{-k}$$

and

$$R_{xx}(k) = \frac{1}{2\pi j} \int S_{xx}(z)z^{k-1}dz \qquad (2.20)$$

Therefore, we can define random signals in terms of their covariances and spectral densities. For example, a purely random or *white-noise* sequence, $e(t)$, is a sequence in which all the $e(t)$ are mutually independent, that is, knowing $e(t)$ in no way can be used to predict $e(t+1)$. A white sequence is called completely random or unpredictable or memoryless (no correlation). Sequences of this type have historically been called white because of their analogy to white light, which possesses all frequencies (constant power spectrum), that is, the *PSD* of white noise is

$$S_{ee}(\Omega) = R_{ee} \qquad (constant)$$

with the corresponding covariance is given by

$$R_{ee}(k) = R_{ee}\delta(k)$$

where R_{ee} is the variance of the noise.

In fact, the white-noise characterization of *random* signals is analogous to the unit impulse representation of *deterministic* signals, that is,

$$R_{ee}\delta(k) \Longleftrightarrow A\delta(t)$$

and

$$S_{ee}(\Omega) = R_{ee} \Longleftrightarrow H(\Omega) = A$$

It is the random counterpart for random systems of the unit impulse excitation for the analysis of linear time-invariant (*LTI*) systems.

Note also that random or white sequences are also Markov since they are uncorrelated:

$$p(e(t) \mid e(t-1), \ldots, e(1)) = p(e(t))$$

If each of the $e(t)$ is also Gaussian-distributed, then as a check for *whiteness* of a sequence, we can perform a statistical hypothesis test (see [3]) using its underlying covariance. The 95% confidence interval or so-called *whiteness test* is given by

$$I = \left[R_{ee}(k) - \frac{1.96R_{ee}(0)}{\sqrt{N}}, R_{ee}(k) + \frac{1.96R_{ee}(0)}{\sqrt{N}}\right] \qquad (2.21)$$

Some properties of the discrete covariance function for stationary processes along with accompanying properties of the *PSD* are given (without proof see [4] for details) in Table 2.1 for reference.

It is important to recognize that these properties are essential to analyze the information available in a discrete random signal. For instance, if we are trying to determine phase information about a particular measured signal, we immediately recognize that the phase is lost in both the auto-covariance and corresponding output power spectrum.

With these properties in mind, we can now reexamine the expression for the *PSD* and decompose it further. From $\mathcal{Z}$-transform relations, we have

$$S_{xx}(z) = \underbrace{\sum_{k=0}^{\infty} R_{xx}(k)z^{-k}}_{S_{xx}^{+}(z)} + \underbrace{\sum_{i=0}^{\infty} R_{xx}(i)z^{i}}_{S_{xx}^{-}(z)} - R_{xx}(0)$$

which is the *sum decomposition* of the *PSD* given by

$$S_{xx}(z) := S_{xx}^{+}(z) + S_{xx}^{-}(z) - R_{xx}(0) \tag{2.22}$$

with the one-sided $\mathcal{Z}$-transform, $S_{xx}^{+}(z) = Z_I\{R_{xx}(k)\}$ and $S_{xx}^{-}(z) = S_{xx}^{+}(z^{-1})$. Note that $S_{xx}^{\pm}(z)$ use the $\pm$ to represent positive and negative time relative to inverse transforms. Also, it can be shown that $S_{xx}^{+}(z)$ has only poles *inside* the unit circle, while $S_{xx}^{-}(z)$ has only those *outside* (see [5] or [6] for details).

Thus, the sum decomposition can be applied to calculate the power spectrum using one-sided transforms of the covariance function. The inverse process is more complicated, since the *PSD* must be decomposed into sums (usually by partial fractions): one having only poles inside the unit circle, the other only poles outside. The corresponding covariance is then determined using inverse $\mathcal{Z}$-transform methods.

In practice, the auto-covariance and power spectrum find most application in the *analysis* of random signals yielding information about spectral content, periodicities, etc., while the cross-covariance and spectrum are used to estimate the properties of two distinct processes (e.g. input and output of a system) to follow.

So we see that the properties of covariance and spectra not only can be used to analyze the information available in random signal data, but also to estimate various signal characteristics. In Section 2.4 we will investigate the properties of linear systems excited by random inputs and use the properties of covariance and spectra to analyze the results.

2.4 Discrete Systems with Random Inputs

When random inputs are applied to linear systems, then covariance and power spectrum techniques must be applied transforming the signal and its Fourier spectrum in deterministic signal theory (see Figure 2.1). In this section, we develop the

Table 2.1 Properties of covariance and spectral functions.

Covariance	Power Spectrum
1. *Average power*:	
$R_{xx}(0) = E\{x^2(t)\}$	$R_{xx}(0) = \frac{1}{2\pi}\int_{2\pi} S_{xx}(z)z^{-1}dz$
2. *Symmetry*:	
$R_{xx}(k) = R_{xx}(-k)$ (even)	$S_{xx}(z) = S_{xx}(z^{-1})$ (even)
$R_{xy}(k) = R_{yx}(-k)$	$S_{xy}(z) = S^*_{yx}(z)$
3. *Maximum*:	
$R_{xx}(0) \geq \lvert R_{xx}(k)\rvert$	$S_{xx}(e^{j\Omega}) \geq 0$
$\frac{1}{2}R_{xx}(0) + \frac{1}{2}R_{yy}(0) \geq \lvert R_{xy}(k)\rvert$	
$R_{xx}(0)R_{yy}(0) \geq \lvert R_{xy}(k)\rvert^2$	
4. *Real*:	
	$S_{xx}(z) = E\{\lvert X(z)\rvert^2\}$ is real
	$S_{xy}(z) = E\{X(z)Y^*(z)\}$ is complex
5. *Sum decomposition*:	
	$S_{xx}(z) = S^+_{xx}(z) + S^-_{xx}(z) - R_{xx}(0)$

relationship between systems and random signals. From linear systems theory, we have the convolution or equivalent frequency relations

$$y(t) = h(t) * x(t) = \sum_{k=0}^{\infty} h(k)x(t-k) \tag{2.23}$$

or taking discrete-time Fourier transforms, we obtain

$$Y(\Omega) = H(\Omega)X(\Omega) \tag{2.24}$$

If we assume that x is a random signal then as we have seen in Section 2.3, we must resort to spectral representations of random processes. Exciting a causal linear system with a zero-mean random signal, we obtain the output power spectrum

$$S_{yy}(\Omega) = E\{Y(\Omega)Y^*(\Omega)\} = E\{(H(\Omega)X(\Omega))\,X^*(\Omega)H^*(\Omega)\} = |H(\Omega)|^2 S_{xx}(\Omega) \tag{2.25}$$

or equivalently

$$S_{yy}(\Omega) = E\{Y(\Omega)Y^*(\Omega)\} = E\{Y(\Omega)\,(X^*(\Omega)H^*(\Omega))\} = S_{yx}(\Omega)H^*(\Omega) \tag{2.26}$$

Table 2.2 Linear system with random inputs: covariance/spectrum relationships.

Covariance	Spectrum
$R_{yy}(k) = h(k) * h(-k) * R_{xx}(k)$	$S_{yy}(z) = H(z)H(z^{-1})S_{xx}(z)$
$R_{yy}(k) = h(k) * R_{xy}(k)$	$S_{yy}(z) = H(z)S_{xy}(z)$
$R_{yx}(k) = h(k) * R_{xx}(k)$	$S_{yx}(z) = H(z)S_{xx}(z)$

where

$$R_{yy}(k) = \frac{1}{2\pi j}\oint S_{yy}(z)z^{k-1} \quad dzS_{yy}(z) = \sum_{k=-\infty}^{\infty} R_{yy}(k)z^{-k}$$

$$R_{xy}(k) = \frac{1}{2\pi j}\oint S_{xy}(z)z^{k-1} \quad dzS_{xy}(z) = \sum_{k=-\infty}^{\infty} R_{xy}(k)z^{-k}$$

Perhaps one of the most important properties that has led to a variety of impulse response identification techniques is

$$S_{yx}(\Omega) = E\{Y(\Omega)X^*(\Omega)\} = E\{H(\Omega)(X(\Omega)X^*(\Omega))\} = H(\Omega)S_{xx}(\Omega) \tag{2.27}$$

Solving for H provides the *Wiener solution* in the frequency domain [7]. Similar results can be obtained for auto and cross-covariances and corresponding spectra. We summarize these linear system relations in Table 2.2.

We can now present an important result that is fundamental for modeling of random signals. Suppose, we have a *LTI* system that is asymptotically stable with rational transfer function. If we excite this system with white noise, we get the so-called *Spectral Factorization Theorem*, which states that

GIVEN a stable, rational, linear system excited by unit variance white noise, then there EXISTS a rational $H(\Omega)$[3] such that:

$$S_{yy}(\Omega) = H(\Omega)H^*(\Omega) = |H(\Omega)|^2 \tag{2.28}$$

where the poles and zeros of $H(\Omega)$ lie within the unit circle.

If we can represent spectral densities in properly factored form, then *all* stationary processes can be thought of as outputs of linear dynamical systems. Synthesizing random signals with given spectra then requires *only* the generation of white-noise sequences. We summarize this discussion with the *Representation Theorem*:[4]

GIVEN a rational spectral density $S_{yy}(\Omega)$, then there EXISTS an asymptotically stable linear system when excited by white noise that produces a stationary output $y(t)$ with this spectrum.

3 Recall that $H(\Omega) = H(e^{j\Omega}) = H(z)|_{z=e^{j\Omega}}$, that is, the $\mathcal{Z}$-transform evaluated along the $j\Omega$-axis.
4 This is really a restricted variant of the famous Wold decomposition for stationary random signals (see [5] for details).

The simulation of random signals with given statistics requires the construction of the power spectrum from the sum decomposition followed by a spectral factorization. This leads to the *spectrum simulation procedure*: (i) calculate $S_{yy}(z)$ from the sum decomposition; (ii) perform the spectral factorization to obtain $H(z)$; (iii) generate a white-noise sequence of variance R_{xx}; and (iv) excite the system $H(z)$ with the sequence [7].

The most difficult part of this procedure is to perform the spectral factorization. For simple systems, the factorization can be performed by equating coefficients of the known $S_{yy}(z)$, obtained in step 1, with the unknown coefficients of the spectral factor and solving the resulting nonlinear algebraic equations, that is,

$$S_{yy}(z) = \begin{cases} S_{yy}^{+}(z) + S_{yy}^{-}(z) - R_{yy}(0) & = \frac{N_s(z,z^{-1})}{D_s(z,z^{-1})} \\ \\ H(z)H(z^{-1})R_{xx} & = \frac{N_H(z)}{D_H(z)}\frac{N_H(z^{-1})}{D_H(z^{-1})}R_{xx} \end{cases} \tag{2.29}$$

$$N_s(z, z^{-1}) = N_H(z)N_H(z^{-1}) \quad \text{and} \quad D_s(z, z^{-1}) = D_H(z)D_H(z^{-1})$$

For higher order systems, more efficient iterative techniques exist including multidimensional systems (see [8–10]). Next, we consider a versatile model that can be applied to represent random signals.

2.5 Classical Spectral Estimation

Spectral estimation techniques have been developed and improved over the years with one major task in mind—the analysis of random data. Based on the previous discussion in this chapter, a majority of the initial effort was focused on applying the Wiener–Khintchine theorem and transform theory, while modern parametric techniques evolved primarily from the speech community. In this section, we begin with the classical (nonparametric) methods that are viable when *large* (long) data records are available and then evolve to the development of a select group of popular parametric techniques [4]. We make no attempt to provide detailed derivations of the algorithms that are available in other texts [4, 11, 12], but just follow a brief outline of the approach and present the final results.

We conclude this chapter with a case study of noisy, sinusoidal signals. With the initial application of Fourier analysis techniques to sun-spot data over 200 years ago, the seeds of spectral estimation were sown by Schuster [13]. Fourier analysis for random signals evolved rapidly after the discovery of the Wiener–Khintchine theorem relating the covariance and power spectrum. Finally with the evolution of the fast Fourier transform (see Cooley and Tukey [14]) and digital computers, all of the essential ingredients were present to establish the classical approach to nonparametric spectral estimation. Classical spectral estimators typically fall into two categories: direct and indirect. The direct methods operate directly on the raw data to transform it into the frequency domain and produce the estimate. Indirect methods first estimate the covariance sequence and then transform it to the frequency domain—an application of the Wiener–Khintchine theorem [4]. First, we develop two basic nonparametric spectral estimation techniques: the correlation method (indirect) and the periodogram method (direct), and then consider a more modern technique—the minimum variance distortionless response (*MVDR*) method.

2.5.1 Correlation Method (Blackman–Tukey)

The *correlation method* or sometimes called the Blackman–Tukey method is simply an implementation of the Wiener–Khintchine theorem: the covariance is obtained using a sample covariance estimator and then the *PSD* is estimated by calculating the *DFT*. The *DFT* transform pair is defined by

$$X(\Omega_m) := DFT\,[x(t)] = \sum_{t=0}^{M-1} x(t)e^{-j\Omega_m t}$$

$$x(t) = IDFT\left[X(\Omega_m)\right] = \frac{1}{M}\sum_{t=0}^{M-1} X(\Omega_m)e^{j\Omega_m t} \tag{2.30}$$

for $\Omega_m = \left(\frac{2\pi}{M}\right) m$ where it can be thought of as the *DtFT* with $\Omega \rightarrow \Omega_m$, that is, the *DtFT* sampled uniformly around the unit circle [4].

Therefore, we have that

$$\hat{S}_{xx}(\Omega_m) = DFT\left[\hat{R}_{xx}(k)\right]$$

$$\hat{R}_{xx}(k) = IDFT\left[\hat{S}_{xx}(\Omega_m)\right]$$

This technique tends to produce a noisy spectral estimate; however, a smoothed estimate can be obtained by multiplying R_{xx} by a *window function*, $\mathcal{W}(k)$ usually called a *lag window*. The window primarily reduces spectral leakage and therefore improves the estimate. It is also interesting to note that a sample covariance estimator does not guarantee the positivity of the *PSD* (auto) when estimated directly from the Wiener–Khintchine theorem [4]. However, if the estimator is implemented directly in the Fourier domain, then it will preserve this property, since it is the *square* of the Fourier spectrum. We summarize the *correlation method* (Blackman–Tukey) of spectral estimation by:[5]

- *Calculate* the *DFT* of $x(t)$, that is, $X(\Omega_m)$
- *Multiply* $X(\Omega_m)$ by its conjugate to obtain, $X(\Omega_m)X^*(\Omega_m)$

5 Note also if we replace X^* by Y^* we can estimate the cross-correlation $\hat{R}_{xy}(k)$ and corresponding cross-spectrum $\hat{S}_{xy}(\Omega_m)$ using this method.

- *Estimate* the covariance from the *IDFT*, $\hat{R}_{xx}(k) = IDFT\left[|X(\Omega_m)|^2\right]$
- *Multiply* the covariance by the lag window, $\mathcal{W}(k)$, and
- *Estimate* the *PSD* from the *DFT* of the windowed covariance, $\hat{S}_{xx}(\Omega_m) = DFT\left[\hat{R}_{xx}(k)\mathcal{W}(k)\right]$

These correlation spectral estimates are statistically improved by using the lag or equivalently spectral window.[6] With practical window selection and long data records, the correlation method can be effectively utilized to estimate the *PSD* (see [4] for more details).

2.5.2 Average Periodogram Method (Welch)

Next, we consider a more direct approach to estimate the *PSD*. We introduce the concept of a periodogram estimator with statistical properties equivalent to the correlation method, then we show how to improve these estimates by statistical averaging and window smoothing leading to *Welch's method* of spectral estimation. The periodogram was devised by statisticians to detect periodicities in noisy data records [13]. The improved method of spectral estimation based on the so-called *periodogram* defined by

$$P_{xx}(\Omega_m) := \frac{1}{N}\left(X(\Omega_m)\, X^*(\Omega_m)\right) = \frac{1}{N}|X(\Omega_m)|^2$$

The samples of P_{xx} are uncorrelated, suggesting that one way of reducing the variance in P_{xx} is to *average* individual periodograms obtained by *sectioning* the original N-point data record into K, L-point sections, that is,

$$\hat{S}_{xx}(\Omega_m) = \frac{1}{K}\sum_{i=1}^{K}\hat{P}_{xx}(\Omega_m, i) \tag{2.31}$$

where $\hat{P}_{xx}(\Omega_m, i)$ is the ith, L-point periodogram. If x is stationary, then it can be shown that this estimate is consistent, since the variance approaches zero as the number of sections becomes infinite [4]. For the periodogram estimator, we have

$$\text{var} \propto \frac{1}{K} \quad \text{and} \quad \text{bias} \propto \frac{K}{N}$$

So we see that for K large, the variance is inversely proportional to K, while the bias is directly proportional. Therefore, for a *fixed* record length N, as the number of periodograms increases, variance decreases, but the bias increases. This is the basic trade-off between precision (variance) and resolution (bias), which can be used to determine a priori the required record length $N = LK$ for an acceptable

6 The window function is called a *lag window* in the time or lag domain, $\mathcal{W}(k)$, and a *spectral window* in the frequency domain, $\mathcal{W}(\Omega_m)$, with its maximum at the origin to match that property of the autocorrelation function; therefore, it is sometimes called a "half" window.

variance. A *full* window, $\mathcal{W}(t)$, can also be applied to obtain a smoothed spectral estimate.

Welch [15] introduced a modification of the original procedure. The data are sectioned into K records of length L; however, the window is applied directly to the segmented records *before* periodogram computation. The modified periodograms are then

$$\hat{P}(\Omega_m, i) = \frac{1}{U}\left|DFT\left[x_i(t)\mathcal{W}_i(t)\right]\right|^2 \quad i = 1, \ldots, K$$

where

$$U = \frac{1}{L}\sum_{t=0}^{L-1}\mathcal{W}^2(t)$$

and

$$\hat{S}_{xx}(\Omega_m) = \frac{1}{K}\sum_{i=1}^{K}\hat{P}(\Omega_m, i)$$

We summarize the *average periodogram method* (Welch's procedure) by:

- *Section* the data, $\{x(t)\}$, $t = 1, \ldots, N$ into K sections, each of length L, where $K = \frac{N}{L}$, that is,

$$x_i(t) = x(t + L(i-1)), \quad i = 1, \ldots, K, \quad t = 0, \ldots, L-1$$

- *Window* the data to obtain, $x_i(t) \times \mathcal{W}_i(t)$
- *Estimate* K periodograms using the DFT as

$$\hat{P}(\Omega_m, i) = \frac{1}{U}\left|DFT\left[x_i(t)\mathcal{W}_i(t)\right]\right|^2 \quad i = 1, \ldots, K$$

  with $U = \frac{1}{L}\sum_{t=0}^{L-1}\mathcal{W}^2(t)$
- *Estimate* the average spectrum using

$$\hat{S}_{xx}(\Omega_m) = \frac{1}{K}\sum_{i=1}^{K}\hat{P}(\Omega_m, i)$$

  with $\text{var}\{\hat{S}_{xx}(\Omega_m)\} \propto \frac{1}{K}$ and $\text{bias}\{\hat{S}_{xx}(\Omega_m)\} \propto \frac{K}{N}$ adjusted for particular windows.

With these metrics in mind, *local bias* can be decreased by decreasing the spectral window main lobe bandwidth making it narrower increasing frequency resolution leading to an asymptotically unbiased estimate. *Broadband bias* evolving from spectral leakage created by the underlying side lobes leaks the broad spectral components into the estimate at a given frequency. Knowledge of these window-based detriments leads to an effective method to design spectral estimators (as before), since

$$S_{xx}(\Omega) = \underbrace{|\mathcal{W}(\Omega)}_{\text{Spectral Window}} * X(\Omega)|^2 \tag{2.32}$$

Therefore, if $\mathcal{W}(\Omega)$ is designed with small side lobes, then leakage is minimized reducing broadband bias, but of course, the trade-off is that decreased side lobes lead to *wider* main lobe bandwidth decreasing *spectral resolution*.

This completes the discussion of the classical or nonparametric methods of spectral estimation. Even though they are considered classical techniques with limited resolution capability, they still can provide us with reasonable information, if we have "long" data records with good signal levels.

Next, we consider another classical methodology that does not directly rely on window length and averaging as in the previously discussed techniques. It is sometimes classified as *parametric* because of its relationship to all-pole models, but here we present it as classical since it is based on estimating the correlation matrix [4].

2.5.3 Minimum Variance Distortionless Response (*MVDR*)

The *MVDR* spectral estimation method relies on estimating the *PSD* by measuring the power through a set of narrowband windows (filters) [16] as introduced in [17] for seismic arrays. It is similar to the classical correlation/periodogram/taper techniques except that the narrowband windows of *MVDR* are different for each frequency, while those windows of the classical techniques are fixed. In this way, the windows of the *MVDR* approach can be thought of as adapting to the process for which *PSD* is desired. It is, therefore, considered a "data adaptive" method. The idea behind the *MVDR* spectral estimator is that for a given discrete frequency, Ω_m, the signal to be extracted is characterized by a complex exponential (narrowband), that is,

$$x(t) = Ae^{j\Omega_m t} \tag{2.33}$$

and

$$X(\Omega) = A\delta(\Omega - \Omega_m)$$

The measured output signal is assumed contaminated with additive white noise

$$y(t) = x(t) + n(t) \tag{2.34}$$

A narrowband window is to be designed so that when the window is convolved with the output, the *variance* is minimized and the frequency under consideration, Ω_m, is passed *undistorted*.

We assume that the window estimator is of the form

$$\hat{y}(t) = h_{\mathcal{W}}(t) * y(t) = \sum_{k=0}^{N} h_{\mathcal{W}}(k)y(t-k) \tag{2.35}$$

Since the window is designed to pass the narrowband signal $x(t)$, we would like to pass $Ae^{j\Omega_m t}$ with unity gain or

$$H_{\mathcal{W}}(\Omega)|_{\Omega=\Omega_m} = 1$$

In order to achieve this, we also require the estimate to be *unbiased*

$$E\{\hat{y}(t)\} = x(t) = Ae^{j\Omega_m t} = \sum_{k=0}^{N} h_{\mathcal{W}}(k)E\{y(t-k)\}$$

but from this relation, we have that

$$Ae^{j\Omega_m t} = \sum_{k=0}^{N} h_{\mathcal{W}}(k)E\{x(t-k) + n(t-k)\}$$

or

$$Ae^{j\Omega_m t} = \sum_{k=0}^{N} h_{\mathcal{W}}(k)\left(Ae^{j\Omega_m(t-k)}\right) \tag{2.36}$$

Dividing, we obtain the desired result

$$1 = \sum_{k=0}^{N} h_{\mathcal{W}}(k)e^{-jk\Omega_m} = H_{\mathcal{W}}(\Omega)|_{\Omega=\Omega_m} \tag{2.37}$$

which implies that the window estimator will pass the complex narrowband signal with unity gain when $\Omega = \Omega_m$. We can express this in vector notation by expanding the sum as

$$\underbrace{\left[1\; e^{-j\Omega_m}\; \dots\; e^{-jN\Omega_m}\right]}_{\mathbf{V}'(\Omega_m)} \underbrace{\begin{bmatrix} h_{\mathcal{W}}(0) \\ h_{\mathcal{W}}(1) \\ \vdots \\ h_{\mathcal{W}}(N) \end{bmatrix}}_{\mathbf{h}_{\mathcal{W}}} = \mathbf{V}'(\Omega_m)\mathbf{h}_{\mathcal{W}} = 1 \tag{2.38}$$

The *MVDR* estimator is obtained by minimizing the variance

$$\min_{\mathbf{h}_{\mathcal{W}}} J = \mathbf{h}'_{\mathcal{W}}\mathbf{R}_{yy}\mathbf{h}_{\mathcal{W}} \tag{2.39}$$

subject to the constraint that $\mathbf{V}'(\Omega_m)\mathbf{h}_{\mathcal{W}} = 1$. This window estimator can be derived using the Lagrange multiplier method of transforming the constrained problem to unconstrained by augmenting the Lagrange multiplier λ into the cost function and minimizing [18]

$$\min_{\mathbf{h}_{\mathcal{W}}} J = \mathbf{h}'_{\mathcal{W}}\mathbf{R}_{yy}\mathbf{h}_{\mathcal{W}} + \lambda(1 - \mathbf{V}'(\Omega_m)\mathbf{h}_{\mathcal{W}}) \tag{2.40}$$

Performing the minimization and substituting for the constraint yields the window weights as

$$\mathbf{h}_{\mathcal{W}} = \frac{\mathbf{R}_{yy}^{-1}\mathbf{V}(\Omega_m)}{\mathbf{V}'(\Omega_m)\mathbf{R}_{yy}^{-1}\mathbf{V}(\Omega_m)} \tag{2.41}$$

Substituting this expression into J gives the *MVDR spectral estimator* at Ω_m as

$$S_{MVDR}(\Omega_m) = \frac{\triangle T}{\mathbf{V}'(\Omega_m)\mathbf{R}_{yy}^{-1}\mathbf{V}(\Omega_m)} = \triangle T\mathbf{h}_{\mathcal{W}}'\mathbf{R}_{yy}\mathbf{h}_{\mathcal{W}} \tag{2.42}$$

which implies that the spectral estimate at a given frequency Ω_m and sampling interval $\triangle T$ is obtained by scaling the output covariance appropriately at the desired frequency. In practice, *MVDR* exhibits more resolution than the classical correlation and periodogram estimators, but less than any of the parametric estimators [19]. We also note that the peak of *MVDR* spectral estimate for a narrowband process is proportional to the power, while for parametric it is proportional to the square of the power.

This completes the discussion of the classical or nonparametric methods of spectral estimation [19]. Even though they are considered classical techniques with limited resolution capability, they still can provide reasonable analytical information, if we have "long" data records with good signal levels. In fact, one other metric that is related to provide an estimate of the quality of the data that employs the spectral estimators is the so-called *coherence function* that is the equivalent of the correlation coefficient in the frequency domain.

2.5.4 Coherence Function

In this section, we develop the coherence function and discuss how it can be applied as a tool to investigate the quality of measured data as well as the linearity of the system under investigation (see [20] for more details). As before, we define the ordinary *squared-coherence function* by

$$\gamma_{xy}^2(\Omega) := \frac{|S_{xy}(\Omega)|^2}{S_{xx}(\Omega)S_{yy}(\Omega)} \tag{2.43}$$

where x is the input and y is the measurement or output. It follows immediately from the properties of the *PSD* that the coherence function satisfies

$$0 \leq \gamma_{xy}^2(\Omega) \leq 1 \tag{2.44}$$

It also follows that if the system under investigation is linear then from the spectral properties

$$S_{xy}(\Omega) = H(\Omega)S_{xx}(\Omega) \quad \text{and} \quad S_{yy}(\Omega) = |H(\Omega)|^2 S_{xx}(\Omega) \tag{2.45}$$

then

$$\gamma_{xy}^2(\Omega) = \frac{|H(\Omega)S_{xx}(\Omega)|^2}{S_{xx}(\Omega)(|H(\Omega)|^2 S_{xx}(\Omega))} = 1 \quad \forall \quad \Omega \tag{2.46}$$

Another way of examining this function follows directly from the first relation, that is,

$$S_{yx}(\Omega) = H^{-1}(\Omega)S_{yy}(\Omega) = S_{xy}^*(\Omega) \tag{2.47}$$

or

$$H^{-1}(\Omega) = \frac{S_{xy}^*(\Omega)}{S_{yy}(\Omega)} \tag{2.48}$$

Clearly, we see that

$$\gamma_{xy}^2(\Omega) = \left[\frac{S_{xy}(\Omega)}{S_{xx}(\Omega)}\right]\left[\frac{S_{xy}^*(\Omega)}{S_{yy}(\Omega)}\right] = H(\Omega)H^{-1}(\Omega) \tag{2.49}$$

and in the ideal case, the constraints must be satisfied. From this relation, we see that the coherence function can be used to determine the frequency band over which data or a system can be considered linear. If it is *perfectly linear*, then the transfer function times its inverse leads to a perfectly *coherent* ($\gamma_{xy}^2(\Omega) = 1$) system. On the other hand, if x and y are uncorrelated then $S_{xy}(\Omega) = 0$ and, therefore, completely *incoherent* ($\gamma_{xy}^2(\Omega) = 0$). We summarize these notions as

$$\gamma_{xy}^2(\Omega) = \begin{cases} 0 & \text{Incoherent} \\ 1 & \text{Coherent} \end{cases} \tag{2.50}$$

In practice, the coherence function ranges between 0 and 1 under the following conditions: (i) measurement noise, (ii) nonlinearities in the system, (iii) biased spectral estimates, and (iv) output due to other inputs. The coherence function can be estimated in a number of ways. We can use any of the spectral estimation techniques of the Section 2.5 for the (auto) *PSD*, but the cross-spectrum is typically estimated with the classical techniques, which utilize both input and output sequences. So, for example, we can use either the correlation or periodogram method to give the estimate

$$\hat{\gamma}_{xy}^2(\Omega) = \frac{|\hat{S}_{xy}(\Omega)|^2}{\hat{S}_{xx}(\Omega)\hat{S}_{yy}(\Omega)} \tag{2.51}$$

or we can use some of the modern parameter estimation techniques, that is,

$$\hat{\gamma}_{xy}^2(\Omega) = \hat{H}(\Omega)\hat{H}^{-1}(\Omega) \tag{2.52}$$

Note that in using this relation the parameter estimation for the inverse transfer function is obtained by reversing the input and output sequences.

$$S_{yy}(\Omega) = E\{Y(\Omega)Y^*(\Omega)\} = |H(\Omega)|^2 S_{xx}(\Omega) + S_{nn}(\Omega) \tag{2.53}$$

and

$$S_{xy}(\Omega) = E\{X(\Omega)Y^*(\Omega)\} = H(\Omega)S_{xx}(\Omega) + S_{xn}(\Omega) = H(\Omega)S_{xx}(\Omega) \tag{2.54}$$

Substituting into the expression for the coherence function, we obtain

$$\gamma_{xy}^2(\Omega) = \frac{|H(\Omega)|^2 S_{xx}^2(\Omega)}{S_{xx}(\Omega)S_{yy}(\Omega)} = \frac{|H(\Omega)|^2 S_{xx}(\Omega)}{|H(\Omega)|^2 S_{xx}(\Omega) + S_{nn}(\Omega)} = \frac{1}{1 + \frac{R_{nn}}{|H(\Omega)|^2 S_{xx}(\Omega)}} \tag{2.55}$$

So we see that for R_{nn} small $\gamma_{xy}^2(\Omega) \to 1$, while large R_{nn} implies that $\gamma_{xy}^2(\Omega) \to 0$. If we are interested in determining that function of the output spectrum linearly due to the input, we can define

$$\begin{aligned} S_{\hat{y}\hat{y}}(\Omega) &= |H(\Omega)|^2 S_{xx}(\Omega) = \left[\frac{S_{xy}(\Omega)}{S_{xx}(\Omega)}\right]^2 S_{xx}(\Omega) \\ &= \left(\frac{|S_{xy}(\Omega)|^2}{S_{xx}(\Omega)S_{yy}(\Omega)}\right) S_{yy}(\Omega) = \gamma_{xy}^2(\Omega) S_{yy}(\Omega) \end{aligned}$$

Here $S_{\hat{y}\hat{y}}(\Omega)$ is called the *coherent output spectrum*. In any case, we see that the coherence function can be utilized to analyze the linearity of a system under investigation as well as the quality of measured data. This completes the section on coherence function estimation and application. We can use the coherence function to determine the quality of our measured data by assuring ourselves that it is reasonably close to unity over the frequency range of interest, similar to the correlation coefficient in statistics over our frequency band of interest. Next, we consider a case study of sinusoids in noise to illustrate the various spectral estimators.

2.6 Case Study: Sinusoids in Noise

Sinusoidal signals occur quite frequently in applications like radar and sonar primarily for target localization and tracking. In this section, we apply the spectral estimation techniques to noisy data generated by filtering a suite of sinusoidal signals (35, 50, 56, and 90 Hz) through a bandpass filter (30–70 Hz) in order to provide a test for each of the methods.

Suppose we generate a 4096-point test signal sampled at 0.001 sec by filtering sinusoids at 35, 50, 56, and 90 Hz contaminated with unit variance white noise

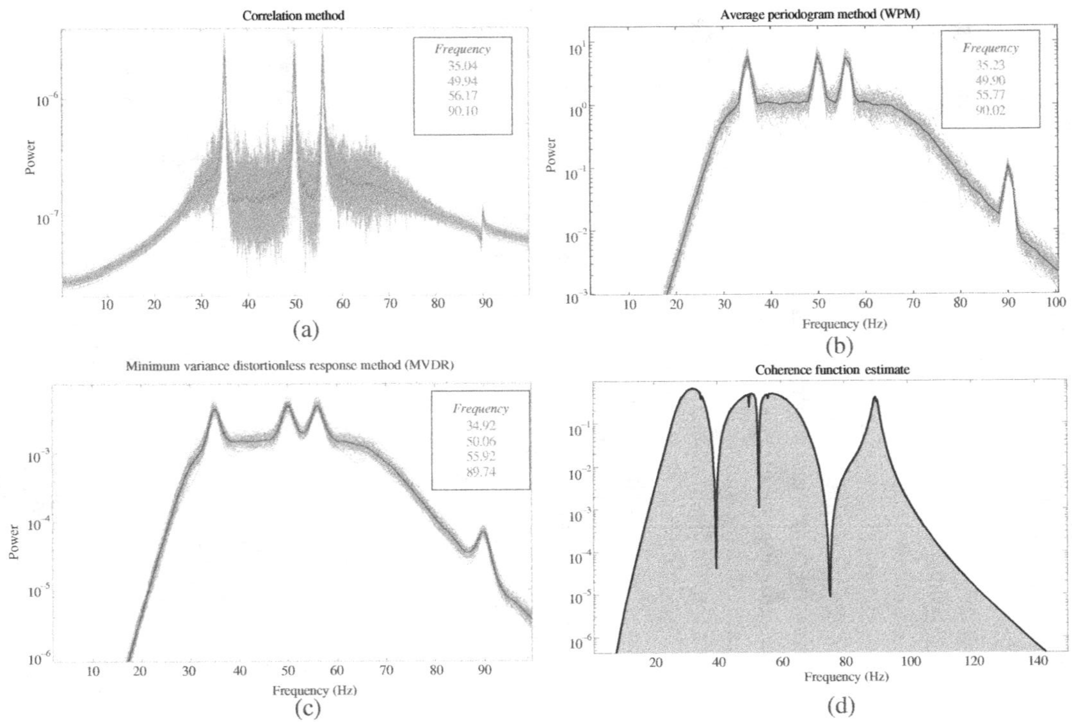

Figure 2.4 Sinusoids in noise (−12 dB) problem (35, 50, 56, 90 Hz) spectral estimation: (a) correlation method (Blackman–Tukey), (b) average periodogram (*WPM*), (c) minimum variance distortionless response (*MVDR*), (d) coherence function (average).

($SNR = -12$ dB)[7] through a bandpass Butterworth (third-order) filter with cutoff frequencies at 30 and 70 Hz. Here, we generate a 100-member ensemble of these bandpass realizations and apply the classical methods of spectral estimation with the results for each shown in Figure 2.4. Here, the *average spectrum* was used to report the final frequency estimates (box insert) using peak detection.

First, we apply the correlation method with the results shown in Figure 2.4a with the thick line of the average spectrum. Here we see a set of sharp peaks with the sinusoids clearly detected in the noisy estimates at spectral peaks (sinusoids) located at (35.04, 49.94, 56.17, 90.10Hz). The average periodogram method (*WPM*) tends to over-smooth the data with the four sinusoids detected at spectral peaks (sinusoids) located at (35.23, 49.90, 55.77, 90.02 Hz) as shown in Figure 2.4b. Next, we investigate the performance of the improved *MVDR* method based on narrowband, adaptive spectral windows for its enhanced performance with its spectral peaks (sinusoids) located at 34.92, 50.06, 55.92, and 89.74 Hz as shown in Figure 2.4c, indicating both a smooth spectrum and good estimates of the sinusoids. The quality of the synthesized data is quite high as confirmed by the estimated average coherence shown in Figure 2.4d where the narrow frequency bands of interest possess coherence amplitudes greater than 0.8.

Summarizing, we see that even though noisy, the classical methods are able to extract the sinusoids with the added advantage of some "smoothing"–a feature that requires long data records. The *MVDR* method is between both classical and parametric methods in performance [19, 20]. Here, the sinusoids are detected along with the bandpass response. This completes the case study.

2.7 Summary

In this chapter, we discussed the fundamental concepts underlying random signal analysis. We discussed the evolution of stochastic processes as models of phenomenological events ranging from coin-flipping to a linear dynamical system. We also discussed special processes (e.g. Gaussian, Markov) and properties of stochastic processes. Under the assumption of stationarity, we developed spectral representations and the concept of simulating a stationary process with given covariance by driving a linear system with white noise. Using these ideas, we developed covariance estimators and employed them in the classical nonparametric spectral estimation schemes: the *correlation* and *periodogram* methods. We then developed the modern classical method of spectral estimation investigating the *MVDR* method. Finally, we concluded with a case study involving the recovery of sinusoids in noise.

7 Define $SNR_{out} := \frac{\text{Signal Energy}}{\text{Noise Variance}}$.

MATLAB® Notes

MATLAB® and its *Signal Processing Toolbox* (**signal**) can be used to design non-parametric spectral estimators. The classical *PSD* methods are available in the periodogram (**periodogram**) and (**pwelch**) along with the coherence function estimator (**cohere**) for a data quality assurance metric. A wealth of model-based spectral estimators are also available to solve for the harmonic temporal as well as spatial models.

References

1 Papoulis, A. (1965). *Probability, Random Variables and Stochastic Processes.* New York: McGraw-Hill.

2 Jazwinski, A. (1970). *Stochastic Processes and Filtering Theory*. New York: Academic Press.

3 Hogg, R. and Craig, A. (1970). *Introduction to Mathematical Statistics*. New York: MacMillan.

4 Candy, J. (1988). *Signal Processing: The Modern Approach*. New York: McGraw-Hill.

5 Astrom, K. (1970). *Introduction to Stochastic Control Theory*. New York: Academic Press.

6 Tretter, S. (1976). *Introduction to Discrete-Time Signal Processing*. New York: Wiley.

7 Candy, J. (2006). *Model-Based Signal Processing*. Hoboken, NJ: Wiley/IEEE Press.

8 Bauer, F. (1955). A direct iterative process for the Hurwitz decomposition of a polynomial. *Arch. Elect. Ubertr.* 9 (1): 285–290.

9 Kailath, T. and Rissanen, J. (1972). Partial realization of random systems. *Automatica* 8 (4): 389–396.

10 Rissanen, J. (1973). Algorithm for the triangular decomposition of block Hankel and Toeplitz matrices with application to factorizing positive matrix polynomials. *Math. Comput.* 17: 147–154.

11 Stoica, P. and Moses, R. (1997). *Introduction to Spectral Estimation*. Englewood Cliffs, NJ: Prentice-Hall.

12 Kay, S. (1988). *Modern Spectral Estimation: Theory and Applications.* Englewood Cliffs, NJ: Prentice-Hall.

13 Schuster, A. (1898). On the investigation of hidden periodicities with application to a supposed 26 day period of meteorological phenomena. *Terr. Magn.* 3 (1): 13–41.

14 Cooley, J. and Tukey, J. (1965). An algorithm for machine calculation of complex Fourier series. *Math. Comput.* 297–301.

15 Welch, P. (1967). The use of fast Fourier transforms for the estimation of power spectra: a method based on time averaging over short modified periodograms. *IEEE Trans. Audio Electroacoust.* AU-15 (2): 70–73.

16 Markel, J. and Gray, A. (1976). *Linear Prediction of Speech*. New York: Springer.

17 Capon, J. (1969). High-resolution frequency-wavenumber spectrum analysis. *Proc. IEEE* 57 (8): 1408–1418.

18 Stewart, G. (1973). *Introduction to Matrix Computations*. New York: Academic Press.

19 Kay, S. and Marple, L. (1981). Spectrum analysis—a modern perspective. *Proc. IEEE* 69 (11): 1380–1419.

20 Bendat, J. and Piersol, A. (1980). *Engineering Applications of Correlation and Spectral Analysis*. New York: Wiley.

Problems

2.1 Suppose the stochastic process $\{y(t)\}$ is generated by

$$y(t) = a\ \exp(-t) + bt, \quad a, b\,\text{random, then}$$

(a) What is the mean of the process?
(b) What is the corresponding covariance?
(c) Is the process stationary, if $E\{a\} = E\{b\} = 0$, and $E\{ab\} = 0$.

2.2 Assume $y(t)$ is a zero mean, ergodic process with covariance $R_{yy}(k)$, calculate the corresponding power spectra, $S_{yy}(z)$ if
(a) $R_{yy}(k) = Ca^{|k|}$.
(b) $R_{yy}(k) = C\ \cos(w|k|)$, $w < \frac{\pi}{2}$.
(c) $R_{yy}(k) = C\ \exp(-a^{|k|})$.

2.3 Verify the covariance-spectral density pairs for the discrete process:
(a) Band limited white noise.
(b) Triangular.

2.4 Let the impulse response of a linear system with random input $u(t)$ be given by $h(t)$, then show that
(a) $R_{yy}(k) = \sum_{m=0}^{\infty} \sum_{i=0}^{\infty} h(m)h(i)R_{uu}(k+i-m)$ and $S_{yy}(z) = H(z)H(z^{-1})$ $S_{uu}(z)$.
(b) $R_{yy}(k) = \sum_{m=0}^{\infty} h(m)R_{yu}(k-m)$ and $S_{yy}(z) = H(z)S_{yu}(z)$.
(c) $R_{uy}(k) = \sum_{m=0}^{\infty} h(m)R_{uu}(k-m)$ and $S_{uy}(z) = H(z)S_{uu}(z)$.

2.5 Derive the *sum decomposition* relation,

$$S_{yy}(z) = S_{yy}^{+}(z) + S_{yy}^{-}(z) - R_{yy}(0)$$

2.6 Develop a computer program to simulate the process

$$y(t) = -ay(t-1) + e(t)$$

where $a = 0.75$, $e \approx N(0, 0.1)$ for 100 data points.
(a) Calculate the analytic covariance $R_{yy}(k)$.
(b) Determine an expression to "recursively" calculate, $R_{yy}(k)$.
(c) Plot the simulated results and construct the $\pm 2\sqrt{R_{yy}(0)}$ bounds.
(d) Do 95% of the samples fall within these bounds?

2.7 Develop the digital filter to simulate a sequence, $y(t)$ with covariance $R_{yy}(k) = 4e^{-3|k|}$.

2.8 Suppose we are given a linear system characterized by transfer function

$$H(z) = \frac{1 - 1/2z^{-1}}{1 - 1/3z^{-1}}$$

which is excited by discrete exponentially correlated noise

$$R_{xx}(k) = (1/2)^{|k|}$$

(a) Determine the output *PSD*, $S_{yy}(z)$;
(b) Determine the output covariance, $R_{yy}(k)$;
(c) Determine the cross-spectrum, $S_{yx}(z)$; and
(d) Determine the cross-covariance, $R_{yx}(k)$.

2.9 Suppose we are given a causal *LTI* system characterized by its impulse response, $h(t)$. If this system is excited by zero mean, unit variance white noise, then
(a) Determine the output variance, $R_{yy}(0)$;
(b) Determine the covariance, $R_{yy}(k)$ for $k > 0$;
(c) Suppose the system transfer function is given by

$$H(z) = \frac{1 + b_0 z^{-1}}{1 + a_1 z^{-1} + a_2 z^{-2}}$$

Find a method to recursively calculate $h(t)$ and therefore $R_{yy}(0)$.

2.10 Given the covariance function

$$R_{yy}(k) = e^{-1/2|k|} \cos \pi|k|$$

Find the digital filter when driven by unit variance white noise produces a sequence $\{y(t)\}$ with these statistics.

2.11 Suppose we have a process characterized by difference equation

$$y(t) = x(t) + 1/2x(t-1) + 1/3x(t-2)$$

(a) Determine a recursion for the output covariance, $R_{yy}(k)$.
(b) If $x(t)$ is white with variance σ_{xx}^2, determine $R_{yy}(k)$.
(c) Determine the output *PSD*, $S_{yy}(z)$.

2.12 Suppose we are given a linear system characterized by the difference equation

$$y(t) - 1/5y(t-1) = \frac{1}{\sqrt{3}}x(t)$$

and the system is excited by
(i) white Gaussian noise, $x \sim N(0, 3)$
(ii) exponentially correlated noise, $R_{ee}(k) = (1/2)^{|k|}$ In both cases, find:
  (a) Output *PSD*, $S_{yy}(z)$;
  (b) Output covariance, $R_{yy}(k)$;
  (c) Cross-spectrum, $S_{ye}(k)$;
  (d) Cross-covariance, $R_{ye}(k)$.

2.13 Suppose we have a process (two-point averager)

$$y(t) = \frac{e(t) + e(t-1)}{2}$$

(a) Develop an expression for $S_{yy}(z)$ when e is white with variance R_{ee}.
(b) Let $z = \exp\{j\Omega\}$ and sketch the "response" of $S_{yy}(e^{j\Omega})$.
(c) Calculate an expression for the covariance, $R_{yy}(k)$ enclosed and recursive form.

2.14 Suppose we are given a zero-mean process with covariance

$$R_{yy}(k) = 10 \exp(-0.5|k|)$$

(a) Determine the digital filter which when driven by white noise will yield a sequence with the above covariance.

(b) Develop a computer program to generate $y(t)$ for 100 points.
(c) Plot the results and determine if 95% of the samples fall within $\pm 2\sqrt{R_{yy}(0)}$.

2.15 Suppose a process is measured using an instrument with uncertainty $v \sim N(0, 4)$ such that

$$y(t) = 2x(t) + v(t)$$

(a) Calculate the output covariance $R_{yy}(k)$.
(b) Develop a computer program to simulate the output.
(c) Plot the process $y(t)$ with the corresponding confidence limits $\pm 2\sqrt{R_{yy}(0)}$ for 100 data points, do 95% of the samples lie within the bounds?

3

Signal Models

3.1 Data-Based Models

When data are collected from a process under investigation, many times phenomenological information is not readily available about the underlying physics generating the data; therefore, a preliminary analysis is necessary to obtain more insight into the underlying problem. Fourier frequency domain techniques aid in unraveling signal and system properties that can be utilized to gain such knowledge. A wide variety of "data-based" signal models evolve based on matrix structures that are constructed from the data, itself. For instance, exciting an unknown system with a sharp transient excitation can provide impulse-like response information that can be used to perform further analysis and lead to characterizing more sophisticated physics-based signal models. In this section, we discuss some of the popular data constructs in the form of *data matrices* that lead to potential nonparametric, nonphysics-based analysis that will be developed.

3.1.1 Data-Based Response Matrices

We start with a simple impulse response matrix that evolves from direct measurements of the transient data, that is,

$$y(t) = h(t) * u(t) = \sum_{k=0}^{N} h(t-k)u(k) \tag{3.1}$$

where u, y are the input excitation, output response, and h the corresponding system response function with $*$ the convolution operator. The data-based signal model evolves from assuming that, $u(t) \approx \delta(t)$, an impulse leading to the output

Signal Processing: An Applied Decomposition Approach, First Edition. James Vincent Candy.

Companion website: www.wiley.com/go/CandySignProc

matrix that contains the collected response data and can be constructed to capture the essence of the system, that is,

$$\mathcal{Y} \approx \mathbf{H}(N) = \underbrace{\begin{bmatrix} h(0) & 0 & \cdots & 0 \\ h(1) & h(0) & \cdots & 0 \\ \vdots & \vdots & \ddots & 0 \\ h(N) & h(N-1) & \cdots & h(0) \end{bmatrix}}_{\text{Impulse Response Matrix}} \tag{3.2}$$

The $(N+1) \times (N+1)$ data-matrix representation $(\mathcal{Y})$ for this signal is therefore the constructed *impulse response matrix* $(\mathbf{H})$ that can be processed to extract the desired information.

Going one step further, suppose we multiply this data matrix by its transpose, then an estimate of the corresponding *correlation* data-matrix results $(\mathbf{R}_{\mathcal{Y}\mathcal{Y}} = \mathcal{Y}' \times \mathcal{Y})$ such that

$$\mathbf{R}_{\mathcal{Y}\mathcal{Y}} \approx \mathcal{Y}' \times \mathcal{Y} = \underbrace{\begin{bmatrix} R_{yy}(0) & R_{yy}(1) & \cdots & R_{yy}(N) \\ R_{yy}(1) & R_{yy}(0) & \cdots & R_{yy}(1) \\ \vdots & \ddots & \ddots & R(0) \\ R_{yy}(N) & R_{yy}(N-1) & \cdots & R_{yy}(0) \end{bmatrix}}_{\text{Correlation Matrix}} \tag{3.3}$$

Note that Fourier transforming the multichannel correlation matrix leads directly to the corresponding multichannel power spectral density matrix from the Wiener–Khintchine theorem [10].

So we see that the data-based signal models can either characterize properties of the signal and can also be used to extract other information that is pertinent to the problem at hand.

From response models, it is apparent that their frequency domain counterparts follow by applying Laplace/Fourier transform techniques to these matrices to obtain both multichannel frequency response function (*FRF*) and transfer function matrices.

The *FRF*-matrix is a transfer function matrix that is simply the Fourier or Laplace transform $(s = j\Omega)$ of the corresponding impulse response matrix, that is,

$$H(\Omega) := H(s)|_{s=j\Omega} \tag{3.4}$$

For a multichannel, multiple input/multiple output (*MIMO*) system with N_y-outputs and N_u-inputs the *FRF*-matrix at frequency Ω_m is defined by

$$\mathbf{H}(\Omega_m) := \begin{bmatrix} h_{11}(\Omega_m) & \cdots & h_{1N_u}(\Omega_m) \\ \vdots & \ddots & \vdots \\ h_{N_y1}(\Omega_m) & \cdots & h_{N_yN_u}(\Omega_m) \end{bmatrix} \in \mathcal{C}^{N_y \times N_u} \tag{3.5}$$

Expanding over a set of discrete frequencies, $m = 0, 1, \ldots, N_{\Omega-1}$, $\Omega = \{\Omega_0, \Omega_1, \ldots, \Omega_{N_\Omega-1}\}$, then a 3D-set of sequential $N_\Omega \times (N_y \times N_u)$-matrices (data cube) result and are given by the *FRF-matrix*

$$\mathbf{H}(\Omega) := \underbrace{\begin{bmatrix} H_{11}(\Omega) & \cdots & H_{1N_u}(\Omega) \\ \vdots & \ddots & \vdots \\ H_{N_y1}(\Omega) & \cdots & H_{N_yN_u}(\Omega) \end{bmatrix}}_{\text{Frequency Response Function Matrix}} \in \mathcal{C}^{N_\Omega \times (N_y \times N_u)} \tag{3.6}$$

Extending this deterministic representation to the stochastic case leads directly to the *power spectral density* as [13–16]

$$S_{hh}(\Omega) := E\left\{H(\Omega)H^{\dagger}(\Omega)\right\} \tag{3.7}$$

where $\dagger$ is the Hermitian conjugate operator with the *MIMO PSD*-matrix *at* frequency Ω_m is defined by

$$\mathbf{S}_{hh}(\Omega_m) := \begin{bmatrix} s_{11}(\Omega_m) & \cdots & s_{1N_y}(\Omega_m) \\ \vdots & \ddots & \vdots \\ s_{N_y1}(\Omega_m) & \cdots & s_{N_yN_y}(\Omega_m) \end{bmatrix} \in \mathcal{R}^{N_y \times N_y} \tag{3.8}$$

Expanding over a set of discrete frequencies, Ω as before, we obtain a 3D set of sequential $N_\Omega \times (N_y \times N_y)$-matrices (data cube) given by the *PSD-matrix* specified by

$$\mathbf{S}_{HH}(\Omega) := \underbrace{\begin{bmatrix} S_{11}(\Omega) & \cdots & S_{1N_y}(\Omega) \\ \vdots & \ddots & \vdots \\ S_{N_y1}(\Omega) & \cdots & S_{N_yN_y}(\Omega) \end{bmatrix}}_{\text{Power Spectral Density Matrix}} \in \mathcal{R}^{N_\Omega \times (N_y \times N_y)} \tag{3.9}$$

completing the data-based spectral representations. Next, consider examining the Toeplitz correlation matrix even further.

3.1.2 Data-Based Toeplitz Matrices

As we readily observe from Eq. (3.3), the correlation matrix has a banded structure—the *Toeplitz matrix* defined by its entries $R(i,j) = R(i-j)$ such that each element along its diagonals is identical. This matrix evolves in a crucial role of problems in the signal processing literature ranging from the well-known Levinson recursion for inversion to the current subspace identification algorithms [10, 12]. The basic properties of this Toeplitz correlation matrix can be exploited to provide insight into the structure of generic signal processing problems.

We list (without proof) the following general *properties* of the Toeplitz (correlation) matrix, $\mathbf{R}$, (see [17, 18] for details) of a wide-sense stationary (*WSS*) process.

1. $R \geq 0$ (positive semi-definite)
2. $R = R'$ (symmetric or Hermitian symmetric for $R \in C^{N \times N}$)
3. $\{\lambda_i\} \geq 0$ for $i = 1, \ldots, N$ (eigenvalues are real and positive)
4. $\mathbf{e}_i'\mathbf{e}_j = 0, \;\; i \neq j$ (eigenvectors are orthogonal or orthonormal for Hermitian)
5. $R = E\Lambda E'$ (eigen-decomposition with E unitary ($EE' = E'E = I$))

Besides the eigen-decomposition property, the Toeplitz matrix can be decomposed using a *Cholesky decomposition* [17] of a positive definite matrix ($R > 0$) defined by

$$R = LDU' = LDL' = \left(LD^{\frac{1}{2}}\right)\left(D^{\frac{T}{2}}L'\right) \tag{3.10}$$

where L is a lower triangular matrix with unity on the diagonal, D is a diagonal matrix with matrix *square roots* defined by $D := D^{\frac{1}{2}}D^{\frac{T}{2}}$ for $D^{\frac{T}{2}} = \left((D')^{\frac{1}{2}}\right)$ and $U = L$ in the real symmetric case.

The structure of the Toeplitz matrix of Eq. (3.3) reveals much about the underlying model-based structure which is exploited to create efficient inversion algorithms. To put this in perspective, the *Levinson–Durbin recursion* follows a linear algebraic approach for ease of comprehension in which prediction error models evolve [17, 18]. So we see that the data-based Toeplitz matrix leads directly to the development of parametric models inherent to a wide variety of processing applications [10]. Next, we consider another important matrix inherent to systems theoretic applications.

3.1.3 Data-Based Hankel Matrices

Just as the data-based Toeplitz matrix plays an intimate role in extracting parametric models from data, the Hankel matrix also plays a role in both extracting multichannel models from both impulse response and input/output data with a wealth of system theoretic implications. The $(N \times N_yN_u) \times (N \times N_yN_u)$ *Hankel matrix* is defined by

$$\mathcal{H}(N) := \begin{bmatrix} H(1) & H(2) & \cdots & H(N) \\ H(2) & H(3) & \cdots & H(N+1) \\ \vdots & \vdots & \cdots & \vdots \\ H(N) & H(N+1) & \cdots & H(2N) \end{bmatrix} \tag{3.11}$$

where $H(t)$ is the corresponding multichannel $(N_y \times N_u)$ impulse response matrix.

Suppose the dimension of the system is N_x, then the Hankel matrix could be constructed such that $N = N_x$ using $2N_x$ impulse response matrices which specifies the minimum number of terms required to extract the Markov parameters. Also the rank, $\rho[\mathcal{H}(N)] = N_x$, which is the dimension of the *minimal realization*. If we did not know the dimension of the system, then we would let (in theory) $N \to \infty$ and determine the rank of $\mathcal{H}(\infty)$. Therefore, the minimal dimension of an "unknown" system is the rank of the Hankel matrix. In order for a system to be *minimal* it must be *completely controllable* and *completely observable*. This can be seen from the fact that the Hankel matrix factors into the observability ($\mathcal{O}$) and controllability ($\mathcal{C}$) matrices as:

$$\mathcal{H}(N) = \mathcal{O}(N) \times \mathcal{C}(N) \tag{3.12}$$

It follows that the $\rho\,[\mathcal{H}(N)] = \min\,[\rho(\mathcal{O}(N)), \rho(\mathcal{C}(N))] = N_x$. We see that the properties of controllability and observability are carefully woven into that of minimality and that testing the rank of the Hankel matrix yields the dimensionality of the underlying dynamic system. This fact will prove crucial when we must "identify" a system from noisy measurement data.

3.2 Parametric-Based Models

In this section, we consider various parametric-based signal representations that are related to decomposition techniques. Starting with time-series models that have evolved from the statistics community we illustrate the connection to signal characterizations. Wave-based structures follow based on so-called lattice representations. Next, *FRF*s or equivalent transfer function models follow along with their related counterparts, harmonic (sinusoidal) models, that appear in a large number of processing applications. Here also the *MIMO* matrix fraction description (*MFD*) model is briefly discussed as well.

3.2.1 ARMAX (AR, ARX, MA, ARMA) Models

Time-series models, that is, models that have evolved directly from statisticians working directly with temporal data such as daily stock market prices were rapidly incorporated by the signal processing community perhaps starting with geophysicists involved in oil exploration applications [19, 20]. In any case, the suite of these models is readily incorporated in signal processing techniques especially in the speech area [21, 22].

The *input–output* or *transfer function* model is familiar to engineers and scientists because it is usually presented in the frequency domain with Laplace transforms. Similarly in the discrete-time case with Z-transforms, it is termed the *pulse transfer function* model and is given by

$$H(z) = \frac{B(z^{-1})}{A(z^{-1})} \tag{3.13}$$

where A and B are polynomials in z or z^{-1}.

$$A(z^{-1}) = 1 + a_1 z^{-1} + \cdots + a_{N_a} z^{-N_a} \tag{3.14}$$

$$B(z^{-1}) = b_0 + b_1 z^{-1} + \cdots + b_{N_b} z^{-N_b} \tag{3.15}$$

If we consider the equivalent discrete-time domain representation, then we have a *difference equation* relating the output sequence $\{y(t)\}$ to the input sequence $\{u(t)\}$.[1] We use the backward shift operator q with the property that $q^{-k}y(t) = y(t-k)$.

$$A(q^{-1})y(t) = B(q^{-1})u(t) \tag{3.16}$$

or,

$$y(t) + a_1 y(t-1) + \cdots + a_{N_a} y(t-N_a) = b_0 u(t) + \cdots b_{N_b} u(t-N_b) \tag{3.17}$$

When the system is excited by random inputs, the time-series models are specified by the *autoregressive-moving average model with exogenous inputs* (*ARMAX*)[2]

$$\underbrace{A(q^{-1})y(t)}_{AR} = \underbrace{B(q^{-1})u(t)}_{X} + \underbrace{C(q^{-1})e(t)}_{MA} \tag{3.18}$$

where A, B, C, are polynomials, and $\{e(t)\}$ is a white noise source, and

$$C(q^{-1}) = 1 + c_1 q^{-1} + \cdots + c_{N_c} q^{-N_c}$$

The *ARMAX* model, usually abbreviated by $ARMAX(N_a, N_b, N_c)$

- Pulse transfer function or infinite impulse response (*IIR*) model: $C(\cdot) = 0$, or $ARMAX(N_a, N_b, 0)$, that is,

$$A(q^{-1})y(t) = B(q^{-1})u(t)$$

- Finite impulse response (*FIR*) model: $A(\cdot) = 1, C(\cdot) = 0$, or $ARMAX(1, N_b, 0)$, that is,

$$y(t) = B(q^{-1})u(t)$$

- Autoregressive (*AR*) model: $B(\cdot) = 0, C(\cdot) = 1$, or $ARMAX(N_a, 0, 1)$, that is,

$$A(q^{-1})y(t) = e(t)$$

- Moving average (*MA*) model: $A(\cdot) = 1, B(\cdot) = 0$, or $ARMAX(1, 0, N_c)$, that is

$$y(t) = C(q^{-1})e(t)$$

---

1 We change from the common signal processing convention of using $x(t)$ for the deterministic excitation to $u(t)$ and we include the b_0 coefficient for generality.

2 The *ARMAX* model can be interpreted in terms of the Wold decomposition of stationary times series, which states that a time series can be decomposed into a predictable or deterministic component ($u(t)$) and nondeterministic or random component ($e(t)$) [17].

- Autoregressive-moving average (*ARMA*) model: $B(\cdot) = 0$, or $ARMAX(N_a, 0, N_c)$, that is,

$$A(q^{-1})y(t) = C(q^{-1})e(t)$$

- Autoregressive model with exogeneous input (*ARX*): $C(\cdot) = 1$, or $ARMAX$ $(N_a, N_b, 1)$, that is,

$$A(q^{-1})y(t) = B(q^{-1})u(t) + e(t)$$

ARMAX models can easily be used for signal processing purposes, since they are basically digital filters with known deterministic ($u(t)$) and random ($e(t)$) excitations. Consider the following example, of spectral shaping using the *ARMAX* model.

Example 3.1 We are asked to model the output sequence $y(t)$ of a 1st-order system driven by Gaussian random noise governed by the parameters: $a = -0.5$ and $\{e(t)\} \sim \mathcal{N}(0, 1)$. We see that this can be accomplished using $ARMAX(1, 0, 0)$, an *AR* model, that is,

$$H(z) = \frac{Y(z)}{E(z)} = \frac{1}{1 + 0.5z^{-1}}$$

Cross-multiplying and substituting the backward shift operation for z, we obtain the difference equation

$$y(t) = -0.5y(t - 1) + e(t)$$

△△△

Since the *ARMAX* model is used to characterize a random signal, we are interested in its statistical properties. The mean value of the output is easily determined by

$$A(q^{-1})E\{y(t)\} = B(q^{-1})E\{u(t)\} + C(q^{-1})E\{e(t)\}$$

or

$$A(q^{-1})m_y(t) = B(q^{-1})u(t) + C(q^{-1})m_e(t) \tag{3.19}$$

Because the first term in the A-polynomial is unity, we can write the *mean propagation recursion* for the *ARMAX* model as

$$m_y(t) = (1 - A(q^{-1}))m_y(t) + B(q^{-1})u(t) + C(q^{-1})m_e(t) \tag{3.20}$$

$$m_y(t) = -\sum_{i=1}^{N_a} a_i m_y(t - i) + \sum_{i=0}^{N_b} b_i u(t - i) + \sum_{i=0}^{N_c} c_i m_e(t - i) \tag{3.21}$$

We note that the mean of the *ARMAX* model is propagated using a recursive digital filter requiring N_a, N_b, N_c past input and output values.

The corresponding variance of the *ARMAX* model is more complex. First, we note that the mean must be removed, that is,

$$\begin{aligned} y(t) - m_y(t) = &\left[(1 - A(q^{-1}))y(t) + B(q^{-1})u(t) + C(q^{-1})e(t)\right] \\ &- \left[(1 - A(q^{-1}))m_y(t) + B(q^{-1})u(t) + C(q^{-1})m_e(t)\right] \end{aligned} \tag{3.22}$$

or

$$y(t) - m_y(t) = (1 - A(q^{-1}))(y(t) - m_y(t)) + C(q^{-1})(e(t) - m_e(t))$$

or finally

$$A(q^{-1})(y(t) - m_y(t)) = C(q^{-1})(e(t) - m_e(t)) \tag{3.23}$$

that is, $y - m_y$ is characterized by an *ARMAX* $(N_a, 0, N_b)$ or equivalently an *ARMA* model.

The covariance of the *ARMAX* model can be calculated utilizing the fact that it is essentially an *IIR* system, that is,

$$\frac{Y(z)}{E(z)} = H(z) = \sum_{t=0}^{\infty} h(t)z^{-t} \tag{3.24}$$

Using this fact and the commutativity of the convolution operator, we have (assuming the mean has been removed)

$$R_{yy}(k) = E\{y(t)y(t+k)\} = E\left\{\sum_{i=0}^{\infty} h(i)e(t-i)\sum_{j=0}^{\infty} h(j+k)e(t-j+k)\right\}$$

or

$$\begin{aligned} R_{yy}(k) = &\sum_{i=0}^{\infty} h(i)h(i+k)E\{e(t-i)e(t-i+k)\} \\ &+ \sum_{i\neq j}\sum h(i)h(j+k)E\{e(t-i)e(t-j+k)\} \end{aligned}$$

The whiteness of $\{e(t)\}$ gives

$$R_{ee}(k) = \begin{cases} R_{ee} & k = 0 \\ 0 & \text{elsewhere} \end{cases}$$

therefore, applying this property above, we have the *covariance* of the *ARMAX* model given by

$$R_{yy}(k) = R_{ee}\sum_{i=0}^{\infty} h(t)h(t+k) \quad \text{for} \quad k \geq 0 \tag{3.25}$$

with corresponding *variance*

$$R_{yy}(0) = R_{ee} \sum_{t=0}^{\infty} h^2(t) \tag{3.26}$$

We note that one property of a stationary signal is that its impulse response is bounded which implies from Eq. (3.25) that the variance is bounded [19]. Clearly, since the variance is characterized by an *ARMA* model (*ARMAX* $(N_a, 0, N_c)$), then we have

$$A(z^{-1})H(z) = C(z^{-1})E(z),\ E(z) = \sqrt{R_{ee}}$$

or taking the inverse $\mathcal{Z}$-transform

$$h(t) = (1 - A(q^{-1}))h(t) + C(q^{-1})\delta(t)$$

or

$$h(t) = -\sum_{i=1}^{N_a} a_i h(t-i) + \sum_{i=0}^{N_c} c_i \delta(t),\ c_0 = 1 \tag{3.27}$$

where $\delta(t)$ is an impulse of weight $\sqrt{R_{ee}}$. So, we see that this recursion coupled with Eq. (3.25) provides a method for calculating the variance of an *ARMAX* model. We summarize these results in Table 3.1 and the following example.

Example 3.2 Suppose we are given the *AR* model of the previous example with $A(z) = 1 + 0.5z^{-1}$, $R_{ee} = 1$ and would like to calculate the variance, then from Eq. (3.27), we have

$$h(t) = -0.5h(t-1) + \delta(t) \longrightarrow (-0.5)^t$$

and

$$R_{yy}(0) = \sum_{i=0}^{\infty} h^2(i) = [1^2 - 0.5^2 + 0.25^2 - 1.25^2 + 0.0875^2 - \cdots] \rightarrow 1.333$$

The covariance is calculated from Eq. (3.25) using *MATLAB*®.

△△△

Before we leave this discussion, let us consider a more complex example to illustrate the use of the *ARMA* model.

Example 3.3 Suppose we would like to simulate an *ARMAX*(2,1,1) model with the following difference equation

$$(1 + 3/4q^{-1} + 1/8q^{-2})y(t) = (1 + 1/8q^{-1})u(t) + (1 + 1/16q^{-1})e(t)$$

Table 3.1 *ARMAX* representation.

Output propagation
$y(t) = (1 - A(q^{-1}))y(t) + B(q^{-1})u(t) + C(q^{-1})e(t)$
Mean propagation
$m_y(t) = (1 - A(q^{-1}))m_y(t) + B(q^{-1})u(t) + C(q^{-1})m_e(t)$
Impulse propagation
$h(t) = (1 - A(q^{-1}))h(t) + C(q^{-1})\delta(t)$
Variance/Covariance propagation
$R_{yy}(k) = R_{ee}\sum_{i=0}^{\infty} h(i)h(i+k) \quad k \geq 0$

y = the output or measurement sequence
u = the input sequence
e = the process (white) noise sequence with variance R_{ee}
h = the impulse response sequence
δ = the impulse input of amplitude $\sqrt{R_{ee}}$
m_y = the mean output or measurement sequence
m_e = the mean process noise sequence
R_{yy} = the stationary output covariance at lag k
A = the N_ath order system characteristic (poles) polynomial
B = the N_bth order input (zeros) polynomial
C = the N_cth order noise (zeros) polynomial

where $u(t) = \sin 2\pi(0.025)t$ and $e \sim \mathcal{N}(1, 0.01)$. Then the corresponding mean propagation equation is

$$(1 + 3/4q^{-1} + 1/8q^{-2})m_y(t) = (1 + 1/8q^{-1})u(t) + (1 + 1/16q^{-1})m_e(t)$$

for $m_e(t) = 1 \,\forall\, t$ and the impulse propagation model is

$$(1 + 3/4q^{-1} + 1/8q^{-2})h(t) = (1 + 1/16q^{-1})\sqrt{R_{ee}}\delta(t) \quad \text{for} \quad \sqrt{R_{ee}} = 0.1$$

and the variance/covariance propagation model is:

$$R_{yy}(k) = R_{ee}\sum_{t=0}^{\infty} h(t)h(t+k) \qquad k \geq 0$$

Using *MATLAB*®, we perform the simulation with $\triangle T = 0.1$ sec and the results are shown in Figure 3.1. In the figure, we see the simulated process, mean (sinusoidal), and corresponding *ARMA* (mean removed) signals. The statistical relations are

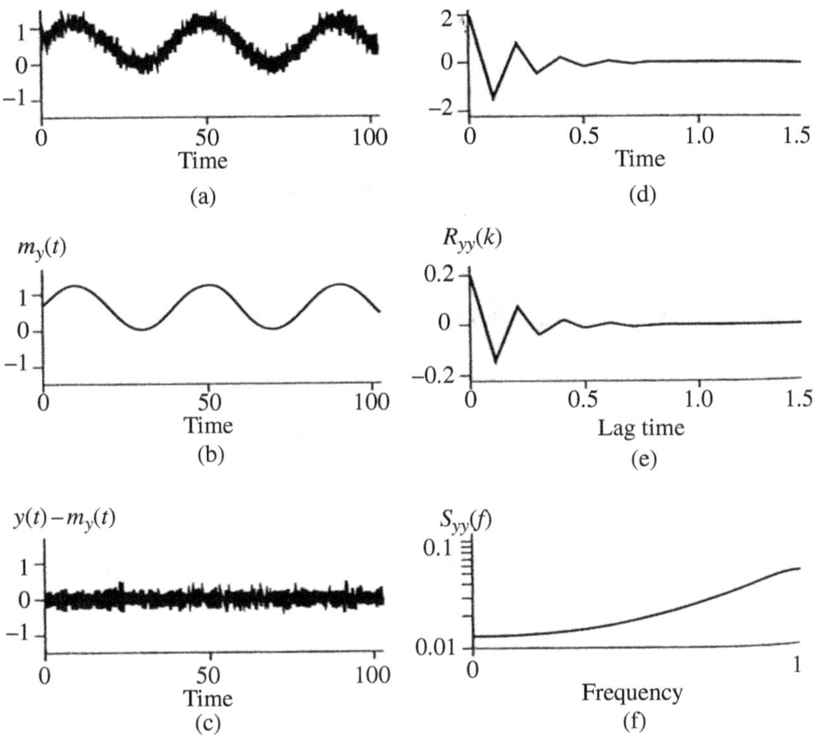

Figure 3.1 *ARMAX*(2,1,1) simulation: (a) simulated signal, (b) mean propagation, (c) *ARMA* (mean removed) signal, (d) impulse response propagation, (e) covariance propagation, and (f) *PSD*.

also shown with the corresponding impulse response and variance propagation models of Eqs. (3.25) and (3.26) and estimated power spectrum. This completes the example.

△△△

It should be noted that for certain special cases of the *ARMAX* model, it is particularly simple to calculate the mean and covariance. For instance, the *MA* model ($ARMAX(1, 0, N_c)$) has *mean*

$$m_y(t) = E\{C(q^{-1})e(t)\} = C(q^{-1})m_e(t) \tag{3.28}$$

and *covariance* (directly from Eq. (3.25) with $h \to c$)

$$R_{yy}(k) = R_{ee}\sum_{i=0}^{N_c} c_i c_{i+k} \quad \text{for} \quad k \geq 0 \tag{3.29}$$

Another special case of interest is the AR ($ARMAX(N_a, 0, 1)$) model with *mean*

$$m_y(t) = (1 - A(q^{-1}))m_y(t) + m_e(t) \tag{3.30}$$

and *covariance* which is easily derived by direct substitution

$$R_{yy}(k) = E\{y(t)y(t+k)\} = (1 - A(q^{-1}))R_{yy}(k) = -\sum_{i=1}^{N_a} a_i R_{yy}(k-i) \quad \text{for} \quad k > 0 \tag{3.31}$$

In fact, the AR covariance model of Eq. (3.31) is essentially a recursive (all-pole) digital filter that can be propagated by exciting it with the variance $R_{yy}(0)$ as initial condition. In this case, the variance is given by

$$R_{yy}(0) = E\{y^2(t)\} = E\left\{\left(-\sum_{i=1}^{N_a} a_i y(t-i) + e(t)\right)y(t)\right\} = -\sum_{i=1}^{N_a} a_i R_{yy}(i) + R_{ee} \tag{3.32}$$

So combining Eqs. (3.31) and (3.32), we have that the *covariance propagation* equations for the AR model are given by

$$R_{yy}(k) = \begin{cases} -\sum_{i=1}^{N_a} a_i R_{yy}(i) + R_{ee} & k = 0 \\ -\sum_{i=1}^{N_a} a_i R_{yy}(k-i) & k > 0 \end{cases}$$

Consider the following example of calculating the statistics of the AR model.

Example 3.4 Consider the AR model of the previous examples and we would like to determine the corresponding mean and variance using the recursions of Eqs. (3.30) and (3.31) with $A(q^{-1}) = 1 + 0.5q^{-1}$. The mean is

$$m_y(t) = -0.5m_y(t-1)$$

and the covariance is

$$R_{yy}(k) = \begin{cases} -0.5R_{yy}(1) + 1 & k = 0 \\ -0.5R_{yy}(k-1) & k > 0 \end{cases}$$

The variance is obtained directly from these recursions, since

$$R_{yy}(1) = -0.5R_{yy}(0)$$

and therefore

$$R_{yy}(0) = -0.5R_{yy}(1) + 1$$

substituting for $R_{yy}(1)$ we obtain

$$R_{yy}(0) = -0.5(-0.5R_{yy}(0)) + 1$$

or

$$R_{yy}(0) = 1.333$$

△△△

This completes the section on *ARMAX* models. In Section 3.2.2 to follow, we develop an alternative but equivalent structure–the lattice model.

3.2.2 Lattice Models

Lattice filters find their roots in analog filter designs and in the design of digital filters as well [23, 24]. They have also evolved as equivalent representations of random signals. Estimation filters based on the lattice design perform well because of their inherent orthogonality properties [24–27]. The (scalar) *lattice model* is defined by the following fundamental propagation equation,

$$\begin{aligned} e_f(t,i) &= e_f(t,i-1) - k_i e_b(t-1,i-1) \\ e_b(t,i) &= e_b(t-1,i-1) - k_i e_f(t,i-1) \end{aligned} \tag{3.33}$$

where $e_f(t,i), e_b(t,i)$ are the respective forward and backward signals of the ith lattice section (or stage) at time t, and k_i is the *reflection coefficient* of the ith section. The lattice model is depicted in Figure 3.2. This model is constructed by a series of connected sections with each section having identical structure with only the values of the reflection coefficients changing.

The lattice model derives its physical origins from the fact that it mathematically represents a *wave model* or *wave* propagating through a layered medium. For instance, lattice models are employed to characterize seismic waves propagating; through a layered earth in the exploration of oil [19], or an acoustic wave propagating through a tube in speech synthesis [21], or an electromagnetic wave

Figure 3.2 Lattice model.

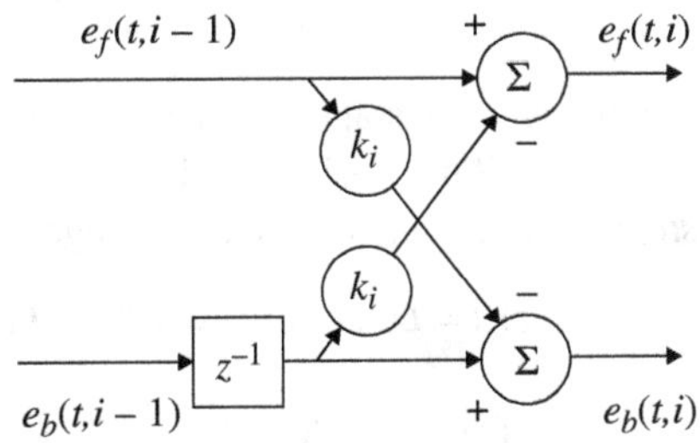

propagating in a transmission line. In each of these cases, we note that the reflection coefficient, which has magnitude between ± 1, indicates the percentage of the wave transmitted and reflected at a layer boundary (or lattice stage).

Structurally the lattice model possesses some very attractive properties from the implementation viewpoint because of its sectional or stage-wise symmetry. Consider a filter of length N consisting of N lattice sections arranged in a feedforward configuration. We first note that the lattice model operation for the ith section can be written in matrix form using the backward shift operator as

$$\begin{bmatrix} e_f(t,i) \\ e_b(t,i) \end{bmatrix} = \begin{bmatrix} 1 & -k_i q^{-1} \\ -k_i & q^{-1} \end{bmatrix} \begin{bmatrix} e_f(t,i-1) \\ e_b(t,i-1) \end{bmatrix} \tag{3.34}$$

Taking the $\mathcal{Z}$-transform of Eq. (3.34) gives

$$\begin{bmatrix} E_f(z,i) \\ E_b(z,i) \end{bmatrix} = \begin{bmatrix} 1 & -k_i z^{-1} \\ -k_i & z^{-1} \end{bmatrix} \begin{bmatrix} E_f(z,i-1) \\ E_b(z,i-1) \end{bmatrix} \tag{3.35}$$

or more compactly

$$\underline{E}(z,i) = L(z,i)\underline{E}(z,i-1) \tag{3.36}$$

where $L(z,i)$ can be thought of as a multichannel transfer function matrix between the ith and $(i-1)$th section of the lattice. In fact, inverting $L(z,i)$ enables us to determine the inverse relation between the ith and the $(i-1)$th stage as

$$\underline{E}(z,i-1) = L^{-1}(z,i)\underline{E}(z,i) \tag{3.37}$$

where

$$L^{-1}(z,i) = \frac{1}{1-k_1^2} \begin{bmatrix} 1 & k_i \\ k_i z & z \end{bmatrix} \tag{3.38}$$

The lattice transfer function is useful for predicting the response of the lattice at various stages. The transfer function for the N-stage lattice can easily be derived using Eq. (3.36) and recursing backward, that is,

$$\begin{aligned} \underline{E}(z,N) &= L(z,N)\underline{E}(z,N-1) \\ &= L(z,N)[L(z,N-1)\underline{E}(z,N-2)] \\ &= L(z,N)L(z,N-1)\ldots L(z,1)\underline{E}(z,0) \end{aligned} \tag{3.39}$$

which gives the stage-wise transfer function. If we start with $(N=1)$ in Eq. (3.38), we can similarly derive relations for the inverse transfer function. Thus, the overall *stage transfer function* of an N-stage lattice and its inverse are specified by

$$L(z) := L(z,N)L(z,N-2)\ldots L(z,1) \tag{3.40}$$

$$L^{-1}(z) := L^{-1}(z,1)L^{-1}(z,2)\ldots L^{-1}(z,N) \tag{3.41}$$

These stage transfer functions must be related to our standard input–output models of Section3.2.1. Before we explore these relations, consider the following example.

Example 3.5 We have a two-stage lattice $\{k_1\ k_2\}$ and would like to determine the overall transfer function $L(z)$ as well as the inverse transfer function $L^{-1}(z)$. From Eq. (3.40), we see that

$$L(z) = L(z,2)L(z,1) = \begin{bmatrix} 1 & -k_2 z^{-1} \\ -k_2 & z^{-1} \end{bmatrix} \begin{bmatrix} 1 & -k_1 z^{-1} \\ -k_1 & z^{-1} \end{bmatrix}$$

or

$$L(z) = \begin{bmatrix} 1 + k_1 k_2 z^{-1} & -k_1 z^{-1} - k_2 z^{-2} \\ -k_2 - k_1 z^{-1} & k_1 k_2 z^{-1} + z^{-2} \end{bmatrix}$$

The inverse transfer function is given by

$$L^{-1}(z) = L^{-1}(z,1)L^{-1}(z,2) = \frac{1}{(1-k_1^2)(1-k_2^2)} \begin{bmatrix} 1 + k_1 k_2 z & k_2 + k_1 z \\ k_1 z + k_2 z^2 & k_1 k_2 z + z^2 \end{bmatrix}$$

△△△

The overall lattice transfer function is related to the standard *IIR* and *FIR* transfer functions of Section 3.2.1. If we examine the *FIR*, *MA* or equivalently all-zero forms of the *ARMAX* model, that is, $ARMAX(1, N_b, 0)$ or $ARMAX(1, 0, N_c)$, then we can develop the required relations. The *FIR* model is characterized by the input–output difference equation

$$y(t) = B(q^{-1})u(t)$$

and equivalently

$$H(z) = \frac{Y(z)}{U(z)} = B(z) \tag{3.42}$$

If we use a feedforward lattice structure, that is,

$$e_f(t,0) = e_b(t,0) := u(t) \quad \text{and} \quad e_f(t,N) := y(t)$$

which is equivalent to exciting both input ports and measuring the output then Eq. (3.38) becomes

$$\underline{E}(z,N) = L(z,N)\ldots L(z,1)\underline{E}(z,0) = L(z)\begin{bmatrix} U(z) \\ U(z) \end{bmatrix} = L(z)\begin{bmatrix} 1 \\ 1 \end{bmatrix} U(z)$$

or dividing both sides by $U(z)$ we obtain

$$\begin{bmatrix} \frac{Y(z)}{U(z)} \\ \frac{Y^R(z)}{U(z)} \end{bmatrix} = \begin{bmatrix} H(z) \\ H^R(z) \end{bmatrix} = \begin{bmatrix} B(z) \\ B^R(z) \end{bmatrix} = L(z)\begin{bmatrix} 1 \\ 1 \end{bmatrix} = \begin{bmatrix} L_{11}(z) + L_{12}(z) \\ L_{21}(z) + L_{22}(z) \end{bmatrix}$$

where $H^R(z)$ is the so-called Nth order *reverse polynomial* [21] given by

$$H^R(z) = z^{-N} H(z^{-1})$$

So we see that if we equate the coefficients $\{b_i\}$ of the all-zero transfer function $H(z)$ to the sum of the row elements of $L(z)$ resulting from the iteration of Eq. (3.40), then we obtain the N-stage recursion (usually called the *Levinson recursion*) [10]

$$\begin{aligned} b(1,1) &= -k_1 \\ b(i,i) &= -k_i \quad \text{for} \quad i = 2, \dots, N_b \\ b(j,i) &= b(j,i-1) - k_i b(i-j,i-1) \quad \text{for} \quad j = 1, \dots, i-1 \end{aligned}$$

where

$$H(z,i) = 1 + \sum_{j=1}^{N_b} b(j,i) z^{-j} \text{ and } H(z,N_b) = 1 + b(1,N_b)z^{-1} + \cdots + b(N_b,N_b)z^{-N_b}.$$

The *ARMAX* polynomial is equivalent to the final stage.

The all-zero *ARMAX* model can also be transformed to the lattice form by using the inverse filtering operation $L^{-1}(z)$ of Eqs. (3.38) and (3.40) which leads to the following recursion[3]

$$\begin{array}{lcll} k_i & = & -b(i,i) & \text{for} \quad i = N_b, \dots, 1 \\ b(j,i-1) & = & \frac{b(j,i) + k_i b(i-j,i)}{1-k_i^2} & \text{for} \quad j = 1, \dots, i-1 \end{array}$$

where we again use the coefficient relations

$$\{b_i\} = \{b(i,N_b)\}$$

The inverse of the feedforward lattice is the feedback lattice. The feedback lattice is an all-pole transfer function, thus, if we apply an input signal to a feedforward lattice and the result is applied to a feedback lattice, with identical coefficients then the original signal will result. From Mason's rule of circuit theory, we know that the reflection coefficients parameterize *both* the feedback as well as feedforward lattices with appropriate changes in signal flow, thus, the conversion from the reflection coefficients to tapped delay line coefficients is identical for the all-zero or all-pole *ARMAX* models where the signal flow specifies which transfer function is implied.

The final useful transformation of the *ARMAX* model is the rational pole-zero form, that is, converting from $ARMAX(N_a, N_b, 0)$ or $ARMAX(N_a, 0, N_b)$ to lattice

3 It should be noted that this procedure is equivalent to the Jury or Schur–Cohn stability test for the roots of the polynomial lying inside the unit circle. In fact, the necessary and sufficient condition for stability is that $|k_i| < 1$.

models, where the tap coefficients are given by $\{g_i\}$ and the pole-zero transfer function is

$$H(z) = H(z, N) = \frac{\sum_{j=0}^{N} b(j, N)z^{-j}}{1 + \sum_{j=1}^{N} a(j, N)z^{-j}} = \frac{\sum_{j=0}^{N} b_j z^{-j}}{1 + \sum_{j=1}^{N} a_j z^{-1}} = \frac{B(z)}{A(z)} \tag{3.43}$$

Using procedures similar to the all-zero case, we obtain the recursion

$$\begin{aligned}
k_i &= -a(i, i) \qquad \text{for} \qquad i = N, \dots, 1 \\
g_i &= b(i, i) \\
a(j, i-1) &= \frac{a(j, i) + k_i a(i-j, i)}{1 - k_i^2} \qquad \text{for} \quad j = 1, \dots, i-1 \\
b(j, i-1) &= b(j, i) - g_i a(i-j, i) \\
b(0, i-1) &= b(0, i) + g_i k_i \\
g_0 &= b(0, 0)
\end{aligned} \tag{3.44}$$

Although these coefficients are related in a nonlinear manner, this recursion is invertible, so that the rational lattice structure can be converted uniquely to a direct-form, pole-zero filter, and vice versa.

Example 3.6 Consider a second-order transfer function with coefficients $\{b_0, b_1, b_2, a_1, a_2\}$ and we would like to calculate the corresponding lattice coefficients $\{k_1, k_2, g_0, g_1, g_2\}$ using the recursion of Eq. (3.44) with $N = 2$

$$\begin{aligned}
i = 2 \qquad & k_2 = -a(2, 2) = -a_2 \\
& g_2 = b(2, 2) = b_2 \\
j = 1 \qquad & a(1, 1) = \frac{a(1, 2) + k_2 a(1, 2)}{1 - k_2^2} = \frac{a_1(1 + k_2)}{1 - k_2^2} = \frac{a_1}{1 - k_2} = \frac{a_1}{1 + a_2} \\
& b(1, 1) = b(1, 2) - g_2 a(1, 2) = b_1 - b_2 a_1 \\
& b(0, 1) = b(0, 2) + g_2 k_2 = b_0 - b_2 a_2 \\
i = 1 \qquad & k_1 = -a(1, 1) \\
& g_1 = b(1, 1) \\
& b(0, 0) = b(0, 1) + g_1 k_1 = b_0 - b_1 \left(\frac{a_1}{1 + a_2}\right) - b_2 \left(a_2 + \frac{a_1^2}{1 + a_2}\right) = g_0
\end{aligned}$$

which completes the example.

△△△

Thus, the lattice structure can be transformed into all-pole (*IIR*, *AR*) all-zero (*FIR*, *MA*), and pole-zero (*IIR*, *ARX*, *ARMA*) models and vice versa. We summarize these useful lattice recursive transformations in Table 3.2.

Table 3.2 *LATTICE* recursions.

Lattice-to-all-pole or all-zero model

$$b(1,1) = -k_1$$
$$b(i,i) = -k_i \qquad i = 2, \ldots, N_b$$
$$b(j,i) = b(j,i-1) - k_i b(i-j, i-1) \qquad j = 1, \ldots, i-1$$

All-pole or all-zero-to-lattice model

$$k_i = -b(i,i) \qquad i = N_b, \ldots, 1$$
$$b(j,i-1) = \frac{b(j,i) + k_i b(i-j,i)}{1-k_i^2} \qquad j = 1, \ldots, i-1$$

where

$$H(z,k) = \sum_{i=0}^{k} b(i,k) z^{-k} \quad \text{and} \quad \{b_i\} = \{b(i, N_b)\}$$

Rational transfer function-to-lattice model

$$k_i = -a(i,i) \qquad i = N, \ldots, 1$$
$$g_i = b(i,i)$$
$$\alpha_{i-1} = \alpha_1(1 + k_i)$$
$$a(j,i-1) = \frac{a(j,i)+k_i a(i-j,i)}{1-k_i^2} \qquad j = 1, \ldots, i-1$$
$$b(j,i-1) = b(j,i) - g_i a(i-j,i)$$
$$b(0,i-1) = b(0,i) + g_i k_i$$
$$g_0 = b(0,0)$$

where

$$H(z,N) = \frac{\sum_{j=0}^{N} b(j,N) z^{-j}}{1 + \sum_{j=1}^{N} a(j,N) z^{-j}} \quad \text{and} \quad \{b_j\} = \{b(j,N)\}$$

It should also be noted that in the stochastic case where we have various versions of the *ARMAX* model, the signal variance can be calculated directly from the white noise variance, R_{ee}, and knowledge of the reflection coefficients, that is, [18]

$$R_{yy}(0) = \text{var}\,\{y(t)\} = \frac{R_{ee}}{\pi_{i=1}^{N}(1 - k_i^2)} \tag{3.45}$$

Compare this result with the recursions developed in Section 3.2.1 for the variance of an *ARMAX* model.

3.2.3 Transfer Function/Frequency Response Function Models

In this section, we discuss transfer functions or equivalently *FRF* models starting with the basic single input/single output (*SISO*), second-order system to provide insight to the multichannel (*MIMO*) *MFDs* to follow. Here, the partial fraction representation is discussed in more detail which will lead to the signal decompositions to follow in Chapter 5.

Suppose we have a simple system driven by an external forcing function. The dynamics of this system can be expressed in terms of a second-order *differential equation* in continuous-time "t" as

$$\ddot{y}(t) + 2\zeta\omega_n\dot{y}(t) + \omega_n^2 y(t) = \omega_n^2 u(t) \tag{3.46}$$

where ω_n is the *undamped natural frequency* ($2\pi f_n$) of the system and ζ is the corresponding *damping ratio*.

This system can also be expressed as a *transfer function* by applying Laplace transform $\mathcal{L}[y_t] := Y(s)$ to give

$$s^2 Y(s) + 2\zeta\omega_n Y(s) + \omega_n^2 Y(s) = \omega_n^2 U(s) \tag{3.47}$$

Factoring out $Y(s)$ and solving, leads to the continuous-time *transfer function* representation

$$\mathcal{H}(s) := \frac{Y(s)}{U(s)} = \frac{\omega_n^2}{s^2 + 2\zeta\omega_n s + \omega_n^2} \tag{3.48}$$

From the second-order system perspective, we can also *factor* the denominator of transfer function of Eq. (3.48) into complex poles and see that it can also be represented equivalently in the generic form with its corresponding partial fraction form as

$$\mathcal{H}(s) = \frac{2\text{Re}[R]s + 2\text{Re}[R]\sigma - 2\omega\text{Im}[R]}{s^2 + 2\sigma s + (\sigma^2 + \omega^2)} = \frac{R}{s + \sigma - j\omega} + \frac{R^*}{s + \sigma + j\omega} \tag{3.49}$$

where Re[·] is the real part of [·], Im[·] is its imaginary part and R its residue.

Performing the inverse Laplace transform of this relation, we obtain the corresponding complex exponential representation

$$\mathcal{H}(t) = Re^{-(\sigma+j\omega)t} + R^* e^{-(\sigma-j\omega)t} \tag{3.50}$$

with poles given by $-\sigma \pm j\omega$ and residues by R.

Relating back to a physical system, the generic second-order system (control theory) has a transfer function of the form

$$\mathcal{H}(s) = \frac{\omega_n^2}{s^2 + 2\zeta\omega_n s + \omega_n^2} = \frac{\omega_n^2}{\left(s + \zeta\omega_n - j\omega_n\sqrt{1-\zeta^2}\right)\left(s + \zeta\omega_n + j\omega_n\sqrt{1-\zeta^2}\right)} \tag{3.51}$$

In terms of Eq. (3.49), we have that

$$\begin{aligned} \sigma_d &= \zeta\omega_n \\ \omega_d &= \omega_n\sqrt{1-\zeta^2} \end{aligned} \tag{3.52}$$

and σ_d is called the *damping coefficient* with ω_d termed the *damped natural frequency*. The system *time constant* is also defined as

$$\tau_c = \frac{1}{\sigma_d} = \frac{1}{\zeta\omega_n}$$

completing the description. Solving the second-order equation using the inverse Laplace transform leads to the real *impulse response* of a damped sinusoidal, *SISO* second-order system.

$$\mathcal{H}(t) = K_t e^{-\sigma_d t} \sin\left(\omega_d t + \phi\right) = K_t e^{-\zeta\omega_n t} \sin\left(\omega_n \sqrt{1-\zeta^2} t + \phi\right) \tag{3.53}$$

for K_t a scale factor.

This relation can easily be extended to define the corresponding *FRF* by investigating

$$F(\omega) := \mathcal{H}(s)|_{s=j\omega}$$

along the $j\omega$-axis, that is, define

$$\begin{aligned} F(\omega) &:= \frac{\omega_n^2}{-\omega^2 + 2j\zeta\omega_n\omega + \omega_n^2} \\ &= \frac{\omega_n^2}{\left(j\omega - \zeta\omega_n - j\omega_n\sqrt{1-\zeta^2}\right)\left(j\omega - \zeta\omega_n + j\omega_n\sqrt{1-\zeta^2}\right)} \end{aligned} \tag{3.54}$$

or more succinctly as

$$F(\omega) = \frac{R}{(j\omega - \lambda)} + \frac{R^*}{(j\omega - \lambda^*)} \quad \text{for} \quad \lambda = -\zeta\omega_n \pm j\omega_n\sqrt{1-\zeta^2} \tag{3.55}$$

Dividing by ω_n^2 gives

$$F(\omega) := \frac{1}{1 + 2j\zeta\left(\frac{\omega}{\omega_n}\right) - \left(\frac{\omega}{\omega_n}\right)^2} \tag{3.56}$$

with corresponding magnitude and phase given by

$$|F(\omega)| := \frac{1}{\sqrt{\left(1 - \left(\frac{\omega}{\omega_n}\right)^2\right)^2 + \left(2\zeta\left(\frac{\omega}{\omega_n}\right)\right)^2}} \quad \text{and} \quad \arg F(\omega) = \tan^{-1}\left(\frac{2\zeta\left(\frac{\omega}{\omega_n}\right)}{1 - \left(\frac{\omega}{\omega_n}\right)^2}\right) \tag{3.57}$$

With this in mind, we can extend these results to *MIMO* transfer functions characterized by the governing multichannel differential equations as

$$\sum_{m=0}^{M} A_m \frac{d^m \mathbf{y}(t)}{dt^m} = \sum_{m=0}^{M} B_m \frac{d^m \mathbf{u}(t)}{dt^m} \tag{3.58}$$

for $A_m \in \mathcal{R}^{N_y \times N_y}$ and $B_m \in \mathcal{R}^{N_y \times N_u}$, the matrix polynomial coefficients. Taking the Laplace transform, the *MFD* of both left and right models evolve as [3, 4]:

$$\underbrace{\mathbf{H}(s) := \mathbf{A}^{-1}(s)\mathbf{B}(s)}_{\text{Left MFD}} \quad \text{and} \quad \underbrace{\mathbf{H}(s) := \mathbf{B}(s)\mathbf{A}^{-1}(s)}_{\text{Right MFD}} \tag{3.59}$$

for $\mathrm{A} \in \mathcal{C}^{N_y \times N_y}$ and $\mathrm{B} \in \mathcal{C}^{N_y \times N_u}$ defined by the matrix polynomials in the complex variable s and

$$\mathbf{A}(s) = \sum_{m=0}^{M} A_m s^m \quad \text{and} \quad \mathbf{B}(s) = \sum_{m=0}^{M} B_m s^m \tag{3.60}$$

completing the descriptions. Note that identical relations exist for discrete-time data with replacing differential equations by difference equations and the S-transform variable s with the $\mathcal{Z}$-transform variable z, that is, and $z \to s$ (more *MFD* details are developed in Refr. [1, 2]).

3.2.4 Harmonic Models

In this section, we develop exponential-based harmonic models starting from a continuous-time system, since they evolve naturally from physical systems, leading directly to the discrete-time version through sampling theory. By "exponential-based" models, we mean starting with the complex exponential signal including the real damped sinusoidal representations as well as harmonic models which are all special cases. Once we complete this development, we discuss harmonic models as a special case of the exponential model set providing a good basis for the models to follow.

We define the *continuous-time exponential model* as

$$x_t = \sum_{i=1}^{N_e} A_i e^{-p_i t} = \sum_{i=1}^{N_e} A_i e^{-(\sigma_i + j\omega_i)t} \tag{3.61}$$

for A_i the complex amplitudes, $p_i = \sigma_i + j\omega_i$ the complex poles with respective damping and angular frequencies given by $\{\sigma_i, \omega_i\}$. Taking the Laplace transform of this expression gives the frequency domain representation

$$X(s) = \sum_{i=1}^{N_e} \frac{A_i}{s + p_i} = \sum_{i=1}^{N_e} \frac{A_i}{s + \sigma_i + j\omega_i} \tag{3.62}$$

Clearly special cases of this model evolve when the poles occur in complex pairs leading to *damped sinusoidal models*, that is,

$$x_t = \sum_{i=1}^{N_e} A_i e^{-p_i t} + A_i^* e^{-p_i^* t} = \sum_{i=1}^{N_e} A_i e^{-(\sigma_i + j\omega_i)t} + A_i^* e^{-(\sigma_i - j\omega_i)t} \tag{3.63}$$

with the frequency domain equivalent

$$X(s) = \sum_{i=1}^{N_e} \frac{A_i}{s+\sigma_i+j\omega_i} + \frac{A_i^*}{s+\sigma_i-j\omega_i} = 2\sum_{i=1}^{N_e} \frac{\mathrm{Re}[A_i](s+\sigma_i)+\mathrm{Im}[A_i]\omega_i}{(s+\sigma_i)^2+\omega_i^2} \tag{3.64}$$

with inverse transforming to obtain the equivalent (real) representation

$$x_t = 2\sum_{i=1}^{N_e} (\mathrm{Re}[A_i])e^{-\sigma_i t}\cos\omega_i t + (\mathrm{Im}[A_i])e^{-\sigma_i t}\sin\omega_i t \tag{3.65}$$

Letting $\omega_i = 0$ gives the real damped exponential model

$$x_t = 2\sum_{i=1}^{N_e} \mathrm{Re}\,[A_i]e^{-\sigma_i t} \tag{3.66}$$

and taking transforms gives the frequency domain representation

$$X(s) = 2\sum_{i=1}^{N_e} \frac{\mathrm{Re}\,[A_i]}{s+\sigma_i} \tag{3.67}$$

Each of these representations is used to develop signal models for a wide variety of applications [10, 23] which makes them very useful in interpreting the physical signals. Next, we develop the discrete-time equivalents.

If we transform the continuous-time model to the discrete-time with sampling interval $\triangle T$ using the *impulse invariant transformation*, $z_i = e^{p_i\triangle T} = e^{(\sigma_i+j\omega_i)\triangle T}$ and $h_t = x_t$ in the model. Thus, we have that $h_{k\triangle T} = h_t|_{t=k\triangle T}$. That is, if we sample the impulse response of Eq. (3.61), then we obtain

$$x(k\triangle T) = \sum_{i=1}^{N_e} A_i e^{-p_i k\triangle T} = \sum_{i=1}^{N_e} A_i e^{-(\sigma_i+j\omega_i)k\triangle T} = \sum_{i=1}^{N_e} \left(A_i e^{-\sigma_i k\triangle T}\right)\left(e^{-j\omega_i}\right)^{k\triangle T}$$

Simplifying, we obtain the *discrete-time-damped exponential model*

$$x(t_k) = \sum_{i=1}^{N_e} \alpha_i(t_k)\beta_i^{t_k} \tag{3.68}$$

where $t_k = k\triangle T$, $\alpha_i(t_k) := A_i e^{-\sigma_i k\triangle T}$ and $\beta_i := e^{-j\omega_i}$.

Thus, we see that the generic exponential model evolves by sampling the continuous-time version and using the impulse invariant transformation. With the relationship between the continuous and discrete-time exponential models in hand, let us examine some of the special discrete signal models that evolve from these representations. Note that we return to the usual discrete-time index notation of t rather than the sampled index of t_k.

Consider the *discrete sinusoidal signal model* given by

$$y(t) = x(t) + n(t) = \sum_{i=1}^{N_s} A_i \sin(\Omega_i t + \phi_i) + n(t) \tag{3.69}$$

where $\Omega_i = 2\pi f_i \triangle T$, n is zero-mean, white with variance R_{nn} and random phase uniformly $\phi \sim \mathrm{U}(-\pi, \pi)$. The covariance of this signal can be calculated analytically as

$$R_{yy}(k) = \sum_{i=1}^{N_s} \frac{A_i^2}{2} \cos k\triangle T\Omega_i + R_{nn}\delta(k) \tag{3.70}$$

or

$$R_{yy}(k) = \begin{cases} R_{nn} + \sum_{i=1}^{N_s} P_i & k = 0 \\ \sum_{i=1}^{N_s} P_i \cos k\triangle T\Omega_i & k \neq 0 \end{cases} \tag{3.71}$$

where $P_i = A_i^2/2$ and $\triangle T$ is the sampling interval.

The corresponding *PSD* is given by

$$S_{yy}(\Omega) = \triangle T \sum_{i=1}^{N_s} 2\pi P_i \left[\delta(\Omega + \Omega_i) + \delta(\Omega - \Omega_i)\right] + R_{nn} \tag{3.72}$$

So we see that the spectrum consists of N_s spectral lines located at $\Omega_i = 2\pi f_i \triangle T$ with power P_i on a base level of R_{nn} and therefore the model is parameterized by $(\{P_i\}, \{\Omega_i\})$ to characterize the sinusoids.

We know from linear systems theory that an individual sinusoid can be characterized by a second-order system or difference equation [10, 25][4]

$$x(t) = -\alpha_1 x(t-1) - \alpha_2 x(t-2) \tag{3.73}$$

where $\alpha_1 = 2\cos\Omega_1$ and $\alpha_2 = 1$. In general, then N_s-sinusoids can be represented by the $2N_s$-order difference equation

$$x(t) = -\sum_{i=1}^{2N_s} \alpha_i x(t-i) \tag{3.74}$$

Equivalently, we can use the sinusoidal model directly

$$y(t) = x(t) + n(t) = -\sum_{i=1}^{2N_s} \alpha_i x(t-i) + n(t) \tag{3.75}$$

4 This evolves from the fact that the $\mathcal{Z}$-transform of a sinusoid is $\left(\frac{z^{-1}\sin\Omega_0}{1+2\cos\Omega_0 z^{-1}+z^{-2}}\right)$ which gives the homogeneous solution.

If we substitute for $x(t-i)$ using Eq. (3.69) above we have

$$y(t) = -\sum_{i=1}^{2N_s} \alpha_i (y(t-i) - n(t-i)) + n(t) \tag{3.76}$$

or

$$y(t) = -\sum_{i=1}^{2N_s} \alpha_i y(t-i) + \sum_{i=0}^{2N_s} \alpha_i n(t-i) \tag{3.77}$$

for $\alpha_0 = 1$ showing that this is a special $2N_s$-th order *ARMA* model.

When the *ARMA* parameters are specified, it is well-known that they can be used to form the system characteristic equation whose roots are the required sinusoidal frequencies, that is [10]

$$z^{2N_s} + \alpha_1 z^{2N_s - 1} + \cdots + \alpha_{2N_s} = \prod_{i=1}^{N_s} (z - z_i)(z - z_i^*) = 0 \tag{3.78}$$

where $z_i = e^{j\Omega_i \triangle T}$.

After determining the roots of the characteristic equation, the sinusoidal frequencies are found from the impulse invariant transformation

$$\Omega_i = \frac{1}{\Delta T} \ln z_i \tag{3.79}$$

Next, we investigate a more complicated yet equivalent version of the sinusoidal model–the *harmonic model*. The sinusoidal model is a special case of this model. We define the *complex harmonic model* by

$$x(t) = \sum_{i=1}^{N_s} A_i e^{j(\Omega_i t + \phi_i)} \tag{3.80}$$

where A_i is complex, $A_i = |A_i| e^{j\theta}$; ϕ_i is random and uniformly distributed as $\phi \sim \mathcal{U}(-\pi, \pi)$ and Ω_i is the harmonic frequency. The set of N_s-complex amplitudes and harmonic frequencies, $[\{A_i\}, \{|\Omega_i\}]$, $i = 1, \ldots, N_s$ are assumed deterministic, but unknown.

The corresponding *measurement model* is characterized by the complex harmonic signal in additive random noise

$$y(t) = x(t) + n(t) \tag{3.81}$$

with $n(t)$, a zero-mean, random sequence with variance, σ_n^2.

Statistically, assuming stationarity, this model is characterized by its autocorrelation function as

$$R_{yy}(k) = E\{y(t)y^*(t+k)\} = R_{xx}(k) + R_{nn}(k) \tag{3.82}$$

since the signal is uncorrelated with the noise. The autocorrelation of the signal for a single harmonic is found by

$$R_{xx}(k) = E\left\{(A_i e^{j(\Omega_i t+\phi_i)})A_i^* e^{-j(\Omega_i(t+k)+\phi_i)}\right\} = E\left\{|A_i|^2 e^{-j\Omega_i k}\right\} \tag{3.83}$$

or simply

$$R_{xx}(k) = P_i e^{-j\Omega_i k} \text{ for } P_i := |A_i|^2 \tag{3.84}$$

Taking Fourier transforms, we obtain the corresponding harmonic power spectrum as

$$P_{xx}(\Omega) = P_i \delta(\Omega - \Omega_i) + \sigma_n^2 \tag{3.85}$$

an impulse or spectral line. Expanding this to the N_s-harmonics, we obtain the final expressions

$$R_{xx}(k) = \sum_{i=1}^{N_s} P_i e^{-j\Omega_i k} + \sigma_n^2 \delta(k) \tag{3.86}$$

or

$$R_{yy}(k) = \begin{cases} \sum_{i=1}^{N_s} P_i + \sigma_n^2 & k = 0 \\ \sum_{i=1}^{N_s} P_i e^{-j\Omega_i k} & k \neq 0 \end{cases} \tag{3.87}$$

This completes the discussion and development of exponentially based models. Next, we consider the set of models corresponding to propagating waves.

3.3 State-space Models

We begin by formally defining the concept of state [1]. The *state* of a system at time t is the "minimum" set of variables (*state variables*) along with the *input* sufficient to uniquely specify the dynamic system behavior for all t over the interval $t \in [t_o, \infty)$. The *state vector* is the collection of state variables into a single vector. The idea of a *minimal set* of state variables is critical and all techniques to define them must ensure that the smallest number of "independent" states have been defined in order to avoid possible violation of some important system theoretic properties [2, 10]. In short, one can think of states mathematically as simply converting an N-th order differential equation into a set of N-first order equations—each of which is a state variable. From a systems perspective, the states are the *internal* variables that may not be measured directly, but provide the critical information about system performance. For instance, measuring only the input/output voltages or currents of an integrated circuit board with a million embedded (internal) transistors—the internal voltages/currents would be the unmeasured states of this system.

3.3.1 Continuous-time State-space Models

Constraining the solution of a set of N_x-th order differential equations completely, we must specify the above-noted functions along with a set of N_x-initial conditions at time t_0 and the input for all $t \geq t_0$. Here N_x is the dimension of the "minimal" set of state variables. If we constrain the state-space representation even further to be linear in the states, then we obtain the generic continuous-time, *linear time-varying state-space* model given by

$$\dot{x}_t = A_t x_t + B_t u_t$$
$$y_t = C_t x_t + D_t u_t \tag{3.88}$$

where $x_t \in \mathcal{R}^{N_x \times 1}$, $u_t \in \mathcal{R}^{N_u \times 1}$, $y_t \in \mathcal{R}^{N_y \times 1}$ and the respective system, input, output and feedthrough matrices are: $A \in \mathcal{R}^{N_x \times N_x}$, $B \in \mathcal{R}^{N_x \times N_u}$, $C \in \mathcal{R}^{N_y \times N_x}$ and $D \in \mathcal{R}^{N_y \times N_u}$.

An interesting property of the state-space representation is to realize that these models represent a complete generic form for almost any physical system. That is, if we have an *RLC* circuit or a *MCK*-mechanical system, their dynamics are governed by the identical set of differential equations—only their coefficients differ. Of course, the physical meaning of the states is different, but this is the idea behind state-space —*many physical systems* can be captured by this *generic* form of differential equations, even though the systems are physically different.

For this development, we constrain the state-space representation above to be a *linear time-invariant (LTI) state-space* model given by

$$\dot{x}_t = A_c x_t + B_c u_t$$
$$y_t = C_c x_t + D_c u_t \tag{3.89}$$

where $A_t \longrightarrow A_c$, $B_t \longrightarrow B_c$, $C_t \longrightarrow C_c$ and $D_t \longrightarrow D_c$ their time-invariant counterparts with the subscript, "c," annotating continuous-time matrices.

This *LTI* model corresponds to the constant coefficient differential equation solutions which can be solved using Laplace transforms. Taking the Laplace transform[5] of these equations and solving for $X(s)$, we have that

$$X(s) = (sI - A_c)^{-1} x_{t_o} + (sI - A_c)^{-1} B_c U(s) \tag{3.90}$$

where $I \in \mathcal{R}^{N_x \times N_x}$ is the identity matrix. The corresponding output is

$$Y(s) = \left[C_c(sI - A_c)^{-1} B_c + D_c\right] U(s) + C_c(sI - A_c)^{-1} x_{t_o} \tag{3.91}$$

From the definition of *transfer function* (zero initial conditions), we have the desired result

$$H(s) \mid_{t_0=0} := \frac{Y(s)}{U(s)} = C_c(sI - A_c)^{-1} B_c + D_c \tag{3.92}$$

---

5 Recall the Laplace transform is defined as $X(s) := \mathcal{L}[x_t] = \int_{-\infty}^{\infty} x_t e^{-st} dt$ with inverse transform given by $x_t = \mathcal{L}^{-1}[X(s)] = \int_{-\infty}^{\infty} X(s) e^{st} ds$ with $s = \sigma + j\omega$, a complex variable.

Taking the inverse Laplace transform of this equation gives us the corresponding *impulse response matrix* of the *LTI* system as [1]

$$H(t, \tau) = C_c e^{A_c(t-\tau)} B_c + D_c \delta(t) \quad \text{for} \quad t \geq \tau \tag{3.93}$$

So we see that the state-space representation enables us to express the input–output relations in terms of the internal variables or states. Note also that this is a multivariable representation as compared to the usual single input-single output (scalar) systems models that frequently appear in the signal processing literature.

Now that we have the multivariable transfer function representation of our *LTI* system, we can solve the state equations directly using inverse transforms to obtain the time-domain solutions. First, we simplify the notation by defining the Laplace transform of the state-transition matrix or the so-called *resolvent matrix* of systems theory [1, 2] as

$$\Phi_c(s) := (sI - A_c)^{-1} \tag{3.94}$$

with

$$\Phi_c(t) = \mathcal{L}^{-1}\left[(sI - A_c)^{-1}\right] = e^{A_c t} \tag{3.95}$$

Therefore, the *state-transition matrix* is a critical component in the solution of the state equations of a *LTI* system and is given by

$$\Phi_c(t, t_0) = e^{A_c(t-t_0)} \qquad t \geq t_0 \tag{3.96}$$

and we can rewrite the transfer function matrix as

$$H(s) = C_c \Phi_c(s) B_c + D_c \tag{3.97}$$

with the corresponding state-input transfer matrix given by

$$X(s) = \Phi_c(s) x_{t_0} + \Phi_c(s) B_c U(s) \tag{3.98}$$

Taking the inverse Laplace transformation gives the time-domain solution

$$x_t = \mathcal{L}^{-1}[X(s)] = \Phi_c(t, t_0) x_{t_0} + \Phi_c(t, t_0) B_c * u_t$$

or

$$x_t = \underbrace{\Phi_c(t, t_0) x_{t_0}}_{\text{zero–input}} + \underbrace{\int_{t_0}^{t} \Phi_c(t, \alpha) B_c u_\alpha d\alpha}_{\text{zero–state}} \tag{3.99}$$

with corresponding output solution

$$y_t = C_c \Phi_c(t, t_0) x_{t_0} + \int_{t_0}^{t} C_c \Phi_c(t, \alpha) B_c u_\alpha d\alpha + D_c \delta(t) \tag{3.100}$$

Revisiting the continuous-time system of Eq. (3.99) and substituting the matrix exponential for the state-transition matrix gives the *LTI* solution as

$$x_t = e^{A_c(t-t_0)}x_{t_0} + \int_{t_0}^{t} e^{A_c(t-\alpha)}B_c u_\alpha d\alpha \tag{3.101}$$

and corresponding measurement system

$$y_t = C_c x_t + D_c u_t \tag{3.102}$$

The transition matrix plays a pivotal role in *LTI* systems theory for the analysis and prediction of the response of linear time-invariant and time-varying systems [1, 2, 10]. For instance, the *poles* of a *LTI* govern such important properties as stability and response time. The poles are the *roots* of the *characteristic* (polynomial) *equation* of A_c, which are found by solving for the roots of the determinant of the resolvent, that is,

$$|\Phi_c(s)| = |(sI - A_c)|_{s=p_i} = 0 \tag{3.103}$$

The poles of the system determine its response as:

$$y_t = \sum_{i=1}^{N_x} K_i e^{-p_i t} \tag{3.104}$$

where K_i is the gain or scale factor and A_c is diagonal which in this case is given by $A_c = \text{diagonal}\left[p_1, p_2, \ldots, p_{N_x}\right]$, the eigenvalues of A_c. Stability is determined by assuring that all of the poles lie within the left half of the S-plane [1].

Next, we consider the sampled-data, state-space representation.

3.3.2 Sampled-data State-space Models

Sampling a continuous-time system is commonplace with the advent of high-speed analog-to-digital converters and modern computers. A sampled-data system lies somewhere between the continuous analog domain (physical system) and the purely discrete domain (stock market prices). Since we are strictly sampling a continuous-time process, we must ensure that all of its properties are preserved. The well-known *Nyquist sampling theorem* precisely expresses the required conditions (twice highest frequency) to achieve "perfect" reconstruction of the process from its samples [10].

Thus, if we have a physical system governed by continuous-time dynamics and we "sample" it at given time instants, then a sampled-data model can be obtained directly from the solution of the continuous-time state-space model. That is, we know from Section 3.3.1 that

$$x_t = \Phi_c(t, t_0)x_{t_0} + \int_{t_0}^{t} \Phi_c(t, \alpha)B_c(\alpha)u_\alpha d\alpha$$

where $\Phi_c(\cdot, \cdot)$ is the continuous-time state-transition matrix that satisfies the matrix differential equation of Eq. (3.16), that is,

$$\dot{\Phi}_c(t, t_0) = A_t \Phi_c(t, t_0), \quad \Phi_c(t_0, t_0) = I, \quad t \geq t_0$$

Sampling this system such that $t \to t_k$ over the interval $(t_k, t_{k-1}]$, then we have the corresponding *sampling interval* defined by $\triangle t_k := t_k - t_{k-1}$. Note this representation need *not* necessarily be equally spaced—another important property of the state-space representation. Thus, the sampled solution becomes

$$x(t_k) = \Phi(t_k, t_{k-1})x(t_{k-1}) + \int_{t_{k-1}}^{t_k} \Phi(t_k, \alpha)B_c(\alpha)u_\alpha d\alpha \tag{3.105}$$

and therefore from the differential equation of Eq. (3.16), we have the solution

$$\Phi(t_k, t_{k-1}) = \int_{t_{k-1}}^{t_k} A(\alpha)\Phi(t_k, \alpha)d\alpha \qquad \text{for} \qquad \Phi(t_0, t_0) = I \tag{3.106}$$

where $\Phi(t_k, t_{k-1})$ is the *sampled-data* state-transition matrix–the critical component in the solution of the state equations enabling us to calculate state evolution in time. Note that for a *LTI* sampled-data system, the state-transition matrix is: $\Phi(t, t_k) = e^{A(t-t_k)}$ where A is the sampled-data system (process) matrix.

If we further assume that the input excitation is *piecewise constant* ($u_\alpha \to u(t_{k-1})$) over the interval $(t_k, t_{k-1}]$, then it can be removed from under the superposition integral in Eq. (3.105) to give

$$x(t_k) = \Phi(t_k, t_{k-1})x(t_{k-1}) + \left(\int_{t_{k-1}}^{t_k} \Phi(t_k, \alpha)B_c(\alpha)d\alpha \right) \times u(t_{k-1}) \tag{3.107}$$

Under this assumption, we can define the sampled-data *input transmission* matrix as

$$B(t_{k-1}) := \int_{t_{k-1}}^{t_k} \Phi(t_k, \alpha)B_c(\alpha)d\alpha \tag{3.108}$$

and therefore the *sampled-data state-space system* with equally or unequally sampled data is given by:

$$\begin{aligned} x(t_k) &= \Phi(t_k, t_{k-1})x(t_{k-1}) + B(t_{k-1})u(t_{k-1}) \\ y(t_k) &= C(t_k)x(t_k) + D(t_k)u(t_k) \end{aligned} \tag{3.109}$$

Computationally, sampled-data systems pose no particular problems when care is taken, especially since reasonable approximation and numerical integration methods exist [28–30]. This completes the discussion of sampled-data systems and approximations, next we consider the discrete-time systems.

3.3.3 Discrete-time State-space Models

Discrete state-space models evolve in two distinct ways: naturally from the problem or from sampling a continuous-time dynamical system. An example of a natural discrete system is the dynamics of balancing our own checkbook. Here the state is the evolving balance given the past balance and the amount of the previous check. There is "no information" between time samples, so this model represents a discrete-time system that evolves naturally from the underlying problem. On the other hand, if we have a physical system governed by continuous-time dynamics, then we "sample" it at given time instants. So we see that discrete-time dynamical systems can evolve from a wide variety of problems both naturally (checkbook) or physically (circuit). In this text we are primarily interested in physical systems (physics-based models), so we will concentrate on sampled systems reducing them to a discrete-time state-space model.

We can use a *first-difference approximation* and apply it to the general *LTI* continuous-time state-space model of Eq. (3.2) to obtain a discrete-time system, that is,

$$\dot{x}_t \approx \frac{x(t) - x(t-1)}{\triangle T} \approx A_c x(t-1) + B_c u(t-1)$$

$$y_t \approx y(t) = C_c x(t) + D_c u(t)$$

Solving for $x(t)$, we obtain

$$\begin{aligned} x(t) &= \left(I + A_c \triangle T\right) x(t-1) + B_c \triangle T u(t-1) \\ y(t) &= C_c x(t) + D_c u(t) \end{aligned} \tag{3.110}$$

Recognizing that the first difference approximation is equivalent to a first-order *Taylor series* approximation of A_c gives the discrete system, input, output, and feedthrough matrices as

$$\begin{aligned} A &\approx I + A_c \triangle T + O(\triangle T^2) \\ B &\approx B_c \triangle T \\ C &\approx C_c \\ D &\approx D_c \end{aligned} \tag{3.111}$$

The discrete *linear time-varying state-space representation* is given by the *system* or *process model* as

$$x(t) = A(t-1)x(t-1) + B(t-1)u(t-1) \tag{3.112}$$

and the corresponding discrete *output* or *measurement model* as

$$y(t) = C(t)x(t) + D(t)u(t) \tag{3.113}$$

where x, u, y are the respective N_x-state, N_u-input, N_y-output and A, B, C, D are the $(N_x \times N_x)$-system, $(N_x \times N_u)$-input, $(N_y \times N_x)$-output and $(N_y \times N_u)$-feedthrough matrices.

The state-space representation for linear, time-invariant (*LTI*), discrete systems is characterized by constant system, input, output, and feedthrough matrices, that is,

$$A(t) \to A, \quad B(t) \to B, \quad \text{and} \quad C(t) \to C, \quad D(t) \to D$$

and is given by the *LTI* system

$$\begin{aligned} x(t) &= Ax(t-1) + Bu(t-1) \\ y(t) &= Cx(t) + Du(t) \end{aligned} \tag{3.114}$$

The discrete system representation replaces the Laplace transform with the $\mathcal{Z}$-transform defined by the transform pair:

$$\begin{aligned} X(z) &:= \sum_{t=0}^{\infty} x(t) z^{-t} \\ x(t) &= \frac{1}{2\pi j} \oint_C X(z) z^{t-1} dz \end{aligned} \tag{3.115}$$

Time-invariant state-space discrete systems can also be represented in *input–output* or *transfer function* form using the $\mathcal{Z}$-transform to give

$$H(z) = C(zI - A)^{-1}B + D = C\left(\frac{\text{adj}(zI - A)}{\det(zI - A)}\right)B + D \tag{3.116}$$

where recall "adj" is the matrix adjoint (transpose of the co-factor matrix) and "det" is the matrix determinant.

We define the *characteristic equation* or *characteristic polynomial* of the N-dimensional system matrix A as

$$\alpha(z) := \det(zI - A) = z^N + \alpha_1 z^{N-1} + \alpha_2 z^{N-2} + \cdots + \alpha_{N-1} z + \alpha_N \tag{3.117}$$

with *roots* corresponding to the *poles* of the underlying system that are also obtained from the *eigenvalues* of A defined by

$$\lambda(A) := \text{eig}(A) = \prod_{n=1}^{N} (z - \lambda_n) \tag{3.118}$$

for λ_n is the nth system pole.

Taking inverse $\mathcal{Z}$-transforms of Eq. (3.116), we obtain the discrete impulse (or pulse) response matrix as

$$\mathcal{H}(t) = CA^{t-1}B + D\delta(t) \tag{3.119}$$

for δ the Kronecker delta function.

The solution to the state-difference equations can easily be derived by induction [7] and is given by

$$x(t) = \Phi(t,k)x(k) + \sum_{i=k+1}^{t} \Phi(t,i)B(i)u(i) \quad \text{for} \quad t > k \tag{3.120}$$

where $\Phi(t,k)$ is the *discrete-time state-transition* matrix. For time-varying systems, it can be shown (by induction) that the state-transition matrix satisfies

$$\Phi(t,k) = A(t-1) \cdot A(t-2) \cdots A(k)$$

while for time-invariant systems, the state-transition matrix is given by

$$\Phi(t,k) = A^{t-k} \quad \text{for} \quad t > k$$

The *discrete-state-transition matrix* possesses properties analogous to its continuous-time counterpart, that is,

1. $\Phi(t,k)$ is uniquely defined [Unique]
2. $\Phi(t,t) = I$ [Identity]
3. $\Phi(t,k)$ satisfies the matrix difference equation:

$$\Phi(t,k) = A(t-1)\Phi(t-1,k), \quad \Phi(k,k) = I, \quad t \geq k+1 \tag{3.121}$$

4. $\Phi(t,k) = \Phi(t,t-1) \times \Phi(t-1,t-2) \times \cdots \times \Phi(k+1,k)$ [Semi-Group]
5. $\Phi^{-1}(t,k) = \Phi(k,t)$ [Inverse]

3.3.4 Gauss–Markov State-space Models

Here we investigate the case when random inputs are applied to a discrete state-space system with random initial conditions. If the excitation is a random signal, then the state is also random. Restricting the input to be deterministic $u(t-1)$ and the noise to be zero-mean, white, Gaussian $w(t-1)$, the *Gauss–Markov model* evolves as

$$x(t) = A(t-1)x(t-1) + B(t-1)u(t-1) + W(t-1)w(t-1) \tag{3.122}$$

where $w \sim \mathcal{N}(0, R_{ww})$ and $x(0) \sim \mathcal{N}(\overline{x}(0), P(0))$.

The solution to the Gauss–Markov equations can easily be obtained by induction to give

$$x(t) = \Phi(t,k)x(k) + \sum_{i=k}^{t-1} \Phi(t,i+1)B(i)u(i) + \sum_{i=k}^{t-1} \Phi(t,i+1)W(i)w(i) \tag{3.123}$$

which is Markov depending only on the previous state. Since $x(t)$ is just a linear transformation of Gaussian processes, it is also Gaussian. Thus, we can represent a Gauss–Markov process easily employing the state-space models.

When the *measurement* model is also included, we have

$$y(t) = C(t)x(t) + D(t)u(t) + v(t) \tag{3.124}$$

where $v \sim \mathcal{N}(0, R_{vv})$.

Since the Gauss–Markov model of Eq. (3.122) is characterized by a Gaussian distribution, it is completely specified (statistically) by its mean and variance. Therefore, if we take the expectation of Eqs. (3.122) and (3.124) respectively, we obtain the *state mean vector* m_x as

$$m_x(t) = A(t-1)m_x(t-1) + B(t-1)u(t-1) \tag{3.125}$$

and the *measurement mean vector* m_y as

$$m_y(t) = C(t)m_x(t) + D(t)u(t) \tag{3.126}$$

The *state variance*[6] $P(t) := \text{var}\{x(t)\}$ is given by the discrete Lyapunov equation:

$$P(t) = A(t-1)P(t-1)A'(t-1) + W(t-1)R_{ww}(t-1)W'(t-1) \tag{3.127}$$

and the measurement variance, $R_{yy}(t) := \text{var}\{y(t)\}$ is

$$R_{yy}(t) = C(t)P(t)C'(t) + R_{vv}(t) \tag{3.128}$$

Similarly, it can be shown that the *state covariance* propagates according to the following equations:

$$P(t,k) = \begin{cases} \Phi(t,k)P(k) & \text{for} \quad t \geq k \\ \\ P(t)\Phi'(t,k) & \text{for} \quad t \leq k \end{cases} \tag{3.129}$$

We summarize the Gauss–Markov and corresponding statistical models in Table 3.3.

If we restrict the Gauss–Markov model to the stationary case, then

$$A(t) \to A, B(t) \to B, C(t) \to C, W(t) \to W, R_{ww}(t) \to R_{ww}, \quad \text{and} \quad R_{vv}(t) \to R_{vv}$$

and the variance equations become

$$P(t) = AP(t-1)A' + WR_{ww}W'$$

with

$$R_{yy}(t) = CP(t)C' + R_{vv} \tag{3.130}$$

At steady-state ($t \to \infty$), we have

$$P(t) = P(t-1) = \cdots = P_{ss} := P$$

---

6 We use the shorthand notation, $P(k) := P_{xx}(k,k) = \text{cov}\{x(k), x(k)\} = \text{var}\{x(k)\}$, throughout this text.

Table 3.3 Gauss–Markov representation.

State propagation

$$x(t) = A(t-1)x(t-1) + B(t-1)u(t-1) + W(t-1)w(t-1)$$

State mean propagation

$$m_x(t) = A(t-1)m_x(t-1) + B(t-1)u(t-1)$$

State variance/covariance propagation

$$P(t) = A(t-1)P(t-1)A'(t-1) + W(t-1)R_{ww}(t-1)W'(t-1)$$

$$P(t,k) = \begin{cases} \Phi(t,k)P(k) & \text{for} \quad t \geq k \\ \\ P(t)\Phi'(t,k) & \text{for} \quad t \leq k \end{cases}$$

Measurement propagation

$$y(t) = C(t)x(t) + v(t)$$

Measurement mean propagation

$$m_y(t) = C(t)m_x(t)$$

Measurement variance/covariance propagation

$$R_{yy}(t) = C(t)P(t)C'(t) + R_{vv}(t)$$

$$R_{yy}(t,k) = C(t)P(t)C'(t) + R_{vv}(t,k)$$

---

and therefore, the measurement covariance relations become

$$R_{yy}(0) = CPC' + R_{vv} \quad \text{for} \quad \text{lag} \quad k = 0 \tag{3.131}$$

By induction, it can be shown that

$$R_{yy}(k) = CA^{|k|}PC' \quad \text{for} \quad k \neq 0 \tag{3.132}$$

The measurement power spectrum is easily obtained by taking the $\mathcal{Z}$-transform of this equation to obtain

$$S_{yy}(z) = CS_{xx}(z)C' + S_{vv}(z) \tag{3.133}$$

where

$$S_{xx}(z) = T(z)S_{ww}(z)T'(z^{-1}) \quad \text{for} \quad T(z) = (zI - A)^{-1}W$$

with

$$S_{ww}(z) = R_{ww} \quad \text{and} \quad S_{vv}(z) = R_{vv}$$

Thus, using $H(z) = CT(z)$ the spectrum is given by

$$S_{yy}(z) = H(z)R_{ww}H'(z^{-1}) + R_{vv} \tag{3.134}$$

So we see that the Gauss–Markov state-space model enables us to have a more general representation of a multichannel stochastic signal. In fact, we are able to easily handle the multichannel and nonstationary statistical cases within this framework. Generalizations are also possible with the vector models, but those forms become quite complicated and require some knowledge of multivariable systems theory and canonical forms (see [1, 2, 10] for details).

3.3.5 Innovations Model

Next, we briefly discuss a special case of the Gauss–Markov representation—the innovations model. The significance of this model will be developed throughout the text, but we take the opportunity now to show its relationship to the basic Gauss–Markov representation. We start by extending the original Gauss–Markov representation to the *correlated* process and measurement noise case and then show how the innovations model is a special case of this structure.

The *standard* Gauss–Markov model for *correlated process and measurement noise* is given by

$$\begin{aligned} x(t) &= Ax(t-1) + Bu(t-1) + W(t-1)w^*(t-1) \\ y(t) &= Cx(t) + v^*(t) \end{aligned} \tag{3.135}$$

where $R^*(t,k) := R^*\delta(t-k)$ and

$$R^* := \left[\begin{array}{c|c} R_{w^*w^*} & R_{w^*v^*} \\ \hline R_{v^*w^*} & R_{v^*v^*} \end{array}\right] = \left[\begin{array}{c|c} WR_{ww}W' & WR_{wv} \\ \hline R_{vw}W' & R_{vv} \end{array}\right]$$

Here we observe that in the standard Gauss–Markov model, the $(N_x + N_v) \times (N_x + N_v)$ block covariance matrix, R^*, is full with cross-covariance matrices $R_{w^*v^*}$ on its off-diagonals. The usual standard model assumes that they are null (uncorrelated). To simulate a system with correlated $w(t)$ and $v(t)$ is more complicated using this form of the Gauss–Markov model because R^* must first be factored such that

$$R^* = \begin{bmatrix} R_1^* \\ R_2^* \end{bmatrix} \begin{bmatrix} R_1^{*\prime} & R_2^{*\prime} \end{bmatrix} \tag{3.136}$$

where R_i^* are matrix square roots [14, 30]. Once the factorization is performed, then the correlated noise is synthesized "coloring" the uncorrelated noise sources,

$w(t)$ and $v(t)$ as

$$\begin{bmatrix} w^*(t) \\ v^*(t) \end{bmatrix} = \begin{bmatrix} R_1^{*\prime} w(t) \\ R_2^{*\prime} v(t) \end{bmatrix} \tag{3.137}$$

The innovations model is a constrained version of the correlated Gauss–Markov characterization. If we assume that $\{e(t)\}$ is a zero-mean, white, Gaussian sequence, that is $e \sim \mathcal{N}(0, R_{ee})$, then the *innovations model* ([10, 31]) evolves as:

$$\begin{aligned} x(t) &= A(t-1)x(t-1) + B(t-1)u(t-1) + K(t-1)e(t-1) \\ y(t) &= C(t)x(t) + e(t) \end{aligned} \tag{3.138}$$

where $e(t)$ is the N_y-dimensional innovations vector and $K(t-1)$ is the $(N_x \times N_y)$ weighting matrix with the innovations covariance specified by:

$$R_{ee}^* := \text{cov}\left(\begin{bmatrix} Ke(t) \\ e(t) \end{bmatrix}\right) = \left[\begin{array}{c|c} KR_{ee}K' & KR_{ee} \\ \hline R_{ee}K' & R_{ee} \end{array}\right] \delta(t-k)$$

It is important to note that the innovations model has implications in Wiener–Kalman filtering (spectral factorization) because R_{ee}^* can be represented in *factored* or *square-root* form ($R := \sqrt{R}\sqrt{R}'$) directly in terms of the weight and innovation covariance matrix as

$$R_{ee}^* := \begin{bmatrix} K\sqrt{R_{ee}} \\ \sqrt{R_{ee}} \end{bmatrix} \begin{bmatrix} \sqrt{R_{ee}}K' & \sqrt{R_{ee}} \end{bmatrix} \delta(t-k) \tag{3.139}$$

Comparing the innovations model to the Gauss–Markov model, we see that they are both equivalent to the case when w and v are correlated. Next, we show the equivalence of the various model sets to this family of state-space representations.

3.3.6 State-space Equivalence Models

In this section, we discuss special state-space structures usually called "canonical forms" in the literature, since they represent unique state constructs that are particularly useful. We will confine the models to *SISO* forms, while the more complicated multivariable structures will be discussed in Chapter 6. Here we will first recall the *ARMAX* model of Section 3.2.1 and then its equivalent representation in the state-space form.

Time-series models are particularly useful representations used frequently by statisticians and signal processors to represent time sequences when no physics is available to employ directly. They form the class of black box or gray box models [10] which are useful in predicting data. These models have an input–output structure, but they can be transformed to an equivalent state-space representation. Each model set has its own advantages: the input–output models are easy to use, while the state-space models are easily generalized and usually evolve when physical phenomenology can be described in a model-based sense [10].

Next, we show the equivalence between the *ARMAX* and state-space models (for scalar processes). That is, we show how to obtain the state-space model given the *ARMAX* models by inspection. We choose particular coordinate systems in the state-space (canonical form) and obtain a relationship between entries of the state-space system to coefficients of the *ARMAX* model. An example is presented that shows how these models can be applied to realize a random signal. First, we consider the *ARMAX* to state-space transformation.

Recall from Eq. (3.18) that the general difference equation form of the *ARMAX* model is given by

$$y(t) = -\sum_{i=1}^{N_a} a_i y(t-i) + \sum_{i=0}^{N_b} b_i u(t-i) + \sum_{i=0}^{N_c} c_i e(t-i) \tag{3.140}$$

or equivalently in the frequency domain as

$$Y(z) = \left(\frac{b_o + b_1 z^{-1} + \cdots + b_{N_b} z^{-N_b}}{1 + a_1 z^{-1} + \cdots + a_{N_a} z^{-N_a}}\right) U(z) + \left(\frac{c_o + c_1 z^{-1} + \cdots + c_{N_c} z^{-N_c}}{1 + a_1 z^{-1} + \cdots + a_{N_a} z^{-N_a}}\right) E(z) \tag{3.141}$$

where $N_a \geq N_b$ and N_c and $\{e(t)\}$ is a zero-mean, white sequence with spectrum given by R_{ee}.

It is straightforward to show (see [10]) that the *ARMAX* model can be represented in *observer canonical form*:

$$\begin{aligned} x(t) &= A_0 x(t-1) + B_0 u(t-1) + W_0 e(t-1) \\ y(t) &= C_0' x(t) + b_0 u(t) + c_0 e(t) \end{aligned} \tag{3.142}$$

where x, u, e, and y are the N_a-state vector, scalar input, noise, and output with

$$A_0 := \left[\begin{array}{c|c} 0 & -a_{N_a} \\ --- & \vdots \\ I_{N_a-1} & -a_1 \end{array}\right] \quad B_0 := \begin{bmatrix} -a_{N_a} b_0 \\ \vdots \\ -a_{N_b+1} b_0 \\ --- \\ b_{N_b} - a_{N_b} b_0 \\ \vdots \\ b_1 - a_1 b_0 \end{bmatrix} \quad W_0 := \begin{bmatrix} -a_{N_a} c_0 \\ \vdots \\ -a_{N_c+1} c_0 \\ --- \\ c_{N_c} - a_{N_c} c_0 \\ \vdots \\ c_1 - a_1 c_0 \end{bmatrix}$$

$$C_0' := [0 \cdots 0 1]$$

Noting this structure we see that each of the matrix or vector elements $\{A_{i,N_a}, B_i, W_i, C_i\}$ $i = 1, \ldots, N_a$ can be determined from the relations

$$\begin{aligned} A_{i,N_a} &= -a_i \qquad i = 1, \ldots, N_a \\ B_i &= b_i - a_i b_0 \\ W_i &= c_i - a_i c_0 \\ C_i &= \delta(N_a - i) \end{aligned} \tag{3.143}$$

where

$$
\begin{aligned}
&b_i = 0 \qquad \text{for} \quad i > N_b \\
&c_i = 0 \qquad \text{for} \quad i > N_c \\
&\delta(i-j) \text{ is the } \textit{Kronecker} \text{ delta}
\end{aligned}
$$

Consider the following example to illustrate these relations.

Example 3.7 Let $N_a = 3$, $N_b = 2$, and $N_c = 1$; then the corresponding *ARMAX* model is

$$
\begin{aligned}
y(t) = &-a_1 y(t-1) - a_2 y(t-2) - a_3 y(t-3) + b_0 u(t) \\
&+ b_1 u(t-1) + b_2 u(t-2) + c_0 e(t) + c_1 e(t-1)
\end{aligned} \tag{3.144}
$$

Using the observer canonical form of Eq. (3.142), we have

$$
x(t) = \left[\begin{array}{cc|c} 0 & 0 & -a_3 \\ \hline 1 & 0 & -a_2 \\ 0 & 1 & -a_1 \end{array}\right] x(t-1) + \left[\begin{array}{c} -a_3 b_0 \\ \hline b_2 - a_2 b_0 \\ b_1 - a_1 b_0 \end{array}\right] u(t-1) + \left[\begin{array}{c} -a_3 c_0 \\ -a_2 c_0 \\ \hline c_1 - a_1 c_0 \end{array}\right] e(t-1)
$$

$$
y(t) = [0 \;\; 0 \;\; 1] x(t) + b_0 u(t) + c_0 e(t)
$$

This completes the example. △△△

It is important to realize that if we assume that $\{e(t)\}$ is Gaussian, then the *ARMAX* model is equivalent to the innovations representation of Section 3.3.5, that is,

$$
\begin{aligned}
x(t) &= Ax(t-1) + Bu(t-1) + We(t-1) \\
y(t) &= C'x(t) + b_0 u(t) + c_0 e(t)
\end{aligned} \tag{3.145}
$$

where in this case, $K \rightarrow W$, $D \rightarrow b_0$, and $1 \rightarrow c_0$. Also, the corresponding covariance matrix becomes:

$$
R_{ee}^* := \text{cov}\left(\begin{bmatrix} We(t) \\ c_0 e(t) \end{bmatrix} \right) = \left[\begin{array}{c|c} WR_{ee}W' & WR_{ee}c_0 \\ \hline c_0 R_{ee} W' & c_0 R_{ee} c_0 \end{array}\right] \delta(t-k)
$$

This completes the discussion on the equivalence of the general *ARMAX* to state-space. Next, let us develop the state-space equivalent models for some of the special cases of the *ARMAX* model presented in Section 2.4.

We begin with the *MA*

$$y(t) = \sum_{i=1}^{N_c} c_i e(t-i) \quad \text{or} \quad Y(z) = C(z)E(z) = (1 + c_1 z^{-1} + \cdots + c_{N_c} z^{-N_c})E(z)$$

Define the state variable as

$$x_i(t-1) := e(t-i-1), \quad i = 1, \ldots, N_c \tag{3.146}$$

and therefore,

$$x_i(t) = e(t-i) = x_{i-1}(t-1), \quad i = 1, \ldots, N_c \tag{3.147}$$

Expanding this expression, we obtain

$$\begin{aligned}
x_1(t) &= e(t-1) \\
x_2(t) &= e(t-2) = x_1(t-1) \\
x_3(t) &= e(t-3) = x_2(t-1) \\
&\vdots \qquad \vdots \qquad \vdots \\
x_{Nc}(t) &= e(t-N_c) = x_{N_c-1}(t-1)
\end{aligned} \tag{3.148}$$

or in vector-matrix form

$$\begin{bmatrix} x_1(t) \\ \\ x_2(t) \\ \vdots \\ x_{N_c}(t) \end{bmatrix} = \left[\begin{array}{ccc|c} 0 & \cdots & 0 & 0 \\ \hline & & & \\ 1 & \cdots & 0 & 0 \\ \vdots & \ddots & \vdots & \vdots \\ 0 & \cdots & 1 & 0 \end{array}\right] \begin{bmatrix} x_1(t-1) \\ \\ x_2(t-1) \\ \vdots \\ x_{N_c}(t-1) \end{bmatrix} + \begin{bmatrix} 1 \\ \\ 0 \\ \vdots \\ 0 \end{bmatrix} e(t-1)$$

$$y(t) = \begin{bmatrix} c_1 & c_2 & \cdots & c_{N_c} \end{bmatrix} \begin{bmatrix} x_1(t-1) \\ x_2(t-1) \\ \vdots \\ x_{N_c}(t-1) \end{bmatrix} + c_0 e(t) \tag{3.149}$$

Thus, the general form for the *MA* state-space is given by

$$\mathbf{x}(t) = \left[\begin{array}{ccc|c} 0 & \cdots & 0 & 0 \\ \hline & & & \\ & \mathbf{I}_{N_c-1} & & \vdots \\ & & & \\ & & & 0 \end{array}\right] \mathbf{x}(t-1) + \mathbf{b}e(t-1)$$

$$y(t) = \mathbf{c}'\mathbf{x}(t) + b_0 e(t) \tag{3.150}$$

with $N_x = N_c$, $\mathbf{b}, \mathbf{c} \in R^{N_x \times 1}$.

Next consider the *autoregressive AR* model (all-pole) given by

$$\sum_{i=1}^{N_a} a_i y(t-i) = \sigma e(t) \quad \text{or} \quad Y(z) = \frac{\sigma E(z)}{1 + a_1 z^{-1} + \cdots + a_{N_a} z^{-N_a}} \tag{3.151}$$

Here the state vector is defined by $x_i(t-1) = y(t-i-1)$ and therefore, $x_i(t) = y(t-i) = x_{i+1}(t-1)$; $i = 1, \ldots, N_a - 1$ with $x_{N_a}(t) = y(t)$. Expanding over i, we obtain the vector-matrix state-space model

$$\begin{bmatrix} x_1(t) \\ x_2(t) \\ \vdots \\ x_{N_a}(t) \end{bmatrix} = \left[\begin{array}{c|cccc} 0 & 1 & 0 & \cdots & 0 \\ 0 & 0 & 1 & \cdots & 0 \\ \vdots & \vdots & \vdots & \ddots & \vdots \\ 0 & 0 & 0 & \cdots & 1 \\ \hline -a_{N_a} & -a_{N_a-1} & -a_{N_a-2} & \cdots & -a_1 \end{array}\right] \begin{bmatrix} x_1(t-1) \\ x_2(t-1) \\ \vdots \\ x_{N_a}(t-1) \end{bmatrix} + \begin{bmatrix} 0 \\ 0 \\ \vdots \\ \sigma \end{bmatrix} e(t-1)$$

$$y(t) = [0 \;\; 0 \;\; \cdots \;\; 1] \begin{bmatrix} x_1(t-1) \\ x_2(t-1) \\ \vdots \\ x_{N_a}(t-1) \end{bmatrix} \tag{3.152}$$

In general, we have the *AR* (*all-pole*) state-space model

$$\mathbf{x}(t) = \left[\begin{array}{c|cccc} 0 & & & & \\ \vdots & & \mathbf{I}_{N_a-1} & & \\ 0 & & & & \\ \hline -a_{N_a} & -a_{N_a-1} & -a_{N_a-2} & \cdots & -a_1 \end{array}\right] \mathbf{x}(t-1) + \mathbf{b}e(t-1)$$

$$y(t) = \mathbf{c}'\mathbf{x}(t) \tag{3.153}$$

with $N_x = N_a$, $\mathbf{b}, \mathbf{c} \in R^{N_x \times 1}$. Consider the following example to demonstrate this form.

Another useful state-space representation is the *normal form* which evolves by performing a partial fraction expansion of a rational discrete transfer function model (*ARMA*) to obtain

$$h(t) = \sum_{i=1}^{N_p} R_i (p_i)^{-t} \quad \text{or} \quad H(z^{-1}) = \frac{Y(z^{-1})}{E(z^{-1})} = \sum_{i=1}^{N_p} \frac{R_i}{1 - p_i z^{-1}} \tag{3.154}$$

for $\{R_i, p_i\}$; $i = 1, \ldots, N_p$ the set of residues and poles of $H(z^{-1})$. Note that the normal form model is the decoupled or parallel system representation based on the following set of relations

$$y_i(t) - p_i y_i(t-1) = e(t), \quad i = 1, \ldots, N_p$$

Defining the state variable as $x_i(t) := y_i(t)$, then equivalently

$$x_i(t) - p_i x_i(t-1) = e(t), \quad i = 1, \dots, N_p \tag{3.155}$$

and therefore, the output is given by

$$y(t) = \sum_{i=1}^{N_p} R_i y_i(t) = \sum_{i=1}^{N_p} R_i x_i(t), \quad i = 1, \dots, N_p \tag{3.156}$$

Expanding these relations over i, we obtain

$$\begin{bmatrix} x_1(t) \\ x_2(t) \\ \vdots \\ x_{N_p}(t) \end{bmatrix} = \begin{bmatrix} p_1 & 0 & \cdots & 0 \\ 0 & p_2 & \cdots & 0 \\ \vdots & \vdots & \ddots & \vdots \\ 0 & 0 & \cdots & p_{N_p} \end{bmatrix} \begin{bmatrix} x_1(t-1) \\ x_2(t-1) \\ \vdots \\ x_{N_a}(t-1) \end{bmatrix} + \begin{bmatrix} 1 \\ 1 \\ \vdots \\ 1 \end{bmatrix} e(t-1)$$

$$y(t) = \begin{bmatrix} R_1 & R_2 & \cdots & R_{N_p} \end{bmatrix} \begin{bmatrix} x_1(t-1) \\ x_2(t-1) \\ \vdots \\ x_{N_p}(t-1) \end{bmatrix} \tag{3.157}$$

Thus, the general decoupled form of the *normal state-space model* is given by

$$\mathbf{x}(t) = \begin{bmatrix} p_1 & 0 & \cdots & 0 \\ 0 & p_2 & \cdots & 0 \\ \vdots & \vdots & \ddots & \vdots \\ 0 & 0 & \cdots & p_{N_p} \end{bmatrix} \mathbf{x}(t-1) + \mathbf{b}e(t-1)$$

$$y(t) = \mathbf{c}'\mathbf{x}(t) \tag{3.158}$$

for $\mathbf{b} \in \mathcal{R}^{N_p \times 1}$ with $\mathbf{b} = \mathbf{1}$, an N_p-vector of unit elements. Here $\mathbf{c} \in \mathcal{R}^{1 \times N_p}$ and $\mathbf{c}' = \begin{bmatrix} R_1 & R_2 & \cdots & R_{N_p} \end{bmatrix}$.

The final model for state-space equivalence is the *MIMO*, *right matrix fraction description* (*RMFD*) that will be applied in frequency domain algorithms to follow.

As before, the governing multichannel differential equation leading to the *RMFD* is

$$\sum_{\ell=0}^{N_a} A_\ell \frac{d^\ell \mathbf{y}(t)}{dt^\ell} = \sum_{\ell=0}^{N_b} B_\ell \frac{d^\ell \mathbf{u}(t)}{dt^\ell}$$

for $A_\ell \in \mathcal{R}^{N_y \times N_y}$ and $B_\ell \in \mathcal{R}^{N_y \times N_u}$, the matrix polynomial coefficients. By Laplace transforming these equations, the *MFD* model evolves as [3]:

$$\mathbf{H}(s) = \mathbf{B}(s)\mathbf{A}^{-1}(s) \tag{3.159}$$

defined by the matrix polynomials (above) in the complex variable s and

$$\mathbf{A}(s) = \sum_{\ell=0}^{N_a} A_\ell s^\ell \quad \text{and} \quad \mathbf{B}(s) = \sum_{\ell=0}^{N_b} B_\ell s^\ell \tag{3.160}$$

It has been shown that the equivalent state-space representation of the *RMFD* $\left(\Sigma_{RMFD} := \{A_R, B_R, C_R, D_R\}\right)$ can be specified by ([see Refr. [3] for proof):

$$\mathbf{A}_R = \begin{bmatrix} -\mathrm{A}_{N_a}^{-1}\mathrm{A}_{N_a-1} & \cdots & -\mathrm{A}_{N_a}^{-1}\mathrm{A}_1 & -\mathrm{A}_{N_a}^{-1}\mathrm{A}_0 \\ \mathbf{I}_{N_a} & \cdots & 0 & 0 \\ \vdots & \ddots & \vdots & \vdots \\ 0 & \cdots & \mathbf{I}_{N_a} & 0 \end{bmatrix};\ \mathbf{B}_R = \begin{bmatrix} \mathrm{A}_{N_a}^{-1} \\ \mathbf{0}_{N_u \times N_a - 1)N_u} \end{bmatrix};$$

$$\mathbf{C}_R = \left[\mathrm{B}_{N_a-1} - \mathrm{B}_{N_a}\mathrm{A}_{N_a}^{-1}\mathrm{A}_{N_a-1} \cdots \mathrm{B}_0 - \mathrm{B}_{N_a}\mathrm{A}_{N_a}^{-1}\mathrm{A}_0\right];\ \mathbf{D}_R = \mathrm{B}_{N_a}\mathrm{A}_{N_a}^{-1} \tag{3.161}$$

It should also be noted that this is a generalization and becomes the familiar *companion matrix* when $(\mathrm{A}_{N_a}^{-1} = \mathbf{I})$.

Another special case of *MFD* models is the *common denominator model* (*CDM*) where each element of the *MIMO*-transfer function has the identical denominator, that is, [3]

$$\mathbf{H}(s) = \frac{\mathbf{B}_c(s)}{\mathrm{a}_C(s)} \tag{3.162}$$

for the scalar $(\mathbf{a}_C(s))$ and matrix $(\mathbf{B}_C(s))$ polynomials

$$\mathbf{a}_C(s) = \sum_{m=0}^{N_a} a_m s^m \quad \text{and} \quad \mathbf{B}_C(s) = \sum_{m=0}^{N_b} B_m s^m \tag{3.163}$$

The *CDM* evolves from the continuous-time, state-space, transfer function [3, 4]

$$\mathbf{H}(s) = \frac{\mathbf{C}\ \mathrm{adj}(s\mathbf{I} - \mathbf{A})\mathbf{B} + \det(s\mathbf{I} - \mathbf{A})\mathbf{D}}{\det(s\mathbf{I} - \mathbf{A})} \tag{3.164}$$

that can be transformed to diagonal form with eigenvalues, $\lambda_\ell; \ell = 1, \dots, N$ such that

$$\mathbf{H}(s) = \frac{\mathbf{B}_c(s)}{\mathrm{a}_C(s)} = \frac{\sum_{\ell=1}^{N} \mathrm{C}_{m,\ell}\mathrm{B}_{m,\ell}\left(\prod_{m=1,m\neq\ell}^{N}(s - \lambda_m)\right) + \left(\prod_{\ell=1}^{N}(s - \lambda_\ell)\right)}{\prod_{\ell=1}^{N}(s - \lambda_\ell)} \tag{3.165}$$

3.4 Summary

In this chapter, we have discussed the development of models that are employed in a variety of signal processing schemes, sometimes dictated by a priori knowledge of the underlying problem physics. Starting with the set of predominant data-based representations that included response matrices of deterministic signals to stochastic representations evolving as correlation and spectral density matrices including Toeplitz and Hankel matrix structures that evolve in many signal processing and multichannel applications. Next, parametric signal representations followed leading to a suite of time-series models including the *ARMAX* model and its variants, followed by the order recursive lattice models. Frequency domain representations were developed starting with the usual second-order, transfer function representation and its various forms as well as its Fourier counterpart the frequency response matrix popular in the structural analysis community. Finally, perhaps the most versatile signal characterization evolved in the form of the state-space model in a variety of forms as continuous-time, sampled-data, and discrete-time state-space models. The stochastic variants of the deterministic models were presented leading to the Gauss–Markov representations for linear systems. The discussion of both the deterministic and stochastic state-space models included a brief development of their second-order statistics were also discussed along with the underlying discrete systems theory where it was shown that the time-series models can easily be represented in state-space form through the use of canonical forms (models). The chapter concluded with a brief development of the multichannel, transfer function representation, the *MFD* and its equivalent state representation.

MATLAB® Notes

MATLAB® has many commands to convert to/from state-space models to other forms useful in signal processing. Many of them reside in the Signal Processing, Control Systems, and System Identification tool-boxes. The matrix exponential is invoked by the **expm** command and is determined from Taylor/Padé approximants using the scaling and squaring approach. Also the commands **expmdemo1**, **expmdemo2**, and **expmdemo3** demonstrate the trade-offs of the Padé, Taylor, and eigenvector approaches to calculating the matrix exponential. The ordinary differential equation method is available using the wide variety of numerical integrators available (**ode***). Converting to/from transfer functions and state space is accomplished using the **ss2tf** and **tf2ss** commands, respectively. *ARMAX* simulations are easily accomplished using the **filter** command with a variety of options converting from armax-to/from transfer functions. The Identification Toolbox converts polynomial-based models to state-space and continuous parameters including Gauss–Markov to discrete parameters (**th2ss**, **thc2thd**, **thd2thc**).

References

1 Kailath, T. (1980). *Linear Systems*. New Jersey: Prentice-Hall.

2 Chen, C. (1984). *Introduction to Linear System Theory*. New York: Holt, Rhinehart, and Winston.

3 Reynders, E. (2012). System identification methods for (operational) modal analysis: review and comparison. *Arch. Comput. Meth. Eng.* 19 (1): 51–124.

4 Rainieri, C. and Fabbrocino, G. (2014). *Operational Modal Analysis of Civil Engineering Structures*. London: Springer.

5 Robinson, E. and Silvia, M. (1979). *Digital Foundations of Time Series Analysis*, vol. 1. San Francisco, CA: Holden-Day.

6 Aoki, M. (1990). *State Space Modeling of Time Series*, 2e. London: Springer.

7 DeCarlo, R. (1989). *Linear Systems: A State Variable Approach*. New Jersey: Prentice-Hall.

8 Juang, J. (1994). *Applied System Identification*. Upper Saddle River, NJ: Prentice-Hall PTR.

9 Maybeck, P. (1979). *Stochastic Models, Estimation and Control*, vol. 1. New York: Academic Press.

10 Candy, J. (2006). *Model-Based Signal Processing*. Hoboken, NJ: Wiley/IEEE Press.

11 Reid, J. (1983). *Linear System Fundamentals: Continuous, and Discrete, Classic and Modern*. New York: McGraw-Hill.

12 van Overschee, P. and De Moor, B. (1996). *Subspace Idenification for Linear Systems: Theory, Implementation, Applications*. Boston, MA: Kluwer Academic Publishers.

13 Papoulis, A. and Pillai, S. (2002). *Probability, Random Variables and Stochastic Processes*, 4e. New York: McGraw-Hill.

14 Jazwinski, A. (1970). *Stochastic Processes and Filtering Theory*. New York: Academic Press.

15 Hogg, R. and Craig, A. (1970). *Introduction to Mathematical Statistics*. New York: Macmillan.

16 MathWorks (1986). *MATLAB Users Manual*. Massachusetts: MathWorks Press.

17 Haykin, S. (1996). *Adaptive Filtering Theory*. New Jersey: Prentice-Hall.

18 Orfanidis, S. (1988). *Optimum Signal Processing*. New York: Macmillan.

19 Robinson, E. and Treitel, S. (1980). *Geophysical Signal Analysis*. New Jersey: Prentice-Hall.

20 Silvia, M. and Robinson, E. (1979). *Deconvolution of Geophysical Time Series in the Exploration of Oil and Natural Gas*. New York: Elsevier.
21 Markel, J. and Gray, A. (1976). *Linear Prediction of Speech*. New York: Springer.
22 Rabiner, L. and Shafer, R. (1978). *Digital Processing of Speech Signals*. New Jersey: Prentice-Hall.
23 Hayes, M. (1996). *Statistical Digital Signal Processing and Modeling*. New York: Wiley.
24 Friedlander, B. (1982). Lattice filters for adaptive processing. *Proc. IEEE* 70 (8): 829—867.
25 Kay, S. (1988). *Modern Spectral Estimation*. New Jersey: Prentice-Hall.
26 Dudgeon, D. and Mersereau, R. (1984). *Multidimensional Digital Signal Processing*. New Jersey: Prentice-Hall.
27 Johnson, D. and Dudgeon, D. (1993). *Array Signal Processing: Concepts and Techniques*. New Jersey: Prentice-Hall.
28 Moler, C. and Van Loan, C. (2003). Nineteen dubious ways to compute the exponential of a matrix, twenty-five years later. *SIAM Rev.* 45 (1): 3–49.
29 Golub, G. and Van Loan, C. (1989). *Matrix Computation*. Virginia: Johns Hopkins University Press.
30 Bierman, G. (1977). *Factorization Methods for Discrete Sequential Estimation*. New York: Academic Press.
31 Candy, J. (2019). *Model-Based Processing: An Applied Subspace Application*. Hoboken, NJ: Wiley/IEEE Press.

Problems

3.1 Suppose we are given the following system

$$y(t) = e^{-t}x(t+1)$$

Determine if it is
(a) time-invariant.
(b) causal.
(c) linear.

3.2 Calculate the response of a linear time-invariant (*LTI*) system with impulse response

$$h(t) = \begin{cases} \alpha^t & 0 \leq t \leq N-1 \\ 0 & \text{elsewhere} \end{cases}$$

when excited by the pulse

$$x(t) = \mu(t) - \mu(t-N)$$

3.3 Suppose we are given a digital filter

$$y(t) = -ay(t-1) + bx(t) \quad \text{for} \quad a < 0,\ b > 0$$

and the values of the impulse response

$$h(1) = 5/2, \quad h(2) = 5/4$$

Find the values of a, b, and $h(0)$.

3.4 Find the impulse response of the system

$$y(t) = -4y(t-1) - \sin \Omega_0 y(t-2) + \ln \Omega_0 x(t-1) \quad \text{for} \quad \Omega_0 > 0$$

3.5 We are given a causal *LTI* system with transfer function

$$T(z) = \frac{4}{1 - 5/2z^{-1} + z^{-2}}$$

(a) Find the unit impulse response of the system.
(b) Find the unit step response.
(c) Is the system stable?
(d) Find the inverse $\mathcal{Z}$-transform of $T(z)$ by division. Calculate $h(t)$, $t = 0, 1, 2$ from (a) and compare.
(e) Sketch the frequency response (magnitude only) of $T(z)$.

3.6 Calculate the frequency response of the three-point average given by the difference equation

$$y(t) = 1/3 \sum_{i=0}^{2} x(t-i)$$

3.7 Suppose we are given the following *LTI* system characterized by the following difference equation

$$y(t) + 3/4y(t-1) + 1/8y(t-2) = x(t) + 1/3x(t-1)$$

(a) Find the frequency response.
(b) Find the impulse response.
(c) Find the response to the excitation

$$x(t) = (1/3)^t \mu(t)$$

and zero initial conditions.

3.8 Calculate the frequency response of the following signal, $x(t) = (1.8)^t \mu(t)$.

3.9 Given the two digital filters, F_1 and F_2 characterized by difference equations

$$F_1: \; y_1(t) - 1/4y_1(t-1) = x_1(t) + 1/2x_1(t-1); \quad y_1(-1) = 4$$
$$F_2: \; y_2(t) - 1/8y_2(t-1) = 4x_2(t); \quad y_2(-1) = -8$$

find the overall performance of the cascaded filters if they are connected in series. Find the overall transfer function and unit impulse response.

3.10 Given the discrete *LTI* system characterized by

$$y(t) + 5/36y(t-1) = 1/36y(t-2) = x(t-1) - 1/2x(t-2)$$

with zero initial conditions, find
(a) Impulse response.
(b) Frequency response.
(c) Simulate the system using *MATLAB* and plot the corresponding frequency response.

3.11 A digital filter designed to operate in a microprocessor need not be causal, suppose we are given the system and REV means time reverse the input $x(t) \rightarrow x(-t)$
(a) Find $h'(t)$, the overall impulse response.
(b) Find the overall frequency response, $H'(\Omega)$. (N.B. This is a technique to remove the phase effect of a filter.)

3.12 Let $h(t)$ be the unit impulse response of a causal *FIR* filter with $h(t)$ real, then the frequency response can be represented in the form

$$H(\Omega) = \hat{H}(\Omega)e^{j\phi(\Omega)} \quad \text{for} \quad \hat{H} \text{ real}$$

Find $\phi(\Omega)$ for $0 \leq \Omega \leq \pi$ when $h(t) = h(N-1-t)$. N even.

3.13 Show that an *FIR* filter is always stable.

3.14 Suppose x, y, z are Gaussian random variables with corresponding means m_x, m_y, m_z and variances R_{xx}, R_{yy}, R_{zz} show that:
(a) If $y = ax + b$, a, b constants, then $y \sim N(am_x + b, a^2 R_{xx})$.
(b) If x and y are uncorrelated, then they are independent.
(c) If $x(i)$ are Gaussian with mean $m(i)$ and variance $R_{xx}(i)$, then for

$$y = \sum_i K_i x(i), \; y \sim N\left(\sum_i K_i m(i), \; \sum_i K_i^2 R_{xx}(i)\right)$$

(d) If x and y are jointly (conditionally) Gaussian, then

$$E\{x|y\} = m_x + R_{xy}R_{yy}^{-1}(y + m_y), \text{ and}$$
$$R_{x|y} = R_{xx} + R_{xy}R_{yy}^{-1}R_{yx}$$

(e) The random variable $x = E\{x|y\}$ is orthogonal to y.

(f) If y and z are independent, then

$$E\{x|y,z\} = E\{x|y\} + E\{x|z\} - m_x$$

(g) If y and z are not independent, show that

$$E\{x|y,z\} = E\{x|y,e\} = E\{x|y\} + E\{x|e\} - m_x$$

for $e = z - E\{x|y\}$.

3.15 Suppose we have the following transfer functions:
(a) $H_1(z) = 1 - 1/8z^{-1}$
(b) $H_2(z) = \frac{1}{1-3/4z^{-1}+1/8z^{-2}}$
(c) $H_3(z) = \frac{1-1/8z^{-1}}{1-3/4z^{-1}+1/8z^{-2}}$ Find the corresponding lattice model for each case, that is, all-zero, all-pole, and pole-zero.

3.16 Suppose we are given the factored power spectrum $S_{yy}(z) = H(z)H(z^{-1})$ with

$$H(z) = \frac{1 + \beta_1 z^{-1} + \beta_2 z^{-2}}{1 + \alpha_1 z^{-1} + \alpha_2 z^{-2}}$$

(a) Develop the *ARMAX* model for the process.
(b) Develop the corresponding Gauss–Markov model for *both* the standard and innovations representation of the process.

3.17 Suppose we are given the following Gauss–Markov model

$$x(t) = 1/3x(t-1) + 1/2w(t-1)$$
$$y(t) = 5x(t) + v(t)$$
$$w \sim N(0,3) \quad v \sim N(0,2)$$

(a) Calculate the state power spectrum, $S_{xx}(z)$.
(b) Calculate the measurement power spectrum, $S_{yy}(z)$.
(c) Calculate the state covariance recursion, $P(t)$.
(d) Calculate the steady-state covariance, $P(t) = \cdots = P = P_{ss}$.
(e) Calculate the output covariance recursion, $R_{yy}(t)$.
(f) Calculate the steady-state output covariance, R_{yy}.

3.18 Suppose we are given the Gauss–Markov process characterized by the state equations

$$x(t) = 0.97x(t-1) + u(t-1) + w(t-1)$$

for $u(t)$ a step of amplitude 0.03 and $w \sim N(0, 10^{-4})$ and $x(0) \sim N(2.5, 10^{-12})$.

(a) Calculate the covariance of x, that is, $P(t) = \mathrm{Cov}(x(t))$.

(b) Since the process is stationary, we know that

$$P(t+k) = P(t+k-1) = \cdots = P(0) = P$$

What is the steady-state covariance, P of this process?

(c) Develop a *MATLAB* program to simulate this process.

(d) Plot the process $x(t)$ with the corresponding confidence limits $\pm 2\sqrt{P(t)}$ for 100 data points, do 95% of the samples lie within the bounds?

3.19 Suppose the process in the problem above is measured using an instrument with uncertainty $v \sim N(0, 4)$ such that

$$y(t) = 2x(t) + v(t)$$

(a) Calculate the output covariance $R_{yy}(k)$.

(b) Develop a *MATLAB* program to simulate the output using the results of the previous problem.

(c) Plot the process $y(t)$ with the corresponding confidence limits $\pm 2\sqrt{R_{yy}(0)}$ for 100 data points, do 95% of the samples lie within the bounds?

3.20 Suppose we are given the *ARMAX* model

$$\begin{aligned} y(t) = &-0.5y(t-1) - 0.7y(t-2) + u(t) + 0.3u(t-1) \\ &+ e(t) + 0.2e(t-1) + 0.4e(t-2) \end{aligned}$$

(a) What is the corresponding innovations model in state-space form for $e \sim N(0, 10)$?

(b) Calculate the corresponding covariance matrix R^*_{ee}.

3.21 Given the following *ARMAX* model

$$A(q^{-1})y(t) = B(q^{-1})u(t) + \frac{C(q^{-1})}{D(q^{-1})}\epsilon(t)$$

for q^{-1} the backward shift (delay) operator such that

$$A(q^{-1}) = 1 + 1.5q^{-1} + 0.7q^{-2}$$
$$B(q^{-1}) = 1 + 0.5q^{-1}$$

$$C(q^{-1}) = 1 + 0.7q^{-1}$$
$$D(q^{-1}) = 1 + 05q^{-1}$$

(a) Find the pulse transfer representation of this process ($C = D = 0$). Convert it to the following equivalent *pole-zero* and *normal* state-space forms. Is the system controllable? Is it observable? Show your calculations.

(b) Find the pole-zero or *ARX* representation of this process ($C = 1, D = 0$). Convert it to the equivalent state-space form.

(c) Find the pole-zero or *ARMAX* representation of this process ($D = 0$). Convert it to the equivalent state-space form.

(d) Find the all-zero or *FIR* representation of this process ($A = 1, C = D = 0$). Convert it to the equivalent state-space form.

(e) Find the all-pole or *IIR* representation of this process ($B = 0, C = 0, D = 0$). Convert it to the equivalent state-space form.

(f) Find the all-zero or *MA* representation of this process ($A = 1, B = 0, D = 0$). Convert it to the equivalent state-space form.

(g) Using the full model above with A, B, C, D polynomials, is it possible to find an equivalent Gauss–Markov representation? If so, find it and convert to the equivalent state-space form. (Hint: Consider the C/D polynomials to be a coloring filter with input $\epsilon(t)$ and output $e(t)$.)

4

Signal Estimation

4.1 Classical Estimation

We have seen that a discrete random signal can be characterized by the equivalent probabilistic concept of a stochastic process. If a measured signal is free from extraneous variations and is repeatable from measurement to measurement, then it is defined as a *deterministic signal*. However, if it varies extraneously and is no longer repeatable, then it is a *random signal*. This chapter is concerned with the development of techniques to extract pertinent information from random signals utilizing any a priori information available. We call these techniques *signal estimation* or *signal enhancement*, and we call a particular algorithm a *signal estimator* or just *estimator*. Symbolically, we use the caret ^ notation throughout this text to annotate an *estimate*, for example, $s \rightarrow \hat{s}$. Sometimes estimators are called filters (e.g. Wiener filter) because they perform the same function as a deterministic (signal) filter except for the fact that the signals are random; that is, they remove unwanted disturbances. Noisy measurements are processed by the estimator to produce "filtered" data. Consider the following example of processing a random signal.

Example 4.1 A random pulse-like signal contaminated by noise is shown in Figure 4.1a. Here we see the measured data along with its Fourier spectrum. We design a signal estimator to extract the desired signal and remove the noise. The processed data are shown in Figure 4.1b where we observe the results of the estimation *filter* and corresponding enhanced spectrum. Here we see how the filtered response has eliminated the random noise.

$\triangle\triangle\triangle$

Estimation can be thought of as a procedure made up of three primary parts:

- Specification of a criterion
- Selection of signal models from a priori knowledge
- Development and implementation of an algorithm

Signal Processing: An Applied Decomposition Approach, First Edition. James Vincent Candy.

Companion website: www.wiley.com/go/CandySignProc

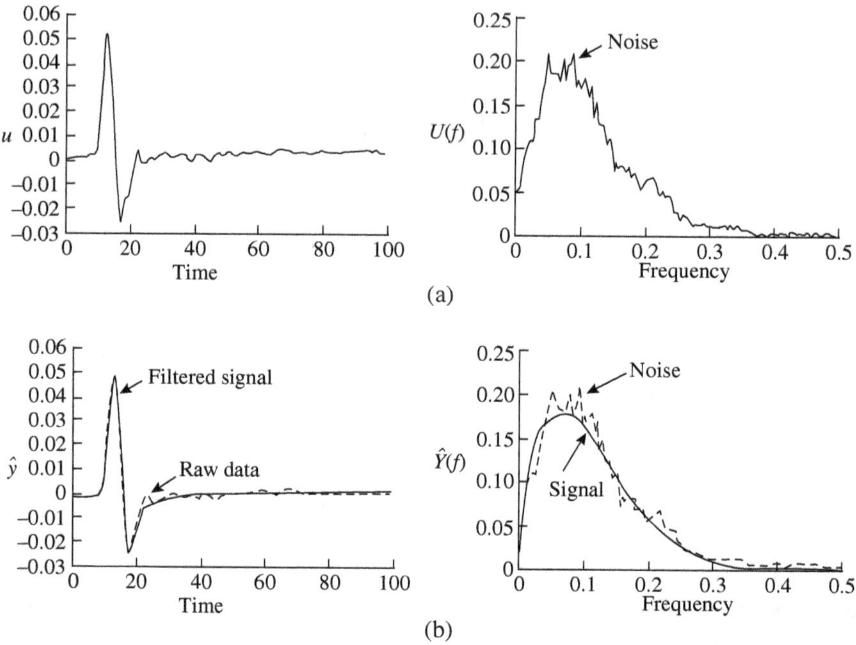

Figure 4.1 Processing of a *random* signal and noise: (a) raw data and Fourier spectrum with noise. (b) Processed data extracting the signal (estimate) and rejecting the extraneous (noise).

The criterion function can take many forms and can also be classified as deterministic or stochastic. Signal models represent a broad class of information formalizing the a priori knowledge about the process generating the signal, measurement instrumentation, noise characterization, underlying probabilistic structure, etc. as discussed in the first chapter. Finally, the algorithm or technique chosen to minimize: (or maximize) the criterion can take many different forms depending on (i) the signal models, (ii) the criterion, and (iii) the choice of solution. For example, one may choose to solve the well-known least-squares problem recursively or with a numerical-optimization algorithm. Another important aspect of most estimation algorithms is that they provide a "measure of quality" of the estimator. Usually what this means is that the estimator also predicts vital statistical information about how well it is performing.

A more formal description of the estimation problem evolves from its inherent structure as depicted in Figure 4.2 [1–3]. The *source* or origin of the problem provides the parameter (Θ) random or deterministic, but unknown, as a point in parameter space. Next, the *probabilistic transition mechanism* ($\Pr(Y|\Theta)$) separates the parameter from the measurements Y. It "knows" the true parameter value

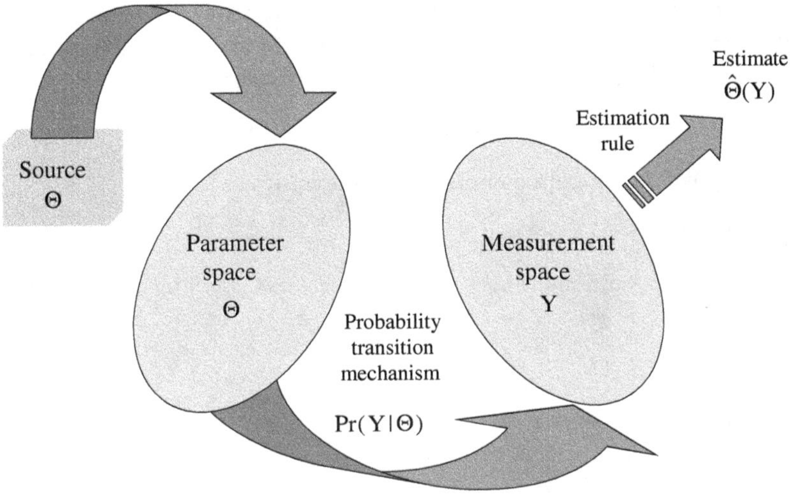

Figure 4.2 The estimation problem: source, probabilistic transition mechanism, measurement space, and estimate.

and generates samples in the *measurement space* according to the underlying probability law. Finally, the *estimation rule* or simply *estimator* ($\Theta(Y)$) is the result providing the "best" (in some sense) parameter value. The estimator is based upon knowledge of a priori distributions (in the random parameter case) and of the inherent conditional probabilities contained in the transition mechanism. Next, we examine the generic properties and performance of estimators.

4.1.1 Estimator Properties

There are many different estimators and algorithms. Suppose we are given two estimators, and asked to evaluate their performance to decide which one is superior. We must have a reasonable measure to make this decision. Thus, we must develop techniques to investigate various properties of estimators as well as a means to decide how well they perform. Sometimes we are interested in estimators that are not optimal for various reasons (e.g. simplicity, implementation ease) and we would like to judge how well they perform compared to the optimal. In this section, we investigate desirable properties of estimators, criteria to evaluate their performance, and investigate some of the most popular schemes.

Thus, the need for agreed-upon metrics to quantify estimator performance is necessary. The two primary statistical measures employed are the estimator mean (accuracy) and the variance (precision). These measures lead to desirable estimator properties, that is, we want estimators that are accurate or unbiased and precise. More formally, an *unbiased estimator* is one whose expected value is identical to the parameter being estimated. Suppose we wish to estimate the random parameter, Θ, then the estimator $\hat{\Theta}$ is *unconditionally unbiased* if

$$E\{\hat{\Theta}\} = E\{\Theta\} \, \forall \Theta \tag{4.1}$$

If Θ is a function of the measurements Y, then it is equivalent to have

$$E_{\hat{\Theta}Y}\{\hat{\Theta}(Y)\} = E\{\Theta\} \, \forall \Theta \tag{4.2}$$

If the estimator is conditioned on Θ, then we desire a *conditionally unbiased* estimator

$$E\{\hat{\Theta}|\Theta = \theta\} = E\{\Theta\} \tag{4.3}$$

A *biased* estimator, then is given by

$$E\{\hat{\Theta}\} = E\{\Theta\} + E\{B(\Theta)\} \tag{4.4}$$

where $B(\cdot)$ is the bias, which may be known or unknown. Known biases are easily removed. Note also that when the parameter Θ is *not* random but merely *unknown*, then the expectation operation results in the parameter itself ($E\{\Theta\} = \Theta$).

An estimator is *consistent*, if the estimate improves as the number of measurements increase, or equivalently it is said to *converge in probability*, if

$$\lim_{t\to\infty} \Pr\left(\left[\Theta - \hat{\Theta}(Y_t)\right] = 0\right) = 1 \tag{4.5}$$

It is said to be *mean-squared convergent* if

$$\lim_{t\to\infty} E\left\{\left[\Theta - \hat{\Theta}(Y_t)\right]' \left[\Theta - \hat{\Theta}(Y_t)\right]\right\} = 0 \tag{4.6}$$

An estimator, $\hat{\Theta}$, is *asymptotically efficient* if its corresponding estimation error defined by $\left(\tilde{\Theta} := \Theta - \hat{\Theta}(Y_t)\right)$ has the smallest (minimum) error variance of all possible estimators. That is, for any other estimator, say $\tilde{\Theta}^*$, then the following condition holds

$$\text{var}(\tilde{\Theta}^*) > \text{var}(\tilde{\Theta}) \tag{4.7}$$

Finally, an estimator is called *sufficient*, if it possesses all of the statistical information contained in the set of measurements regarding the parameter to be estimated. More formally, a statistic, $\hat{\Theta}$ is sufficient for Θ, if the conditional probability, $\Pr(Y|\hat{\Theta})$ does not depend on Θ. The underlying theory of this result is based on being able to factor the probability such that $\Pr(Y, \Theta) = f(\hat{\Theta}, \Theta)g(Y)$ (Neyman–Fisher factorization theorem) as discussed in more detail in [4] or [5]. These properties are desirable in an estimator and are determined in its evaluation.

4.1.2 Estimator Performance

The quality of an estimator is usually measured in terms of its *estimation error*,

$$\tilde{\Theta} = \Theta - \hat{\Theta} \tag{4.8}$$

A common measure of estimator quality is called the Cramér–Rao (*CRLB*) lower bound. The *CRLB* bound offers a means of assessing estimator quality prior to processing the measured data. We restrict discussion of the *CRLB* bound to the case of unbiased estimates, $\hat{\Theta}$, of a "nonrandom" parameter Θ. The bound is easily extended to more complex cases for biased estimates as well as random parameters [1, 6]. The *Cramér–Rao bound*[1] for any unbiased estimate $\hat{\Theta}$ of Θ based on the measurement, Y, is given by

$$R_{\tilde{\Theta}|\Theta} = \text{cov}(\Theta - \hat{\Theta}(Y)|\Theta = \theta) \geq \mathcal{I}^{-1} \tag{4.9}$$

where $\mathcal{I}$ the $N_\theta \times N_\theta$ *information matrix* defined by

$$\mathcal{I} := -E_y\left\{\nabla_\Theta\left(\nabla_\Theta \ln \Pr(Y|\Theta)\right)'\right\} \tag{4.10}$$

with the *gradient vector* $\nabla_\Theta \in R^{N_\theta \times 1}$ defined by

$$\nabla_\Theta := \left[\frac{\partial}{\partial\Theta_1} \cdots \frac{\partial}{\partial\Theta_{N_\theta}}\right]' \tag{4.11}$$

Any estimator satisfying the *CRLB* bound with equality is called *efficient*. The bound is easily calculated using the *chain rule* from vector calculus [7] defined by

$$\nabla_\Theta(a'b) := (\nabla_\Theta a')b + (\nabla_\Theta b')a \qquad a, b \in R^{N_\theta \times 1} \tag{4.12}$$

where a, b are functions of Θ. Consider the following example illustrating the calculation of the *CRLB* bound.

Example 4.2 Suppose we would like to estimate a nonrandom but unknown parameter, Θ, from a measurement y contaminated by additive Gaussian noise, that is,

$$y = \Theta + v$$

where $v \sim \mathcal{N}(0, R_{vv})$ and Θ is unknown. Thus, we have that

$$E\{Y|\Theta\} = E\{\Theta + v|\Theta\} = \Theta$$

and

$$\text{var}(Y|\Theta) = E\{(y - E\{Y|\Theta\})^2|\Theta\} = E\{v^2|\Theta\} = R_{vv}$$

---

1 We choose the matrix-vector version, since parameter estimators are typically vector estimates.

which gives

$$\Pr(Y|\Theta) \sim \mathcal{N}(\Theta, R_{vv})$$

and therefore

$$\ln \Pr(Y|\Theta) = -\frac{1}{2} \ln 2\pi\sqrt{R_{vv}} - \frac{1}{2}\frac{(y-\Theta)^2}{R_{vv}}$$

Differentiating, according to the chain rule of Eq. (4.12), and taking the expectation we obtain

$$\mathcal{I} = -E\left\{\frac{\partial^2}{\partial\Theta^2}\ln \Pr(Y|\Theta)\right\} = -E\left\{\frac{\partial}{\partial\Theta}\frac{(y-\Theta)}{R_{vv}}(-1)\right\} = \frac{1}{R_{vv}}$$

and therefore the *CRLB* bound is

$$R_{\tilde{\Theta}|\Theta} \geq R_{vv}$$

△△△

The utility of the *CRLB* bound is twofold: (i) it enables us to measure estimator quality because it indicates the "best" (minimum error covariance) that any estimator can achieve, and (ii) it allows us to decide whether or not the designed estimator is efficient, a desirable statistical property.

In summary, the properties of an estimator can be calculated prior to estimation (in some cases) and these properties can be used to answer the question "how well does this estimator perform." In subsequent sections, we investigate some popular estimators and their associated performance.

4.2 Minimum Variance (MV) Estimation

In this section, we review the concepts of minimum variance estimation and use it as a foundation to develop both Bayesian and likelihood estimators. The general techniques developed will be applied to develop various processors in subsequent sections.

Suppose we are asked to obtain an estimate of a N_θ-parameter vector $\mathbf{\Theta}$ from a set of noisy measurements characterized by the N_y-measurement vector $\mathbf{Y}$. We would like to construct an estimate that is best or optimal in some sense. The most natural criterion to consider is one that minimizes the error between the true parameter and its estimate based on the measured data. The *error variance criterion* defined by

$$J(\Theta) = E_\Theta\left\{[\Theta - \hat{\Theta}(Y)]'[\Theta - \hat{\Theta}(Y)]|Y\right\} \tag{4.13}$$

where

Θ is the true random N_θ-vector

Y is the measured random N_y-vector (data) and
$\hat{\Theta}$ is the estimate of Θ given Y

whose minimization leads to the *minimum variance estimator* [8]. Thus, if we minimize $J(\Theta)$ using the chain rule of Eq. (4.12), then we have that

$$\begin{aligned}\nabla_{\hat{\Theta}} J(\Theta) &= E_\theta\left\{\nabla_{\hat{\Theta}}\left(\Theta - \hat{\Theta}(Y)\right)'\left(\Theta - \hat{\Theta}(Y)\right)|Y\right\}\\ &= E_\theta\left\{-\left(\Theta - \hat{\Theta}(Y)\right) - \left(\Theta - \hat{\Theta}(Y)\right)|Y\right\}\\ &= -2\left[E_\theta\{\Theta - \hat{\Theta}(Y)|Y\}\right]\end{aligned}$$

performing the conditional expectation operation gives

$$\nabla_{\hat{\Theta}} J(\Theta) = -2\left[E_\theta\{\Theta|Y\} - \hat{\Theta}(Y)\right] \tag{4.14}$$

and setting this equation to zero and solving yields the minimum variance estimate as

$$\hat{\Theta}_{MV} = \hat{\Theta}(Y) = E_\theta\{\Theta|Y\} \tag{4.15}$$

We check for the global minimum by examining the sign (positive) of the second derivative, since

$$\nabla_{\hat{\Theta}}\left(\nabla_{\hat{\Theta}} J(\Theta)\right)' = 2I \tag{4.16}$$

indicating the unique minimum.

Thus, the minimum variance estimator is the *conditional mean*. The associated error variance is determined by substituting for $\hat{\Theta}(Y)$ into Eq. (4.13) giving

$$J_{MV} = R_{\tilde{\Theta}|Y} = E_\theta\{(\Theta - E\{\Theta|Y\})'(\Theta - E\{\Theta|Y\})|Y\} = E_\theta\{\Theta'\Theta|Y\} - E_\theta^2\{\Theta|Y\} \tag{4.17}$$

The minimum variance estimator can be summarized as:

Criterion:	$J = E\{\tilde{\Theta}'\tilde{\Theta}\}$
Model:	$\Pr(\Theta|Y)$
Algorithm:	$\hat{\Theta}_{MV} = E_\theta\{\Theta|Y\}$
Quality:	$R_{\tilde{\Theta}|Y}$

Note also that this estimator is linear, unconditionally, and conditionally unbiased and possesses general orthogonality properties. To see this, we investigate the estimation error defined by

$$\tilde{\Theta} = \Theta - \hat{\Theta}(Y) \tag{4.18}$$

Taking expectations, we have the fact that the estimator is unconditionally unbiased

$$E_\theta\{\tilde{\Theta}\} = E_\theta\{\Theta - \hat{\Theta}(Y)\} = E_\theta\{\Theta\} - E_\theta\{E_\theta\{\Theta|Y\}\} = E_\theta\{\Theta\} - E_\theta\{\Theta\} = 0 \tag{4.19}$$

as well as conditionally unbiased, since

$$\begin{aligned} E_\theta\{\tilde{\Theta}|Y\} &= E_\theta\{\Theta - \hat{\Theta}(Y)|Y\} = E_\theta\{\Theta|Y\} - E_\theta\{E_\theta\{\Theta|Y\}|Y\} \\ &= E_\theta\{\Theta|Y\} - E_\theta\{\Theta|Y\} = 0 \end{aligned} \tag{4.20}$$

Another important property of the minimum variance estimator is that the estimation error is orthogonal to *any* function, say $f(\cdot)$, of the data vector Y [9], that is,

$$E_{\theta Y}\{f(Y)\tilde{\Theta}'\} = 0 \tag{4.21}$$

Also,

$$E_\theta\{f(Y)\tilde{\Theta}'|Y\} = 0 \tag{4.22}$$

This is the well-known orthogonality condition. To see this, we substitute for the error

$$\begin{aligned} E_\theta\{f(Y)\tilde{\Theta}'|Y\} &= E_\theta\left\{f(Y)\left(\Theta - \hat{\Theta}(Y)\right)'|Y\right\} \\ &= f(Y)E_\theta\left\{\left(\Theta - \hat{\Theta}(Y)\right)'|Y\right\} \\ &= f(Y)\left(E_\theta\{\Theta|Y\} - \hat{\Theta}(Y)'\right) = 0 \end{aligned}$$

Taking the expectation over Y proves the unconditional result as well. Thus, the estimation error is orthogonal to *any* function of Y, a fact that is used in proofs throughout the literature for both linear and nonlinear estimators.

Let us now investigate the special case of the *linear* minimum variance estimator. The *estimation error* is *orthogonal* to all past data Y, that is,

$$E_{\Theta Y}\{Y\tilde{\Theta}'\} = 0 \tag{4.23}$$

or as before

$$E_\Theta\{Y\tilde{\Theta}'|Y\} = 0 \tag{4.24}$$

This is the well-known minimum variance estimator in the linear case [2]. For a linear function of the parameter, we have that

$$y = C\Theta + v \tag{4.25}$$

where $y, v, \in R^{N_y \times 1}, \Theta \in R^{N_\theta \times 1}, C \in R^{N_y \times N_\theta}$ and v is zero-mean, white with R_{vv}. The mean-squared error criterion

$$J(\Theta) = E\{\tilde{\Theta}'\tilde{\Theta}\} \tag{4.26}$$

is minimized to determine the estimate. The minimization results in the orthogonality condition of Eq. (4.23) which we write as

$$E\{y\tilde{\Theta}'\} = E\{y\Theta'\} - E\{y\hat{\Theta}'_{MV}\} = 0 \tag{4.27}$$

for $\hat{\Theta}_{MV} = K_{MV}y$, a linear function of the data vector. Substituting for y and $\hat{\Theta}$ in this equation gives

$$K_{MV} = R_{\Theta\Theta}C'(CR_{\Theta\Theta}C' + R_{vv})^{-1} \tag{4.28}$$

The corresponding quality is obtained as

$$R_{\tilde{\Theta}\tilde{\Theta}} = (R_{\Theta\Theta}^{-1} + C'R_{vv}^{-1}C)^{-1} \tag{4.29}$$

So we see that the minimum variance estimator for the linear case is:

Criterion:	$J = E\{\tilde{\Theta}'\tilde{\Theta}\}$
Model:	$y = C\Theta$
Algorithm:	$\hat{\Theta}_{MV} = K_{MV}y$
Quality:	$R_{\tilde{\Theta}\tilde{\Theta}}$

It is also interesting to note that the fundamental Wiener result [7] is easily obtained from the orthogonality condition of Eq. (4.23), that is,

$$E\{y\tilde{\Theta}'\} = E\{y\Theta'\} - E\{yy'\}K'_{MV} = R_{y\Theta} - R_{yy}K'_{MV} = 0 \tag{4.30}$$

which is called the *discrete Wiener–Hopf equation*. Solving for K_{MV}, we obtain the Wiener solution for a linear (batch) estimation scheme, that is,

$$K_{MV} = R_{\Theta y}R_{yy}^{-1} \tag{4.31}$$

We also mention in passing that *least-squares estimation* is similar to that of minimum variance except that no statistical information is assumed to be known about the process, that is, the least-squares estimator, $\hat{y}_{LS}$ minimizes the squared error criterion

$$\min_{\hat{y}} \; J = \tilde{y}'\tilde{y} \tag{4.32}$$

for $\tilde{y} = y - \hat{y}_{LS}$.

In contrast to the least-squares approach requiring *no* statistical information, we introduce two popular estimators: the "most probable" or maximum a posteriori (*MAP*) estimator and the maximum likelihood (*ML*) estimator, which is a special case of *MAP*. We show their relationship and analyze their performance.

4.3 Maximum A-Posteriori (MAP) Estimation

Suppose we are trying to estimate a random parameter, say Θ, from data $Y = y$. Then the associated conditional density $\Pr(\Theta|Y = y)$ is called the *posterior density* because the estimate is conditioned "after (*post*) the measurements" have been acquired. Estimators based on the a posteriori density are usually called *Bayesian* because they are constructed from Bayes' rule, since $\Pr(\Theta|Y)$ is difficult to obtain directly. That is,

$$\Pr(\Theta|Y) = \Pr(Y|\Theta)\frac{\Pr(\Theta)}{\Pr(Y)} \tag{4.33}$$

where $\Pr(\Theta)$ is called the *prior density* (before measurement), $\Pr(Y|\Theta)$ is called the likelihood (more likely to be true), and $\Pr(Y)$ is called the *evidence* (scales the posterior to assure its integral is unity). Bayesian methods view the sought-after parameter as random possessing a "known" a priori density. As measurements are made, the *prior* is converted to the *posterior density* function adjusting the parameter estimates. Thus, the result of increasing the number of measurements is to improve the a posteriori density resulting in a sharper peak closer to the true parameter.

To solve the estimation problem, the first step requires the determination of the a posteriori density. A logical solution to this problem leads us to find the "most probable" value of $\Pr(\Theta|Y)$ that is, its *maximum*. The *maximum a posteriori* (*MAP*) *estimate* is the value that yields the maximum value of the a posteriori density. The optimization is carried out in the usual manner by

$$\nabla_\Theta \Pr(\Theta|Y)|_{\Theta=\hat{\Theta}_{MAP}} = 0 \tag{4.34}$$

Because many problems are based on the exponential class of densities, the $\ln \Pr(\Theta|Y)$ is considered instead. Since the logarithm is a monotonic function, the maximum of $\Pr(\Theta|Y)$ and $\ln \Pr(\Theta|Y)$ occur at the same value of Θ. Therefore, the *MAP equation* is

$$\nabla_\Theta \ln \Pr(\Theta|Y)|_{\Theta=\hat{\Theta}_{MAP}} = 0 \tag{4.35}$$

Now, if we apply Bayes' rule to Eq. (4.33), then

$$\ln \Pr(\Theta|Y) = \ln \Pr(Y|\Theta) + \ln \Pr(\Theta) - \ln \Pr(Y) \tag{4.36}$$

Since $\Pr(Y)$ is not a function of the parameter Θ, the *MAP* equation can be written

$$\nabla_\Theta \ln \Pr(\Theta|Y)\big|_{\Theta=\hat{\Theta}_{MAP}} = \nabla_\Theta(\ln \Pr(Y|\Theta) + \ln \Pr(\Theta))\big|_{\Theta=\hat{\Theta}_{MAP}} = 0 \tag{4.37}$$

Of course, the second derivative (negative) must be checked to ensure that a global maximum is obtained. It can be shown that for a variety of circumstances (cost function convex, symmetrical) the minimum variance estimate of the Section 4.2 is in fact identical to the *MAP* estimate [2]. Next, we consider the case when the parameter Θ is not random leading to the maximum likelihood estimator.

4.4 Maximum Likelihood (ML) Estimation

In contrast to the Bayesian approach, the likelihood method views the parameter as deterministic, but *unknown*. It produces the "best" estimate as the value that maximizes the probability of the measurements given that the parameter value is "most likely" true. In the estimation problem, the measurement data are given along with the underlying structure of the probability density function (as in the Bayesian case), but the parameters of the density are unknown and must be determined from the measurements; therefore, the maximum likelihood estimate can be considered heuristically as that value of the parameter that best "explains" the measured data giving the most likely estimation.

More formally, let $\boldsymbol{\Theta}$ be a vector of unknown parameters, $\boldsymbol{\Theta} \in R^{N_\theta \times 1}$ and the corresponding set of N-independent measurements, $Y(N) := \{\mathbf{y}(1) \cdots \mathbf{y}(N)\}$ for $\mathbf{y} \in R^{N_y \times 1}$. The *likelihood* of Θ, given the measurements is defined to be proportional to the value of the probability density of the measurements given the parameters, that is,

$$\mathcal{L}(Y(N); \Theta) \propto \Pr(Y(N)|\Theta) = \Pr(\mathbf{y}(1) \cdots \mathbf{y}(N)|\Theta) = \Pi_{i=1}^{N} \Pr(\mathbf{y}(i)|\Theta) \quad (4.38)$$

where $\mathcal{L}$ is the likelihood function and $\Pr(Y|\Theta)$ is the joint probability density functions of the measurements given the unknown parameters. This expression indicates the usefulness of the likelihood function in the sense that in many applications measurements are available and are assumed drawn as a sample from a "known" or assumed known probability density function with unknown parameters (e.g. Poisson with unknown mean). Once we have the measurements (given) and the likelihood function, we would like to find the best estimate of the parameters. If we search through parameter space over various values of Θ, say Θ_i, then we select the value of $\hat{\Theta}$ that most likely (most probable) specifies the underlying probability function that the measurement sample was drawn from, that is, suppose we have two estimates, $\hat{\Theta}_i$ and $\hat{\Theta}_j$ for which

$$\Pr(Y|\Theta_i) > \Pr(Y|\Theta_j) \quad (4.39)$$

Thus, it is "more likely" that the $Y(N)$ was drawn for parameter value $\hat{\Theta}_i$ than $\hat{\Theta}_j$, since, equivalently, $\mathcal{L}(Y; \hat{\Theta}_i) > \mathcal{L}(Y; \hat{\Theta}_j)$. Searching over all Θ and selecting the value of Θ that is maximum (most probable) leads to the maximum likelihood estimate (*ML*) given by

$$\hat{\Theta}_{ML}(Y) = \underset{\Theta}{\arg\max} \ \Pr(Y|\Theta) \quad (4.40)$$

As noted before for the Bayesian estimator, many problems are characterized by the class of exponential densities making it simpler to use the natural logarithm function; therefore, we define the *log-likelihood function* as

$$\Lambda(Y(N)|\Theta) := \ln\ \mathcal{L}(Y;\Theta) = \ln \Pr(Y(N)|\Theta) \tag{4.41}$$

Since the logarithm is monotonic, it preserves the maximum of the likelihood providing the identical result, that is,

$$\hat{\Theta}_{ML}(Y) = \underset{\Theta}{\arg\max}\ \ln \Pr(Y|\Theta) \tag{4.42}$$

What makes the *ML* estimator popular is the fact that it enjoys some very desirable properties that we list without proof (see [5] for details).

1. *ML* estimates are *consistent.*
2. *ML* estimates are *asymptotically efficient* with $R_{\tilde{\Theta}|\Theta} = \mathcal{I}^{-1}$ (Information matrix of Eq. (4.9)).
3. *ML* estimates are *asymptotically Gaussian* with $N(\Theta, R_{\tilde{\Theta}\tilde{\Theta}})$.
4. *ML* estimates are *invariant*, that is, if $\hat{\Theta}_{ML}$, then any function of the *ML* estimate is the *ML* estimate of the function, $\hat{f}_{ML} = f\left(\hat{\Theta}_{ML}\right)$.
5. *ML* estimates of the *sufficient statistic* are equivalent to the *ML* estimates over the original data.

These properties are asymptotic and therefore imply that a large amount of data must be available for processing.

Mathematically, the relationship between the *MAP* and *ML* estimators is clear even though philosophically they are quite different in construct. If we take the *MAP* equation of Eq. (4.33) and ignore the a priori density $\Pr(\Theta)$ (assume Θ is unknown but deterministic), then the maximum likelihood estimator is only a special case of *MAP*. Using the same arguments as before, we use the $\ln \Pr(\Theta|Y)$ instead of $\Pr(\Theta|Y)$. We obtain the maximum likelihood estimate by solving the *log-likelihood equation* and checking for the existence of a maximum; that is,

$$\nabla_\Theta \ln \Pr(\Theta|Y)|_{\Theta=\hat{\Theta}_{ML}} = 0 \tag{4.43}$$

where recall that the $\Pr(\Theta|Y)$ is the likelihood of Θ. Of course, to check for a maximum we have that $\nabla_\Theta(\nabla_\Theta \ln \Pr(\Theta|Y)') < 0$. Again applying Bayes' rule as in Eq. (4.33) and ignoring $\Pr(\Theta)$, we have

$$\nabla_\Theta \ln \Pr(\Theta|Y) = \nabla_\Theta \ln \Pr(Y|\Theta)\big|_{\Theta=\hat{\Theta}_{ML}} \tag{4.44}$$

Consider the following example to demonstrate this relationship between *MAP* and *ML*.

Example 4.3 Consider estimating an unknown *constant*, from a noisy measurement as in the previous example. Further assume that the noise is an independent

Gaussian random variable such that $v \sim N(0, R_{vv})$ and the measurement model is given by

$$y = \Theta + v \tag{4.45}$$

First, we assume no "a priori" statistical knowledge of Θ just that it is an unknown, nonrandom constant. Thus, we require the maximum likelihood estimate, since no prior information is assumed about $\Pr(\Theta)$. The associated conditional density is

$$\Pr(Y|\Theta) = \frac{1}{\sqrt{2\pi R_{vv}}} e^{-\frac{1}{2}\frac{(y-\Theta)^2}{R_{vv}}} \tag{4.46}$$

The maximum likelihood estimate of Θ is found by solving the log-likelihood equation:

$$\nabla_\Theta \ln \Pr(Y|\Theta)\big|_{\Theta=\hat{\Theta}_{ML}} = 0| \tag{4.47}$$

or

$$\begin{aligned}\frac{\partial}{\partial\Theta} \ln \Pr(Y|\Theta) &= \frac{\partial}{\partial\Theta}\left\{-\frac{1}{2}\ln 2\pi R_{vv} - \frac{1}{2R_{vv}}(y-\Theta)^2\right\}\\ &= \frac{1}{R_{vv}}(y-\Theta)\end{aligned}$$

Setting this expressiorn to zero and solving for Θ, we obtain

$$\hat{\Theta}_{ML} = y \tag{4.48}$$

that is, the best estimate of Θ in a maximum likelihood sense is the raw data y. The corresponding error variance is easily calculated (as before)

$$R_{\tilde{\Theta}|\Theta} = R_{vv} \tag{4.49}$$

Next, we model Θ as a random variable with Gaussian prior, that is, $\Theta \sim N(\overline{\Theta}, R_{\Theta\Theta})$ and desire the maximum a posteriori estimate. The *MAP* equation is

$$\mathcal{J}_{MAP} = \nabla_\Theta(\ln \Pr(Y|\Theta) + \ln \Pr(\Theta))$$

$$\mathcal{J}_{MAP} = \frac{\partial}{\partial\theta}\left\{-\frac{1}{2}\ln 2\pi R_{vv} - \frac{1}{2R_{vv}}(y-\Theta)^2 - \frac{1}{2}\ln 2\pi R_{\Theta\Theta} - \frac{\frac{1}{2}(\Theta-\overline{\Theta})^2}{R_{\Theta\Theta}}\right\}$$

or

$$\mathcal{J}_{MAP} = \frac{1}{R_{vv}}(y-\Theta) - \frac{1}{R_{\Theta\Theta}}(\Theta-\overline{\Theta})$$

Setting this expression to zero and solving for $\Theta = \hat{\Theta}_{MAP}$, we obtain

$$\hat{\Theta}_{MAP} = \frac{y + \frac{R_{vv}}{R_{\Theta\Theta}}\overline{\Theta}}{1 + \frac{R_{vv}}{R_{\Theta\Theta}}}$$

It can be shown that the corresponding error variance is

$$R_{\tilde{\Theta}|\Theta} = \frac{R_{vv}}{1 + \frac{R_{vv}}{R_{\Theta\Theta}}}$$

Examining the results of this example, we see that when the parameter variance is large ($R_{\Theta\Theta} \gg R_{vv}$), the *MAP* and *ML* estimates perform equivalently. However, when the variance is small, the *MAP* estimator performs better because the corresponding error variance is smaller. This completes the example.

△△△

The main point to note is that the *MAP* estimate provides a mechanism to incorporate the a priori information, while the *ML* estimate does not. Therefore, for some problems, *MAP* is the efficient estimator. In this example, if Θ were actually Gaussian, then the *ML* solution, which models Θ as an unknown parameter, is not an efficient estimator, while the *MAP* solution incorporates this information by using $\Pr(\Theta)$.

This completes the introduction to minimum variance, maximum a posteriori, and maximum likelihood estimation that will be applied to the decomposition approaches to follow. Next, we consider the nonstatistical least-squares approach.

4.5 Least-squares (LS) Estimation

The richness of the least-squares estimation problem is sometimes overlooked. It provides an interesting and full description of optimal signal processing techniques which have been exploited throughout the literature, especially in model-based approaches [7, 10] and many of the array signal processing methods [11–13] involving subspace processing.

In this section, we discuss least-squares (*LS*) estimation techniques and relate them to the minimum variance estimator discussed previously. Weighted *LS* estimators in the strict sense are estimators that minimize the sum-squared error criterion

$$\mathcal{J}_{LS} = \frac{1}{2}\tilde{y}' W^{-1}\tilde{y} \tag{4.50}$$

where $\tilde{y} := y - \hat{y}$, $y \in \mathcal{R}^{N_y \times 1}$, $W \in \mathcal{R}^{N_y \times N_y}$ is the symmetric, positive definite, *weighting matrix*. No probabilistic or statistical description of the estimation problem is assumed. In fact, we can consider the solution to the least-squares estimation problem to be a "deterministic" optimization problem.

4.5.1 Batch Least Squares

In this section, we investigate "batch" solutions to the least-squares estimation problem. We consider the specific problem of estimating an N_θ-parameter vector

Θ from data and assume that the *LS* estimator of Θ is linearly related to the *LS* estimate of y by

$$\hat{Y}_{LS} = C\hat{\Theta}_{LS} \tag{4.51}$$

where $C \in \mathcal{R}^{N_y \times N_\theta}$ is a full rank matrix, that is, $\rho(C) = N_\theta$, $N_\theta < N_y$. The *LS* estimator is found by minimizing $\mathcal{J}_{LS}$ with respect to $\hat{\Theta}$, by taking the derivative (gradient), setting the result to zero, and solving for $\hat{\Theta} = \hat{\Theta}_{LS}$; that is,

$$\nabla_{\hat{\Theta}} \mathcal{J}_{LS}|_{\hat{\Theta}=\hat{\Theta}_{LS}} = \nabla_{\hat{\Theta}} \left(\frac{1}{2}\tilde{y}' W^{-1}\tilde{y}\right) = \frac{1}{2}\nabla_{\hat{\Theta}}(y - C\Theta)' W^{-1}(y - C\Theta) = 0 \tag{4.52}$$

Applying the chain rule of Eq. (4.12) with $a' = (y - C\Theta)$ and $b = W^{-1}(y - C\Theta)$, we obtain

$$\nabla_{\hat{\Theta}} \mathcal{J}_{LS} = \frac{1}{2}\left[-C' W^{-1}(y - C\Theta) - C' W^{-1}(y - C\Theta)\right] = -C' W^{-1}(y - C\Theta) = 0$$

Solving this equation for $\hat{\Theta}$, we obtain the *weighted least-squares estimate*:

$$\hat{\Theta}_{LS} = K_{LS}y = \left(C' W^{-1} C\right)^{-1} C' W^{-1} y \tag{4.53}$$

If we further assume additive measurement noise with variance R_{vv}, then

$$y = C\theta + v \tag{4.54}$$

and therefore

$$R_{\tilde{\Theta}\tilde{\Theta}} = \text{cov}\ \tilde{\Theta} = K_{LS} R_{vv} K_{LS}' \tag{4.55}$$

To investigate "how good" the least-squares estimator is for this problem, we must calculate the *CRLB*—assuming some prior statistics for the underlying process. First, we assume that the noise is $v \sim \mathcal{N}(0, R_{vv})$, then $\Pr(Y|\Theta) \sim \mathcal{N}(C\Theta, R_{vv})$ and

$$\ln \Pr(Y|\Theta) = -\frac{1}{2}\ \ln\ (2\pi)_y^N |R_{vv}|^{-\frac{1}{2}} - \frac{1}{2}(y - C\Theta)' R_{vv}^{-1}(y - C\Theta)$$

Again, using the chain rule with $a' = (y - C\Theta)'$ and $b = R_{vv}^{-1}(y - C\Theta)$, we have that the information matrix $\mathcal{I}$ is

$$\mathcal{I} = -E\{\nabla_\theta(\nabla_\theta\ \ln \Pr(Y|\Theta))'\} = -\nabla_\theta(C' R_{vv}^{-1}(y - C\Theta)) = \left(C' R_{vv}^{-1} C\right)$$

Thus, the *CRLB* is given by

$$R_{\tilde{\Theta}|\Theta} = \left(C' R_{vv}^{-1} C\right)^{-1}$$

Note also in passing that if we solve for the minimum variance estimator (set gradient to zero), then we obtain

$$\hat{\Theta}_{MV} = \left(C' R_{vv}^{-1} C\right)^{-1} C' R_{vv}^{-1} y$$

Comparing this result with the least-squares estimator, we see that the weighted *LS* estimator is identical to the *MV* estimator, if the weighting matrix $W = R_{vv}$, that is, both estimators satisfy the *CRLB* with *equality* (efficient) for estimating a random vector from noisy data. To see this substitute, $W = R_{vv}$ for K_{LS} in Eq. (4.55) to obtain

$$R_{\tilde{\theta}\tilde{\theta}} = \left[(C'R_{vv}^{-1}C)^{-1}C'R_{vv}^{-1}R_{vv}R_{vv}^{-1}C(C'R_{vv}^{-1}C)^{-1}\right] = (C'R_{vv}^{-1}C)^{-1} \tag{4.56}$$

This estimator is also considered a "batch" estimator, since it processes all of the N_y-measurements (C is $N_y \times N_\theta$) in one batch to obtain the estimate. Other techniques to solve this problem sequentially or recursively yield the identical results but update the estimate only as a new data sample becomes available. Note that the batch approach would require the entire problem to be recalculated each time as the next measurement becomes available, that is, C becomes a $(N_y + 1) \times N_\theta$ matrix.

We summarize the batch *weighted least-squares (WLS) estimator* as follows:

Criterion: $J_{WLS} = \frac{1}{2}\tilde{y}'W^{-1}\tilde{y}$
Signal Model: $y = C\Theta$
Algorithm: $\hat{\Theta}_{LS} = K_{LS}y$
Quality: $R_{\tilde{\Theta}\tilde{\Theta}} = K_{LS}R_{vv}K_{LS}'$

assuming that $y = C\Theta + v$ and cov $v = R_{vv}$.

Finally, we consider the batch weighted *LS* estimator for a common estimation problem—trend removal.

Example 4.4 Consider the problem of removing a trend from noisy measurement data. A typical problem application is tracking a flight vehicle using a radar system [14] and compensating for the known ballistic trajectory. The basic motion compensation problem is that of removing the trajectory or trend from the measured data. In general, suppose we are given N measurements and would like to develop the compensation algorithm. Our underlying scalar measurement model is

$$y(t) = s(t) + \Theta(t) + n(t)$$

where $\Theta(t)$ is the unknown trajectory (trend) and y, s, and n are the respective measurement, signal, and random noise. The *trend estimation* problem is to estimate the trend, $\hat{\Theta}(t)$ and remove it from the data. Generally, a temporal polynomial is used as the model, that is,

$$\Theta(t) = \theta_0 + \theta_1 t + \cdots + \theta_{N_\theta-1}t^{N_\theta-1} = \mathbf{c}'(t)\theta$$

where $\mathbf{c}'(t) = [1 \; t \; \cdots \; t^{N_\theta-1}]$ and $\theta = [\theta_0 \; \theta_1 \; \cdots \; \theta^{N_\theta-1}]'$. The measurement model at a given time, t, is therefore

$$y(t) = s(t) + \Theta(t) + n(t) = s(t) + \mathbf{c}'(t)\theta + n(t)$$

If we expand over the N_y measurements, we obtain

$$\mathbf{y} = \mathbf{s} + C(t)\theta + \mathbf{n}$$

with $\mathbf{y} = [y(1) \cdots y(N_y)]'$, $\mathbf{s} = [s(1) \cdots s(N_y)]'$, and $\mathbf{n} = [n(1) \cdots n(N_y)]'$ for the $N_y \times N_\theta$ measurement matrix, $C(t) = [\mathbf{c}(1) \cdots \mathbf{c}(N_\theta)]'$, $\mathbf{c} \in \mathcal{R}^{N_y \times 1}$. For this problem,

$$J_{LS} = \frac{1}{2}(\mathbf{y} - \hat{\mathbf{y}})' W^{-1} (\mathbf{y} - \hat{\mathbf{y}})$$

and the *WLS* estimator is of the form

$$\hat{Y}_{LS} = C\hat{\mathbf{\Theta}}_{LS}$$

From Eq. (4.53), we obtain the *WLS* estimator

$$\hat{\theta}_{LS} = K_{LS}\mathbf{y} = \left(C'W^{-1}C\right)^{-1}C'W^{-1}\mathbf{y}$$

and therefore,

$$\hat{\mathbf{\Theta}}(t) = \mathbf{c}'(t)\hat{\theta}_{LS}$$

Now if we assume that n is zero-mean, white with covariance, R_{nn}, then the quality of the estimator is

$$R_{\tilde{\theta}\tilde{\theta}} = K_{LS} R_{nn} K'_{LS}$$

Also note (as before) that if we choose the weighting matrix to be $W = R_{nn}$, then we obtain the minimum variance estimate. Once $\hat{\mathbf{\Theta}}(t)$ is estimated, it can then be removed from the measurement data to obtain

$$z(t) = y(t) - \hat{\mathbf{\Theta}}(t) = s(t) + \left[\Theta(t) - \hat{\mathbf{\Theta}}(t)\right] + n(t) \approx s(t) + n(t)$$

Now in terms of our motion compensation application [14], our polynomial model has *physical* interpretation as

$$\Theta(t) = R_o + v_R t + a_R t^2 = [1 \;\; t \;\; t^2] \begin{bmatrix} R_o \\ v_R \\ a_R \end{bmatrix} = \mathbf{c}'(t)\theta$$

with R_o the nominal range, v_R the velocity, and a_R the acceleration of the vehicle. We show the results of applying the *WLS* estimator to measured trajectory data in Figure 4.3. Here we see the effect is to remove the ballistic trajectory from the measurement enabling further processing to extract the pertinent signals such as the power spectrum shown in Figure 4.3c.

Summarizing we have

Criterion: $J_{LS} = \frac{1}{2}(\mathbf{y} - \hat{\mathbf{y}})' W^{-1} (\mathbf{y} - \hat{\mathbf{y}})$

Signal Models:

Measurement: $\mathbf{y} = \mathbf{s} + C(t)\theta + \mathbf{n}$

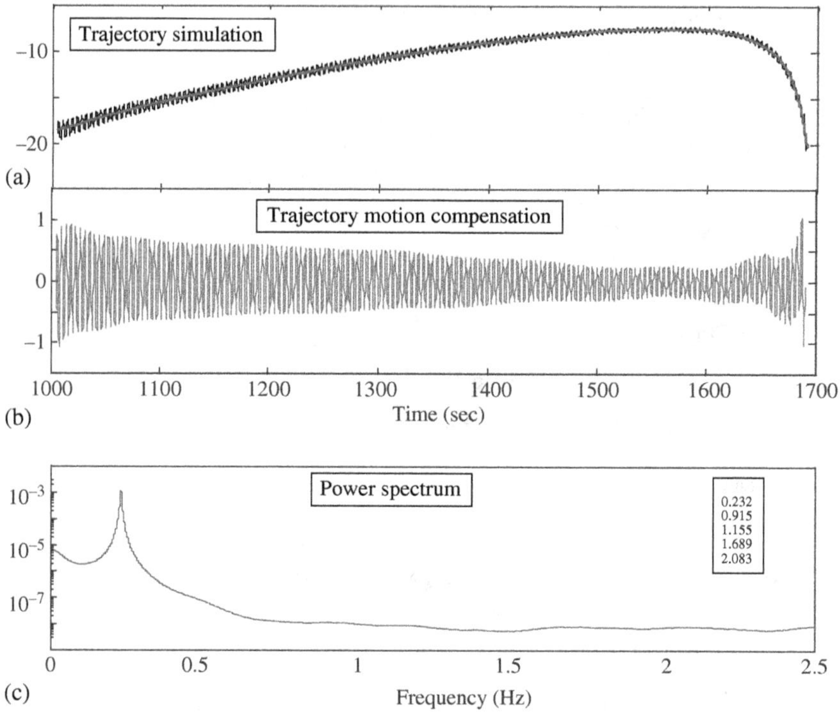

Figure 4.3 Trajectory motion compensation using the *WLS* estimator: (a) raw data and *WLS* trajectory estimate. (b) Compensated (trend removed) data. (c) Power spectral density estimate.

Signal: $s(t)$
Noise: $n(t)$ and cov $\mathbf{n} = R_{nn}$
Trend: $\Theta(t) = \mathbf{c}'(t)\theta$
Algorithm: $\hat{\Theta}_{LS} = K_{LS}\mathbf{y}$ for $K_{LS} = \left(C'W^{-1}C\right)^{-1}C'W^{-1}$
Quality: $R_{\tilde{\Theta}\tilde{\Theta}} = K_{LS}R_{nn}K'_{LS}$

△△△

4.5.2 Recursive Least-squares

In this section, we briefly investigate "recursive" or equivalently "sequential" solutions to the least-squares estimation problem. It is important to realize that the recursive least-squares solution is identical to the batch solution after it converges, so there is *no gain* in estimator performance properties (e.g. *CRLB*); however, the number of computations is significantly less than the equivalent batch technique.

Many processors can be placed in a recursive form with various subtleties emerging in the calculation of the current estimate ($\hat{\Theta}_{\text{old}}$). The standard technique employed is based on correcting or updating the current estimate as a new measurement data sample becomes available. The estimates generally take the *recursive form*:

$$\hat{\Theta}_{\text{new}} = \hat{\Theta}_{\text{old}} + KE_{\text{new}} \tag{4.57}$$

where

$$E_{\text{new}} = Y - \hat{Y}_{\text{old}} = Y - C\hat{\Theta}_{\text{old}}$$

Here we see that the new estimate is obtained by correcting the old estimate with a K-weighted error. The error term E_{new} is the new information or innovation; that is, it is the difference between the actual measurement and the predicted measurement ($\hat{Y}_{\text{old}}$) based on the old estimate ($\hat{\Theta}_{\text{old}}$). The computation of the weight matrix K depends on the criterion used (e.g. mean-squared error, absolute error, etc.).

Consider the following example that demonstrates how to *recursively* estimate the sample mean.

Example 4.5 The sample mean estimator can easily be put in recursive form. The estimator is given by

$$\hat{X}(N) = \frac{1}{N}\sum_{t=1}^{N} y(t)$$

Extracting the Nth term from the sum, we obtain

$$\hat{X}(N) = \frac{1}{N}y(N) + \frac{1}{N}\sum_{t=1}^{N-1} y(t)$$

Identify $\hat{X}(N-1)$ from the last term, that is,

$$\hat{X}(N) = \frac{1}{N}y(N) + \left(\frac{N-1}{N}\right)\hat{X}(N-1)$$

Multiplying and grouping terms the recursive form is obtained as

$$\underbrace{\hat{X}(N)}_{\text{NEW}} = \underbrace{\hat{X}(N-1)}_{\text{OLD}} + \underbrace{\frac{1}{N}}_{\text{WEIGHT}} \underbrace{[y(N) - \hat{X}(N-1)]}_{\text{ERROR}}$$

△△△

This procedure to develop the "recursive form" is very important and can be applied to a multitude of processors. Note the steps in its determination:

1. remove the Nth term from the summation;
2. identify the previous estimate in terms of the $N-1$ remaining terms; and
3. perform the algebra to determine the gain factor and place the estimator in the recursive form of Eq. (4.57) for a *scalar* measurement.

Note that the vector case follows the "identical" steps with weighting matrix, $W \in \mathcal{R}^{N_y \times N_\theta}$, the identity matrix instead of unity in the scalar equations (see [2] for details).

Next, we briefly develop the general recursive form from the "batch" least squares estimator of Eq. (4.53) for a scalar measurement $y(t)$ by defining

$$\mathbf{y}(t-1) := \begin{bmatrix} y(1) \\ \vdots \\ y(t-1) \end{bmatrix}, \boldsymbol{\Theta}(t-1) := \begin{bmatrix} \theta_1 \\ \vdots \\ \theta_{N_\theta} \end{bmatrix}, C(t-1) := \begin{bmatrix} \mathbf{c}'(1) \\ \vdots \\ \mathbf{c}'(t-1) \end{bmatrix}$$

$\boldsymbol{\Theta} \in \mathcal{R}^{N_\Theta \times 1}, \mathbf{c}' \in \mathcal{R}^{1 \times N_\theta}, \mathbf{y} \in \mathcal{R}^{(t-1)\times 1}, C \in \mathcal{R}^{(t-1)\times N_\theta}$

The batch estimate of θ based on the set of past data up to $t-1$ is obtained ($W = 1$) from Eq. (4.53) as

$$\hat{\Theta}_{LS}(t-1) = P(t-1)C'(t-1)\mathbf{y}(t-1) \tag{4.58}$$

where $P(t-1) = [C'(t-1)C(t-1)]^{-1}$. Now suppose we would like to add a new (scalar) measurement data sample, $y(t)$, then we have

$$\Theta_{LS}(t) = P(t)C'(t)\mathbf{y}(t) \tag{4.59}$$

Now, the matrix, C' can be augmented in terms of the past and new data samples as

$$C'(t) = [\ \mathbf{c}(1)\ \cdots\ \mathbf{c}(t-1)|\mathbf{c}(t)\ ] = \left[\ C'(t-1)|\mathbf{c}(t)\ \right] \tag{4.60}$$

with the corresponding augmented data vector

$$\mathbf{y}(t) = \left[\ \mathbf{y}(t-1)|y(t)\ \right] \tag{4.61}$$

Therefore substituting these relations into Eq. (4.59) and multiplying, we obtain

$$\begin{aligned} \Theta_{LS}(t) &= P(t)\left[\ C'(t-1)|\mathbf{c}(t)\ \right]\begin{bmatrix} \mathbf{y}(t-1) \\ -- \\ y(t) \end{bmatrix} \\ &= P(t)\left[\ C'(t-1)\mathbf{y}(t-1) + \mathbf{c}(t)y(t)\ \right] \end{aligned} \tag{4.62}$$

Also $P(t)$ can be expressed in terms of C' of Eq. (4.60) as

$$P^{-1}(t+1) = C'(t+1)C(t+1) = \left[\ C'(t-1)|\mathbf{c}(t)\ \right]\begin{bmatrix} C(t-1) \\ -- \\ \mathbf{c}'(t) \end{bmatrix}$$

or

$$P(t) = \left[P^{-1}(t-1) + \mathbf{c}(t)\mathbf{c}'(t)\right]^{-1} \tag{4.63}$$

Applying the *matrix inversion lemma*[2] with $A = P^{-1}(t-1)$, $B = \mathbf{c}(t)$, $C = 1$, and $D = \mathbf{c}'(t)$, we obtain

$$P(t) = P(t-1) - P(t-1)\mathbf{c}(t)\left[\mathbf{c}'(t)P(t-1)\mathbf{c}(t)+1\right]^{-1}\mathbf{c}'(t)P(t-1) \quad (4.64)$$

Define the *weight* or *gain* matrix as

$$K(t) := P(t-1)\mathbf{c}(t)\left[\mathbf{c}'(t)P(t-1)\mathbf{c}(t)+1\right]^{-1}$$

Therefore,

$$P(t) = \left[I - K(t)\mathbf{c}'(t)\right]P(t-1) \quad (4.65)$$

and substituting into Eq. (4.62) for $P(t)$, we obtain

$$\Theta_{LS}(t) = \left[I - K(t)\mathbf{c}'(t)\right]P(t-1)\left[\,C'(t-1)\mathbf{y}(t-1) + \mathbf{c}(t)y(t)\,\right] \quad (4.66)$$

Thus, using Eq. (4.58), the *LS* estimate of Eq. (4.59) can be written as

$$\Theta_{LS}(t) = \left[I - K(t)\mathbf{c}'(t)\right]\Theta_{LS}(t-1) + \left[I - K(t)\mathbf{c}'(t)\right]P(t-1)\mathbf{c}(t)y(t) \quad (4.67)$$

Concentrating on the *last term* of this expression, we perform the indicated multiplication to obtain

$$\left[I - K(t)\mathbf{c}'(t)\right]P(t-1)\mathbf{c}(t)y(t) = \left[P(t-1)\mathbf{c}(t) - K(t)\mathbf{c}'(t)P(t-1)\mathbf{c}(t)\right]y(t)$$

and substituting for $K(t)$, we have

$$\left[P(t-1)\mathbf{c}(t) - P(t-1)\mathbf{c}(t)\left[\mathbf{c}'(t)P(t-1)\mathbf{c}(t)+1\right]^{-1}\mathbf{c}'(t)P(t-1)\mathbf{c}(t)\right]y(t)$$

factoring $P(t)\mathbf{c}(t)$ on the left gives

$$P(t-1)\mathbf{c}(t)\left[I - [\mathbf{c}'(t)P(t-1)\mathbf{c}(t)+1]^{-1}\mathbf{c}'(t)P(t-1)\mathbf{c}(t)\right]y(t)$$

Factoring the inverse from the left of this expression then gives

$$P(t-1)\mathbf{c}(t)\left[\mathbf{c}'(t)P(t-1)\mathbf{c}(t)+1\right]^{-1}\left\{[\mathbf{c}'(t)P(t-1)\mathbf{c}(t)+1] - \mathbf{c}'(t)P(t-1)\mathbf{c}(t)\right\}y(t) \quad (4.68)$$

But this term is just $K(t)y(t)$; therefore, substituting this result into Eq. (4.67), we obtain

$$\Theta_{LS}(t) = \left[I - K(t)\mathbf{c}'(t)\right]\Theta_{LS}(t-1) + K(t)y(t) \quad (4.69)$$

which by multiplying and combining like terms gives the *recursive LS estimate*

$$\underbrace{\Theta_{LS}(t)}_{\text{NEW}} = \underbrace{\Theta_{LS}(t-1)}_{\text{OLD}} + \underbrace{K(t)}_{\text{WT}}\underbrace{\left[y(t) - \mathbf{c}'(t)\Theta_{LS}(t-1)\right]}_{\text{ERROR}} \quad (4.70)$$

We summarize the results in Table 4.1 for the scalar measurement case.

2 The matrix inversion lemma is given by $[A+BCD]^{-1} = A^{-1} - A^{-1}B[DA^{-1}B + C^{-1}]^{-1}DA^{-1}$ [2].

Table 4.1 Recursive least-squares (scalar) algorithm.

Parameter correction

$$\underbrace{\hat{\Theta}(t)}_{\text{NEW}} = \underbrace{\hat{\Theta}(t-1)}_{\text{OLD}} + \underbrace{K(t)}_{\text{WT}} \underbrace{e(t)}_{\text{ERROR}}$$

Error

$$e(t) = y(t) - \mathbf{c}'(t)\hat{\Theta}(t-1)$$

Gain

$$K(t) = P(t-1)\mathbf{c}(t)\left[\mathbf{c}'(t)P(t-1)\mathbf{c}(t) + 1\right]^{-1}$$

Covariance correction

$$P(t) = \left[I - K(t)\mathbf{c}'(t)\right] P(t-1)$$

Initial conditions

$$\hat{\Theta}(0), \qquad P(0)$$

$$\Theta \in \mathcal{R}^{N_\theta \times 1}, y \in \mathcal{R}^{1\times 1}, \mathbf{c}' \in \mathcal{R}^{1\times N_\theta} \textit{ and } K \in \mathcal{R}^{N_\theta \times 1}, P \in \mathcal{R}^{N_\theta \times N_\theta}$$

Consider the following example of formulating the trend removal problem using the recursive rather than batch approach.

Example 4.6 Starting with the trajectory representation of Example 4.4, we have that the underlying scalar measurement model is

$$y(t) = s(t) + \Theta(t) + n(t) \text{ for } t = 1, \ldots, N$$

where $\Theta(t)$ is the unknown trajectory (trend) and y, s, and n are the respective measurement, signal, and random noise. The problem is to estimate the trend, $\hat{\Theta}(t)$, and remove it from the data using a temporal polynomial as the model, that is,

$$\Theta(t) = \theta_0 + \theta_1 t + \cdots + \theta_{N_\theta - 1} t^{N_\theta - 1} = \mathbf{c}'(t)\theta$$

where $\mathbf{c}'(t) = [1 \; t \; \cdots \; t^{N_\theta - 1}]$ and $\boldsymbol{\theta} = [\theta_0 \; \theta_1 \; \cdots \; \theta^{N_\theta - 1}]'$. The dynamic measurement model at time t is therefore

$$y(t) = s(t) + \Theta(t) + n(t) = s(t) + \mathbf{c}'(t)\theta + n(t)$$

Now in the recursive formulation, $\boldsymbol{\theta}(t) \longrightarrow \boldsymbol{\theta}$, and at each point in time, the estimate of $\boldsymbol{\theta}(t)$ is accomplished using the *RLS* algorithm of Table 4.1, that is (simply),

$$\hat{\theta}(t) = \hat{\theta}(t-1) + K(t)e(t)$$

is used to estimate the polynomial coefficients which provide the desired trend estimate

$$\hat{\Theta}(t) = \mathbf{c}'(t)\hat{\theta}(t)$$

that is removed from the measurements to yield the "trend-free" data

$$z(t) = y(t) - \hat{\Theta}(t)$$

This completes the example. △△△

4.6 Optimal Signal Estimation

In this section, we discuss the concept of optimal signal estimation, that is, we estimate signals that exist in a random (vector) space and introduce the concept of signal estimation/decomposition into orthogonal vector spaces. Here, we represent a random signal as a vector and produce an optimal estimate by minimizing the error between the noisy signal and its estimate. This is accomplished by performing a linear transformation of the vector into an orthogonal space representing this error—a *projection* operation. We start with the signal vector and then proceed to its decomposition into components leading to the idea of a projection.

A signal can be considered a set of time samples, $\{s(t)\}; t = 0, 1, \ldots, N_t$ or an N_t-vector defined by $\mathbf{s}(t) = [s(0)\ s(1), \ \ldots, \ s(N_t)]'$. The decomposition of this vector, say "**s**," into meaningful components evolves from the fundamental concept of orthogonality. By definition, for a real vector space $\mathcal{S} \in \mathcal{R}^{N_t \times 1}$ with vectors, $\mathbf{s}_m$ and $\mathbf{s}_n$, the *inner product* or equivalently *dot product* is defined by

$$\langle \mathbf{s}_m, \mathbf{s}_n \rangle := \mathbf{s}_m \bullet \mathbf{s}_n = \mathbf{s}'_m \mathbf{s}_n = |\mathbf{s}_m||\mathbf{s}_n| \cos\theta \tag{4.71}$$

If the vectors are orthogonal, then $\theta = \frac{\pi}{2}$ or 90° and therefore $\langle \mathbf{s}_m, \mathbf{s}_n \rangle = 0$. Also, it follows that the *angle* between two vectors is given by

$$\theta = \arccos \frac{\langle \mathbf{s}_m, \mathbf{s}_n \rangle}{|\mathbf{s}_m||\mathbf{s}_n|} \tag{4.72}$$

Signals are typically captured in terms of their signal vectors and corresponding vector (function) space, either in time, $\mathbf{s}(t)$, or frequency, $\mathbf{S}(\Omega)$, with the corresponding space defined in either domain and *orthogonality* specified by

$$\langle \mathbf{s}_m(t), \mathbf{s}_n(t) \rangle := \mathbf{s}_m(t) \bullet \mathbf{s}_n(t) = \mathbf{s}'_m(t)\mathbf{s}_n(t) = 0 \ (m \neq n)$$

or

$$\langle \mathbf{S}_m(\Omega), \mathbf{S}_n(\Omega) \rangle := \mathbf{S}_m(\Omega) \bullet \mathbf{S}_n(\Omega) = \mathbf{S}'_m(\Omega)\mathbf{S}_n(\Omega) = 0 \ (m \neq n) \tag{4.73}$$

Suppose we are given a set of orthogonal *basis* vectors, $\{\mathbf{o}_k\}; k = 1, \ldots, N$ such that $\langle \mathbf{o}_m, \mathbf{o}_n \rangle = 0$ for $m \neq n$, then any vector in this space $\mathbf{s} \in \mathcal{S}^N$ can be uniquely expressed or decomposed in terms of this basis, that is,

$$\mathbf{s} = \sum_{k=1}^{N} \alpha_k \mathbf{o}_k \tag{4.74}$$

Further, let the signal vector be uniquely "decomposed" into its parallel ($\mathbf{s}^{\parallel}$) projection components and its perpendicular ($\mathbf{s}^{\perp}$) or orthogonal components such that

$$\mathbf{s} = \mathbf{s}^{\parallel} + \mathbf{s}^{\perp} \tag{4.75}$$

This unique decomposition is the *orthogonal projection* ($\mathbf{s}^{\parallel} \in \mathcal{S}$) of $\mathbf{s}$ onto $\mathcal{S}$ defined by the projection operator, $\mathcal{P}_{\mathbf{s}|\mathcal{S}}$.

If we further assume that the set of basis vectors is *orthonormal* ($\langle \mathbf{o}_m, \mathbf{o}_n \rangle = 1$ for $m = n$), then the projection can be expressed simply in terms of the inner product as

$$\mathcal{P}_{\mathbf{s}|\mathcal{S}} = \sum_{k=1}^{N} \langle \mathbf{o}_k, \mathbf{s} \rangle \times \mathbf{o}_k = (\mathbf{o}_1' \mathbf{s})\, \mathbf{o}_1 + \cdots + (\mathbf{o}_N' \mathbf{s})\, \mathbf{o}_N \tag{4.76}$$

The subspace consisting of the set of orthogonal vectors is termed the *orthogonal complement* and designated by $\mathcal{S}^{\perp}$ such that $\mathcal{S}^{\perp} = \{\mathbf{s} \in \mathcal{S}^N : \mathbf{o} \bullet \mathbf{s} = 0\ \forall\ \mathbf{o} \in \mathcal{S}^N\}$.

A well-known procedure to construct an orthonormal set of basis vectors ($\mathbf{o}_j$) from a set of independent vectors spanning a signal space, $\mathcal{S} \in \mathcal{R}^N$ is given by the Gram–Schmidt (*G-S*) process [15]. Returning to the generic signal space, we would like to construct a signal space composed of the set of orthonormal vectors such that, $\mathcal{O} \in \mathcal{S}^N$. Assuming the set of independent vectors spanning $\mathcal{S}$ is defined by, $\{\mathbf{s}_n\}; n = 1, \ldots, N$, then by construction the first normalized vector is defined by

$$\mathbf{o}_1 = \frac{1}{\| \mathbf{s}_1 \|} \mathbf{s}_1$$

for $\| \bullet \|$ the Euclidean norm [15].

Next, a vector $\mathbf{o}_2$ is constructed that is orthogonal to $\mathbf{o}_1$ such that (see Eq. (4.75))

$$\mathbf{s}_2^{\perp} = \mathbf{s}_2 - \mathbf{s}_2^{\parallel} = \mathbf{s}_2 - \mathcal{P}_{\mathbf{s}_2|\mathbf{o}_1} = \mathbf{s}_2 - \langle \mathbf{o}_1, \mathbf{s}_2 \rangle \mathbf{o}_1$$

and normalizing, we obtain

$$\mathbf{o}_2 = \frac{1}{\| \mathbf{s}_2^{\perp} \|} \mathbf{s}_2^{\perp}$$

Proceeding, the generic *G-S* process for the *k*th-vector is given by

$$\mathbf{o}_k = \frac{1}{\| \mathbf{s}_k^{\perp} \|} \mathbf{s}_k^{\perp} = \mathbf{s}_k - \underbrace{\sum_{i=1}^{k-1} \langle \mathbf{o}_i, \mathbf{s}_k \rangle \mathbf{o}_i}_{\mathcal{P}_{\mathbf{s}_k|\mathbf{o}_i}} \tag{4.77}$$

These orthogonality properties extend directly to signal (function) spaces capturing all of the concepts and properties replacing $\mathbf{s} \rightarrow \mathbf{s}(t)$. As an example, consider the Fourier decomposition of a signal vector.

Example 4.7 Consider the discrete Fourier transform in the M-dimensional complex signal space $\mathcal{S} \in \mathcal{C}^{M\times 1}$ with inner product defined by

$$\langle \mathbf{s}(t), \mathbf{o}(t) \rangle = \mathbf{s}(t) \bullet \mathbf{o}^*(t) = \mathbf{s}'(t) \times \mathbf{o}^*(t) = \sum_{t=0}^{M-1} \mathbf{s}(t)\mathbf{o}^*(t)$$

where we define the orthogonal basis vectors as $\{\mathbf{o}_m(t)\}, m = 0, 1, \ldots, M-1$

$$\mathbf{o}_m(t) := e^{j\Omega_m t} = e^{j\left(\frac{2\pi m}{M}\right)\times t}$$

with $\left\langle \mathbf{o}_m(t), \mathbf{o}_n(t) \right\rangle = \mathbf{o}_m^{\dagger}(t) \times \mathbf{o}_n(t) = 0$ for $m \neq n$ and * denotes the complex conjugate.

Therefore, the *decomposition* of the signal vector over the orthogonal basis $\mathbf{o}_m(t)$ into constituent frequency components, $\{\mathbf{S}(\Omega_m)\}; m = 0, 1, \ldots, M-1$ is

$$\mathbf{s}(t) = \sum_{m=0}^{M-1} \mathbf{S}(\Omega_m)\mathbf{o}_m(t) = \frac{1}{M}\sum_{m=0}^{M-1} \mathbf{S}(\Omega_m)e^{j\Omega_m t}$$

or

$$\mathbf{s}(t) = \frac{1}{M}\left(\mathbf{S}(\Omega_0)\underbrace{e^{j\Omega_0 t}}_{\mathbf{o}_0(t)} + \mathbf{S}(\Omega_1)\underbrace{e^{j\Omega_1 t}}_{\mathbf{o}_1(t)} + \cdots + \mathbf{S}(\Omega_{M-1})\underbrace{e^{j\Omega_{M-1} t}}_{\mathbf{o}_{M-1}(t)}\right)$$

So we see that even this simple relationship can be expressed in vector space concepts quite easily.

△△△

Thus, optimal signal estimation, when viewed from the vector space perspective, can be seen as a *projection* of the data into two distinct spaces: one parallel (correlated) and one orthogonal (uncorrelated). To see this, consider the problem of developing an optimal estimate from a noisy signal. The criterion function of most interest in signal processing literature is that of the mean-squared error defined by

$$J(t) := E\{e^2(t)\} \tag{4.78}$$

where the error is derived directly from the measurement and estimated as

$$e(t) = s(t) - \hat{s}(t) \tag{4.79}$$

The optimal estimate is that which produces the smallest error. Geometrically, we can think of this error sequence, $\{e(t)\}$ as the result of projecting the current

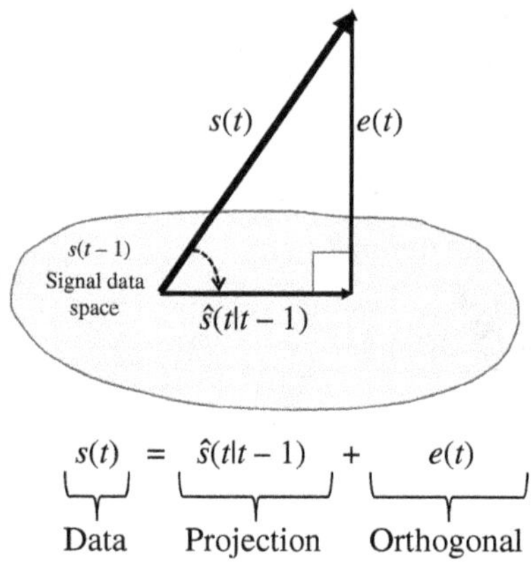

Figure 4.4 Orthogonal projection of the optimal signal estimate.

data onto a space of past data. Intuitively, the minimum error occurs when the *projection* is perpendicular or *orthogonal* to the space, that is,

$$e(t) \perp S(t-1) \tag{4.80}$$

where $S(t-1) = \{s(t-1), \ldots, s(0)\}$ is the past signal (data) space. We depict this relationship in Figure 4.4. We can think of the error or more commonly the *innovation* (new information [16]) as the orthogonal projection of the signal (data) vector $s(t)$ onto the past signal space $S(t-1)$ spanned by

$$\hat{s}(t|t-1) = E\{s(t)|S(t-1)\} \tag{4.81}$$

So we see that the optimal estimator can be thought of as the orthogonal projection of $s(t)$ onto $S(t-1)$.

Another interpretation of this result is to think of $\hat{s}(t|t-1)$ as an estimate of that part of $s(t)$ *correlated* with past data $S(t-1)$ and $e(t)$ as the uncorrelated or orthogonal part, that is,

$$\underbrace{s(t)}_{\text{Data}} = \underbrace{e(t)}_{\text{Uncorrelated}} + \underbrace{\hat{s}(t|t-1)}_{\text{Correlated}} \tag{4.82}$$

We know that the optimal estimator satisfies the *orthogonality condition* which is equivalent to Eq. (4.80)

$$E\{e(t)s(T)\} = 0 \quad \text{for} \quad T < t \tag{4.83}$$

In fact, we take the correlation-canceling approach of [17] and consider the following example.

Example 4.8 Suppose the signal data are composed of two parts: one correlated with the past and one not, that is,

$$s(t) = s_1(t) + s_2(t)$$

for

$$s_1(t) \in S(t-1) \quad \text{and} \quad s_2(t) \perp S(t-1)$$

and we are asked to analyze the performance of an estimator. The optimal estimator is an estimate *only* of the correlated part

$$\hat{s}(t|t-1) = s_1(t)$$

and therefore the residual or innovation is

$$e(t) = s(t) - \hat{s}(t|t-1) = (s_1(t) + s_2(t)) - s_1(t) = s_2(t)$$

the uncorrelated part of the data. Thus, we see that the results of canceling or removing the correlated part of the measurement with the optimal estimate are the uncorrelated part or innovation. This completes the example.

△△△

This example emphasizes a very important point that is crucial to analyzing the performance of *any* optimal estimator. If the estimator is performing properly and we have estimated all of the information in the currently available data, then the *error sequence must be zero-mean and white.* We will see when we design estimators whether parametric, signal, spectral, or model-based that checking the statistical properties of the corresponding residual or innovation sequence is the first step in analyzing proper performance.

4.7 Projection Theory

Projection theory plays an important role in signal estimation and signal decomposition primarily because signal vectors are created by transforming or "projecting" a vector into a lower dimensional space—the subspace [15, 18–21]. We are primarily interested in two projection operators: (i) orthogonal and (ii) oblique or parallel. The *orthogonal projection operator* "projects" • onto $\mathcal{Y}$ as $\mathcal{P}_{\bullet|\mathcal{Y}}$ or its complement $\mathcal{P}^{\perp}_{\bullet|\mathcal{Y}}$, while the *oblique projection operator* "projects" • onto $\mathcal{Y}$ "along" $\mathcal{Z}$ as $\mathcal{P}^{\|}_{\bullet|\mathcal{Y}\circ\mathcal{Z}}$. For instance, orthogonally projecting the vector $\mathbf{s}$, we have $\mathcal{P}_{\mathbf{s}|\mathcal{Y}}$ or $\mathcal{P}^{\perp}_{\mathbf{s}|\mathcal{Y}}$, while obliquely projecting $\mathbf{s}$ along $\mathcal{Z}$ is $\mathcal{P}_{\mathbf{s}|\mathcal{Y}\circ\mathcal{Z}}$. Note that more details of projection theory both on vectors and matrices are developed in Appendix B.

4.7.1 Orthogonal Projections: A Geometric Decomposition Perspective

Perhaps the simplest way to motivate a geometric approach (as in [18]) is to consider the problem that we are given a vector $\mathbf{y}$ in an N_y-dimensional vector space

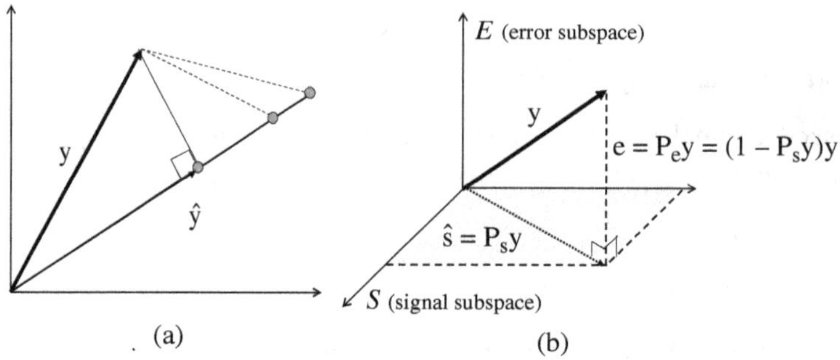

Figure 4.5 Geometric perspective of least squares: (a) projection of point **y** onto line $\hat{\mathbf{y}}$. (b) *LS* estimation.

whose tip defines a point and we would like to find the distance from that point to a line that lies in a certain direction defined by the vector $\hat{\mathbf{y}}$ as shown in Figure 4.5a. We search along this line (vector) for the point closest to **y**. We know from geometry that the solution is a perpendicular line drawn from the tip of **y** or equivalently the point to the line defined by $\hat{\mathbf{y}}$. Knowledge of this fact allows us to find not only the point on the line defining $\hat{\mathbf{y}}$, but also its distance from the tip of **y**. If we replace the line with a plane or more generally a subspace $\mathcal{S}$, as in Figure 4.5b then the problem becomes that of finding the location on the subspace that is "closest to **y**." The intersection point defines the vector $\hat{\mathbf{y}}$ which is called the *projection* of **y** onto $\mathcal{S}$.

Now let us assume that the N_y-dimensional data space, $\mathcal{Y}$, is decomposed into two orthogonal subspaces, $\mathcal{S}$ and $\mathcal{E}$ of respective dimensions N_s and N_e [18] such that

$$\mathcal{Y} = \mathcal{S} \oplus \mathcal{E}$$

Therefore, any vector in $\mathcal{Y}$, $\mathbf{y} \in \mathcal{R}^{N_y \times 1}$ can be decomposed as

$$\mathbf{y} = P_s\mathbf{y} + P_e\mathbf{y} \text{ for } \mathcal{S} \perp \mathcal{E} \tag{4.84}$$

where $P_s\mathbf{y}$ represents the *projection* with P_s the corresponding *projection matrix* of **y** onto the subspace $\mathcal{S}$. This subspace is defined as the "signal" subspace, while $P_e\mathbf{y}$ is the projection with P_e the corresponding projection matrix of **y** onto the orthogonal "error" or "noise" subspace $\mathcal{E}$ as shown in Figure 4.3b.

Since this is an *orthogonal decomposition*, the orthogonal projection matrices have the following properties:

1. $P_s = P_s'$, $P_e = P_e'$ [Symmetric]
2. $P_s = P_s^2$, $P_e = P_e^2$ [Idempotent]
3. $P_e = I - P_s$ [Identity]
4. $P_sP_e = P_eP_s = 0$ [Orthogonal]

Example 4.9 Next, let us apply the orthogonal decomposition and projection concepts to the batch least squares problem. Recall the least-squares estimation problem of Eq. (4.50) with $W = I$ for simplicity. Here, we are tasked with finding the best estimate of the signal defined by

$$\mathbf{s} = C\boldsymbol{\Theta} = \sum_{i=1}^{N_\theta} \theta_i \mathbf{c}_i \text{ for } C \in \mathcal{R}^{N_y \times N_\theta}, \boldsymbol{\Theta} \in \mathcal{R}^{N_\theta \times 1}, \mathbf{c}_i \in \mathcal{R}^{N_y \times 1}, i = 1, \ldots, N_\theta \tag{4.85}$$

where $\mathbf{c}_i$ are the set of N_θ-column vectors of C spanning the signal subspace. Another way of stating the same problem is to choose that value of Θ to minimize the square of the *error* or *noise* vector defined by

$$\mathbf{e} = \mathbf{y} - \mathbf{s} = \mathbf{y} - C\boldsymbol{\Theta} \tag{4.86}$$

in the least squares sense of Eq. (4.50). The solution to this problem then provides the closest point than any other point spanned by the column vectors of C. Thus, minimizing the squared error is equivalent to finding the orthogonal projection of $\mathbf{y}$ onto the signal subspace defined spanned by the $\mathbf{c}_i$ vectors. Therefore, we solve for the unknown parameter vector, $\boldsymbol{\Theta}$, that defines the signal vector using orthogonal projection theory to gain the geometric insight we seek. In the least squares problem, we have that the projection of $\mathbf{y}$ onto $\mathcal{S}$ gives the signal vector

$$\hat{\mathbf{s}} = P_c\mathbf{y} = C\hat{\boldsymbol{\Theta}}_{LS} = C(C'C)^{-1}C'\mathbf{y} \tag{4.87}$$

where $P_c := C(C'C)^{-1}C'$ is the orthogonal projection matrix defined on the signal subspace. Similarly, the projection of $\mathbf{y}$ onto $\mathcal{E}$ (orthogonal complement) gives the error vector

$$\mathbf{e} = P_e\mathbf{y} = \mathbf{y} - \hat{\mathbf{s}} = \mathbf{y} - C\hat{\boldsymbol{\Theta}}_{LS} = \mathbf{y} - P_c\mathbf{y} = (I - P_c)\,\mathbf{y} \tag{4.88}$$

Thus, we see that the orthogonal projection of the data $\mathbf{y}$ onto the signal subspace results in a linear transformation of the least squares parameter estimate $\hat{\boldsymbol{\Theta}}_{LS}$, while the orthogonal projection of the data vector onto the error subspace (orthogonal complement) results in the error vector. The orthogonality of the error vector to the signal vector defines the *orthogonality condition*

$$\mathbf{e}'\hat{\mathbf{s}} = \mathbf{0} \tag{4.89}$$

which states that the error vector is orthogonal to the space spanned by the signal vectors. To see this we use the projection matrices and the error vector

$$\mathbf{e}'\hat{\mathbf{s}} = (P_e\mathbf{y})'(P_c\mathbf{y}) = \mathbf{y}'P_eP_c\mathbf{y} = \mathbf{y}'(I - P_c)P_c\mathbf{y} = \mathbf{y}'(P_c - P_c^2)P_c\mathbf{y} = \mathbf{0} \tag{4.90}$$

since the projection matrix is idempotent. The corresponding cost function is also defined in terms of the projections by

$$\mathcal{J}_{LS} = \frac{1}{2}\mathbf{e}'\mathbf{e} = \mathbf{y}'P_e'P_e\mathbf{y} = \mathbf{y}'P_e\mathbf{y} = \mathbf{y}'(I - P_c)\mathbf{y} = \mathbf{y}'\mathbf{y} - (P_c\mathbf{y})'(P_c\mathbf{y}) = \mathbf{y}'\mathbf{y} - \hat{\mathbf{s}}'\hat{\mathbf{s}}. \tag{4.91}$$

In summary, we have the batch "geometric" description of the LS estimator as:

Criterion: $J_{LS} = \frac{1}{2}\mathbf{e}'\mathbf{e} = \frac{1}{2}\mathbf{y}'P_e\mathbf{y}$
Models: $s = P_c\mathbf{y} = C\Theta$

Algorithm: $\hat{\Theta}_{LS} = (C'C)^{-1}C'\mathbf{y};\ \hat{s} = P_c\mathbf{y} = C\hat{\Theta}_{LS}$

Quality: $\mathcal{J}_{LS} = \mathbf{y}'\mathbf{y} - \hat{\mathbf{s}}'\hat{\mathbf{s}}$

△△△

4.7.2 Orthogonal Projections: Singular Value Decomposition

It is interesting to see that the least-squares estimation problem when viewed from the geometric perspective leads to the idea of orthogonal projections on subspaces defined by the signal and associated error vectors. Next let us investigate the same ideas cast into a computational problem which can be attacked using the singular value decomposition [19] of linear algebra.

The *singular value decomposition* (*SVD*) of a (rectangular) matrix, say $Y \in \mathcal{R}^{N_y \times M_y}$ is defined by

$$Y = U\Sigma V' \tag{4.92}$$

where $U \in \mathcal{R}^{N_y \times N_y}$, $V \in \mathcal{R}^{M_y \times M_y}$, U, V are *unitary* matrices ($U' = U^{-1}$) and $\Sigma \in \mathcal{R}^{N_y \times M_y}$ is a matrix with elements σ_{ii} arranged in descending order of magnitude such that $\sigma_{11} \geq \sigma_{22}, \ldots, \geq \sigma_{\min(N_y, M_y)}$. The matrices U and V are called the left and right *singular vectors* of Y and are the respective eigenvectors of YY' and $Y'Y$. The singular values $\{\sigma_{ii}\}$ are the square roots of the eigenvalues of YY' or $Y'Y$.

In general, the *SVD* of Y can be written as

$$Y = [U_1 | U_2] \begin{bmatrix} \Sigma_1 & | & 0 \\ -- & -- & -- \\ 0 & | & \Sigma_2 \end{bmatrix} \begin{bmatrix} V_1' \\ -- \\ V_1' \end{bmatrix} = U_1\Sigma_1 V_1' + U_2\Sigma_2 V_2' \tag{4.93}$$

with $Y \in \mathcal{R}^{N_y \times M_y}$, $U_1 \in \mathcal{R}^{N_y \times N_1}$, $U_2 \in \mathcal{R}^{N_y \times N_2}$, $V_1 \in \mathcal{R}^{N_1 \times M_y}$, $V_2 \in \mathcal{R}^{N_2 \times M_y}$, $\Sigma_1 \in \mathcal{R}^{N_1 \times N_1}$, and $\Sigma_2 \in \mathcal{R}^{N_2 \times N_2}$. Thus, the *SVD* explicitly demonstrates the orthogonal decomposition of Y into its constituent subspaces.

Next, let us return to the idea of the orthogonal decomposition of the data space by defining two sets of independent vectors: one set spanning the *signal subspace*, $\mathcal{S}$, such that $\mathbf{ci} \in \mathcal{R}^{N_y \times 1}; i =, \ldots, N_\theta$, and the other set spanning the orthogonal *error subspace*, $\mathcal{E}$, such that $\epsilon_{\mathbf{i}} \in \mathcal{R}^{N_y \times 1}; i =, \ldots, N_y - N_\theta$. Collecting the column vectors, we construct two full-rank matrices,

$$C = \begin{bmatrix} \mathbf{c}_1 & \mathbf{c}_2 & \cdots & \mathbf{c}_{N_\theta} \end{bmatrix} \in \mathcal{R}^{N_y \times N_\theta} \text{ and } \mathcal{E} = \begin{bmatrix} \epsilon_1 & \epsilon_2 & \cdots & \epsilon_{N_y - N_\theta} \end{bmatrix} \in \mathcal{R}^{N_y \times (N_y - N_\theta)}$$

such that the orthogonality condition holds

$$\mathcal{E}'C = \mathbf{0}$$

establishing the orthogonal decomposition of the N_y-dimensional data space (see Figure 4.3b).

If $\rho(Y) = n \leq \min(N_y, M_y)$, then it can be shown [15, 19–22] that

$$\begin{cases} \sigma_{kk} > 0 & k = 1, \ldots, n \\ \sigma_{kk} = 0 & k = n+1, \ldots, \min(N_y, M_y) \end{cases} \tag{4.94}$$

Therefore, the rank n matrix Y can be written as

$$Y = [U_1 | U_2] \begin{bmatrix} \Sigma_1 & | & 0 \\ -- & -- & -- \\ 0 & | & 0 \end{bmatrix} \begin{bmatrix} V_1' \\ -- \\ V_2' \end{bmatrix} = U_1 \Sigma_1 V_1' \tag{4.95}$$

where $\Sigma_1 \in \mathcal{R}^{n \times n}$ is nonsingular and U_1, V_1 are an orthonormal basis for S while U_2 and V_2 are an orthonormal basis for the $\mathcal{E}$.

It is also important to realize that once the *SVD* of a matrix has been performed, its *inverse* (in the square matrix case) is trivial. Let $X = U\Sigma V'$ be the *SVD* of $X \in \mathcal{R}^{N_x \times N_x}$, then

$$X^{-1} = (V^{-1})' \Sigma^{-1} U^{-1} = (V')' \Sigma^{-1} U' = V \Sigma^{-1} U' \tag{4.96}$$

Thus, since Σ is diagonal, we merely take reciprocals and multiply to construct the inverse. Next we use the *SVD* to revisit the batch least squares problem.

Example 4.10 With this information in mind, let us now construct the data matrix as

$$Y = [C | E] \in \mathcal{R}^{N_y \times N_y}$$

with C and E defined (above) as the signal and error subspaces. Recall that the batch least squares estimator is obtained by solving the *normal equations* given by

$$(C'C)\, \hat{\Theta}_{LS} = C'\mathbf{y} \tag{4.97}$$

or using the *SVD* of $C = U_C \Sigma_C V_C'$ and substituting, we have

$$(V_C \Sigma_C U_C')\,(U_C \Sigma_C V_C')\, \hat{\Theta}_{LS} = (V_C \Sigma_C \Sigma_C V_C')\, \hat{\Theta}_{LS} = (V_C \Sigma_C U_C')\, \mathbf{y} \tag{4.98}$$

or inverting

$$\hat{\Theta}_{LS} = (V_C \Sigma_C^{-1} \Sigma_C^{-1} V_C')\,(V_C \Sigma_C U_C')\, \mathbf{y} = V_C \Sigma_C^{-1} U_C' \mathbf{y} = \sum_{i=1}^{N_\theta} \left(\frac{\mathbf{v}_i \mathbf{u}_i'}{\sigma_i} \right) \mathbf{y} \tag{4.99}$$

Thus, the signal estimate is

$$\hat{\mathbf{s}} = C \hat{\Theta}_{LS} = (U_C \Sigma_C V_C')\,(V_C \Sigma_C^{-1} U_C')\, \mathbf{y} = U_C U_C' \mathbf{y} = \sum_{i=1}^{N_\theta} (\mathbf{u}_i \mathbf{u}_i')\, \mathbf{y} \tag{4.100}$$

The corresponding projection matrices are given by

$$P_c = C(C'C)^{-1}C' = U_C U_C' \text{ and } P_e = (I - P_c) = I - U_C U_C' \tag{4.101}$$

Thus, in summary, we have that the *SVD* description of the batch least squares problem yields

- Criterion: $J_{LS} = \frac{1}{2}\mathbf{e}'\mathbf{e} = \frac{1}{2}\mathbf{y}'P_e\mathbf{y} = \mathbf{y}'(I - U_C U_C')\mathbf{y}$
- Models: $\underline{s} = U_C \Sigma_C V_C' \Theta$
- Algorithm: $\hat{\Theta}_{LS} = V_C \Sigma_C^{-1} U_C' \mathbf{y}$; $\hat{\underline{s}} = U_C U_C' \mathbf{y}$
- Quality: $\mathcal{J}_{LS} = \mathbf{y}'\mathbf{y} - \hat{\mathbf{s}}'\hat{\mathbf{s}}$

△△△

This completes the geometric perspective of projection offering insight into the idea of optimal estimation and orthogonal decompositions, projections, and computations using the *SVD* (see Appendix B for more details).

4.8 Summary

In this chapter, we have discussed the fundamental concepts underlying random signal estimation. Starting with the statistical concepts of estimation, we developed desirable properties of an estimator and showed how to compare and evaluate its performance. Next, we developed the minimum variance, maximum a posteriori, and maximum likelihood estimators and showed how they are related. However, we concentrated on least-squares estimation starting with the well-known batch solutions and progressing to the sequential or recursive approach. These ideas were applied to a trajectory estimation problem to illustrate the approach. Finally, the vector-matrix approach to signal estimation was developed based on projection theory. Here both orthogonal and oblique projections were discussed leading to basic optimal signal estimation. Finally both estimation and decomposition concepts were discussed from a geometric perspective leading to the development of an optimal signal estimator and the decomposition approach based on the singular value decomposition. An understanding of this approach is fundamental to the decomposition techniques to follow.

MATLAB® Notes

MATLAB® has a wealth of linear algebraic commands to *solve* the various estimation problems discussed in this chapter. The **svd** command performs the singular value decomposition providing the order set of singular values (diag Σ) and the unitary matrices of the decomposition (U, V). The **eig** command performs the eigen-decomposition of a matrix as well, along with the efficient, numerically stable **qr** decomposition (Gram–Schmidt) and its variants (see [18–20], for details). Polynomial coefficient estimation is performed using the *SVD*-based, **polyfit** command, while the *RLS* algorithms require the *MATLAB System Identification* toolbox. The optimal Wiener solution is performed in the frequency domain using the transfer function estimation command **tfestimate**.

References

1 Van Trees, H. (1968). *Detection, Estimation and Modulation Theory*. New York: Wiley.

2 Sage, A. and Melsa, J. (1971). *Estimation Theory with Applications to Communications and Control*. New York: McGraw-Hill.

3 Srinath, M., Rajasekaran, P., and Viswanathan, R. (1996). *Introduction to Statistical Signal Processing with Applications*. New Jersey: Prentice-Hall.

4 Hogg, R. and Craig, A. (1970). *Introduction to Mathematical Statistics*. New York: Macmillan.

5 Porat, B. (1994). *Digital Processing of Random Signals: Theory and Methods*. New Jersey: Prentice-Hall.

6 Candy, J. (1977). The Cramer–Rao Bound: A Measure of Estimator Quality. Lawrence Livermore National Laboratory Report, *UCID-17660*.

7 Candy, J. (1986). *Signal Processing: The Model-Based Approach*. New York: McGraw-Hill.

8 Jazwinski, A. (1970). *Stochastic Processes and Filtering Theory*. New York: Academic Press.

9 Rhodes, I. (1971). A tutorial introduction to estimation and filtering. *IEEE Trans. Autom. Control* AC-16 (6): 688–706.

10 Mendel, J. (1995). *Lessons in Estimation Theory for Signal Processing, Communications, and Control*. New Jersey: Prentice-Hall.

11 Haykin, S. (ed.) (1984). *Array Signal Processing*. New Jersey: Prentice-Hall.

12 Johnson, D. and Dudgeon, D. (1993). *Array Signal Processing: Concepts and Techniques*. New Jersey: Prentice-Hall.

13 Van Trees, H. (2002). *Optimum Array Processing*. New York: Wiley-Interscience.

14 Candy, J. (1999). Processing of RV signatures from radar tracking data. RTO Set Symposium on High Resolution Radar Techniques, RTO MP-40, pp. 38-1 to 38-12.

15 Bretscher, O. (2015). *Linear Algebra with Applications*, 5e. Upper Saddle River, NJ: Pearson Prentice-Hall.

16 Kailath, T. (1981). *Lectures on Kalman and Wiener Filtering Theory*. New York: Springer.

17 Orfanidis, S. (1985). *Optimum Signal Processing*. New York: Macmillan.

18 Strang, G. (1976). *Linear Algebra and Its Applications*. New York: Academic Press.

19 Golub, G. and Van Loan, C. (1989). *Matrix Computations*. Baltimore, MD: Johns Hopkins University Press.

20 Lawson, C.L. and Hanson, R.J. (1974). *Solving Least Squares Problems*. New Jersey: Prentice-Hall.

21 Stewart, G. (1973). *Introduction to Matrix Computations*. New York: Academic Press.

22 Scharf, L. (1991). *Statistical Signal Processing: Detection, Estimation, and Time Series Analysis*. Massachusetts: Addison-Wesley.

Problems

4.1 Suppose we use the covariance estimator

$$\hat{R}_{xx}(k) = \frac{1}{N-k-1} \sum_{t=0}^{N-k-1} x(t)x(t+k) \quad \text{for } k \geq 0$$

(a) Is this a biased estimate for $R_{xx}(k) = E\{x(t)x(t+k)\}$?
(b) What is the variance of this estimator?

4.2 Derive the following properties of conditional expectations:
(a) $E_x\{X|Y\} = E\{X\}$ if X and Y are independent.
(b) $E\{X\} = E_y\{E\{X|Y\}\}$.
(c) $E_x\{g(Y)X\} = E_y\{g(Y)E\{X|Y\}\}$.
(d) $E_{xy}\{g(Y)X\} = E_y\{g(Y)E\{X|Y\}\}$.
(e) $E_x\{c|Y\} = c$.
(f) $E_x\{g(Y)|Y\} = g(Y)$.
(g) $E_{xy}\{cX + dY|Z\} = cE\{X|Z\} + dE\{Y|Z\}$.

4.3 Verify the following properties:
(a) $\nabla_x(a'b) = (\nabla_x a')b + (\nabla_x b')a$, for $a, b \epsilon R^n$ and functions of x.
(b) $\nabla_x(b'x) = b$.
(c) $\nabla_x(x'C) = C, C\epsilon R^{n\times m}$.
(d) $\nabla_x(x') = I$.
(e) $\nabla_x(x'x) = 2x$.
(f) $\nabla_x(x'Ax) = Ax + A'x$, for A not a function of x.

4.4 Show that for any unbiased estimate of $\hat{\mathbf{x}}(y)$ of the parameter vector $\mathbf{x}$ the *Cramér–Rao bound* is

$$\mathrm{Cov}(\tilde{x}|x) \geq \mathcal{I}^{-1} \quad \text{for} \quad \tilde{x} = x = \hat{x}(y) \quad \text{and} \quad \mathbf{x} \in \mathcal{R}^{n}, \mathcal{I} \in \mathcal{R}^{n\times n}$$

where the *information matrix* is defined by

$$\mathcal{I} := -E_y\{\nabla_x(\nabla_x \ \ln \Pr(Y|X))'\}$$

4.5 Let $x(t)$ be an unknown dynamic parameter obtained from the linear measurement system

$$y(t) = C(t)x(t) + v(t)$$

where $v \sim \mathcal{N}(0, R_{vv}(t))$ with $y, v \in \mathcal{R}^p$ and x is governed by the state transition mechanism

$$x(t+1) = \Phi(t+1, t)x(t) \quad \text{for} \quad \Phi \in \mathcal{R}^{n\times n}$$

Calculate the Cramér Rao bound for the case of estimating the initial state $x(0)$.

4.6 Suppose we have the following signal in additive Gaussian noise:

$$y = x + n \quad \text{with} \quad n \sim N(0, R_{nn})$$

(a) Find the Cramér–Rao bound if x is assumed unknown.
(b) Find the Cramér–Rao bound if $p(x) = xe^{-x^2/R_{nn}}$, $x \geq 0$.

4.7 Suppose we have two jointly distributed random vectors x and y with know means and variances. Find the linear minimum variance estimator. That is, find

$$\hat{x}_{MV} = \hat{A}y + \hat{b}$$

and the corresponding error covariance $\mathrm{cov}\tilde{x}$

4.8 We would like to estimate a vector of unknown parameters $p\epsilon R^n$ from a sequence of N-noisy measurements given by

$$y = Cp + n$$

where $C\epsilon R^{p\times n}, y, n\epsilon R^p$, and v is zero-mean and white covariance R_{nn}.
(a) Find the minimum variance estimate $\hat{p}_{MV}$.
(b) Find the corresponding quality $\mathrm{cov}(p - \hat{p}_{MV})$.

4.9 Suppose we have N samples $\{x(0) \cdots x(N-1)\}$ of a process $x(t)$ to be estimated by N complex sinusoids of arbitrary frequencies $\{f_0, \ldots, f_{n-1}\}$. Then

$$x(k\triangle T) = \sum_{m=0}^{N-1} a_m \exp(j2\pi f_m k\triangle T) \quad \text{for} \quad k = 0, \ldots, N-1$$

Find the least-squares estimate $\hat{a}_{LS}$ of $\{a_m\}$.

4.10 Suppose we are given a measurement modeled by

$$y(t) = s + n(t)$$

where s is random and zero-mean with variance $\sigma_s^2 = 4$ and n is zero-mean and white with a unit variance. Find the two-weight minimum variance estimate of s, that is,

$$\hat{s}_{MV} = \sum_{i=1}^{2} w_i y(i)$$

that minimizes

$$J = E\{(s - \hat{s})^2\}$$

4.11 Find the maximum likelihood and maximum a posteriori estimate of the parameter x is

$$p(x) = \alpha e^{-\alpha x} \quad x \geq 0, \quad \alpha \geq 0$$

and

$$p(y|x) = xe^{-xy} \quad x \geq 0, \quad y \geq 0$$

4.12 Suppose we are given a measurement system

$$y(t) = x(t) + v(t) \quad t = 1, \ldots, N$$

where $v(t) \sim N(0, 1)$.

(a) Find the maximum likelihood estimate of $x(t)$, that is, $\hat{x}_{ML}(t)$, for $t = 1, \ldots, N$.

(b) Find the maximum a posteriori estimate of $x(t)$ that is, $\hat{x}_{MAP}(t)$, if $p(x) = e^{-x}$.

4.13 Suppose we have a simple AM receiver with signal

$$y(t) = \Theta s(t) + v(t) \quad t = 1, \ldots, N$$

where Θ is a random amplitude, s is the known carrier, and $v \sim N(0, R_{vv})$.

(a) Find the maximum likelihood estimate $\hat{\Theta}_{ML}$.

(b) Assume $\Theta \sim N(\Theta_0, R_{\Theta\Theta})$. Find the maximum a posteriori estimate $\hat{\Theta}_{MAP}$.

(c) Assume Θ is Rayleigh-distributed (a common assumption). Find $\hat{\Theta}_{MAP}$.

4.14 We would like to estimate a signal from a noisy measurement

$$y = s + v$$

where $v \sim N(0, 3)$ and s is Rayleigh-distributed

$$p(s) = se^{-s^2/2}$$

with

$$p(y|s) = \frac{1}{\sqrt{6\pi}} e^{-\frac{(y-s)^2}{6}}$$

(a) Find the maximum likelihood estimate.
(b) Find the maximum a posteriori estimate.
(c) Calculate the Cramér–Rao bound (ignoring $p(s)$).

4.15 Assume that a planet travels an elliptical orbit centered about a known point. Suppose we make M observations.

(a) Find the best estimate of the orbit and the corresponding quality, that is, for the elliptical orbit (gaussian problem)

$$\beta_1 x^2 + \beta_2 y^2 = 1$$

Find $\hat{\beta} = [\hat{\beta}_1 \ \hat{\beta}_2]'$.

(b) Suppose we are given the following measurements: $(x, y) = \{(2, 2), (0, 2), (-1, 1), \text{ and } (-1, 2)\}$, find $\hat{\beta}$ and $\hat{J}$, the cost function.

4.16 Find the parameters, β_1, β_2, and β_3 such that

$$f(t) = \beta_1 t + \beta_2 t^2 + \beta_3 \sin t$$

fits $(t, f(t))\colon = \{(0, 1), (\pi/4, 2), (\pi/2, 3), \text{ and } (\pi, 4)\}$ with corresponding quality estimate, $\hat{J}$.

4.17 Calculate the batch and sequential least-squares estimate of the parameter vector x based on two measurements $y(1)$ and $y(2)$ where

$$y(1) = C(1)x + v(1) = \begin{bmatrix} 2 \\ 1 \end{bmatrix}$$

$$y(2) = c'x + v(2) = 4$$

$$C = \begin{bmatrix} 1 & 1 \\ 0 & 1 \end{bmatrix}, c'(1) = [1 \ 2], W = I$$

4.18 Suppose we have a two-measurement system given by

$$y = \begin{bmatrix} 3 \\ 4 \end{bmatrix} + v$$

where $R_{vv} = \text{diag}[1,\ 0.1]$.

(a) What is the batch least-squares estimate ($W = I$) of the parameter x, if $y = [7\ 21]'$?

(b) What is the batch-weighted least-squares estimate of the parameter x with W selected for minimum variance estimation?

4.19 Given a sequence of random variables $\{y(t)\}$, define

$$e(t) = y(t) - \hat{y}(t|t-1) \quad \hat{y}(1|0) = 0$$

$$\hat{y}(t|t-1) = E\{y(t)|Y(t-1)\}$$

is the minimum variance estimate. Then

(a) Show that the $\{e(t)\}$ are orthogonal.

(b) Show that if $E\{e^2(t)\} > 0$, then

$$\underline{Y}(N) = \begin{bmatrix} y(1) \\ \vdots \\ y(N) \end{bmatrix} \quad \text{and} \quad \underline{e}(N) = \begin{bmatrix} e(1) \\ \vdots \\ e(N) \end{bmatrix}$$

are related by

$$\underline{Y}(N) = L(\underline{N})\underline{e}(N)$$

(c) If $W(\underline{N})$ is a linear operation that yields $\hat{\underline{Y}}(N)$ from $\underline{Y}(N)$, show that

$$W(\underline{N}) = I - L^{-1}(\underline{N}) \quad \text{and} \quad \hat{\underline{Y}}(N) = \underline{Y}(N) - \underline{e}(N)$$

(d) Show that $L(\underline{N}) = R_{ye}(\underline{N})R_{ee}^{-1}(\underline{N})$

(e) Show that $R_{yy}(\underline{N}) = L(\underline{N})R_{ee}(\underline{N})L'(\underline{N})$

(f) If x is a related vector, show that

$$\hat{x} = R_{xy}(\underline{N})R_{yy}^{-1}(\underline{N})\underline{Y}(N) = R_{xe}(\underline{N})R_{ee}^{-1}(\underline{N})\underline{e}(N)$$

(g) If $R_{yy}(i,j) = \alpha^{|i-j|}$, for $0 < \alpha < 1$, find $L(\underline{N})$ and $R_{yy}^{-1}(\underline{N})$.

4.20 Construct an innovation sequence $\{e(1), e(2)\}$ from the measurement sequence

$$y(1) = \begin{bmatrix} 1 \\ 4 \end{bmatrix},\ y(2) = \begin{bmatrix} 2 \\ 3 \end{bmatrix} \text{ if } R_{ee} = \text{diag}\{1, 2\} \quad \text{and} \quad R_{ye}(2,1) = \begin{bmatrix} 2 & 1 \\ 1 & 6 \end{bmatrix}$$

5

Signal Decomposition

5.1 Introduction

Signal decomposition techniques evolve from various constructs ranging from the statistical, signal processing, and controls communities. In statistics, processing the raw data with little or no prior information leads directly to the popular approaches such as principal component analysis (*PCA*), independent component analysis (*ICA*), and the singular spectral analysis (*SSA*), while in signal processing emphasis is on the processing of either data or correlation matrices leading to spectral decompositions. Here some decompositions such as the multiple taper method (*MTM*) based on prolate spheroidal function approximations and subspace decomposition techniques based on harmonic functions as the Pisaranko harmonic decomposition (*PHD*) method and the multiple signal classification (*MUSIC*) method decompose the underlying signal data matrices. In control systems, model-based decomposition approaches rely on parametrically estimating the various components that make up the signal model itself based on the decomposition of a selected model structure. Here, we investigate the damped exponential (*DExp*), order-recursive lattice (*LATTICE*), and state-space (*SSP*) model decompositions. In this section, we discuss each of these primary approaches, statistical, processing, and controls detailing the major techniques that have evolved from these disciplines. We summarize a road map of the subsequent discussions in Figure 5.1 along with each particular method annotated.

5.2 Data-Based Decompositions

Measuring phenomena occurs in a wide variety of scientific as well as industrial and military applications resulting in large datasets that must be investigated to extract the desired information. Typically, models, that is, signal models are

Signal Processing: An Applied Decomposition Approach, First Edition. James Vincent Candy.

Companion website: www.wiley.com/go/CandySignProc

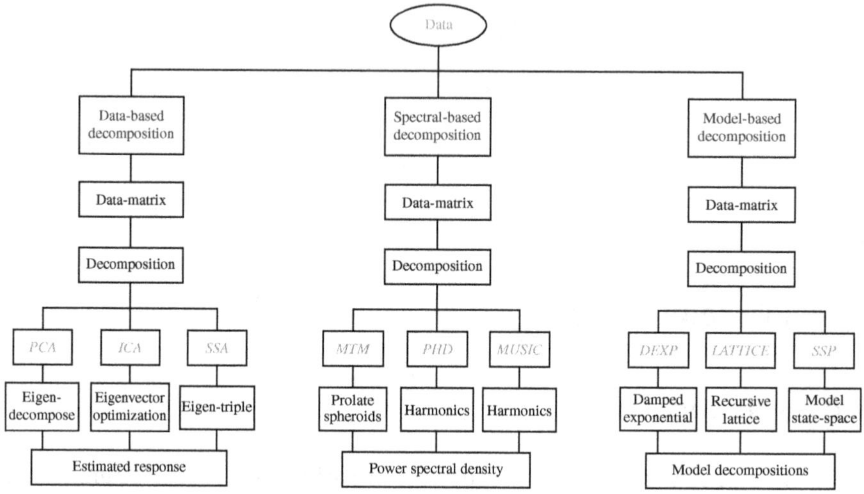

Figure 5.1 Signal decomposition approach: data-based (*PCA, ICA, SSA*), spectral-based (*MTM, PHD, MUSIC*), model-based (*DExp, LATTICE, SSP*).

employed by engineers and scientists to aid in understanding and separating the "best" possible information available in a comprehensive manner. Data-based signal models basically represent, perhaps, the lowest level of phenomenological understanding, but just as a Fourier transform reveals fundamental frequency information to a controls engineer, data-based representations can reveal information to the phenomenologist that leads to the development of more sophisticated models and a better understanding of the process to solve the underlying problems at hand. The usual vector-matrix formulation of the data enable the development of a variety of formats leading to matrix structures that can be exploited to extract the desired information. The data, assumed viable, can lead to a deeper understanding of problem at hand. One of the most obvious approaches in signal processing is the decomposition of a signal into constituent components that are easily comprehended. For instance, consider the basic Fourier estimation/decomposition of a signal in terms of its transform, it is the coefficients of the transform that provide the information in the frequency domain that enable engineers and scientists to understand a variety of physical phenomena.

In this section, we explore three of the most popular data (signal) decomposition methodologies: (i) *PCA*; (ii) *ICA*; and (iii) singular spectrum analysis (*SSA*) all evolving primarily from the statistical computation community [1–11]. In *PCA*, large datasets are orthogonally decomposed based on their inherent energy (variance) starting with the largest and progressing to the smallest, while *ICA* seeks to create a decomposition of the data into completely independent sets. The *SSA*-approach sections the dataset based on an eigen-decomposition.

5.2.1 Data Decomposition: Principal Component Analysis (*PCA*)

When large datasets exist that must be processed to extract pertinent information, statistical data-based methods evolve as a potential solution, especially when exploring data that may lack known phenomenological explanation [8–10]. *PCA* is a statistical data-based technique to decompose datasets into constituent components that explain the existing variations in the data. The first component is a linear combination of the original data variables that explains the largest variance, while a second component explains the residual variance remaining after the first is removed and so on. This results in a set of orthonormal (independent) basis vectors spanning the data space resulting in a linear representation of the underlying signal and noise. Thus, the extracted components are uncorrelated and independent of each other forming the basis that explains the phenomenological data set. The algorithmic approach is a linear projection method that decomposes highly correlated variables into a new set of uncorrelated (orthogonal) variables. Historically, *PCA* evolved as a methodology to characterize random vectors in an orthogonal basis termed the *Karhunen–Loeve transformation* [11–13]. Therefore, *PCA* can be defined (simply) as the orthogonal projection of data onto a lower dimensional (reduced) linear vector space—the *principal subspace* maximizing the variance of the projected data [11, 14–17].

Principal component theory is based on developing a set of orthonormal basis vectors spanning the measurement or data space such that each basis vector maximizes its own variance in its principal direction, that is, the direction defined by its inherent angle. The set of basis vectors is ordered in terms of decreasing variance (by construction) similar to the Gram–Schmidt (*G-S*) process developed previously. Simply stated the *PCA* problem is:

GIVEN a set of vectors, $\{\mathbf{y}_n\}$, spanning the data space, $\mathcal{Y}$ with covariance matrix, $\mathbf{R}_{yy}$, FIND a set of ordered orthonormal basis vectors $\{\mathbf{o}_i\}; i = 1, \ldots, N$ that *maximize* their variance in their principal directions, that is,

$$\max_{|\mathbf{o}|} \text{var}(\mathbf{o}) \ni \mathbf{o}'\mathbf{o} = 1$$

Suppose the data space $\mathcal{Y}^{N\times1}$ is spanned by $\{\mathbf{y}_n\}; n = 1, \ldots, N$ and we would like to construct a set of orthonormal vectors, $\{\mathbf{o}_n\}$, spanning the space, such that each component is specified by

$$\mathbf{o}_i = \sum_{i=1}^{N} \alpha_{in}\mathbf{y}_n \tag{5.1}$$

Expanding over i, we have

$$\mathbf{o} = \alpha'\mathbf{y} \tag{5.2}$$

Similar to the *G-S* process of Eq. (4.77), we select the first vector $\mathbf{o}_1$ to satisfy

$$\mathbf{o}_1 = \boldsymbol{\alpha}_1'\mathbf{y} = \sum_{i=1}^{N} \alpha_{1n}\mathbf{y}_n \tag{5.3}$$

and maximize the variance of $\mathbf{o}_1$ subject to the constraint that the coefficients of $\boldsymbol{\alpha}_1 = [\alpha_{11}\alpha_{12},\ldots,\alpha_{1N}]'$ satisfy $\boldsymbol{\alpha}_1'\boldsymbol{\alpha}_1 = 1$. Therefore, the variance (assuming zero mean) is

$$\text{var}\left(\mathbf{o}_1\right) = E\{\mathbf{o}_1\mathbf{o}_1'\} = \boldsymbol{\alpha}_1'\underbrace{E\{\mathbf{y}\mathbf{y}'\}}_{\mathbf{R}_{yy}}\boldsymbol{\alpha}_1 = \boldsymbol{\alpha}_1'\mathbf{R}_{yy}\boldsymbol{\alpha}_1 \tag{5.4}$$

The fundamental problem for the "first" component can be formulated as

$$\max_{\|\boldsymbol{\alpha}_1\| = 1} \quad \mathcal{J}(\boldsymbol{\alpha}_1) = \boldsymbol{\alpha}_1'\mathbf{R}_{yy}\boldsymbol{\alpha}_1 - \lambda_1(\boldsymbol{\alpha}_1'\boldsymbol{\alpha}_1 - 1) \tag{5.5}$$

which is a constrained optimization problem solved easily by the Lagrange multiplier method [17, 18]. Differentiating, setting to zero and solving, that is,

$$\nabla_{\boldsymbol{\alpha}_1}\mathcal{J}(\boldsymbol{\alpha}_1) = 2\mathbf{R}_{yy}\boldsymbol{\alpha}_1 - 2\lambda_1\boldsymbol{\alpha}_1 = 0 \tag{5.6}$$

or

$$\mathbf{R}_{yy}\boldsymbol{\alpha}_1 = \lambda_1\boldsymbol{\alpha}_1 \tag{5.7}$$

an eigenvalue problem. Pre-multiplying this expression by $\boldsymbol{\alpha}_1'$ yields the variance of $\mathbf{o}_1$ as the corresponding eigenvalue, that is,

$$\underbrace{\boldsymbol{\alpha}_1' \times \mathbf{R}_{yy}\boldsymbol{\alpha}_1}_{\text{var}\left(\mathbf{o}_1\right)} = \lambda_1\underbrace{\boldsymbol{\alpha}_1' \times \boldsymbol{\alpha}_1}_{1} = \lambda_1 \tag{5.8}$$

implying that the first principal component ($\mathbf{o}_1$) has the largest variance of $\{\mathbf{y}_n\}$ and $\boldsymbol{\alpha}_1$ is an *eigenvector* of $\mathbf{R}_{yy}$ with eigenvalue λ_1.

Next, the second principal component, $\mathbf{o}_2 = \boldsymbol{\alpha}_2'\mathbf{y}$ is constructed to maximize the remaining variance after removing var($\mathbf{o}_1$) constrained by $\|\boldsymbol{\alpha}_2\| = 1$ as before, but now including the additional orthogonality constraint, $\boldsymbol{\alpha}_2'\boldsymbol{\alpha}_1 = 0$ and therefore, $\boldsymbol{\alpha}_2'\mathbf{R}_{yy}\boldsymbol{\alpha}_1 = 0$. The criterion function for this constrained optimization problem is

$$\mathcal{J}(\boldsymbol{\alpha}_2) = \underbrace{\boldsymbol{\alpha}_2'\mathbf{R}_{yy}\boldsymbol{\alpha}_2}_{\text{var}(\mathbf{o})} - \lambda_2\underbrace{\left(\boldsymbol{\alpha}_2'\boldsymbol{\alpha}_2 - 1\right)}_{\|\boldsymbol{\alpha}_2\| = 1} - \eta\ \underbrace{\boldsymbol{\alpha}_2'\boldsymbol{\alpha}_1}_{\boldsymbol{\alpha}_2 \perp \boldsymbol{\alpha}_1} \tag{5.9}$$

with corresponding gradient

$$\nabla_{\boldsymbol{\alpha}_2}\mathcal{J}(\boldsymbol{\alpha}_2) = 2\mathbf{R}_{yy}\boldsymbol{\alpha}_2 - 2\lambda_2\boldsymbol{\alpha}_2 - \eta\boldsymbol{\alpha}_1 = 0 \tag{5.10}$$

Pre-multiplying this expression by $\boldsymbol{\alpha}_1'$ yields

$$\boldsymbol{\alpha}_1'\nabla_{\boldsymbol{\alpha}_2}\mathcal{J}(\boldsymbol{\alpha}_2) = \underbrace{2\boldsymbol{\alpha}_1'\mathbf{R}_{yy}\boldsymbol{\alpha}_2}_{0} - 2\lambda_2 \underbrace{\boldsymbol{\alpha}_1'\boldsymbol{\alpha}_2}_{0} - \eta \underbrace{\boldsymbol{\alpha}_1'\boldsymbol{\alpha}_1}_{1} = 0 \tag{5.11}$$

but the orthogonality condition implies that the first term in this expression is zero leading to $\eta = 0$ and therefore

$$\mathbf{R}_{yy}\boldsymbol{\alpha}_2 = \lambda_2\boldsymbol{\alpha}_2 \tag{5.12}$$

so that $\boldsymbol{\alpha}_2$ is an eigenvector of $\mathbf{R}_{yy}$ orthogonal to $\boldsymbol{\alpha}_1$ with λ_2 the second largest eigenvalue (variance).

Continuing with this construction, then the ℓth principal component satisfies

$$\mathbf{o}_\ell = \boldsymbol{\alpha}_\ell'\mathbf{y}$$

with α_ℓ the eigenvector corresponding to the ℓth largest eigenvalue (variance) of $\mathbf{R}_{yy}$.

So we see from Eq. (5.2) that

$$\mathbf{o} = \mathcal{A}'\mathbf{y} = \left[\boldsymbol{\alpha}_1\boldsymbol{\alpha}_2 \mid \cdots \mid \boldsymbol{\alpha}_N\right]\mathbf{y} \tag{5.13}$$

with the columns of $\mathcal{A}$ the eigenvectors of $\mathbf{R}_{yy}$ with corresponding eigenvalues (variances) in descending order (largest to smallest). Thus, these eigenvalues offer some explanation of the data, since the *total variance* of the data is equal to sum of the eigenvalues

$$\sum_{n=1}^{N} \text{var}\left(\mathbf{o}_n\right) = \sum_{n=1}^{N} \lambda_n \tag{5.14}$$

It follows that selecting the first K-principal components account for the corresponding %var($\mathbf{y}$) of the total data variance, that is,

$$\%\text{var}(\mathbf{y}) = \frac{\sum_{k=1}^{K} \lambda_k}{\sum_{n=1}^{N} \lambda_n} \times 100 \tag{5.15}$$

In fact, one way to reduce the size of the dataset is to approximate with its first K-principal components such that

$$\mathbf{o}_K = \mathcal{A}_K'\mathbf{y} = \left[\boldsymbol{\alpha}_1\ \boldsymbol{\alpha}_2 \mid \cdots \mid \boldsymbol{\alpha}_N\right]\mathbf{y} \tag{5.16}$$

for $A_K' \in R^{K\times N}$ and $\mathbf{o}_K \in R^{K\times 1}$. In the signal processing literature, this is known as an eigen-filter [16, 19–21]. Numerically, an eigen-decomposition directly of the data covariance matrix leads to a single-step solution of the principal component problem, that is,

$$\mathbf{R}_{YY} \times \mathcal{A} = \mathcal{A} \times \Lambda \quad \text{for} \quad \Lambda = \text{diag}\left[\lambda_1\ \lambda_2\ \cdots\ \lambda_N\right] \quad \text{with} \quad \lambda_1 > \lambda_2 > \cdots \lambda_N \tag{5.17}$$

Another popular approach to develop and implement *PCA*, that is, numerically stable, is to minimize the "reconstruction error" [22]. Starting with the set of desired bases vectors $\{\mathbf{o}\}$ (as before) such that

$$\mathbf{o} = \mathcal{A}'\mathbf{y}$$

the objective is to estimate the transformation matrix, $\mathcal{A} \in R^{N\times K}$, by constructing orthonormal column vectors such that the following error criterion is minimized, that is,

$$\mathcal{J}(\mathcal{A}) = \begin{matrix} \min \\ \mathcal{A} \end{matrix} \sum_{i=1}^{N} \|\mathbf{y}_i - \underbrace{\mathcal{A}\mathcal{A}'}_{P_{\mathbf{o}|\mathcal{Y}}}\mathbf{y}_i\|^2 \tag{5.18}$$

Stacking the data into an $N \times K$-data matrix, $\mathbf{Y}$, a singular value decomposition (*SVD*) is performed to yield

$$\mathbf{Y} = \mathbf{U} \times \mathbf{\Sigma} \times \mathbf{V}' \tag{5.19}$$

with $\mathbf{U} \in R^{N\times K}$ an orthogonal matrix ($\mathbf{U}\mathbf{U}' = \mathbf{I}_K$) with corresponding left singular vectors, $\mathbf{u}_k$, $\mathbf{V} \in R^{K\times K}$ also an orthogonal matrix of right singular values, $\mathbf{v}_k$ and $\mathbf{\Sigma} \in R^{K\times K}$ a diagonal matrix of singular values, $\{\sigma_k\}$ with $\sigma_k = \sqrt{\lambda_k}$. Here, the *columns* of the product $\mathbf{U} \times \mathbf{\Sigma}$ are the *principal components* and the data "decomposition" is

$$\mathbf{Y} = \sum_{k=1}^{K} \sigma_k \mathbf{u_k} \times \mathbf{v}'_k \tag{5.20}$$

where $\mathbf{u}_k$ are the eigenvectors of $\mathbf{YY}'$ and $\mathbf{v}_k$ are the eigenvectors of $\mathbf{Y}'\mathbf{Y}$ with Eq. (5.20) providing the decomposition of Y [14, 15].

The *SVD* minimizes the reconstruction or equivalently approximation error

$$\epsilon^2 = \sum_{k=1}^{K} \lambda_k$$

providing the best rank approximation of $\mathbf{Y}$ (for $K < N$) [23]. So, we see that the *PCA* can be estimated using the data matrix directly in contrast to the covariance matrix decomposition by using the *SVD*. Note also that the *SVD* minimizes the approximation error while the previous method minimized the mean-squared error (e.g. discrete Karhunen–Loeve decomposition [11]). Next we consider an example based on actual sensor data without an explicit physics-based solution to relate two critical variables.

The *PCA* algorithm is performed by:

- *Construct* data matrix $\mathbf{Y}$
- *Perform* a *SVD*, svd($\mathbf{Y}$)
- *Estimate* the "best" rank of $\rho(\mathbf{Y})$

- *Perform* the best rank approximation of the data matrix $\hat{\mathbf{Y}}$, and
- *Extract* the corresponding set of principal components, $\{\mathbf{o}_\ell\}; \ell = 1,\ldots,K$.

Example 5.1 Consider the three-mass mechanical system of Example 1.7 where we know there exists three-complex poles governing the system. However, *PCA* has no a priori knowledge of the underlying *MCK* system. The best we can hope for is to estimate the number of dominant poles and their decomposition of the overall data space and no more. We generate 256-samples of the *MCK* system ($t = 20$ msec) and provide the deterministic impulse response data as input to the *PCA* algorithm. We know a priori from the physics and relationship of the spectral response that two to three poles dominate. The *PCA* of these data is shown in Figure 5.2a where a 3D plot of the original data vectors (Yv_1, Yv_2, Yv_3) are shown (dashed lines), and the principal components (rotated) (Ev_1, Ev_2, Ev_3) are shown in the thick solid lines expressed in this new coordinate system. Examining the 3D plot and Figure 5.2a we see that much of the data is clustered in the (Ev_1 and Ev_2), directions with maximum spread in Ev_1, this is confirmed by investigating the 2D-plot in Figure 5.2b where we see how (Ev_1 and Ev_2) capture the majority of the data (spread) and confirm this visualization by the explaining approximately 94% of the data (%var). The extracted component signals applying the *PCA* are shown in Figure 5.3 where the estimated signals are illustrated along with their corresponding power spectra. From this figure, we see that the *PCA* decomposition is able to obtain the 0.123 and 0.345 Hz individual signals but cannot extract the weaker component at 0.49 Hz. We note that *PCA* can be useful in decomposing

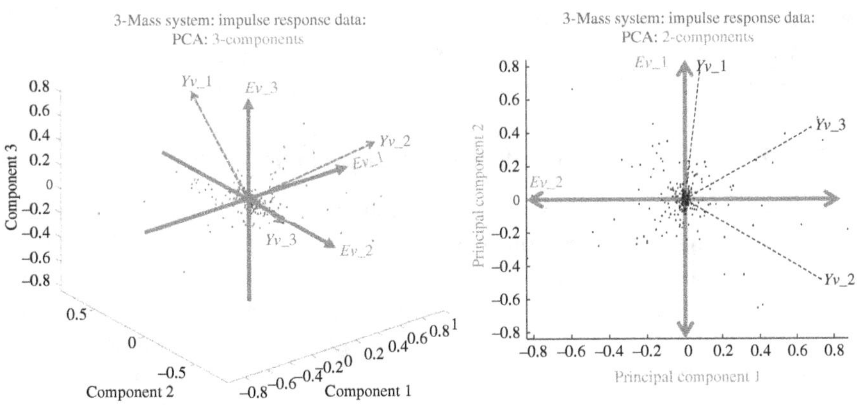

(a) (b)

Figure 5.2 Impulse response data from 3-mass *MCK*-system: (a) 3*D*-scatter plot with data vectors (Yv) and eigenvectors (Ev) indicating spread and directions and (b) 2*D*-scatter plot with data vectors (Yv) and eigenvectors (Ev) indicating spread and directions.

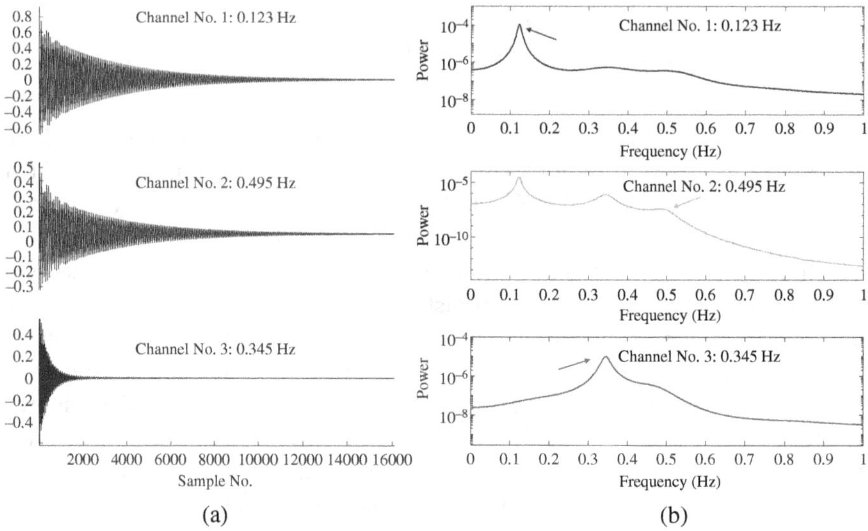

Figure 5.3 Impulse response data from 3-mass *MCK*-system: (a) extracted (separated) channel responses using *PCA* and (b) corresponding power spectra with estimated resonant peaks illustrating potential separations.

the raw data even with little or no prior information and provide some insight into its decomposition and its information content.

△△△

This completes the primary deterministic development of the *PCA* technique; however, there exist a variety of advances in *PCA* research leading to the iterative-*PCA* and probabilistic *PCA* for instance [14, 17]. Next, we consider a similar data-based decomposition approach based on independent signals.

5.2.2 Data Decomposition: Independent Component Analysis (*ICA*)

Noisy measurement data can be corrupted by many external disturbances besides those created by instrumentation. For instance, the well-known cocktail noise problem in signal processing is where one person is speaking at the same time as others. Here the problem is separating the speech of each. Other applications of similar problems are those of separating neuron responses in medical electroencephalogram (EEG) measurements or separating corrupted facial images in image processing or simply extracting individual photos from corrupted multiple exposures in photography. In any case, *ICA* has evolved in a variety of disciplines to attack the signal separation/extraction problem [24–28].

In contrast to the *PCA*, *ICA* has as its primary objective to decompose or separate a combined measurement into individual responses of the underlying system.

In essence, *ICA* is a computational methodology used to extract source signals from a composite measurement containing a mixture of these sources and noise [29–33]. Principal components are focused on maximizing signal variances, while independent components are focused on maximizing signal independence. In fact, the usual first step in *ICA* methods is to perform a *PCA* to "decorrelate" or "whiten" the measured data.

To see this, we start out with what is termed the "mixing model" consisting of known component signals, mixing matrix, and measured data specified by

$$\mathbf{y}(t) = \mathcal{M} \times \mathbf{s}(t) = \sum_{i=1}^{N_s} \mathbf{m}_i \, s_i(t) \tag{5.21}$$

for $\mathbf{y} \in \mathbf{R}^{N_y \times 1}$, $\mathbf{s} \in \mathbf{R}^{N_s \times 1}$, and $\mathcal{M} \in \mathbf{R}^{N_y \times N_s}$—the full-rank, *mixing (coefficient) matrix*.

To recover the individual component channels, it is necessary to invert the non-singular mixing matrix such that

$$\hat{\mathbf{s}}(t) = \underbrace{\mathcal{U}}_{\mathcal{M}^{-1}} \times \mathbf{y}(t) = \sum_{i=1}^{N_y} \mathbf{u}_i \, y_i(t) \tag{5.22}$$

for $\mathcal{U} \in \mathbf{R}^{N_s \times N_y}$—the *UN-mixing matrix* with $\mathcal{M}^{-1}$ the usual matrix inverse, if $N_y = N_s$ or the pseudo-inverse, $\mathcal{M}^{\#} := \left(\mathcal{M}'\mathcal{M}\right)^{-1}\mathcal{M}'$ if $N_y > N_s$.

To motivate the decomposition approach to the *ICA* problem, we investigate the second-order moments of the measurement vector assuming a *zero-mean* process and ignoring *time*, that is,

$$\text{Cov}\,(\mathbf{y}) = \text{E}\{\mathbf{y}\mathbf{y}'\} = \text{E}\{\mathcal{M}\,\mathbf{s}(\mathcal{M}\,\mathbf{s})'\} = \mathcal{M}\,\underbrace{\text{E}\{\mathbf{s}\,\mathbf{s}'\}}_{\text{Cov}\,(\mathbf{s})}\mathcal{M}' \tag{5.23}$$

Further, assuming the sources uncorrelated and have been standardized (normalized), then we have $\text{Cov}\,(\mathbf{s}) = \mathbf{I}$.

Therefore, if we were to perform a decomposition (*SVD*) of the mixing matrix so that

$$\mathcal{M} = U \times \Sigma \times V' \tag{5.24}$$

where U and V are orthogonal rotation (direction) matrices and Σ provides scaling [33–35]. Under this transformation, the measurement covariance becomes (sources standardized)

$$\text{Cov}\,(\mathbf{y}) = \mathcal{M} \times \mathcal{M}' = \left(U\Sigma V'\right) \times \left(U\Sigma V'\right)' = U\,\Sigma^2\,U' \tag{5.25}$$

just a function of the data, U and Σ, *not* "**s**" or "V."

For *ICA*, we seek an unmixing matrix that is the inverse of $\mathcal{M}$, that is,

$$\mathcal{U} = \mathcal{M}^{-1} = \left(U\Sigma V'\right)^{-1} = \underbrace{\left(V'\right)^{-1}}_{V} \times \Sigma^{-1} \times \underbrace{U^{-1}}_{U'} = V \times \Sigma^{-1} \times U' \quad (5.26)$$

that follows since U and V are orthogonal and unitary ($UU' = I$).

Formally, we can now state the source *signal separation/extraction problem* as: GIVEN a set of measurements $\{\mathbf{y}(t)\}; t = 0, 1, \ldots, N_t$ composed of a set of mixed, *independent* source signals, $\{\mathbf{s}(t)\}$, FIND the set of "best" signal estimates of $\{\hat{\mathbf{s}}_i(t)\}; i = 0, 1, \ldots, N_s$, that is, separate/extract the set of optimal individual source signals.

In order to achieve a reasonable solution to this problem certain assumptions about the properties of the underlying source signals are required:

1. *Independent* (statistically) component signals; and
2. *Non-Gaussian* signals.

Statistical independence simply means the joint distribution of the source signals factors into each of the marginal distributions, that is,

$$\Pr[\mathbf{s}(t)] = \Pr[s_1(t), \ldots, s_{N_s}(t)] = \Pr[s_1(t)] \times \cdots \times \Pr[s_{N_s}(t)] \quad \text{[Independence]} \quad (5.27)$$

while non-Gaussian implies that the joint or marginal distributions are not Gaussian

$$\Pr[\mathbf{s}(\mathbf{t})] \nsim \mathcal{N}(\mathbf{M}, \mathbf{\Sigma}) \quad \text{[Non-Gaussian]} \quad (5.28)$$

for mean vector, $\mathbf{M}$, and covariance matrix $\mathbf{\Sigma}$.

Recall that the *Central Limit Theorem* in statistics states that the distribution of the sum of independent random variables (signals) of finite variances tends to be Gaussian; therefore, this theorem *not* being satisfied is a critical requirement for *ICA* to succeed.

For *ICA*, the first step will actually be a *PCA* to "whiten" or standardize the data by first *removing* its mean (centralizing) and then *normalizing* its variance to unity or the identity matrix. That is, no particular principal directions exist leaving the pre-processed data rotationally symmetric illustrating why Gaussian signals are indistinguishable and disabling the independent component decomposition (separation) approach [29, 33].

$$\mathbf{z}(t) = \mathbf{y}(t) - \boldsymbol{\mu}_y \quad \text{for} \quad \mathrm{E}\{\mathbf{s}\} = \mathcal{M}^{-1}\mu_y \qquad \text{[Mean Removal]} \quad (5.29)$$

followed by an eigen-decomposition of the centralized covariance (symmetric) matrix,

$$\mathrm{Cov}\,(\mathbf{z}) = \mathrm{E}\left\{\mathbf{z}\mathbf{z}'\right\} = \mathbf{E} \times \mathbf{\Lambda} \times \mathbf{E}' \qquad \text{[Eigen-Decomposition]} \quad (5.30)$$

Here **E** is the orthogonal eigenvector matrix and **Λ** is a diagonal matrix of eigenvalues, $\{\lambda_i\}; i = 1, \ldots, N_z$; therefore, the "whitened" measurement is obtained by

$$\tilde{\mathbf{z}}(t) = \mathbf{\Lambda}^{-\frac{1}{2}}\mathbf{E}' \times \mathbf{z}(t) = \underbrace{\left(\mathbf{\Lambda}^{-\frac{1}{2}}\mathbf{E}'\right)\mathcal{M}}_{\tilde{\mathcal{M}}} \times \mathbf{s}(t) = \tilde{\mathcal{M}}\mathbf{s}(t) \qquad \text{[Whitening]} \quad (5.31)$$

with $\tilde{\mathcal{M}}$ orthogonal.

Therefore, we have a set of "decorrelated" (normalized) measurements as

$$\mathrm{E}\{\tilde{\mathbf{z}}(t)\} = \mathbf{0} \quad \text{and} \quad \mathrm{E}\left\{\tilde{\mathbf{z}}(t)\tilde{\mathbf{z}}'(t)\right\} = \mathbf{I}$$

Comparing this "whitening" measurement decomposition to the original approach of decomposing the mixing matrix (Eq. 5.24) and inverting (Eq. 5.26) using the *SVD*, we see that performing *ICA* becomes simplified requiring only an estimation of the remaining rotation matrix to provide an estimate of the source signals as

$$\hat{\mathbf{s}}(t) = V \times \tilde{\mathbf{z}}(t) = \underbrace{\left(V\mathbf{\Lambda}^{-\frac{1}{2}}\mathbf{E}'\right)}_{\mathcal{U}}\mathbf{z}(t) \qquad (5.32)$$

We note that since the transformed data $\tilde{\mathbf{z}}$ are now rotationally symmetric, then so is the recovered set of source signals $\{\hat{s}_i(t)\}$ consistent with the original *ICA* assumptions of Eq. (5.23) above.

Perhaps the primary key to *ICA* is non-Gaussian signals, that is, maximizing the non-Gaussian property of the independent source signals yields the required independent components. In fact, the main objective of *ICA* is to estimate the unmixing matrix $\mathcal{U}$ in order to extract the source signals $\hat{\mathbf{s}}(t) = \mathcal{U}\mathbf{z}(t)$, that is, we must find the optimal $\mathcal{U}$ to maximize the non-Gaussian source signals. Therefore, after whitening the data reduce the *ICA* problem to finding the optimal "V", or equivalently, finding the column vectors of $\tilde{\mathcal{M}}$ that maximize this non-Gaussian property corresponding to $\tilde{\mathcal{M}}\mathbf{s}(t)$—an optimization problem [33].

There have been a number of viable approaches in *ICA* to solve this problem leading to a number of measures to quantify non-Gaussian signals such as higher order moments (kurtosis), information-theory metrics (negentropy, mutual information), and estimation (maximum likelihood) to name a few. Each approach defines a criterion (or so-called *contrast function*) to be maximized (or minimized) converging to a solution of the signal separation/extraction problem. Next, we briefly discuss each of these approaches (without proof) and then apply *ICA* to our three-mass problem [36].

5.2.2.1 Higher Order Statistics

The classical metric of statistics to quantify the Gaussian nature of data is based on the fourth-order cumulant—kurtosis [28]. Recall that for a Gaussian distributed random variable the kurtosis is zero, $\mathcal{K}(\cdot) = 0$, while for non-Gaussian variables, $\mathcal{K}(\cdot) > 0$ (super-Gaussian) are spikey with heavy tails (e.g. Laplace distribution) and $\mathcal{K}(\cdot) < 0$ (sub-Gaussian) are flat (e.g. uniform distribution) where

$$\mathcal{K}(z) = \text{E}\{z^4\} - 3 \times \left(\text{E}\{z^2\}\right)^2 \qquad \text{[Kurtosis]} \tag{5.33}$$

Kurtosis as a metric that can theoretically be employed as an optimization criterion, when $\mathcal{K}(z)$ reaches a maximum, then $\mathbf{z}(t) = \mathbf{s}(t)$. This can be achieved when the unmixing matrix is orthogonal (e.g. using *G-S* orthogonalization). A typical implementation utilizes a gradient-based descent technique based on the recursion (see Ref. [30] for details)

$$\mathcal{U}_{i+1} = \mathcal{U}_i - \Delta\text{E}\{\mathbf{z}(t)\left(\mathcal{U}_i'\mathbf{z}(t)\right)^3\} \qquad \text{[Gradient]} \tag{5.34}$$

Unfortunately, kurtosis estimation is sensitive to outliers limiting its robustness as an *ICA* processor [37].

5.2.2.2 Information Theory: Negentropy

Negentropy is a metric that has evolved from information theory, that is, for continuous random variables the *differential entropy* is defined by

$$\mathcal{H}(z) := -\int \Pr[z] \times \log_2 \Pr[z] dz \qquad \text{[Differential Entropy]} \tag{5.35}$$

Gaussian random variables possess the property that they have the largest entropy of all random variables of equal variances [5] leading to this measure of non-Gaussian signals for *ICA*. That is, using a Gaussian as a reference, the estimation criterion is [29, 30]

$$g(z) - \frac{1}{\alpha}\log(\cosh(\alpha z)) \quad \text{for} \quad 1 \le \alpha \le 2; \quad \text{or} \quad g(z) = -e^{-z^2/2}$$

$$\mathcal{J}(z) := \mathcal{H}(\mathbf{z}_{gauss}(t)) - \mathcal{H}(\mathbf{z}(t)) \approx \left(\text{E}\left\{g(\mathbf{z}_{gauss}(t))\right\} - \text{E}\left\{g(\tilde{\mathbf{z}}(T))\right\}\right)^2 \quad \text{[Negentropy]} \tag{5.36}$$

where $\mathbf{z} \sim (0, \mathbf{I})$ and $\tilde{\mathbf{z}}$ is zero mean/unit variance and $g(\cdot)$ is a non-quadratic function [25]. Typical uses for $g(\cdot)$ are

$$g(z) = \log(\cosh(\alpha z)) \quad \text{for} \quad 1 \le \alpha \le 2; \quad \text{or} \quad g(z) = -e^{-z^2/2}$$

Another approximation based on higher-order moments is [25, 32]

$$\mathcal{J}(z) = \frac{1}{12}\left(\text{E}\{z^3\}\right)^2 + \frac{1}{48}\mathcal{K}^2(z) \tag{5.37}$$

which still suffers from robustness issues due to the kurtosis estimation problem as before (see Ref. [30] for more details and techniques).

An approximate negentropy approach, after whitening the data, is given in [30] and is quite similar in structure to the "FastICA" (maximum likelihood) to follow.

5.2.2.3 Information Theory: Mutual Information

One of the major issues in *ICA* is how to determine whether or not one of its basic assumptions is satisfied—the *independence* of the underlying sources. Information theory can provide a reasonable metric to determine independence as well as provide an *ICA* methodology to solve the problem. The major problem then is to determine when a multivariate or "multi" distribution can be factored or equivalently decomposed into a set of independent variables.

Mutual information is defined in terms of the information available in the multivariate source vector **s** and its joint distribution

$$\mathcal{I}\,(\mathbf{s}(t)) := \int \Pr[\mathbf{s}(t)]\log_2\left(\frac{\Pr[\mathbf{s}(t)]}{\prod_{i=1}^{N_s}\Pr[s_i(t)]}\right)d\mathbf{s} \quad \text{[Mutual Information]} \tag{5.38}$$

Clearly, if the joint distribution is factorable, then the ratio is unity and $\mathcal{I}\,(\mathbf{s}(t)) = 0$ indicating independent variates satisfying the *ICA* premise. So, the metric for *ICA* is to *minimize* the mutual information, that is, *iff* the variables are statistically independent. Mutual information is a measure of the dependence between variables equivalent to the Kullbach–Liebler divergence function for independence [26]. Thus, it implies that there is *no* information about each other variate, thereby indicating potential separability.

Returning to the whitened data, we can define the information-related *entropy*, a metric that provides a measure of uncertainty of the underlying distribution, by

$$\mathcal{H}(\tilde{\mathbf{z}}(t)) := -\int \Pr[\tilde{\mathbf{z}}(t)]\log_2\ \Pr[\tilde{\mathbf{z}}(t)]d\tilde{\mathbf{z}} \tag{5.39}$$

Using this metric, the multivariate information is defined in terms of differential entropy as

$$\mathcal{I}(\tilde{\mathbf{z}}(t)) := \sum_i \mathcal{H}(\tilde{z}_i(t)) - \mathcal{H}(\tilde{\mathbf{z}}(t)) \quad \text{[Multivariate Information]} \tag{5.40}$$

Applying these metrics to the estimated source of Eq. (5.32) of the *ICA* problem yields the multivariate information as

$$\mathcal{I}(\hat{\mathbf{s}}(t)) := \sum_i \mathcal{H}(\hat{s}_i(t)) - \mathcal{H}(\hat{\mathbf{s}}(t)) = \sum_i \mathcal{H}(V\tilde{z}_i(t)) - \mathcal{H}(V\tilde{\mathbf{z}}(t)) \tag{5.41}$$

Using the definition of entropy (last term) provides a simplified expression for the mutual information criterion as [29, 33] yielding a reasonable criterion as

$$\mathbf{V} = \arg\min_{V}\ \textstyle\sum_i \mathcal{H}(V\tilde{z}_i(t))$$

which is equivalent to estimating a V that minimizes the sum of $\hat{\mathbf{s}}(t)$ or maximizing their statistical independence [34]. Thus, extracting independent source signals is based on minimizing mutual information in order to locate the most non-Gaussian directions.

To summarize, *ICA* first performs a *PCA* eliminating (whitening) second-order correlations followed by estimating the rotation matrix $\mathbf{V}$ to remove higher-order correlations for source independence [18, 33]. Next, we consider an *ICA*-approach based on estimation theory—the maximum likelihood approach.

5.2.2.4 Estimation Theory: Maximum Likelihood

The estimation theory approach is based on the fundamental maximum likelihood estimator that incorporates a required source signal distribution—in the *ICA* case the source signals are specified by $\Pr[\mathbf{s}(t)]$. The main idea is to estimate the unmixing matrix $\mathcal{U}$ such that [29, 30]

$$\hat{\mathbf{s}}(t) = \mathcal{U}\mathbf{y}(t)$$

Applying the usual likelihood estimator construct of Eq. (4.40) to the *ICA* problem, we obtain [5]

$$\mathcal{L}(\mathcal{U}) := \Pr{}_s\left[\mathcal{U}\ \mathbf{y}(t)\right] \times |\det\ \mathcal{U}| \tag{5.42}$$

The basic maximum likelihood estimation problem for *ICA* is
GIVEN a set of measurements $\{\mathbf{y}(t)\}; t = 0, 1, \ldots, N_t$ composed of a set of mixed, *independent* source signals, $\{\mathbf{s}(t)\}$, FIND the "best" estimate of the unmixing matrix $\hat{\mathcal{U}}$ maximizing the likelihood $\mathcal{L}(\mathcal{U})$ to separate/extract the set of optimal individual source signals, $\{\hat{\mathbf{s}}_i(t)\}; i = 0, 1, \ldots, N_s$.

Since the logarithm is a monotonic function, it preserves the maximum of the likelihood providing the identical result, that is, $\Lambda(\mathcal{U}) := \ln\ \mathcal{L}(\mathcal{U})$

$$\Lambda(\mathcal{U}) = \ln \Pr{}_s\left[\mathcal{U}\ \mathbf{y}(t)\right] + \ln|\det\ \mathcal{U}| = \sum_{i=1}^{M}\sum_{t=1}^{N_t} \ln \Pr{}_s\left[\mathbf{u}_i'\ \mathbf{y}(t)\right] + N_t \ln|\det\ \mathcal{U}| \tag{5.43}$$

A typical distribution model for the source signals is [30]

$$\Pr{}_s\left[\mathcal{U}\ \mathbf{y}(t)\right] = 1 - \tanh\left(\mathbf{s}(t)\right)^2 \tag{5.44}$$

and therefore, the log-likelihood searches for the $\mathcal{U}$ that maximizes

$$\Lambda(\mathcal{U}) = \frac{1}{N_t}\sum_{i=1}^{M}\sum_{t=1}^{N_t} \ln\ 1 - \tanh\left(\mathbf{u}_i'\ \mathbf{y}(t)\right)^2 + \ln|\det\ \mathcal{U}| \tag{5.45}$$

providing a potential solution to the *ICA* problem.

A version of the popular so-called FastICA implementation for a single signal estimate of this (scalar) approach [29] is:

- Select a random weight $\mathbf{u}$
- Let $\hat{\mathbf{u}} = \mathrm{E}\left\{\mathbf{y}(t)g(\mathbf{u}'\mathbf{y}(t))\right\} - \mathrm{E}\left\{g'(\mathbf{u}'\mathbf{y}(t))\right\} \times \mathbf{u}$
- $\mathbf{u} = \frac{\hat{\mathbf{u}}}{\|\hat{\mathbf{u}}\|}$

where $g(u) = -\frac{1}{\alpha}\log(\tanh(\alpha u))$ for $1 \le \alpha \le 2$; or $g(u) = -u\, e^{-u^2/2}$. For a more detailed development and extension of the FastICA methodology to the multivariate problem, see Hyvarinen for details [29, 30].

Next, we apply a *multivariate ICA* technique to the multichannel three-mass *MCK* problem.

Example 5.2 Reconsider the three-mass mechanical system of Example 1.7 where we know there exists three-complex poles governing the system. As in *PCA*, the *ICA* has no a priori knowledge of the underlying *MCK* system. The best we can hope for is to estimate the each of dominant pole responses and their decomposition of the overall data space. We generate 16*K* samples of the *MCK* system ($t = 20$ msec) and provide the deterministic impulse response data to the *ICA* algorithm. We know a priori from the physics and relationship of the spectral response that two to three poles dominate. The *ICA* of these data is shown in Figure 5.4a where a plot of the extracted individual signal vectors is shown along with their corresponding power spectra. From the figure, we see that the *ICA* is able to separate the channel No. 1 and No. 2 (upper and lower plots) as indicated by the strong spectral peaks at 0.12 and 0.34 Hz, but not able to extract the source signal at 0.49 Hz (center plot). This is because this mode is much weaker than the initial two modes and leads to results similar to those of the *PCA* approach for this dataset.

△△△

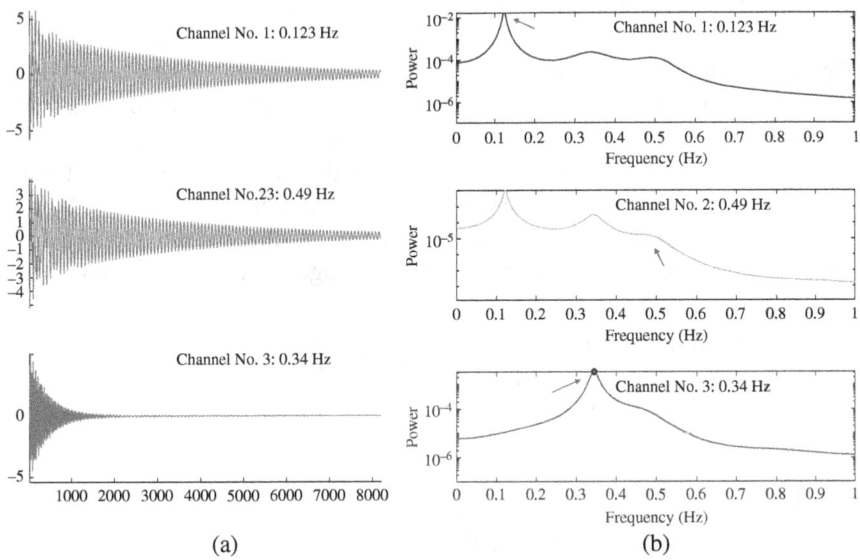

Figure 5.4 Estimated (*ICA*) individual response data from three-mass *MCK*-system: (a) extracted (separated) channel responses using *ICA* and (b) corresponding power spectra with estimated resonant peaks illustrating potential separations.

The *ICA* technique is able to capture the majority of each individual channel source signal along with their corresponding power spectra confirming the signal extraction. Thus, *ICA* can be useful in not only decomposing the raw data even with little or no prior information but also providing some insight into its decomposition and information content. We note in closing that in this section *ICA* is introduced and motivated as another method of decomposing data with little or no a priori information available. An outstanding reference on this approach with detailed mathematics, proofs, and practical algorithms for the interested reader is Ref. [30].

5.2.3 Data Decomposition: Singular Spectral Analysis (*SSA*)

SSA is a nonparametric signal decomposition method employed heavily in time-series analysis based on the eigen-decomposition of a data matrix created by sectioning the signal [22, 38–47]. The methodology decomposes the signal into trends, harmonics (sinusoids), and noise enabling its separation into individual components similar to *ICA*. Trends are the smooth portions of the data with long-term variance, while the harmonic components exhibit periodicities and the remaining uncertainties are captured by the noise component. Its name evolves from the "singular" representation (spectral decomposition) based on the eigen-expansion into eigenvalues and eigenvectors [45–47]. In this section, we develop the *SSA*-technique and apply it to our on-going three-mass problem for comparison.

We start with the signal time series representation (scalar case) and embed it to construct the data matrix termed the *trajectory matrix*. Let the N_t-length dataset be defined by $Y_t := \{y(t)\}, t = 0, \ldots, N_t - 1$ and a sectioning window of length L $(1 < L < N_t)$ such that the data are K-sectioned with $K = N_t - L + 1$. Next, define a set of sections or column vectors, say $\mathbf{Y}_m :=$ $[y(m)\ y(m+1)\ldots,\ y(m+L-1)]' \in \mathrm{R}^{L\times 1}$; $m = 1, \ldots, K$–the so-called *elementary matrices*. Using these sections of *embedded* data (lagged by one-sample) construct the $(L \times L)$-trajectory matrix, $\mathbf{Y}$ as

$$\mathbf{Y} := \underbrace{\begin{bmatrix} y(0) & y(1) & \cdots & y(N_t - L) \\ y(1) & y(2) & \cdots & y(N_t - L - +1) \\ \vdots & \vdots & \vdots & \vdots \\ y(L-1) & y(L) & \cdots & y(N_t - 1) \\ \end{bmatrix}}_{\mathbf{Y}_i;i=0,\ldots,N_t-1} = \begin{bmatrix} \mathbf{Y}_0\ \mathbf{Y}_1 \cdots \mathbf{Y}_{N_t-1} \end{bmatrix} \tag{5.46}$$

a *Hankel matrix* with anti-diagonals defined by elements, $y_{ij} = y_{i+k,j-k}\ \forall k = 0, \ldots, j - i$ assuming $i \leq j$. Note also another variant of the *SSA* approach creates the covariance matrix $\mathbf{C} = \mathbf{Y}' \times \mathbf{Y}$ evolving to a Toeplitz rather than Hankel structure and proceeding with its decomposition; however, the *SVD* of each matrix results in the same *SSA* technique [44].

Once the trajectory matrix is created, it is decomposed by the *SVD* and can be expressed in terms of its elementary matrices

$$\mathbf{Y} = \mathbf{U} \times \boldsymbol{\Sigma} \times \mathbf{V}' = \sum_{i=0}^{N_\sigma-1} \mathbf{Y}_i = \sum_{i=0}^{N_\sigma-1} \underbrace{\sigma_i \mathbf{u}_i \mathbf{v}_i'}_{\mathbf{Y}_i} \tag{5.47}$$

where $\mathbf{U} \in \mathrm{R}^{L\times L}$ is the unitary matrix consisting of the set of orthonormal left singular vectors, $\mathbf{V} \in \mathrm{R}^{K\times K}$ the unitary matrix consisting of the set of orthonormal right singular vectors with $\boldsymbol{\Sigma} \in \mathrm{R}^{L\times K}$ the rectangular diagonal matrix consisting of the set of singular values, $\{\sigma_i\}$; $i = 1, \ldots, N_\sigma$ for $N_\sigma < L$ the rank of the trajectory matrix. Here in *SSA*, the set of singular values and vectors are called the *eigen-triple*, $\{\sigma_i, \mathbf{u}_i, \mathbf{v}_i\}$.

It is interesting to note that the columns of the left singular matrix $\mathbf{U}$ are an orthonormal basis characterizing the column data (time series) of the trajectory matrix. Likewise, the rows of the right singular matrix $\mathbf{V}$ are an orthonormal basis characterizing the corresponding row data (time series) of the trajectory matrix and the set of ordered singular values $\{\sigma_i\}$ are scale factors specifying the relative importance of the ith eigen-triple corresponding to the elementary matrix, $\mathbf{Y}_i$ [46, 47].

The singular values can be used to determine the rank of the trajectory matrix. The so-called *best rank approximation* is calculated by the ratio of the summation of singular values up to the selected order over all the values, that is, β (percentage) based on the ratio of singular values, [4]

$$\beta(\hat{N}_\sigma) := \left(\frac{\sum_{k=1}^{\hat{N}_\sigma} \sigma_k}{\sum_{k=1}^{K} \sigma_k} \right) \times 100 \tag{5.48}$$

$\beta(\%)$ is the percentage of the original trajectory matrix (Hankel) approximated by choosing $\hat{N}_\sigma$. A threshold, τ_x can be selected to search over k for that particular $\hat{N}_x$ that "best" approximates $\mathbf{Y}$ up to the threshold, that is,

$$\beta_n(\hat{N}_\sigma) \geq \tau_\sigma(\%) \qquad \text{for} \qquad n = 1, \ldots, \hat{N}_\sigma \tag{5.49}$$

After decomposing the trajectory matrix and determining its rank (N_σ), the next step in *SSA* is to "approximately" reconstruct the original data based on a subset of elementary $L \times K$-matrices that capture the decomposition. To extract the signal (time series) from the data (trajectory) matrix assuming separability, the elementary matrices can be grouped, say N-groups, into the various disjoint data components (e.g. trends, harmonics, etc.) given by

$$\mathcal{Y} := \bigcup_{n \in \mathcal{Y}} \mathcal{Y}_n = \mathcal{Y}_0 \cup \cdots \cup \mathcal{Y}_{N_\cup - 1} \tag{5.50}$$

such that the various sets of grouped elementary matrices $\{\mathbf{Y}(n)\}$ combine to yield

$$\mathbf{Y} = \sum_{n=0}^{N_U - 1} \mathbf{Y}(n) \tag{5.51}$$

The grouped elementary matrices are then transformed to have a Hankel matrix structure $\mathbf{Y}(n) \overbrace{\rightarrow}^{\mathcal{H}} \tilde{\mathbf{Y}}(n)$ for $\mathcal{H}$ the Hankel operator [47].

Once the Hankel transformation is accomplished, the *SSA* employs a *diagonal averaging* method by averaging the anti-diagonals of the grouped Hankel matrices $\tilde{\mathbf{Y}}$. The (m, n)th element of $\tilde{\mathbf{Y}}$ is obtained for $\ell = m + n$

$$\tilde{y}_{m,n} = \begin{cases} \frac{1}{\ell+1} \sum_{k=0}^{\ell} y_{k,\ell-k} & 0 \le \ell \le L-1 \\ \frac{1}{L-1} \sum_{k=0}^{L-1} y_{k,\ell-k} & L \le \ell \le K-1 \\ \frac{1}{K+L-\ell-1} \sum_{k=\ell-K+1}^{L} y_{k,\ell-k} & K \le \ell \le K+L-2 \end{cases} \tag{5.52}$$

This calculation of $\tilde{y}_{m,n}$ is simply averaging the elements of the anti-diagonals of $\tilde{\mathbf{Y}}$ based on the location of m and n leading to the *decomposition* of the rank N_σ-trajectory matrix, that is,

$$\mathbf{Y} = \sum_{i=0}^{N_\sigma} \tilde{\mathbf{Y}}(i) \tag{5.53}$$

Defining the groupings relies on the observation of the individual sections as trends, harmonics, and noise. Rather than this approach, a more formal quantization metric has evolved based on the so-called weighted-correlation method (see Refs. [39, 47] for more details). In any case, this decomposition technique relies primarily on the window length L and rank of the trajectory matrix N_σ to parametrically adapt the algorithm to the time series data that can be validated by the quality of the resulting estimates produced by the reconstruction.

We summarize the steps of the *SSA* method as:

- *Construct* the trajectory (Hankel) matrix $\mathbf{Y}$ with selected window length L as in Eq. (5.46)
- *Decompose* $\mathbf{Y}$ using a *SVD* to obtain the decomposition in Eq. (5.47)
- *Best Rank* using the extracted singular values estimate the rank of $\mathbf{Y}$ in Eq. (5.48)
- *Construct* the set of elementary matrices $\{\mathbf{Y}_i\}$ from the eigen-triple $\{\sigma_i, \mathbf{u}_i, \mathbf{v}_i\}$ as in Eq. (5.50)
- *Group* the set of elementary matrices and perform a Hankel transform of each to obtain $\{\tilde{\mathbf{Y}}_i\}$
- *Diagonally Average* the set of grouped elementary matrices $\{\tilde{\mathbf{Y}}\}$ as Eq. (5.52)

- *Reconstruct* or approximate the data based on the decomposed, diagonally averaged, grouped matrices as specified by Eq. (5.53)

Next, we apply the *SSA* technique to the three-mass problem.

Example 5.3 Again we reconsider the three-mass mechanical system of Example 1.7 where we know there exist three-complex poles governing the system. As in *PCA* and *ICA*, no a priori knowledge of the underlying *MCK* system is assumed. The best we can hope for is to estimate responses and their decomposition of the overall data space. As before. we generated turn 16*K* samples of the *MCK* system ($t = 20$ msec) and provide the *average* deterministic impulse response data to the *SSA* algorithm. We know from the physics and the relationship of the spectral response that two to three poles dominate. The application of the *SSA* technique to these data is shown in Figure 5.5a where a plot of the extracted individual signal components is shown along with their corresponding power spectra in Figure 5.5b. From the figure, we see that the estimated responses

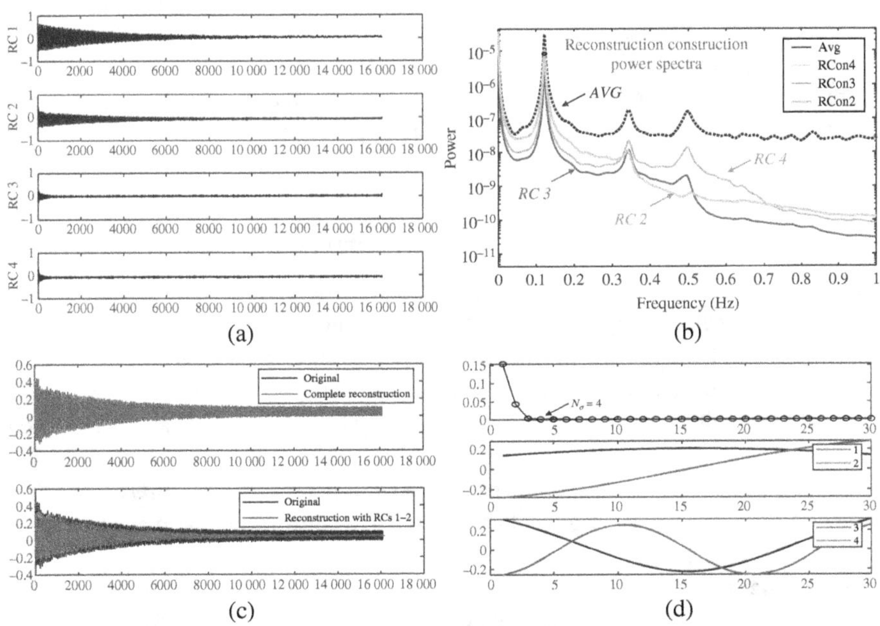

Figure 5.5 *SSA* processing of 3-mass *MCK*-system data: (a) decomposed (reconstructed) component time series: *RC*1–*RC*4, (b) decomposed component power spectra: average (*RC*1) and *RC*2–*RC*4, (c) time series estimation (reconstruction): total components and *RC*1/*RC*2 estimates, and (d) singular values order estimation with $N_\sigma = 4$ shown and *RC*1/*RC*2 periodic/harmonic decomposition components.

capture the majority of the spectrum based on the approximate reconstruction of the first four grouped signals. Comparing the estimated data (average) using all available reconstructed components shows an almost perfect reconstruction. Here the time series estimation is limited to the first two components ($RC1/RC2$) to illustrate the reasonable approximation. The actual estimated order is shown in the singular values plot and both the trends and harmonic estimates in Figure 5.5d. Although a somewhat different approach to signal "decomposition," the *SSA* is a popular methodology in areas of climatology, economics, and biomedical applications [40, 42, 44].

$\triangle\triangle\triangle$

5.3 Spectral-Based Decompositions

The decomposition of spectra into individual frequency bins is inherent in basic Fourier techniques; however, other methods have evolved usually employing a variety of methods based on selecting sets of orthogonal basis functions and ranging to parametric schemes directly aimed at fitting harmonics to data [48]. In this section, we start with a classical spectral estimation approach that decomposes data by fitting sets of orthogonal spheroidal functions followed by more parametric techniques that employ robust numerical methods to decompose the data, similar to principal component theory, but based on a priori assumption of an underlying class of models to be fit.

5.3.1 Spectral Decomposition: Multitaper Method (*MTM*)

The Multitaper Method (*MTM*) is a classical spectral estimator that decomposes data and reduces bias by obtaining statistically independent estimates that can effectively be averaged to reduce uncertainty, similar to the *WPM* method of Section 2.5. Each window of *MTM* is pairwise *orthogonal* to all other windows enabling the statistically independent set of spectral estimates that are averaged (weighted) to provide the final spectrum. The orthogonal windows are *Slepian* discrete prolate spheroidal sequences (*DPSS*) that possess the required properties such as minimizing spectral leakage, while concentrating the power in tight bands [48–51].

Recall that frequency resolution is specified by the number of samples, N, and the corresponding sampling interval Δt determined by the highest frequency supported by the data, that is, the sampling frequency $f_s = \frac{1}{\Delta t}$ with corresponding Nyquist frequency, f_{NYQ}. Therefore, the *frequency resolution*, Δf is

$$\Delta f := \frac{f_s}{N} = \frac{1}{N\Delta t} \quad \text{and} \quad f_{NYQ} = \frac{f_s}{2} \tag{5.54}$$

yielding the minimal frequency band that distinct sinusoids can be resolved [50].

MTM estimation, in its simplest form, is given by

$$\hat{S}_{MTM}(\Omega) = \frac{1}{K}\sum_{k=0}^{K-1}\hat{S}_{MTM}(\Omega, k) \tag{5.55}$$

and

$$\hat{S}_{MTM}(\Omega, k) = \frac{\Delta t}{K}\sum_{k=0}^{K-1} = \Big|\underbrace{\mathcal{W}_k(t)}_{DPSS}\, y(t)e^{-j\Omega t}\Big|^2 \tag{5.56}$$

with $\mathcal{W}_k(t)$ the kth *DPSS* of length N or window associated with the corresponding eigen-spectrum $\hat{S}_{MTM}(\Omega, k)$ [48–51]. The *spectral* window

$$\mathcal{W}_k(\Omega) = \Delta t\sum_{t=0}^{N-1}\mathcal{W}_k(t)e^{-j\Omega t} \tag{5.57}$$

is given by

$$\hat{\hat{S}}_{MTM}(\Omega) = E\left\{\hat{S}_{MTM}(\Omega)\right\} = \frac{1}{K}\int_{-f_{NYQ}}^{f_{NYQ}} \overline{\mathcal{W}}_k(\Omega - \Omega')\hat{S}_{MTM}(\Omega')d\Omega' \tag{5.58}$$

for

$$\overline{\mathcal{W}}_k(\Omega) = \frac{1}{K}\sum_{k=0}^{K-1}\mathcal{W}_k(\Omega)$$

For *MTM*, the windows are the orthogonal set of *DPSS* $\left\{\mathcal{W}_k(t)\right\}, k = 0, 1, \ldots, K-1; t = 0, 1, \ldots, N-1$ that are solutions of the so-called concentration problem.[1] Here lies the *key* to the *MTM*—the design of spectral windows that solve the concentration problem, that is,

Among ALL window sequences $\left\{\mathcal{W}_k\right\}$ for a given duration $N\Delta t$ and pre-specified bandwidth $\left[-f_{BW}, f_{BW}\right]$, FIND the sequence that maximizes the spectral concentration within this band or equivalently obtain the optimal sequence that minimizes the side-lobe energy *outside* the band.

That is, find a sequence such that the *spectral concentration* defined by $\lambda(N\Delta t, f_{BW})$ on the interval $\left[-f_{BW}, f_{BW}\right]$ is maximum where

$$\lambda(N\Delta t, f_{BW}) := \frac{\int_{-f_{BW}}^{f_{BW}} |\mathcal{W}(\Omega)|^2 d\Omega}{\int_{-f_{NYQ}}^{f_{NYQ}} |\mathcal{W}(\Omega)|^2 d\Omega} \tag{5.59}$$

then, FIND $\left\{\mathcal{W}_k(t)\right\}, k = 0, 1, \ldots, K-1; t = 0, 1, \ldots, N-1$ such that $\lambda(N\Delta t, f_{BW})$ is a maximum.

1 The spectral concentration problem consists of finding a finite length window sequence whose spectrum is most localized in a given frequency interval specified by $\pm f_{BW}$.

The solution to this constrained optimization problem is available in [48–51] and becomes an eigenvalue problem defined by:

$$\mathcal{D}(f_{BW}) \times \underline{\mathcal{W}}(t') = \underbrace{\lambda(N\Delta t, f_{BW})}_{\text{eigenvalues}} \times \underbrace{\underline{\mathcal{W}}(t)}_{\text{eigenvectors}} \tag{5.60}$$

where the matrix components are $\mathcal{D}_{t,t'} = \frac{\sin 2\pi f_{BW}(t-t')}{\pi(t-t')} \in \mathcal{R}^{N \times K}$.

Since this matrix $\mathcal{D}$ is positive definite, the largest eigenvalue corresponds to the largest spectral concentration and therefore the corresponding eigenvector sequence, $\underline{\mathcal{W}}_0(t)$ is termed the 0th order *DPSS* (Slepian) [51]. The number of dominant eigenvalues of $\mathcal{D}$ are all close to unity and correspond to

$$\mathcal{S} := 2f_{BW} \times N\Delta t \qquad \text{[Shannon Number]} \tag{5.61}$$

with $\mathcal{S}$ defined as the *Shannon number* [50] with $\lambda_0 > \lambda_1, \ldots, \lambda_{K-1}$ where the Kth-order window offers the best side-lobe suppression. In fact, the first $\mathcal{S} - 1$ eigenvalues are close to unity indicating relatively *low* side-lobe energy (leakage).

We illustrate the calculation of the *DPSS* windows in Figure 5.6a where the first five functions are shown along with their corresponding spectra and functional

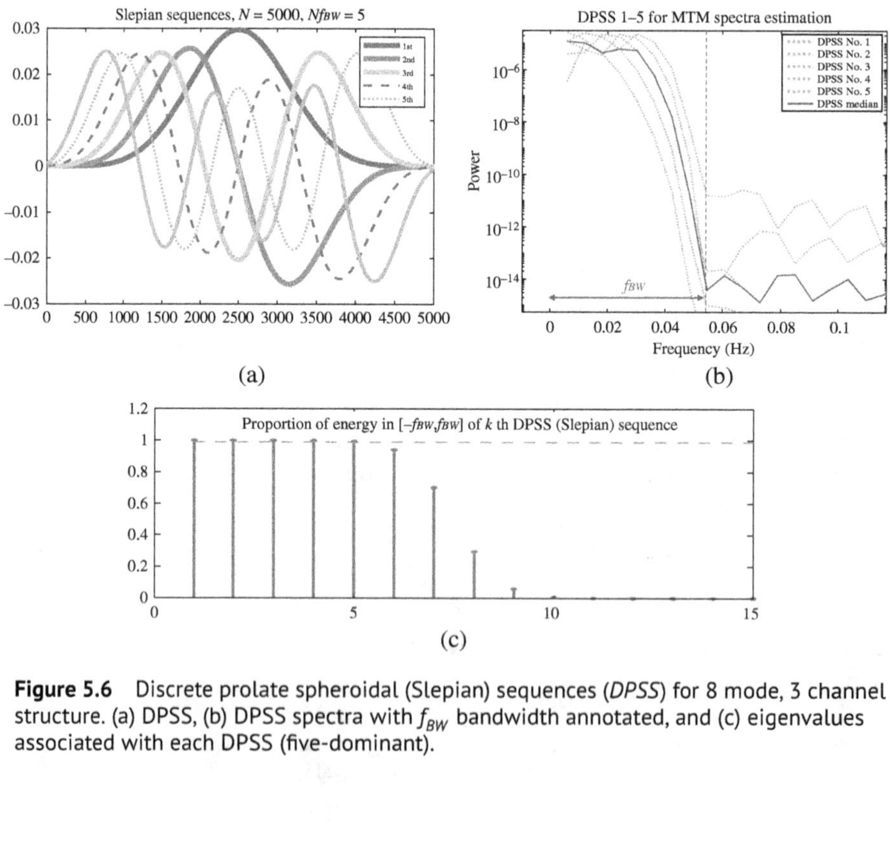

Figure 5.6 Discrete prolate spheroidal (Slepian) sequences (*DPSS*) for 8 mode, 3 channel structure. (a) DPSS, (b) DPSS spectra with f_{BW} bandwidth annotated, and (c) eigenvalues associated with each DPSS (five-dominant).

bandwidth in Figure 5.6b. An application to a synthesized eight-mode mechanical system demonstrates the corresponding *DPSS* eigenvalues in descending order in Figure 5.6c. By imposing a unity threshold, it is clear that the "optimal" *number* of *DPSS*s or equivalently *order* is 5, since their values are either at or above 1. This represents a convenient method for selection.

In the *MTM*, the length *N*-*DPSS* windows of orders $k = 0, 1, \ldots, (K-1)$ are usually applied in practice with $K \leq S - 1$. *Fixed* N and f_{BW}, then lead to smaller values of K generally reduce leakage with larger values reducing variance. *Leakage* can be approximated by

$$\mathcal{L} = \frac{1}{K}\sum_{k=0}^{K-1}\left(1 - \lambda_k(N\triangle t, f_{BW})\right) \qquad \text{[Leakage]} \tag{5.62}$$

and the corresponding *variance* is inversely proportional to the number of windows employed

$$\text{var} = \frac{1}{K} \qquad \text{[Variance]} \tag{5.63}$$

with frequency *resolution* specified by the design bandwidth, that is,

$$\text{res} = \triangle f = f_{BW} \qquad \text{[Resolution]} \tag{5.64}$$

Also, the *time-bandwidth* product is given by

$$\mathcal{T}_{BW} := \frac{S}{2} := f_{BW} \times N\triangle t \tag{5.65}$$

It is this expression coupled with the number of windows that define the variance of *MTM* estimates. Choosing $\mathcal{T}_{BW}$ and K provide a trade-off of spectral resolution, bias, and variance. Bias is primarily controlled by the largest eigenvalue, $\lambda_0(N\triangle t, f_{BW})$ since

$$\lambda_0(N\triangle t, f_{BW}) \approx 1 - 4\pi\sqrt{\mathcal{T}_{BW}}e^{-2\pi\mathcal{T}_{BW}} \qquad \text{[Total Side Lobe Energy]} \tag{5.66}$$

the leakage outside $[-f_{BW}, f_{BW}]$ that decreases rapidly with increasing $\mathcal{T}_{BW}$ [52–55].

We summarize these properties below in Table 5.1.

Table 5.1 MTM LEAKAGE (Window) properties.

MTM DPSS-window (Slepian) leakage properties		
Time-bandwidth ($\mathcal{T}_{BW}$)	**Asymptotic leakage ($1 - \lambda_0$)**	**Leakage (dB)**
4	3.05×10^{-10}	−95
6	1.31×10^{-15}	−149
8	5.26×10^{-21}	−203
10	2.05×10^{-26}	−257

Source: Adapted from [54].

So we see that it is in fact the optimal window design employing the *DPSS* that enables the solution to the spectral concentration problem. Therefore, in summary, the major *properties* of the *MTM* are:

- *Bias* is *local* characterized by the window main lobe width (resolution) and *broadband* characterized by the side lobes (leakage);
- *Local Bias* is *small* if the spectral window for $\hat{S}_{MTM}(\Omega)$ closely approximates $\frac{1}{2\times f_{BW}}$;
- *Broadband Bias* is *bounded* $\left(1-\frac{1}{K}\sum_{k=0}^{K-1}\lambda_k(N\Delta t, f_{BW})\right)$ for $K-$ *DPSS* windows;
- *Variance* is controlled by the *number* of *DPSS* windows given by $var \approx \frac{1}{K}$;
- *Resolution* is specified (a priori) by *twice* the design *bandwidth* of $2\times f_{BW}$;
- *Mean-Squared Error* is *negligible*, if $k \leq S-1$ and $\{\lambda_k)(N\Delta t, f_{BW}) \approx 1$;
- *MTM* is a *direct method* of spectral estimation; and
- *MTM* can be applied to problems with *missing* or *irregularly-spaced* datasets.

The *MTM* technique can be summarized as:

MULTITAPER METHOD (*MTM*) Spectral Decomposition:

1. *Select* the frequency *resolution*: $f_{BW} < f_{NYQ}$ $Res = 2\times f_{BW}$;
2. *Calculate* the time–bandwidth product (Shannon number): $S = 2f_{BW}\times N\Delta t$;
3. *Calculate* the *number* of *DPSS* functions (eigenvectors): $K \leq S-1$;
4. *Calculate* the *DPSS* weights: $\mathcal{W}_k(t), k = 0, 1, \dots, (K-1); t = 0, 1, \dots, (N-1)$;
5. *Calculate* the individual eigen-spectra: $\hat{S}_{MTM}(\Omega, k) = \sum_{t=0}^{N-1}\mathcal{W}_k(t)y(t)e^{-j\Omega t}$;
6. *Estimate* the final *MTM* spectrum (averaging): $\hat{S}_{MTM}(\Omega) = \sum_{k=0}^{K-1}\hat{S}_{MTM}(\Omega, k)$;

Next, we investigate extensions of the *MTM* spectral estimator using these Slepian sequences.

A reasonable *unbiased MTM* spectral estimate (*eigenvalue weighting method*) can be scaled by the associated eigenvalues such that

$$\hat{S}_{MTM}(\Omega) = \frac{\sum_{k=1}^{K-1}\lambda_k(N\Delta t, f_{BW})\times|Y_k(\Omega)|^2}{\sum_{k=1}^{K-1}\lambda_k(N\Delta t, f_{BW})} \tag{5.67}$$

the denominator ensures an unbiased estimate, while the individual eigenvalues (<1) decrease the weighting of the larger spectra in the final estimate, since $\lambda_0 > \lambda_1 > \cdots > \lambda_K < 1$. That is, as the number of windows increase ($K\uparrow; \lambda\downarrow$), the corresponding eignenvalues decrease, mitigating more spectral leakage.

Another extension to the *MTM* proposed by Thomson is *adaptive weighting MTM* where a set of adaptive weights $\{\alpha_k(\Omega)\}$ is employed to downweight the higher-order spectra using an iterative approach [50]. Here a weighted average of the eigen-spectra from individual *DPSSs* is performed leading to an iterative technique that adapts (adjusts) the weights for each band to provide a balance between broadband bias ($\mathcal{L}_k$) and the variance of the estimates. The following set of relations provide the iterations (over i):

$$\hat{S}_{MTM}(\Omega, i+1) = \frac{\sum_{k=0}^{K-1} |\alpha_k(\Omega, i)|^2 \times \hat{S}_{MTM}(\Omega, k; i)}{\sum_{k=0}^{K-1} |\alpha_k(\Omega, i)|^2}$$

$$\alpha_k(\Omega, i) = \frac{\sqrt{\lambda_k(N\Delta t, f_{BW})} \times S(\Omega, i)}{\lambda_k(N\Delta t, f_{BW})S(\Omega, i) + \mathcal{L}_k} \tag{5.68}$$

where $S(\Omega, 0)$ is the initial estimate of the power spectrum and $\mathcal{L}_k$ is the leakage that the kth *DPSS* bank has at frequency Ω. The technique iterates over i converging to the desired solution completing the spectral estimation.

5.3.2 Spectral Decomposition: Subspace Method

The fundamental subspace approach to spectral decomposition is to decompose either the data or associated covariance matrix into two disjoint subspaces: one for the signal and the other for noise. The technique is to first perform the associated matrix decomposition and identify the individual subspaces by determining their rank. Once extracted, the signal or the noise subspace is utilized to perform the spectral estimation. Here, we first investigate the subspace decomposition using the *SVD* and then extract the spectral based on an assumed harmonic model. With the signal and noise subspaces defined in terms of a *SVD* decomposition of the underlying measurement covariance matrix, the harmonic signal representation is used to estimate the power spectrum.

The *complex harmonic model* can be defined by

$$s(t) = \sum_{i=1}^{N_s} A_i e^{j\Omega_i t + \phi_i} \tag{5.69}$$

where A_i is complex, $A_i = |A_i|e^{-j\theta}$; ϕ_i is random and uniformly distributed as $\phi \sim \mathcal{U}(-\pi, \pi)$ and Ω_i is the harmonic frequency. The set of N_s-complex amplitudes and harmonic frequencies, $\{A_i, \Omega_i\}$, $i = 1, \ldots, N_s$ is assumed deterministic, but unknown.

The corresponding *measurement model* is characterized by the complex harmonic signal in additive random noise as

$$y(t) = s(t) + n(t) \tag{5.70}$$

with $n(t)$, a zero-mean, random sequence with variance, σ_n^2. Therefore, since the signal is uncorrelated with the noise, the autocorrelation function is given by

$$R_{yy}(k) = R_{ss}(k) + R_{nn}(k) = \sum_{i=1}^{N_s} P_i e^{j\Omega_i k} + \sigma_n^2 \quad \text{for} \quad P_i := |A_i|^2 \tag{5.71}$$

Expanding this relation to incorporate the N_s-harmonics, we obtain

$$R_{yy}(k) = \begin{cases} \sum_{i=1}^{N_s} P_i + \sigma_n^2 \delta(k) & k = 0 \\ \sum_{i=1}^{N_s} P_i e^{j\Omega_i k} & k \neq 0 \end{cases} \tag{5.72}$$

We apply the *power method* to develop this time-domain spectral decomposition, that is, the cost function is based on maximizing the output power in the signal where the resulting function is related to its power spectrum [56]. All peaks of the spectrum locate the decomposed harmonic frequencies of the underlying model.

Performing the expansion of Eq. (5.72) over k, we obtain

$$R_{yy}(N) := R_{ss} + R_{nn} = V(\Omega) P V^{\dagger}(\Omega) + \sigma_n^2 I \tag{5.73}$$

where $V(\Omega) \in C^{N \times N_s}$ is the harmonic signal matrix with corresponding power matrix, $P \in C^{N_s \times N_s}$ and $P = \text{diag}\left[P_1 \ \cdots \ P_{N_s}\right]$.

The *signal matrix* is constructed from the harmonic model as

$$V(\Omega) = \left[\mathbf{v}_1(\Omega) \mid \cdots \mid \mathbf{v}_{N_s}(\Omega)\right], i = 1, \ldots, N_s \ \text{ with } \mathbf{v}_i(\Omega) := \left[1 \ e^{j\Omega_i} \ \cdots \ e^{j(N-1)\Omega_i}\right]' \tag{5.74}$$

Therefore, the output measurement covariance matrix can be *decomposed* into an constituent eigenvalue (Λ)-eigenvector (E) system as

$$R_{yy}(N) = E \, \Lambda \, E^{\dagger} \tag{5.75}$$

or substituting the harmonic model, we have

$$R_{yy}(N) = E \begin{bmatrix} \lambda_1 & & 0 \\ & \ddots & \\ 0 & & \lambda_N \end{bmatrix} E^{\dagger} = V(\Omega) \begin{bmatrix} \lambda_1^s & & & 0 \\ & \ddots & & \\ & & \lambda_{N_s}^s & \\ 0 & & & 0 \end{bmatrix} V^{\dagger}(\Omega) + \sigma_n^2 \mathbf{I} \tag{5.76}$$

which follows from the fact that there are N_s-harmonic signals. Therefore, we have that $\lambda_1 > \lambda_2 > \cdots > \lambda_N$ and

$$\lambda_i = \begin{cases} \lambda_i^s + \sigma_n^2 & i = 1, \ldots, N_s \\ \sigma_n^2 & i = N_s + 1, \ldots, N \end{cases} \tag{5.77}$$

The output covariance can also be expressed in partitioned form as

$$R_{yy}(N) = \left[E(N_s) \mid E(N - N_s)\right] \begin{bmatrix} \Lambda(N_s) & | & 0 \\ --- & | & --- \\ 0 & | & \Lambda(N - N_s) \end{bmatrix} \begin{bmatrix} E^{\dagger}(N_s) \\ --- \\ E^{\dagger}(N - N_s) \end{bmatrix} \tag{5.78}$$

where both $E(N_s) \in C^{N \times N_s}$ and $E(N - N_s) \in C^{N \times (N-N_s)}$ are eigenvector matrices. Multiplying the partitions gives the relation

$$R_{yy}(N) = E(N_s)\Lambda(N_s)E^{\dagger}(N_s) + E(N - N_s)\Lambda(N - N_s)E^{\dagger}(N - N_s) \tag{5.79}$$

or

$$R_{yy}(N) = \begin{bmatrix} \mathbf{e}_1 & \cdots & \mathbf{e}_{N_s} \end{bmatrix} \begin{bmatrix} \lambda_1^s + \sigma_n^2 & & 0 \\ & \ddots & \\ 0 & & \lambda_{N_s}^s + \sigma_n^2 \end{bmatrix} \begin{bmatrix} \mathbf{e}_1^{\dagger} \\ \vdots \\ \mathbf{e}_{N_s}^{\dagger} \end{bmatrix}$$
$$+ \begin{bmatrix} \mathbf{e}_{N_s+1} & \cdots & \mathbf{e}_N \end{bmatrix} \begin{bmatrix} \sigma_n^2 & & 0 \\ & \ddots & \\ 0 & & \sigma_n^2 \end{bmatrix} \begin{bmatrix} \mathbf{e}_{N_s+1}^{\dagger} \\ \vdots \\ \mathbf{e}_N^{\dagger} \end{bmatrix} \tag{5.80}$$

which implies the general *spectral decomposition* as

$$R_{yy}(N) = \sum_{i=1}^{N_s} \left(\lambda_i^s + \sigma_n^2\right) \mathbf{e}_i \mathbf{e}_i^{\dagger} + \sum_{i=N_s+1}^{N} \sigma_n^2 \mathbf{e}_i \mathbf{e}_i^{\dagger} \tag{5.81}$$

From this *decomposition*, it is possible to define various subspaces as:

1. The *signal subspace* is defined as the N_s-dimensional subspace spanned by the *signal eigenvectors*, $\{\mathbf{e}_i\}$; $i = 1, \ldots, N_s$, corresponding to the N_s-largest eigenvalues.
2. The *noise subspace* is defined as the $(N - N_s)$-dimensional subspace spanned by the *noise eigenvectors*, $\{\mathbf{e}_i\}$; $i = N_s + 1, \ldots, N$, corresponding to the remaining $(N - N_s)$-eigenvalues.
3. The *signal* and *noise subspaces* are *orthogonal*.

Applying this decomposition to the harmonic estimation problem, we see that the *harmonic signal vectors*, $\mathbf{v}_i(\Omega)$ lie in the signal subspace spanned by the signal eigenvectors, $\{\mathbf{e}_i\}$; $i = 1, \ldots, N_s$, and the corresponding *noise vectors* lie in the orthogonal noise subspace spanned by $\{\mathbf{e}_i\}$; $i = N_s + 1, \ldots, N$. Thus, the harmonic signal vectors lie in the subspace spanned by the first N_s-eigenvectors of $R_{yy}(N)$ and are *orthogonal* to each noise eigenvector, that is,

$$\mathbf{v}_i^{\dagger}(\Omega)\mathbf{e}_j = 0 \quad \text{for} \quad i = 1, \ldots, N_s; \quad j = N_s + 1, \ldots, N \tag{5.82}$$

which also implies that

$$V^{\dagger}(\Omega)\mathbf{e}_j = 0 \quad \text{for} \quad j = N_s + 1, \ldots, N \quad \text{and} \quad V(\Omega) = \begin{bmatrix} \mathbf{v}_1(\Omega) \mid \cdots \; \mathbf{v}_{N_s}(\Omega) \end{bmatrix} \tag{5.83}$$

A harmonic frequency *power estimator* [57] follows directly from the orthogonality property of Eq. (5.82) as

$$P(\Omega) := \frac{1}{\sum_{j=N_s+1}^{N} \left|\mathbf{v}^{\dagger}(\Omega)\mathbf{e}_j\right|^2} \quad \text{for} \quad P(\Omega) \longrightarrow \infty$$
$$\text{whenever} \quad \Omega = \Omega_i, \quad i = 1, \ldots, N_s \tag{5.84}$$

where we have defined the *frequency vector*, $\mathbf{v}(\Omega)$, to represent the corresponding temporal Fourier frequencies.

Thus, if we calculate the power estimator over a set of frequencies, $\{\Omega\}$, then whenever this estimator passes through a harmonic frequency, $\Omega = \Omega_i$, a *peak* will occur in the power function. Simple peak detection then enables the estimation of the desired harmonic frequencies. An alternative to peak picking is to create the corresponding characteristic polynomial and extracting its roots (poles) directly [48].

5.3.3 Spectral Decomposition: Pisarenko Harmonic Decomposition (*PHD*) Method

The Pisarenko Harmonic Decomposition (*PHD*) method is based on the assumption that the *number* of N_s-harmonics in white additive noise are *known* a priori and it is desired to extract them from the data. It is further assumed that the dimension of the noise subspace is unity so that the data length is $N = N_s + 1$. Thus, the N_s largest eigenvalues correspond to the signal eigenvectors, $\mathbf{e}_i$, $i = 1, \ldots, N_s$ and the smallest or minimum eigenvalue $(N_s + 1)$ corresponds to the noise eigenvector, $\mathbf{e}_{N_s+1}$ with $\lambda_{\min} = \sigma_N^2$. The signal eigenvectors are orthogonal to the noise eigenvector, that is,

$$\mathbf{v}_i(\Omega) \perp \mathbf{e}_{N_s+1} = \mathbf{v}_i^H(\Omega)\mathbf{e}_{N_s+1} = 0 \tag{5.85}$$

which can also be written as the *DtFT* relation of Eq. (5.84) to obtain $E_{\min}(\Omega)$. Using the power method, we have that

$$P_{PHD}(\Omega) := \frac{1}{|\mathbf{v}^H(\Omega)\mathbf{e}_{N_s+1}|^2} = \frac{1}{|E_{\min}(\Omega)|^2} \quad \text{for}$$
$$P(\Omega) \longrightarrow \infty, \quad \Omega = \Omega_i, \quad i = 1, \ldots, N_s \tag{5.86}$$

with the peaks corresponding to a set of harmonic frequencies or equivalently finding the roots of Eq. (5.86).

The corresponding power coefficients are then found by solving a set of linear equations developed from the eigen-equations

$$R_{yy}\mathbf{e}_i = \lambda_i \mathbf{e}_i, \quad i = 1, \ldots, N_s \tag{5.87}$$

Multiplying both sides by $\mathbf{e}_i^H$ and substituting Eq. (5.73), we have that

$$\mathbf{e}_i^H R_{yy}\mathbf{e}_i = \mathbf{e}_i^H \left(\sum_{k=1}^{N_s} P_k \mathbf{v}_k(\Omega)\mathbf{v}_k^H(\Omega) + \sigma_n^2 I \right) \mathbf{e}_i = \mathbf{e}_i^H \lambda_i \mathbf{e}_i = \lambda_i, \quad i = 1, \ldots, N_s$$
$$= \sum_{k=1}^{N_s} P_k \mathbf{e}_i^H \mathbf{v}_k(\Omega)\mathbf{v}_k^H(\Omega)\mathbf{e}_i + \sigma_n^2 \mathbf{e}_i^H \mathbf{e}_i = \sum_{k=1}^{N_s} P_k |\mathbf{v}_k^H(\Omega)\mathbf{e}_i|^2 + \sigma_n^2 = \lambda_i$$

or solving, we have

$$\sum_{k=1}^{N_s} P_k |\mathbf{v}_k^H(\Omega)\mathbf{e}_i|^2 = \lambda_i - \sigma_n^2 \tag{5.88}$$

Defining $H_k(\Omega_i) = \mathbf{v}_k^H(\Omega)\mathbf{e}_i$ and expanding over i and k we obtain

$$\begin{bmatrix} |H_1(\Omega_1)|^2 & \cdots & |H_1(\Omega_{N_s})|^2 \\ \vdots & \ddots & \vdots \\ |H_{N_s}(\Omega_1)|^2 & \cdots & |H_{N_s}(\Omega_{N_s})|^2 \end{bmatrix} \begin{bmatrix} P_1 \\ \vdots \\ P_{N_s} \end{bmatrix} = \begin{bmatrix} \lambda_1 - \sigma_n^2 \\ \vdots \\ \lambda_{N_s} - \sigma_n^2 \end{bmatrix}$$

or compactly

$$H\mathbf{P} = \Lambda - \sigma_n^2 I \tag{5.89}$$

which is easily solved.

This is the *PHD* technique of estimating harmonics in white noise. This method can be generalized further by incorporating more of the noise eigenvectors and averaging which we discuss next. Note, however, that the *PHD* method provides the foundation for the eigen-decomposition approaches to harmonic retrieval in noise.

5.3.4 Spectral Decomposition: Multiple Signal Classification (*MUSIC*) Method

The *MUSIC* method is a well-known spectral decomposition method in signal processing. It was developed by Schmidt [57] as an extension of the original Pisarenko harmonic estimation technique [56] evolving from spatial (array) signal processing [58]. It is also based on attempting to extract harmonics from noisy measurement data incorporating the *SVD* decomposition of the underlying covariance matrix [57]. Similar to the *PHD* approach of solving for the sinusoids in noise, the *MUSIC* methodology generalizes this extraction problem and utilizes the *SVD* to decompose or separate the signals from the noise.

It follows directly from the *decomposition* of the measurement covariance matrix with the steps summarized as:

1. The eigen-decomposition of the autocorrelation matrix is performed, $R_{yy} = E\Lambda E^{\dagger}$.
2. The number of harmonics, N_s, is determined using the *order estimation* or *best rank approximation* methods [59] with the noise subspace dimensioned as $N - N_s$.
3. The *signal subspace* is defined by the set of largest eigenvalues and corresponding eigenvectors, $[\{\lambda_i\}, \{\mathbf{e}_i\}]$, $i = 1, \ldots, N_s$.
4. The *noise subspace* is defined by the set of the remaining (smallest) eigenvalues and corresponding eigenvectors, $[\{\lambda_i\}, \{\mathbf{e}_i\}]$, $i = N_s + 1, \ldots, N$.

5. The *power estimator* of Eq. (5.90) (to follow) is based on averaging over all of the *noise eigenvectors* to reduce the effects of spurious peaks, $|\mathbf{v}_i^\dagger(\Omega)\mathbf{e}_j|^2 \longrightarrow \sum_{j=N_s+1}^{N} |\mathbf{v}^\dagger(\Omega)\mathbf{e}_j|^2$.
6. The power is found by solving the linear least-squares problem of Eq. (5.91) (below).

Using the noise averaging, we have that the *MUSIC*-power estimator is

$$P_{MUSIC}(\Omega) := \frac{1}{\mathbf{v}^\dagger(\Omega)\left[E(N-N_s)E^\dagger(N-N_s)\right]\mathbf{v}(\Omega)} = \frac{1}{\sum_{j=N_s+1}^{N} |\mathbf{v}^\dagger(\Omega)\mathbf{e}_j|^2} \tag{5.90}$$

The power is found by solving Eq. (5.73) using least-squares for P as

$$\hat{P} = \left[V^\dagger(\Omega)V(\Omega)\right]^{-1} V^\dagger(\Omega)\left(R_{yy} - \sigma_N^2 I\right) V(\Omega)\left[V^\dagger(\Omega)V(\Omega)\right]^{-\dagger} \tag{5.91}$$

where † is the Hermitian conjugate operation completing this spectral decomposition approach.

Next, we apply these spectral decomposition techniques to a problem consisting of multiple sinusoids contaminated with noise.

Example 5.4 Each of the spectral decomposition methods *MTM, PHD, MUSIC* are applied to the deterministic 3-mass MCK-system response data previously generated from the underlying complex resonant modes. The deterministic data along with its frequency response functions (Fourier spectra) are shown in Figure 5.7a. It is clear from the figure and peaks that the decomposition using these techniques is able to perform the spectral decomposition quite reasonably along with their corresponding power spectra shown in Figure 5.7b, It is clear that the set of true modal frequencies, $\{0.123, 0.343, 0.497\,\text{Hz}\}$, are well-estimated as:

MTM/MUSIC ($\{0.123, 0.344, 0.496\,\text{Hz}\}$) and PHD ($\{0.123, 0.345, 0.491\,\text{Hz}\}$) each along with their damping estimates to produce a reasonable signal decomposition completing this example.

△△△

From the data-based and spectral-based decomposition methods, it appears that the most logical approach to decomposing a complex signal is based on its *SVD* or its variants in one form or another. Typically, data can be arranged in a data or covariance matrix where decomposition operations are performed to extract the desired signal components while mitigating disturbances and noise. A "reduced" form or equivalently enhanced signal under scrutiny can be achieved through these decomposition methods [60–65].

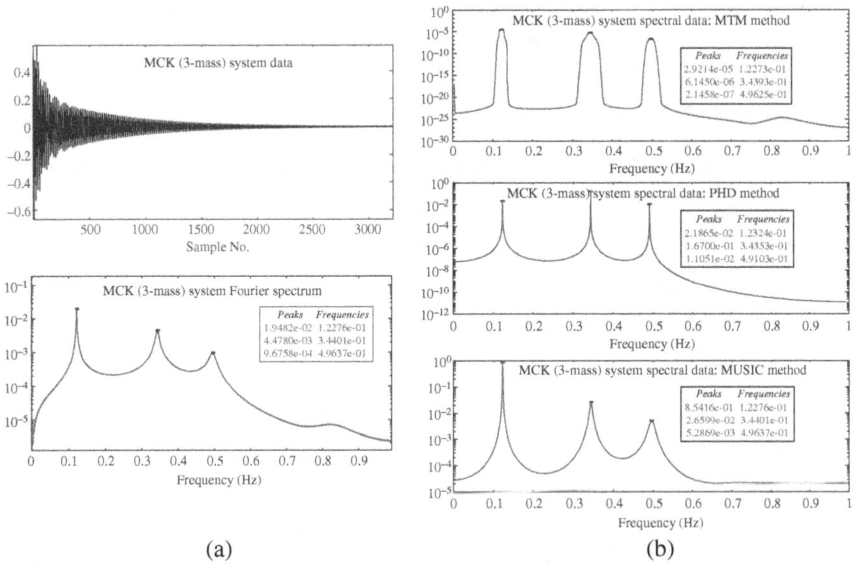

Figure 5.7 Spectral decomposition of 3-mass MCK response data: (a) deterministic response data and Fourier spectrum and (b) decomposed power spectra: MTM, PHD, and MUSIC methods.

5.4 Model-Based Decomposition

Model-based decomposition methods are based on estimating an assumed model structure, usually motivated by the underlying physics, applying a decomposition technique to the data or covariance matrices, fitting this model to the data, and then possibly performing the another decomposition directly on the extracted model. These ideas evolve from the design of decoupled (modal) control systems and can be quite successful in extracting subsystems for subsequent analysis and control. In fact, this approach can be thought of as a "divide and conquer" system theoretical approach to decomposing a system first and then extracting the corresponding signals associated with each decoupled subsystem. We shall observe this direct model decomposition approach for multichannel systems in Chapter 6 using *SSP* representations; however, here we concentrate on *scalar* models that are useful in measured signal representations.

We begin with the development of techniques aimed at *DExp* signal models of which the harmonic representations of the spectral decompositions of Section 5.3 are special cases. Next, we proceed to the approach of characterizing signals using the order-recursive lattice (*LATTICE*) models performing the required matrix decompositions. Finally, we consider an approach of the *SVD*-based *SSP* models estimated from noisy input/output data.

5.4.1 Model-Based Decomposition: Damped Exponential Method

Let us assume that the measured data are characterized by a set of N_e complex exponential signals in additive random noise given by

$$y(t) = s(t) + n(t) \quad \text{for} \quad t = 0, \ldots, N-1 \tag{5.92}$$

with the *DExp* signal model

$$s(t) = \sum_{i=1}^{N_e} A_i e^{p_i t} \tag{5.93}$$

where A_i is complex and p_i is a complex pole defined by

$$p_i := d_i + j\Omega_i$$

for d_i the damping and Ω_i the angular frequency.

The signal can be decomposed into a set of parametrically dependent sub-signals, $\{s_i(t;\theta)\}$. The basic problem reduces to first estimating the number of complex exponentials required to approximate the signal–the order estimation problem in the system identification literature [59]. To complicate the problem even further, the measured data are usually contaminated with noise and disturbances as in Eq. (5.92) and therefore $\hat{N}_e$ must be estimated from the noisy data. This is where the *SVD*-approach evolves in signal modeling, similar to the *MUSIC*-approach for spectral decomposition or the *SSP* approach that follows can be applied [63, 66].

This representation can be employed to decompose the signal space into its constituent components, that is,

$$s(t;\Theta) = \sum_{i=1}^{N_e} s_i(t;\theta) = \underbrace{A_1 e^{p_1 t}}_{s_1(t;\theta_1)} + \underbrace{A_2 e^{p_2 t}}_{s_1(t;\theta_2)} + \cdots + \underbrace{A_{N_e} e^{p_{N_e} t}}_{s_{N_e}(t;\theta_{N_e})} \tag{5.94}$$

where the parameter set is defined by $\Theta := \{N_e, A_i, d_i, \Omega_i\}; i = 1, \ldots, N_e$, the respective unknown model order, complex amplitude, damping, and frequency.

Thus, the basic approach is to decompose the data to estimate the number of exponentials ($\hat{N}_e$) and then estimate the underlying component signals based on Eq. (5.94) performing the desired decomposition.

An optimization approach can be used to solve the parameter estimation problem based on the sum-squared (model) error (*SSQE*) criterion between the data and the signal model, that is,

$$\mathcal{J}(\Theta) = \sum_{t=0}^{N\_1} (y(t) - \hat{s}(t;\Theta))^2 \tag{5.95}$$

for $s(t;\hat{\Theta}) = \hat{s}(t)$. Direct minimization of the *SSQE* is a nonlinear optimization problem that can be computationally intensive; therefore, a popular alternative is to use a suboptimal technique based on the Prony method [48, 67] that involves solving two sets of linear equations with intermediate polynomial rooting which is the nonlinear component of the problem.

The crucial ingredient leading to the Prony technique is to realize that Eq. (5.93) is the solution of a homogeneous linear difference equation with appropriate initial conditions $\{\hat{s}(0),\dots,\hat{s}(N_e-1)\}$, that is,

$$\sum_{i=0}^{N_e} a_i\hat{s}(t-i) = 0 \quad \text{for} \quad a_0 = 1 \quad \text{and} \quad N_e \le t \le N-1 \tag{5.96}$$

Taking $\mathcal{Z}$-transforms, leads directly to the characteristic equation of the system relating coefficients of the difference equation to the roots or equivalently, the complex exponentials,

$$A(z) = \sum_{i=0}^{N_e} a_i z^{-i} = \prod_{i=1}^{N_e}(1-p_i z^{-i}), \quad a_0 = 1 \qquad \text{[Characteristic Equation]} \tag{5.97}$$

or $A(z)$ has $\{p_i\}$ as its roots and coefficients $\{a_i\}$ when expanded as above. Based on this representation, the extended Prony technique can be developed.

We first define the *modeling error* as

$$e(t) := y(t) - \hat{s}(t) \quad \text{for} \quad 0 \le t \le N-1 \tag{5.98}$$

Solving for $\hat{s}(t)$ and substituting, we have

$$e(t) = y(t) + \sum_{i=1}^{N_e} a_i\hat{s}(t-i) = y(t) + \sum_{i=1}^{N_e} a_i\,(y(t-i) - e(t-i))$$

Now solving this expression for $y(t)$, we obtain

$$y(t) = -\sum_{i=1}^{N_e} a_i y(t-i) + \sum_{i=0}^{N_e} a_i e(t-i) \quad \text{for} \quad N_e \le t \le N-1 \tag{5.99}$$

or combining like variables give in terms of the backward shift operator q with the property that $q^{-k}y(t) = y(t-k)$ gives

$$A(q^{-1})y(t) = A(q^{-1})e(t) \tag{5.100}$$

which is an equivalent *ARMA* model, (*ARMAX* $(N_a, 0, N_a)$). Note the indices on t which follow directly from the limits of Eq. (5.96). If we attempt to construct an estimator for this model, it will also be nonlinear [67]; therefore, we develop a suboptimal approach called the *extended Prony technique* where the prediction error is defined as

$$\varepsilon(t) := \sum_{i=0}^{N_e} a_i e(t-i) \quad \text{for} \quad t = N_e, \ldots, N-1 \tag{5.101}$$

Substituting into Eq. (5.99) gives an *AR* model

$$A(q^{-1})y(t) = \sum_{i=0}^{N_e} a_i y(t-i) = \varepsilon(t) \tag{5.102}$$

Expanding this relation over $N_e \le t \le N-1$, gives the partitioned matrix

$$\underbrace{\begin{bmatrix} \varepsilon(N_e) \\ \vdots \\ \varepsilon(N-1) \end{bmatrix}}_{\mathcal{E}} = \underbrace{\begin{bmatrix} y(N_e) & | & y(N_e+1) & \cdots & y(0) \\ \vdots & | & & \ddots & \vdots \\ y(N-1) & | & y(N-2) & \cdots & y(N-1-N_e) \end{bmatrix}}_{\mathbf{y}\,|\,\mathbf{Y}} \underbrace{\begin{bmatrix} 1 \\ -- \\ a_1 \\ \vdots \\ a_{N_e} \end{bmatrix}}_{\mathbf{a}} \tag{5.103}$$

or in compact vector-matrix form as

$$\mathcal{E} = [\mathbf{y} \mid \mathbf{Y}] \begin{bmatrix} 1 \\ -- \\ -\mathbf{a} \end{bmatrix} = \mathbf{y} - \mathbf{Y}\mathbf{a} \tag{5.104}$$

where $\mathbf{y} \in \mathcal{R}^{N-1-N_e \times 1}$, $\mathbf{Y} \in \mathcal{R}^{N-1-N_e \times N_e+1}$, $\mathbf{a} \in \mathcal{R}^{N_e \times 1}$.

The optimal least-squares parameter estimate is found by minimizing the sum-squared prediction error criterion over $N_e \le t \le N-1$, that is,

$$J_{\min} := \min_{\mathbf{a}} J(t) = \mathcal{E}'\mathcal{E} = \sum_{t=N_e}^{N-1} \varepsilon^2(t) \tag{5.105}$$

Let us return to Eq. (5.104)

$$\mathbf{y} = \mathbf{Y}\mathbf{a} + \mathcal{E} \tag{5.106}$$

The *SVD* approach is to choose a larger order N_e (where $N_e > N/3$) for N the number of data samples and then perform an *SVD* of the data matrix to give [62]

$$\mathbf{Y} := \mathbf{U} \times \boldsymbol{\Sigma} \times \mathbf{V}' \tag{5.107}$$

and therefore, we have

$$\left(\mathbf{U}\boldsymbol{\Sigma}\mathbf{V}'\right)\mathbf{a} = \epsilon \tag{5.108}$$

where and $\epsilon := \mathbf{y} - \mathcal{E}$.

Inverting the data matrix using the *SVD*, we obtain the predictor coefficients

$$\hat{\mathbf{a}} = \left(\mathbf{V}\mathbf{\Sigma}^{-1}\mathbf{U}'\right)\epsilon \tag{5.109}$$

The idea is to estimate the order or the number of exponentials (N_e) by the largest of the singular values characterizing the data matrix, since

$$\Sigma = \text{diag}\left[\ \sigma_1\ \sigma_2\ \cdots\ \sigma_{N_e}\ \mid\ \sigma_{N_e+1}\ \cdots\ \sigma_{N_\mathcal{E}}\ \right]$$

One technique suggested by Golub and Van Loan [4] is the "rank approximation" of $\mathbf{Y}$ defined by the ratio

$$\rho(N_e) = \frac{|\hat{Y}(N_e)|}{|Y(N_\mathcal{E})|} = \sqrt{\frac{\sum_{i=1}^{N_e}\sigma_i}{\sum_{i=1}^{N_\mathcal{E}}\sigma_i}} \tag{5.110}$$

Clearly, when $N_e \approx N_\mathcal{E}$, $\rho(N_e) \approx 1$ and the matrix $\hat{Y}(N_e) \longrightarrow Y(N_\mathcal{E})$ with high accuracy. The idea is to select a value of N_e such that $\hat{Y}(N_e)$ approximates the data matrix. A reasonable choice of N_e is to choose a desired or target value of $\rho(N_e) = \rho^*$ for $0 < \rho^* < 1$ and continue to increase the estimated order $\hat{N}_e$ until $\rho(\hat{N}_e) \geq \rho^*$ is achieved. Once the number of exponentials is estimated, the predictor coefficients can then be estimated using the "truncated" *SVD* as [62, 64]

$$\hat{\mathbf{a}}(\hat{N}_e) = V(\hat{N}_e)\Sigma^{-1}(\hat{N}_e)U'(\hat{N}_e)\epsilon = \sum_{i=1}^{\hat{N}_e}\frac{\mathbf{v}_i}{\sigma_i}\mathbf{u}_i'\epsilon_i \tag{5.111}$$

With the $\{\hat{a}_i\}$ estimated, the complex exponentials $\{p_i\}$ are determined from the roots of the characteristic equation, $A(z)$ of Eq. (5.97).

The set of complex gains $\{A_i\}$ can be estimated by expanding Eq. (5.93) with $i = 1,\ldots,N_e$, that is, the filtered estimate, $\hat{y}(t)$, from the *AR* model approximates the estimated signal, $\hat{y}(t) \approx \hat{s}(t)$, that is,

$$\hat{y}(t) = \begin{bmatrix} e^{p_1 k\triangle T} & e^{p_2 k\triangle T} & \cdots & e^{p_{N_e} k\triangle T}\end{bmatrix}\begin{bmatrix} A_1 \\ A_2 \\ \vdots \\ A_{N_e}\end{bmatrix} \quad \text{for} \quad t = k\triangle T; \quad k = 0,\ldots, N-1 \tag{5.112}$$

to obtain

$$\mathbf{PA} = \hat{\mathbf{y}} \tag{5.113}$$

where the $\mathbf{P} \in R^{N\times N_e}$ is a Van der Monde matrix of exponentials given by

$$\mathbf{P} = \begin{bmatrix} 1 & \cdots & 1 \\ e^{p_1\triangle T} & & e^{p_{N_e}\triangle T} \\ \vdots & & \vdots \\ e^{p_1(N-1)\triangle T} & \cdots & e^{p_{N_e}(N-1)\triangle T}\end{bmatrix}$$

with

$$\mathbf{A}' = [A_1 \ldots A_{N_e}], \quad \hat{\mathbf{y}}' = [\hat{y}(1) \ldots \hat{y}(N_e)]$$

which can also be solved using least squares (pseudo-inverse) techniques implemented using the *SVD* decomposition to give

$$\hat{\mathbf{A}} = \mathbf{P}^{\#}\hat{\mathbf{y}} = \left(\mathbf{V}_p\mathbf{\Sigma}_p^{-1}\mathbf{U}_p'\right)\mathbf{y} \tag{5.114}$$

completing the set of required set of amplitude parameters $\{\hat{A}_i\}$ to enable the decomposition of $s(t)$ estimates into the set of *DExps* of Eq. (5.94).

The effect of employing the *SVD* of the data matrix and truncating it to $\hat{N}_e$ is to increase the *SNR* of the data. In low *SNR* measurements, its performance can be improved even further by using the correlation matrix instead of the data matrix [64]. That is, multiplying Eq. (5.108) by Y' and defining $R := Y \times Y'$ as before gives

$$R\mathbf{a} = Y'\epsilon \tag{5.115}$$

and therefore

$$\hat{\mathbf{a}} = R^{-1}Y'\epsilon \tag{5.116}$$

This method is called the autocorrelation matrix algorithm by Cadzow et al. [64]. It can be improved as before in the data matrix case by estimating the number of exponentials using the rank approximation and truncating the correlation matrices appropriately to give

$$\hat{\mathbf{a}}(N_e) = R^{-1}(N_e)Y'(N_e)\epsilon(N_e) \tag{5.117}$$

Again the predictor coefficients are estimated using the *SVD* approach, the resulting polynomial is factored into it roots and the corresponding amplitudes are estimated by solving the least-squares problem. We summarize the *DExp* decomposition algorithm as:

Criterion: $\mathcal{J} = \mathcal{E}'\mathcal{E}$
Models:
  Measurement: $\mathbf{y} = \mathbf{Ya} + \mathcal{E}$
  Signal: $\mathbf{s} := \mathbf{Ya} = \sum_{i=1}^{N_e} A_i e^{p_i t}$
  Noise: $R_{\epsilon\epsilon}$
Algorithm: $\hat{\mathbf{a}}(\hat{N}_e) = \sum_{i=1}^{\hat{N}_e} \frac{\mathbf{v}_i}{\sigma_i}\mathbf{u}_i'\epsilon_i$
Quality: $\mathcal{J}_{\min}(\hat{\mathbf{a}}) = \mathcal{E}'\mathcal{E}$

Example 5.5 Using the deterministic 3-mass MCK-system response data previously generated from the underlying complex resonant modes, we investigate the model-based signal decomposition based on the *DExp* model. The deterministic data along with its frequency response functions (Fourier spectra) are shown in Figure 5.7a. The SVD-based (Prony) approach is applied to data, that is, the extended Prony technique is used to estimate the set of model parameters (Θ)

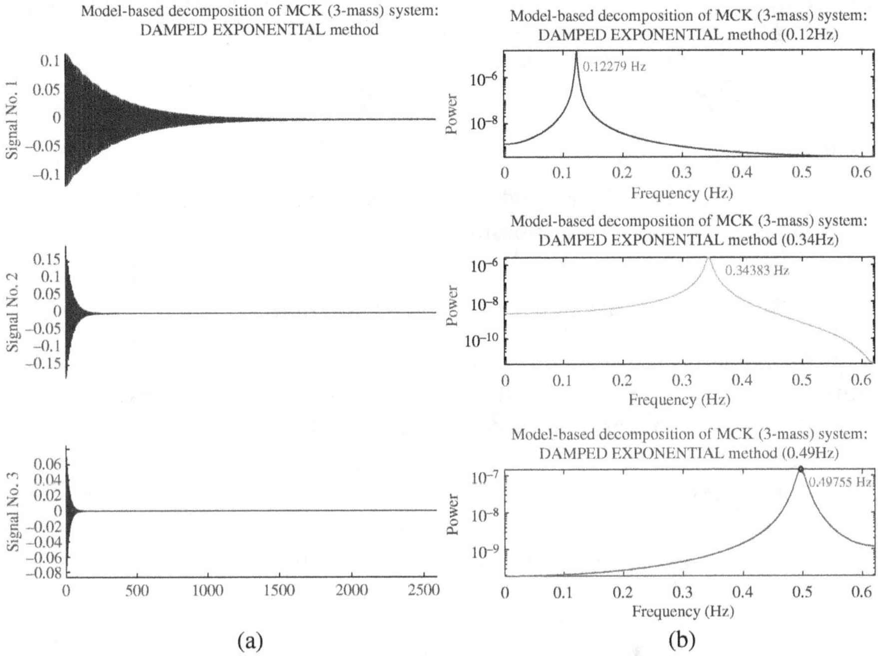

Figure 5.8 Damped exponential method of model-based decomposition for deterministic 3-mass MCK response data: (a) estimated signal components and (b) decomposed signal component power spectra: (MVDR, Order = 100).

after the "rank approximation" method is used to estimate the number of exponentials (poles) to capture the data ($\hat{N}_e = 6$). With these estimated parameters available, the individual component signal response estimates are synthesized with the results shown in Figure 5.8a along with their corresponding power spectra shown in Figure 5.8b, It is clear that the set of true modal frequencies, $\{0.123, 0.343, 0.497\,\text{Hz}\}$, are well-estimated as $\{0.1228, 0.3438, 0.4976\,\text{Hz}\}$ along with the damping estimates to produce a reasonable signal decomposition completing this example.

△△△

We summarize the Prony model-based parameter estimation technique below:

Criterion: $\mathcal{J}(\Theta) = \sum_{t=0}^{N-1} (y(t) - \hat{s}(t;\Theta))^2$

Models:

Measurement: $y(t) = s(t) + e(t)$

Signal: $s(t) = \sum_{i=1}^{N_e} A_i e^{p_i t}$

Noise: R_{ee}

Algorithm:

$$\hat{\mathbf{a}} = \left[\mathbf{Y}'\mathbf{Y}\right]^{-1}\mathbf{Y}'\mathbf{y}$$
$$\hat{\mathbf{A}} = \left[\mathbf{P}'\mathbf{P}\right]^{-1}\mathbf{P}'\hat{\mathbf{y}}$$

Quality: $\mathcal{J}_{\min} = \mathcal{E}'\mathcal{E}$

Next, we consider a similar and related method based on an order-recursive, lattice approach to signal decomposition.

5.4.2 Model-Based Decomposition: Lattice Method

The lattice approach uses its basic structure to decompose the signal into individual orthogonal, stage-wise components where the data are processed initially to estimate its parameters (reflection coefficients). By its inherent structure and order-recursive property, an implied decomposition is available. However, this result can be interpreted as a "composition" rather than a decomposition, since subsystem responses are not obvious in contrast to previous decomposition approaches. Single lattice stages capture an exponential response (single pole), while with two stages a sinusoidal response (two poles) and so on (multiple poles).

By parameter estimation and construction, the lattice provides a decomposition representation that is recursive "in-order" where the orthogonality properties of each stage are characterized uniquely by its particular reflection coefficient. Using this construct, the overall response of the system is captured, stage-by-stage; however, individual subsystem responses are not as obvious as in the *DExp* decomposition, for example. Here, we simply extract the lattice parameters and investigate the sequential construction of the responses leading to the overall system response.

Since its roots were established in [68–73], the lattice filter and lattice techniques have evolved to a high degree of popularity for many applications. The lattice method combines the orthogonal projection ideas in *both* time and parameter spaces [60, 72].

As before, we are interested in developing an estimate of the signal $s(t)$ based on past measurements and/or past excitations, that is,

$$\hat{s}(t) = \hat{s}(t|Y_{past} \quad \text{or} \quad X_{past})$$

Taking the model-based approach using the parametric lattice model, we know that given the lattice parameters, we can construct all-pole or all-zero signal estimators [60].

Recall the *lattice recursion* for the ith-section at time t is given by

$$\begin{aligned} e_f(t,i) &= e_f(t,i-1) - k_i e_b(t-1,i-1) \\ e_b(t,i) &= e_b(t-1,i-1) - k_i e_f(t,i-1) \end{aligned} \tag{5.118}$$

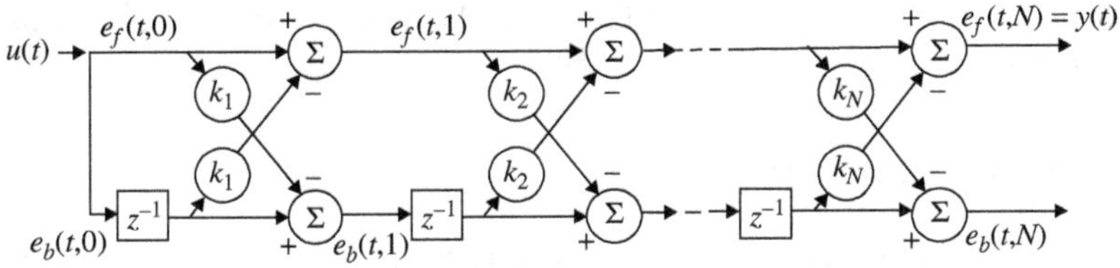

Figure 5.9 Feed forward lattice (all-zero) structure.

where $e_f(t, i)$ is the *forward* prediction error, $e_b(t, i)$ is the *backward* prediction error, k_i is the reflection coefficient of the ith-stage.

Consider a filter consisting of N-lattice sections arranged in a *feedforward* configuration as shown in Figure 5.9. We first note that the lattice model operation for the ith-section can be written in matrix form using the delay operator ($q^{-i}y(t) = y(t - i)$) as

$$\begin{bmatrix} e_f(t, i) \\ e_b(t, i) \end{bmatrix} = \begin{bmatrix} 1 & -k_i q^{-1} \\ -k_i & q^{-1} \end{bmatrix} \begin{bmatrix} e_f(t, i-1) \\ e_b(t, i-1) \end{bmatrix} \tag{5.119}$$

The $\mathcal{Z}$-transform yields a vector-matrix relation

$$\underline{E}(z, i) = L(z, i)\underline{E}(z, i-1) \tag{5.120}$$

or in terms of the reflection coefficients as

$$\begin{bmatrix} E_f(z, i) \\ E_b(z, i) \end{bmatrix} = \begin{bmatrix} 1 & -k_i z^{-1} \\ -k_i & z^{-1} \end{bmatrix} \begin{bmatrix} E_f(z, i-1) \\ E_b(z, i-1) \end{bmatrix} \tag{5.121}$$

where $L(z, i)$ can be thought of as a stage-wise transfer function matrix[2] between the ith and $(i - 1)$th-section of the lattice. In fact, inverting $L(z, i)$ enables us to determine the inverse relation between the ith and the $(i - 1)$th-stage as

$$\underline{E}(z, i-1) = L^{-1}(z, i)\underline{E}(z, i) \tag{5.122}$$

where

$$L^{-1}(z, i) = \frac{1}{1 - k_1^2} \begin{bmatrix} 1 & k_i \\ k_i z & z \end{bmatrix}$$

This transfer function is useful for predicting the response of the lattice at various stages–a very important feature from the composition/decomposition perspective, since it is naturally provided by its internal structure. Thus, the

2 We note that the lattice filter is a two-port network and $L(z, i)$ is the so-called transfer matrix from input-to-output [60].

overall transfer function for the N-stage lattice can be decomposed in terms of its individual stages as

$$\begin{aligned}\underline{E}(z,N) &= \prod_{n=1}^{N} L(z,n) \times \underline{E}(z,n-1) \\ &= \underbrace{L(z,N)}_{\text{Stage } N} \times \underbrace{L(z,N-1)}_{\text{Stage } (N-1)} \times \cdots \times \underbrace{L(z,1)}_{\text{Stage } 1} \times \underline{E}(z,0)\end{aligned} \tag{5.123}$$

which gives the stage-wise transfer function. We can similarly derive relations for the inverse transfer function. Thus, the overall *stage transfer function* of an N-stage lattice and its inverse are given by

$$\begin{aligned}L(z) &:= L(z,N) \times L(z,N-2) \times \cdots \times L(z,1) \\ L^{-1}(z) &:= L^{-1}(z,1) \times L^{-1}(z,2) \times \cdots \times L^{-1}(z,N)\end{aligned} \tag{5.124}$$

Consider the following example that illustrates how to construct the usual transfer functions in terms of the underlying reflection coefficients.

Example 5.6 We are given a three-stage feedforward lattice structure, and we would like to convert it to the all-zero model,

$$L(z,i) = 1 + \sum_{j=1}^{N_\alpha} \alpha(j,i) z^{-j} \quad \text{and} \quad L(z,N_\alpha) = 1 + \alpha(1,N_\alpha)z^{-1} + \cdots + \alpha(N_\alpha,N_\alpha)z^{-N_\alpha}$$

Applying the N-stage Levinson recursion (see Section 3.3) [48]

$$\begin{aligned}\alpha(1,1) &= -k_1 \\ \alpha(i,i) &= -k_i \quad \text{for} \quad i = 2,\ldots,N_\alpha \\ \alpha(j,i) &= \alpha(j,i-1) - k_i\alpha(i-j,i-1) \quad \text{for} \quad j = 1,\ldots,i-1\end{aligned}$$

$i = 1$

$$\alpha(1,1) = -k_1 \qquad L(z,1) = 1 + \alpha(1,1)z^{-1} = 1 - k_1 z^{-1}$$

$i = 2$

$$\begin{aligned}\alpha(1,2) &= \alpha(1,1) - k_2\alpha(1,1) = -k_1(1-k_2) & L(z,2) &= 1 + \alpha(1,2)z^{-1} + \alpha(2,2)z^{-2} \\ \alpha(2,2) &= -k_2 & &= 1 - k_1(1-k_2)z^{-1} - k_2 z^{-2}\end{aligned}$$

$i = 3$

$$\begin{aligned}\alpha(1,3) &= \alpha(1,2) - k_3\alpha(2,2) = -k_1(1-k_2) + k_2k_3 & L(z,3) &= 1 + \alpha(1,2)z^{-1} + \alpha(2,2)z^{-2} \\ \alpha(2,3) &= \alpha(2,2) - k_3\alpha(1,2) = -k_2 + k_1k_3(1-k_2) & &= 1 - [k_1(1-k_2) - k_2k_3]z^{-1} \\ & & &\quad - [k_2 - k_1k_3(1-k_2)]z^{-2} \\ \alpha(3,3) &= -k_3 & &\quad - k_3 z^{-3}\end{aligned}$$

The resulting all-zero filter is simply found by inspection of the final stage coefficients, that is,

$$\{\alpha_i\} = \{\alpha(i, N_\alpha)\} \quad \text{with} \quad \alpha_0 = 1$$

which completes the example.

△△△

Thus, the order-recursive lattice structure provides a "natural" set of stages that can decompose a signal or system in terms of these orthogonal sections. Grouping of lattice stages to create viable subsystems still appears to be an open question. Here, we just accept the stage-wise order and sequentially catenate each of the stages eventually leading to the overall transfer function, $L(z)$. From a decomposition perspective, we see that the lattice already provides an extraction of each subsystem by its very structure. So we can easily extract the component responses from the inverse lattice-section signals.

Let us return to the basic inverse section governed by [67]

$$\begin{aligned} e_f(t, i-1) &= e_f(t, i) + k_i e_b(t-1, i-1) \\ e_b(t, i) &= e_b(t-1, i-1) - k_i e_f(t, i-1) \end{aligned} \tag{5.125}$$

With this in mind, let us now construct an estimator for the lattice model. The model-based (parametric) problem is characterized by minimizing the joint error function in terms of the lattice parameters as

$$\min_{\mathbf{k}} \mathcal{J}(t) = E\{e_f^2(t) + e_b^2(t)\} \tag{5.126}$$

Performing the indicated minimization of Eq. (5.126), we have the orthogonality condition

$$\frac{\partial J}{\partial k_i} = \frac{\partial}{\partial k_i} E\{e_f^2(t, i) + e_b^2(t, i)\} = 2E\left\{ e_f(t, i)\frac{\partial e_f(t, i)}{\partial k_i} + e_b(t, i)\frac{\partial e_b(t, i)}{\partial k_i} \right\} = 0 \tag{5.127}$$

The error gradients can be calculated directly from the model of Eq. (5.118) as

$$\frac{\partial e_f}{\partial k_i} = -e_b(t-1, i-1)$$

and similarly we have

$$\frac{\partial e_b}{\partial k_i} = -e_f(t, i-1)$$

Substituting these gradients into Eq. (5.127) gives

$$\frac{\partial J}{\partial k_i} = -2E\{e_f(t, i)e_b(t-1, i-1) + e_b(t, i)e_f(t, i-1)\} = 0$$

Using the model recursion of Eq. (5.118), we obtain

$$\frac{\partial J}{\partial k_i} = -2E\{e_f(t,i-1)e_b(t-1,i-1) - k_i(e_f^2(t,i-1) + e_b^2(t-1,i-1))\} = 0 \tag{5.128}$$

Solving this equation for k_i[3] gives

$$k_i = \frac{2E\{e_f(t,i-1)e_b(t-1,i-1)\}}{E\{e_f^2(t,i-1) + e_b^2(t-1,i-1)\}} \tag{5.129}$$

The variance of the lattice recursion can be calculated directly from the recursion, since

$$\begin{aligned} R_{ee}(i) = E\{e_f^2(t,i)\} &= E\{(e_f(t,i-1) - k_i e_b(t-1,i-1))^2\} \\ &= E\{e_f^2(t,i-1)\} - 2k_i E\{e_f(t,i-1)e_b(t-1,i-1)\} + k_i^2 E\{e_b^2(t-1,i-1)\} \end{aligned} \tag{5.130}$$

Solving Eq. (5.129) for the numerator term and substituting, we obtain

$$\begin{aligned} R_{ee}(i) = E\{e_f^2(t,i-1) &- k_i^2(E\{e_f^2(t,i-1)\} + E\{e_b^2(t-1,i-1)\}) \\ &+ k_i^2 E\{e_b^2(t-1,i-1)\} \end{aligned}$$

or

$$R_{ee}(i) = (1-k_i^2)R_{ee}(i-1) \tag{5.131}$$

where $R_{ee}(0) = R_{yy}(0) = \text{var}(y(t))$. Expanding over an N-stage lattice we see that

$$R_{ee} = R_{ee}(N) = \prod_{i=1}^{N}(1-k_i^2)R_{ee}(0) \tag{5.132}$$

If we assume the data are ergodic, then Burg [60, 67] proposed replacing the ensemble averages of Eq. (5.129) with time averages using only the unwindowed data record, that is,

$$k_i = \frac{2\sum_{t=N_a}^{N-1} e_f(t,i-1)e_b(t-1,i-1)}{\sum_{t=N_a}^{N-1} e_f^2(t,i-1) + e_b^2(t-1,i-1)} \quad \text{for} \quad i = 1,\ldots,N_a \tag{5.133}$$

Once the reflection coefficients are estimated from the data using Eqs. (5.118) and (5.133) for each stage, the prediction error variance is obtained from Eq. (5.132).

The latest decomposition into subsystems is based on a superposition of lattice stages or sections in which each individual stage is augmented to the previous stage (order recursive) constructing a subsystem consisting of it and the previous stages. To illustrate this, consider the application of the lattice decomposition to the 3-mass MCK problem.

---

3 The reflection coefficient evolves from the Schur–Cohn stability technique and is used to test stability. If the reflection coefficient lies between ± 1 the system is *stable*, that is, $|k_i| \leq 1$ for stability.

Example 5.7 Again using the deterministic 3-DOF system response data (Figure 5.7) to illustrate the model-based decomposition based on the lattice method applied to single-channel response data. As before, with the model-based decomposition approach, the underlying lattice model parameters are first estimated by applying the parametric estimation technique (Burg) to the synthesized data to estimate the set of parameters (Θ) after the "rank approximation" method is used to determine the number of lattice stages required to capture the response ($\hat{N} = 6$). With these estimated lattice parameters available, the individual component subsystem signal response estimates are synthesized by transforming the reflection coefficients to a transfer function representation via the Levinson method (see Example 5.6) with the results shown in Figure 5.10a along with their corresponding power spectra shown in Figure 5.10b.

Here the first stage is equivalent to a simple single pole (*DExp*) system while the combination of the first and second stages corresponds to a second-order or quadratic system that captures a sinusoidal response, Adding the third-stage provides another pole (*DExp*) while the addition of the fourth-stage corresponds to combined sinusoidal response. Continuing with the superposition of lattice stages leads to the full 3-mass modal frequency response that captures the entire response of the MCK system. Here the output of each section or the combined (superposed)

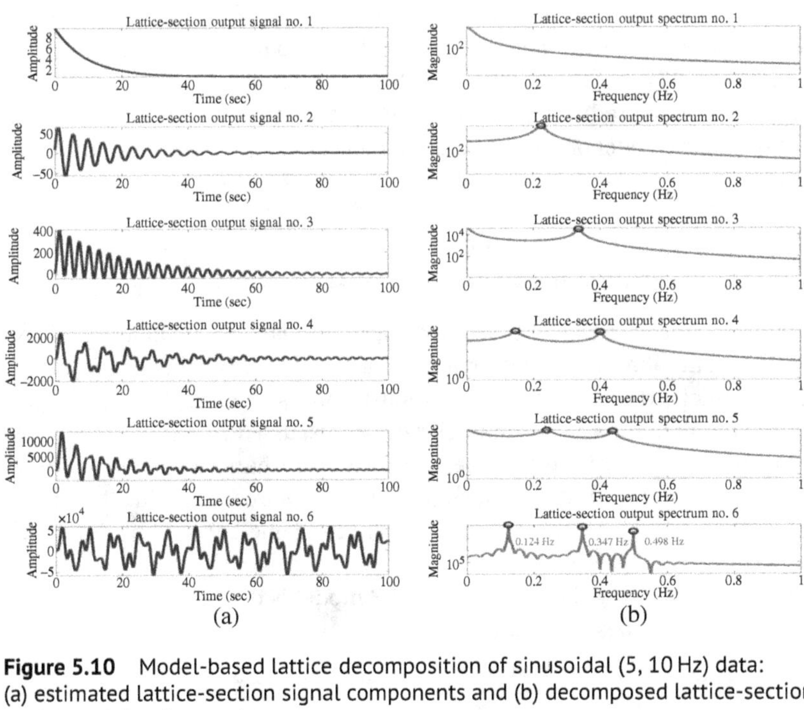

Figure 5.10 Model-based lattice decomposition of sinusoidal (5, 10 Hz) data: (a) estimated lattice-section signal components and (b) decomposed lattice-section component Fourier spectra.

stages is shown

$$1, (1-2), (1-2-3), \ldots, (1-2-3-4-5-6)$$

with the final section capturing the overall transfer function response. The set of estimated reflection coefficients are given by:

$$\hat{k}_i = \{-0.9896, 0.9873, -0.9863, 0.9862, -0.9738, 1.0000\}$$

The corresponding individual all-pole transfer functions with the set of coefficients specified by $\{b_0, a_{11}, a_{21}, \ldots, a_{1N_a}\}$ for this decomposition are given by:

$$\underbrace{\{3.14\times 10^{-5}, -1, 1\}}_{\text{Stage No. 1}}, \underbrace{\{4.15\times 10^{-5}, 1, -1.98, 1\}}_{\text{Stage No. (1-2)}}, \underbrace{\{5.47\times 10^{-5}, 1, 2.96, -2.96, 1\}}_{\text{Stage No. (1-2-3)}},$$

$$\underbrace{\{6.8\times 10^{-5}, 1, -3.93, 5.85, -3.93, 1\}}_{\text{Stage No. (1-2-3-4)}}, \underbrace{\{1.47\times 10^{-4}, -1, 4.9, -9.7, 9.7, -4.9, 1\}}_{\text{Stage No. (1-2-3-4-5)}},$$

$$\underbrace{\{1.66\times 10^{-6}, 1, -5.85, 14.4, -19.1, 14.4, -5.85, 1\}}_{\text{Stage No. (1-2-3-4-5-6)}}$$

The subsystem responses are quite different than that of the other approaches with the decomposition becoming available directly from the lattice structure itself. In any case, the final estimated modal frequencies match those of the system reasonably well, since the set of true modal frequencies given by $\{0.123, 0.343, 0.497\,\text{Hz}\}$ are estimated as $\{0.124, 0.347, 0.498\,\text{Hz}\}$ along with the damping estimates to produce a reasonable signal decomposition almost identical to that of the *DExp* model of the previous example (Example 5.5) completing this example.

△△△

5.4.3 Model-Based Decomposition: State-Space Method

An alternate approach to model-based decomposition is rooted fundamentally in the *SSP* representation of the underlying signals/systems. Although not limited to single channels, the *SSP* provides a robust approach to capturing complex signals/systems including multichannel data as discussed in detail in the next chapter. Here, we *limit* this development to the single channel representation in order to compare it with the previous model-based approaches with the understanding that more sophisticated techniques are available and will be developed to follow.

Recall that a linear time-invariant (*LTI*) *SSP* model is defined by both its state vector $\mathbf{x}(t)$ and *scalar* measurement $y(t)$ as

$$\begin{aligned} \mathbf{x}(t) &= \mathbf{A}\mathbf{x}(t-1) + \mathbf{b}u(t-1) \qquad &&\text{[State Equations]} \\ y(t) &= \mathbf{c}'\mathbf{x}(\mathbf{t}) &&\text{[Measurement Equations]} \end{aligned} \tag{5.134}$$

where $\mathbf{x} \in R^{N_x \times 1}$, $u, y \in R^{1\times 1}$ with system, input, measurement matrices: $\mathbf{A} \in R^{N_x \times N_x}$, $\mathbf{b} \in R^{N_x \times 1}$ and $\mathbf{c}' \in R^{1\times N_x}$

For the decomposition problem solution, it is necessary to solve the parameter estimation problem first (as before). Parameter estimation for the *SSP* model is based on creating a *Hankel matrix* with its desirable system theoretic properties that are exploited to extract the parameters from a set of measurements.

In systems theory, the Hankel matrix can be constructed from impulse response data, that is, we define the $(K \times K)$ *Hankel matrix* $(K \to \infty)$ by

$$\mathcal{H}_{K,K} := \begin{bmatrix} h_1 & h_2 & \cdots & h_K \\ h_2 & h_3 & \cdots & h_{K+1} \\ \vdots & \vdots & \cdots & \vdots \\ h_K & h_{K+1} & \cdots & h_{2K-1} \end{bmatrix} \tag{5.135}$$

where $h_k := \mathbf{c}'\mathbf{A}^{k-1}\mathbf{b}$; $k = 1, 2, \ldots, K$ is the impulse response with $\Sigma_{ABC} := \{\mathbf{A}, \mathbf{b}, \mathbf{c}\}$ defined as the system with its set of Markov parameters.

If Σ_{ABC} is a realization of $\{h_t\}$, then based on its Markov parameters, the Hankel matrix can be *factored* as the product of the controllability and observability matrices, that is,

$$\mathcal{H}_{K,K} = \begin{bmatrix} \mathbf{c}'b & \mathbf{c}'\mathbf{A}\mathbf{b} & \cdots & \mathbf{c}'\mathbf{A}^{K-1}\mathbf{b} \\ \mathbf{c}'\mathbf{A}\mathbf{b} & \mathbf{c}'\mathbf{A}^2\mathbf{b} & \cdots & \mathbf{c}'\mathbf{A}^{K}\mathbf{b} \\ \vdots & \vdots & \cdots & \vdots \\ \mathbf{c}'\mathbf{A}^{K-1}\mathbf{b} & \mathbf{c}'\mathbf{A}^{K-2}\mathbf{b} & \cdots & \mathbf{c}'\mathbf{A}^{2K-2}\mathbf{b} \end{bmatrix}$$

$$= \underbrace{\begin{bmatrix} \mathbf{c}' \\ \mathbf{c}'\mathbf{A} \\ \vdots \\ \mathbf{c}'\mathbf{A}^{K-1} \end{bmatrix}}_{\mathcal{O}_K} \underbrace{\left[\mathbf{b} \mid \mathbf{A}\mathbf{b} \mid \cdots \mid \mathbf{A}^{K-1}\mathbf{b}\right]}_{\mathcal{C}_K} \tag{5.136}$$

or simply

$$\mathcal{H}_{K,K} = \mathcal{O}_K \times \mathcal{C}_K \tag{5.137}$$

Another interesting property of the Hankel matrix utilized by realization algorithms is its inherent *shift-invariant property*, that is, multiplication by the system matrix A shifts the Hankel matrix "up" ($^{\uparrow}$) or "left" ($^{\leftarrow}$) since

$$\mathcal{O}_K^{\uparrow} = \underbrace{\begin{bmatrix} \mathbf{c}' \\ \mathbf{c}'A \\ \vdots \\ \mathbf{c}'A^{K-1} \end{bmatrix}}_{\mathcal{O}_K} \times A = \underbrace{\begin{bmatrix} \mathbf{c}'A \\ \mathbf{c}'A^2 \\ \vdots \\ \mathbf{c}'A^{K} \end{bmatrix}}_{\mathcal{O}_K^{\uparrow}} \tag{5.138}$$

$$\mathcal{C}_K^{\leftarrow} = A \times \underbrace{[\mathbf{b}\ A\mathbf{b} \quad \cdots \quad A^{K-1}\mathbf{b}]}_{\mathcal{C}_K} = \underbrace{[A\mathbf{b}\ A^2\mathbf{b} \quad \cdots \quad A^K\mathbf{b}]}_{\mathcal{C}_K^{\leftarrow}} \tag{5.139}$$

therefore

$$\mathcal{H}_{K,K}^{\uparrow} := \mathcal{O}_K^{\uparrow}\mathcal{C}_K = \underbrace{(\mathcal{O}_K \times A)}_{\mathcal{O}_K^{\uparrow}} \times \mathcal{C}_K \iff \mathcal{H}_{K,K}^{\leftarrow} := \mathcal{O}_K\mathcal{C}_K^{\leftarrow} = \mathcal{O}_K \times \underbrace{(A \times \mathcal{C}_K)}_{\mathcal{C}_K^{\leftarrow}} \tag{5.140}$$

Perhaps, the most important property of the Hankel matrix is that its rank provides an estimate of the number of states and therefore the number of poles. Knowledge of the system order N_x indicates the minimum number of terms required to exactly "match" the Markov sequence and extract the Markov parameters. It follows that the rank of the Hankel matrix must be N_x, that is, $\rho\left(\mathcal{H}_{K,K}\right) = N_x$ which is the dimension of the *minimal realization*. This rank property of the Hankel matrix is fundamental to linear systems theory [74]. In order for a system to be *minimal*, it must be *completely controllable* and *completely observable*. Thus, testing the rank of the Hankel matrix yields the dimensionality of the underlying dynamic system. With this in mind, it is clear that providing estimates of the system and output matrices enables us to decompose the system by performing change of coordinates through a similarity transformation [59]. If this transformation is selected to diagonalize the system matrix, it is deemed a "modal transformation" and extracting the subsystem parameters is possible [75].

An *SVD* approach can be used to decompose the Hankel matrix and utilize its underlying properties to extract the system matrices by taking advantage of the Hankel shift property. It is also clear that the eigenvalues are complex for resonant systems enabling us to extract the system poles, while the amplitude is available from the output data. For this reason, once the available system is transformed to modal form, then the subsystems are easily extracted.

$$\mathcal{H}_{m,K} = \left[U_{N_x} \mid U_N\right]\begin{bmatrix} \Sigma_{N_x} & | & 0 \\ - & - & - \\ 0 & | & 0 \end{bmatrix}\begin{bmatrix} V'_{N_x} \\ -- \\ V'_N \end{bmatrix} = U_{N_x}\Sigma_{N_x}V'_{N_x} \tag{5.141}$$

for $\Sigma_{N_x} = \text{diag}\left[\ \sigma_1,\ \sigma_2,\ \ldots,\ \sigma_{N_x}\right]$

From this factorization of the Hankel matrix, we have that

$$\mathcal{H}_{m,K} = \mathcal{O}_m \times \mathcal{C}_K = \underbrace{\left(U_{N_x}\Sigma_{N_x}^{1/2}\right)}_{\mathcal{O}_{N_x}}\underbrace{\left((\Sigma'_{N_x})^{1/2}V'_{N_x}\right)}_{\mathcal{C}_{N_x}} \tag{5.142}$$

where $\Sigma_{N_x}^{1/2}$ is the matrix square root.

Now, from shift-invariant property (see Eq. 5.140) of the observability matrix, we can extract the system matrix from

$$\mathcal{O}_{N_x} \times \mathbf{A} = \mathcal{O}^{\uparrow}_{N_x} \tag{5.143}$$

Therefore; solving by performing the pseudo-inverse[4] $\left(Z^{\#} := (Z'Z)^{-1}Z'\right)$ yields

$$\mathbf{A} = \mathcal{O}^{\#}_{N_x} \times \mathcal{O}^{\uparrow}_{N_x} = \mathcal{O}^{\#}_{N_x}\mathcal{O}^{\uparrow}_{N_x} = \left(\Sigma^{-1/2}_{N_x}U'_{N_x}\right) \times \mathcal{O}^{\uparrow}_{N_x} \tag{5.144}$$

with the input and output matrices extracted directly from the controllability and observability matrices as

$$\mathbf{b} = \mathcal{C}(1 : N_x, 1) \quad \text{and} \quad \mathbf{c}' = \mathcal{O}(1, 1 : N_x) \tag{5.145}$$

Another interesting system theoretic property that evolves from the observability, controllability, and Hankel matrices is the concept of a *balanced realization*. Simply stated a balanced realization is one in which "every state is as controllable as it is observable" [76]. It is inherited from by *SVD* approach.

With the *SSP* system model, Σ_{ABC}, available, the transformation (similarity) to modal coordinates is performed to obtain the system decomposition (see Chapter 6 for more details):

Perhaps, one of the most expository representations of a system is its modal representation [77], where its internal structure and response to various excitations are exposed. The modal *decomposition* of a system can easily be determined from *SSP* systems by transforming its current coordinates to modal space which is accomplished through a *similarity transformation* constructed from an eigen-decomposition. That is, the system matrices $\Sigma := \{A, B, C\}$ are transformed to modal coordinates by the transformation matrix T_M constructed of the eigenvectors of the underlying system [67, 74, 78–81], that is,

$$\Sigma \xrightarrow{T_M} \Sigma_M$$

where we have

$$\{\mathbf{A}, \mathbf{b}, \mathbf{c}'\} \longrightarrow \{\mathbf{A}_M, \mathbf{b}_M, \mathbf{c}'_M,\} := \{T_M\mathbf{A}T_M^{-1},\ T_M\mathbf{b},\ \mathbf{c}'T_M^{-1}\}$$

In this case, the system matrix can be decomposed, as before, using the eigen-decomposition which now yields complex eigen pairs along with the corresponding complex eigenvectors, that is,

$$\begin{aligned} &\lambda_i = -\sigma_i \pm j\omega_i \\ &\mathbf{t}_i = [\xi_i \mid \xi_i^*] \quad \text{for} \quad i = 1, 2, \ldots, N_d \quad \text{for} \quad N_x = 2 \times N_d \end{aligned} \tag{5.146}$$

for N_x the number of the complex poles and $\mathbf{t}_i$ the ith-column vector of T_M with σ_i is the *damping coefficient*, ω_i the *damped natural frequency*, and ϕ_i the *phase*.

---

4 Recall the *SVD* of $Z = U\Sigma V'$ efficiently provides $Z^{\#} = V\Sigma^{-1}U' = V(\Sigma^{-1/2})'\Sigma^{-1/2}U'$.

We also have that the modal transformation matrix becomes

$$T_M = \left[\xi_1\ \xi_1^* \mid \xi_2\ \xi_2^* \mid \cdots \mid \xi_{N_x/2}\ \xi_{N_x/2}^*\right] \tag{5.147}$$

Here T_M generates an "equivalent" system from an input/output perspective, that is, the transfer functions are identical.

Applying this transformation to the system matrix **A**, we obtain Λ

$$\Lambda = T_M \mathbf{A} T_M^{-1} = \begin{bmatrix} \sigma_1 & \omega_1 & \cdots & 0 & 0 \\ -\omega_1 & \sigma_1 & \cdots & 0 & 0 \\ \vdots & \vdots & \ddots & \vdots & \vdots \\ 0 & \cdots & 0 & \sigma_{N_x} & \omega_{N_x} \\ 0 & \cdots & 0 & -\omega_{N_x} & \sigma_{N_x} \end{bmatrix} \tag{5.148}$$

with

$$\mathbf{b}_M = T_M \times \mathbf{b} \quad \text{and} \quad \mathbf{c}'_M = \mathbf{c}' \times T_M^{-1} \tag{5.149}$$

Thus, the complex modal *SSP* system is given by

$$\begin{aligned} \dot{\mathbf{x}}(t) &= \mathbf{A}_M \mathbf{x}(t) + \mathbf{b}_M \mathbf{u}(t) \\ \mathbf{y}(t) &= \mathbf{c}'_M \mathbf{x}(t) \end{aligned} \tag{5.150}$$

where the system matrices become

$$\mathbf{A}_M = \begin{bmatrix} A_{M_1} & 0 & \cdots & 0 \\ 0 & A_{M_2} & \cdots & 0 \\ \vdots & \vdots & \ddots & 0 \\ 0 & 0 & \cdots & A_{M_{N_x}} \end{bmatrix}; \quad \mathbf{b}_M = \begin{bmatrix} \mathbf{b}_{M_1} \\ --- \\ \vdots \\ --- \\ \mathbf{b}_{M_{N_x}} \end{bmatrix}; \quad \mathbf{c}'_M = \left[\mathbf{c}'_{M_1} \mid \cdots \mid \mathbf{c}'_{M_{N_x}}\right]$$

with $A_{M_i} = \Lambda_i$.

From this decomposition, each subsystem can be characterized, that is, the *i*th-subsystem is

$$\begin{aligned} \dot{\mathbf{x}}_i(t) &= \mathbf{A}_i \mathbf{x}_i(t) + \mathbf{b}_i \mathbf{u}(t) \\ \mathbf{y}_i(t) &= \mathbf{c}'_i \mathbf{x}_i(t) \end{aligned} \tag{5.151}$$

and therefore, for the *i*th modal subsystem with state vector $\mathbf{x}_i := \left[x_{i1} \mid x_{i2}\right]'$, we have

$$\frac{d}{dt}\begin{bmatrix} x_{i1}(t) \\ --- \\ x_{i2}(t) \end{bmatrix} = \underbrace{\begin{bmatrix} \sigma_i & | & \omega_i \\ - & - & - \\ -\omega_i & | & \sigma_i \end{bmatrix}}_{\mathbf{A}_i} \begin{bmatrix} x_{i1}(t) \\ --- \\ x_{i2}(t) \end{bmatrix} + \underbrace{\begin{bmatrix} b_{i1}(t) \\ --- \\ b_{i2}(t) \end{bmatrix}}_{\mathbf{b}_i} u_i(t)$$

$$y_i(t) = \underbrace{\left[c_{i1} \mid c_{i2}\right]}_{\mathbf{c}'_i} \begin{bmatrix} x_{i1}(t) \\ --- \\ x_{i2}(t) \end{bmatrix} \tag{5.152}$$

or expanding

$$
\begin{aligned}
\dot{\mathbf{x}}_{i1}(t) &= \sigma_i \mathbf{x}_{i1}(t) + \omega_i \mathbf{x}_{i2}(t) + \mathbf{b}_{i1} u_i(t) \\
\dot{\mathbf{x}}_{i2}(t) &= -\sigma_i \mathbf{x}_{i1}(t) + \omega_i \mathbf{x}_{i2}(t) + \mathbf{b}_{i2} u_i(t) \\
\mathbf{y}_i(t) &= \mathbf{c}'_{i1} \mathbf{x}_{i1}(t) + \mathbf{c}'_{i2} \mathbf{x}_{i2}(t)
\end{aligned} \tag{5.153}
$$

which can be used to extract the individual modal responses.

We summarize the *SSP decomposition* approach as:

1. *Construct* the Hankel matrix as Eq. (5.135)
2. *Perform* the *SVD* decomposition as Eq. (5.141)
3. *Extract* the system Σ_{ABC}
4. *Perform* the modal transformation as Eq. (5.148)
5. *Extract* the sub-systems from $\Sigma_{A_M B_M C_M}$ as Eq. (5.153)

This completes the development of the *SSP* approach to model-based decomposition, next we revisit the harmonic example and apply this approach.

Example 5.8 The deterministic 3-DOF system response data in Figure 5.7 is provided to illustrate the model-based signal decomposition based on the

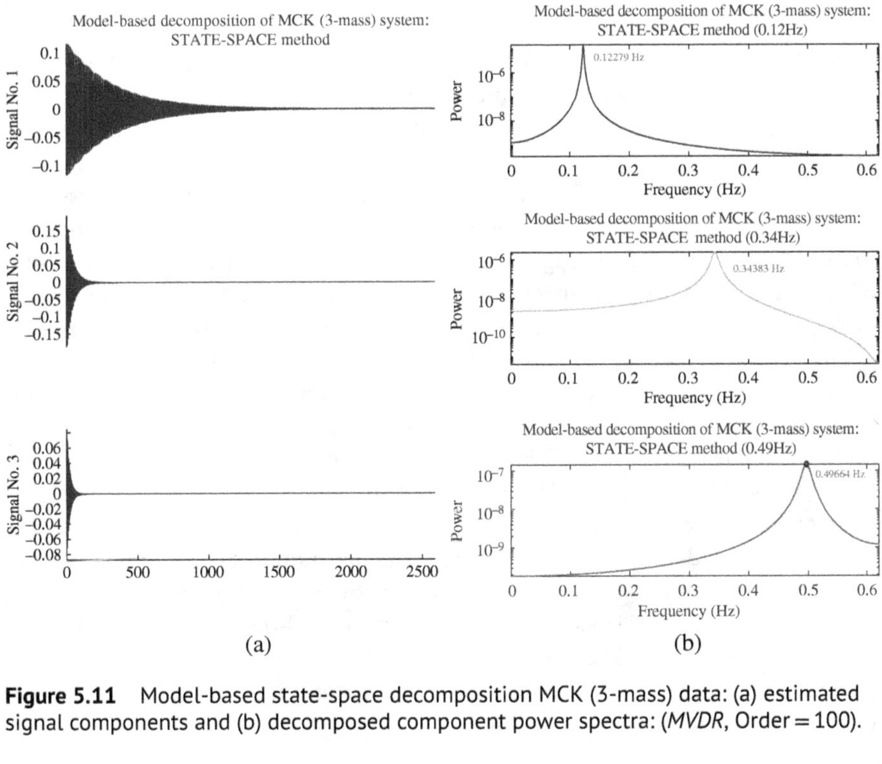

Figure 5.11 Model-based state-space decomposition MCK (3-mass) data: (a) estimated signal components and (b) decomposed component power spectra: (*MVDR*, Order = 100).

SSP modal model. The SVD-based realization approach was applied to the single-channel data. As before with the model-based decomposition approach, the underlying MCK model parameters are first estimated using the Hankel matrix *SVD*-decomposition to estimate the set of modal parameters (Θ) after the "rank approximation" method is used to estimate the number of modes required to capture the data ($\hat{N}_x = 6$). With these estimated parameters available, the individual component subsystem signal response estimates are synthesized with the results shown in Figure 5.11a along with their corresponding power spectra shown in Figure 5.11b, It is clear that the set of true modal frequencies, $\{0.123, 0.343, 0.497\,\text{Hz}\}$, are well-estimated as $\{0.12279, 0.34383, 0.49664\,\text{Hz}\}$ along with the damping estimates to produce a reasonable signal decomposition almost identical to that of the *DExp* model (Example 5.5) completing this example.

$\triangle\triangle\triangle$

5.5 Case Study: Harmonics in Noise

In this case study, we investigate the application of various signal decomposition methods applied to a set of harmonics synthesized in noisy data at a *SNR* of 0 dB. The signal data consists of 1, 2, 5, 10, and 20 Hz harmonic signals sampled at 0.02 sec for 5000 samples.

Starting with the application of the spectral decomposition method of *MTM*, *PHD*, and *MUSIC* followed by the model-based decomposition methods into a set of *DExps* and *SSP* modal decompositions techniques are all applied to decompose and extract the harmonics from noisy data. The application of the applied spectral decomposition methods is shown in Figure 5.12a where we observe the noisy time series and the Fourier spectrum followed by spectral decomposition methods of *MTM*, *PHD*, and *MUSIC* in Figure 5.12b. Each of the methods is a decomposition of the harmonic power spectrum extracting the resonant peaks using simple peak peaking methods. Next, we observe the results of the model-based decomposition.

The model-based *DExp*, lattice, and *SSP* processors are illustrated in the subsequent figures with the decomposed sub-signals shown along with their accompanying power spectra to confirm their properties. The *DExp* results are quite good with the estimated harmonic frequencies illustrated in Figure 5.13a where we see the extracted time series and the power spectrum of each signal component in Figure 5.13b.

The lattice model-based decomposition of the harmonic data is shown in Figure 5.14 with the decomposed subsystem data shown in Figure 5.14a illustrating the extracted time series after transforming to second-order system coordinates $(2, 4, 6, 8, 10)$ with the corresponding spectral response using the *MVDR*-estimator (Order = 100) shown in Figure 5.14b. The composite representation shown confirms a successful decomposition.

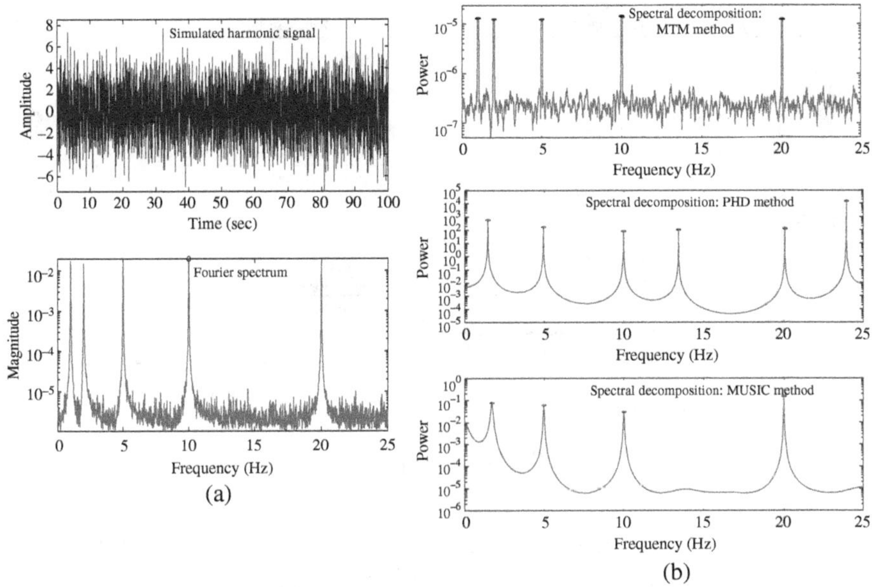

Figure 5.12 Spectral decomposition of harmonic data: (a) deterministic response data and Fourier spectrum and (b) decomposed power spectra: *MTM*, *PHD*, and *MUSIC* methods.

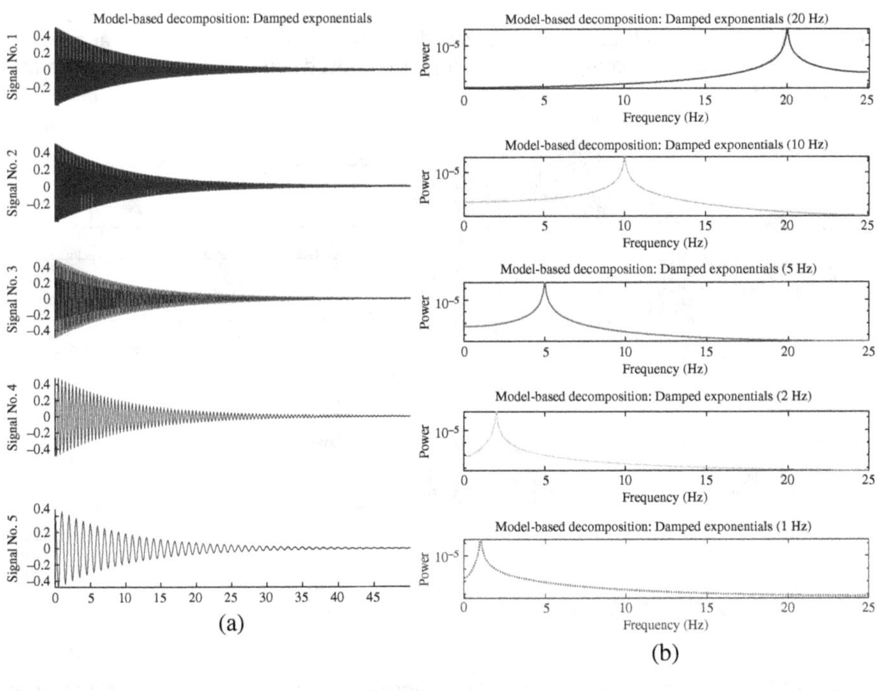

Figure 5.13 Damped exponential method of model-based decomposition of harmonic data at a *SNR* of 0 dB: (a) noisy signal data and Fourier spectrum and (b) decomposed spectra.

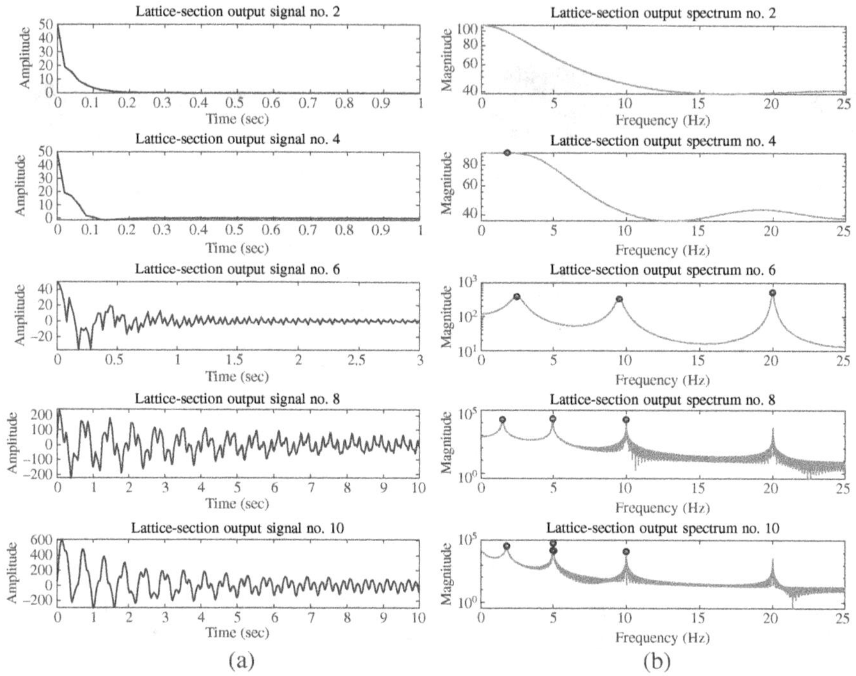

Figure 5.14 Lattice method of model-based decomposition of harmonic data at a *SNR* of 0 dB: (a) estimated signal components and (b) decomposed component power spectra at 2, 5, 10, 20 Hz.

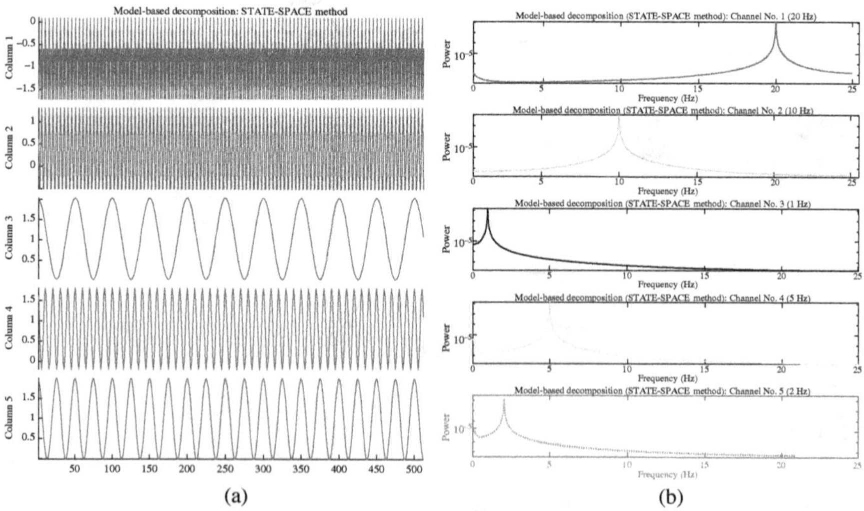

Figure 5.15 Model-based state-space decomposition of harmonic data at a *SNR* of 0 dB: (a) estimated signal components and (b) decomposed component power spectra at 20, 10, 1, 5, 2 Hz.

The *SSP* model-based decomposition of the harmonic data is shown in Figure 5.15. The decomposed subsystem data are illustrated in Figure 5.15a showing the extracted time series after transforming to modal coordinates with the corresponding spectral response using the *MVDR*-estimator (Order = 100) shown in Figure 5.15b. Each of the decomposed harmonic components are estimated and extracted successfully.

5.6 Summary

This chapter encompasses a variety of signal/system decomposition techniques grouped into three categories: data-based, spectral-based, and model-based decomposition methods. Data-based methods incorporating principal component and independent component as well as the singular spectral spectrum analysis approaches are shown to provide decomposition methods of a signal with no prior knowledge of its origins. Spectral-based methods incorporate robust methods of representing a signal in terms of special functions (e.g. prolate spheroidal functions) of the MTM and subspace decompositions based on underlying harmonic model assumptions consisting of the *PHD* and the *MUSIC* methods. Finally, specific model-based methods of *DExp*, lattice, and *SSP* approaches are discussed. All of these approaches are applied to the 3-mass MCK problem for comparative purposes as well as a case study based on multiple harmonic signals in noise aimed at investigating both the spectral and model-based decomposition methodologies.

MATLAB® Notes

MATLAB® and its *Signal Processing Toolbox* (**signal**) can be used to design nonparametric spectral estimators. The classical *PSD* methods are available in the periodogram (**periodogram** and (**pwelch**)) and the multiple taper window method using prolate spheroidal functions (**pmtm**). The transfer function is estimated using the (**tfe**) command which solves the Wiener filter problem in the frequency domain. A wealth of model-based spectral estimators (decomposition) are also available to solve for the harmonic temporal as well as spatial models incorporating the Pisaranko, **phd**, lattice **pburg**, **levinson**, and music commands **pmusic** and **rootmusic**. *DExp* model-based decomposition can be performed using the **prony** command.

MATLAB® offers a wealth of linear algebraic algorithms that can be applied to the subspace realization problem including commands to create the **hankel** and

toeplitz matrices and their decompositions using the singular value decomposition (**svd**), the Gram–Schmidt or **QR** decomposition that is easily transformed to the lower triangular/orthogonal LQ transformation used throughout. Other techniques such as the **Cholesky** factorization for covariance matrices as well as the **LDU** decomposition methods extend these methods even further.

The *MATLAB® Statistics and Machine Learning Toolbox* offer a wealth of computational statistics. The **pca**, **ppca**, **pcacov**, **pcares** offer a set of *PCA* tools enabling the calculation of principal components including the probabilistic pca tool. The accompanying covariance and residual analysis are also available. Independent component (reconstruction) analysis is available using the **risa** coupled with the **kurtosis** command as well as *SSA* capability applying the **trenddecomp** command. Other commands available in this toolbox are the maximum likelihood and corresponding covariance analysis tools **mle** and **mlecov** [18, 21, 24–35].

The *System Identification Toolbox* (**ident**) offers a large variety of both subspace (**n4sid**) and prediction-error methods for *SSP* models including *black-box* as well as *gray-box* representations. Analysis command enables the performance analysis of the identified models as well as well-founded validation methods. Using the embedded GUI allows the user to easily import the input/output data, perform preliminary pre-processing (e.g. mean and trend removal), partition the data for both identification and validation as well as enabling instantaneous analysis of results performing signal estimation plots, residual analysis, frequency domain plots, and transient response analysis. This is quite a powerful toolbox for this problem set. Other implementations of subspace algorithms exist and can be extracted from the texts of [77] and [79] (disk).

References

1 Bretscher, O. (2009). *Linear Algebra with Applications*, 4e. Upper Saddle River, NJ: Pearson Prentice-Hall.

2 Stewart, G. (1973). *Introduction to Matrix Computations*. New York: Academic Press.

3 Strang, G. (1976). *Linear Algebra and Its Applications*. New York: Academic Press.

4 Golub, G. and Van Loan, C. (1989). *Matrix Computations*. Baltimore, MD: Johns Hopkins University Press.

5 Papoulis, A. (1965). *Probability, Random Variables and Stochastic Processes*. New York: McGraw-Hill.

6 Jazwinski, A. (1970). *Stochastic Processes and Filtering Theory*. New York: Academic Press.

7 Hogg, R. and Craig, A. (1970). *Introduction to Mathematical Statistics*. New York: Macmillan.
8 Abdi, H. and Williams, L. (2010). Principal component analysis. *WIREs Comput. Stat.* 2: 433–459.
9 Jolliffe, I. (2002). *Principal Component Analysis*. New York: Springer.
10 Jackson, J. (1991). *A User's Guide to Principal Components*. Hoboken, NJ: Wiley.
11 Duda, R., Hart, P., and Stork, D. (2001). *Pattern Classification*, 2e. Hoboken, NJ: Wiley.
12 Scharf, L. (1991). *Statistical Signal Processing: Detection, Estimation, and Time Series Analysis*. Reading, MA: Addison-Wesley.
13 Lawson, C. and Hanson, R. (1974). *Solving Least Squares Problems*. New Jersey: Prentice-Hall.
14 Bishop, C. (2006). *Pattern Recognition and Machine Learning*. New York: Springer.
15 Fukunaga, K. (1990). *Statistical Pattern Recognition*, 2e. New York: Academic Press.
16 Murphy, K. (2014). *Machine Learning*. Cambridge, MA: MIT Press.
17 Webb, A. (2003). *Statistical Pattern Classification*, 2e. Hoboken, NJ: Wiley.
18 Shlens, J. (2014). *A Tutorial on Principal Component Analysis*, vol. 7, 1–12. Google Research.
19 Behrens, R. and Scharf, L. (1994). Signal processing applications of oblique projection operators. *IEEE Trans. Signal Process.* 42 (6): 1413–1424.
20 Verhaegen, M. and Verdult, V. (2007). *Filtering and System Identification: A Least-Squares Approach*. Cambridge: Cambridge University Press.
21 Nabney, I. (2002). *NETLAB: Algorithms for Pattern Recognition*. New York: Springer.
22 Cadzow, J. (1988). Signal enhancement - a composite property mapping algorithm. *IEEE Trans. Acoust. Speech Signal Process.* 36 (1): 49–62.
23 Tangirala, A. (2015). *Principles of System Identification: Theory and Practice*. Boca Raton, FL: CRC Press.
24 Haykin, S. (2009). *Neural Networks and Learning Machines*. New Jersey: Pearson Inc.
25 Comon, P. (1994). Independent component analysis, a new concept? *Signal Process.* 36: 287–314.
26 Cardoso, J. (1989). Sources separation using higher order moments. Proceedings of International Conference on Acoustics, Speech, and Signal Processing, 2109–2112.
27 Cardoso, J. (1997). Infomax and maximum likelihood for source separation. *IEEE Signal Process Lett.* 4: 112–114.
28 Kendall, M. and Stuart, A. (1977). *The Advanced Theory of Statistics*, vol. 1. New York: MacMillan Press.

29 Hyvarinen, A. and Aapo, E. (2000). Independent component analysis: algorithms and applications. *Neural Networks* 13 (4–5): 411–430.

30 Hyvarinen, A., Karhunen, J., and Oja, E. (2001). *Independent Component Analysis*. New York: Wiley.

31 Stone, J. (2004). *Independent Component Analysis: A Tutorial Introduction*. Cambridge, MA: MIT Press.

32 Bell, A. and Sejnowski, D. (1995). An information-maximization approach to blind separation and blind deconvolution. *Neural Comput.* 7: 1129–1159.

33 Shlens, J. (2014). *A Tutorial on Independent Component Analysis*, vol. 7, 1–13. Google Research.

34 Lee, T., Girolami, M., and Sejinowski, T. (1999). Independent component analysis using an extended infomax algorithm for mixed sub-Gaussian and super-Gaussian sources. *Neural Comput.* 11 (2): 417–441.

35 Pham, D., Garrat, P., and Jutten, C. (1992). Separation of a mixture of independent sources through a maximum likelihood approach. *Proc. EUSIPCO* 771–774.

36 Gawronski, W. (2004). *Advanced Structural Dynamics and Active Control of Structures*. New York: Springer.

37 Huber, P. (1985). Projection Pursuit. *Ann. Stat.* 13 (2): 435–475.

38 Broomhead, D. and King, G. (1986). Extracting qualitative dynamics from experimental data. *Physica D* 20: 217–236.

39 Elsner, J. and Tsonis, A. (1996). *Singular Spectral Analysis: A New Tool in Time Series Analysis*. New York: Plenum Press.

40 Allen, M. and Robertson, A. (1996). Distinguishing modulated oscillation from colored noise in multivariate datasets. *Clim. Dyn.* 12: 775–784.

41 Golyandina, N. and Osipov, E. (2007). The "Caterpiller"-SSA method for analysis of time series with missing values. *J. Stat. Plann. Inference* 137 (8): 2642–2653.

42 Zhigljavsky, A. (2010). Special issue on theory and practice in singular spectral analysis. *Stat. Interface* 20: 255–258.

43 Gillard, J. (2010). Cadzow's basic algorithm, alternating projections and singular spectral analysis. *Stat. Interface* 3: 335–343.

44 Hassani, H. and Thomakos, D. (2010). A review on singular spectrum analysis for economic and financial time series. *Stat. Interface* 3: 377–397.

45 Golyandina, N., Nekrutkin, V., and Zhigljavsky, A.A. (2013). *Analysis of Time Series Structure: SSA and Related Techniques*. New York: Chapman & Hall CRC Press.

46 Golyandina, N. and Zhigljavsky, A. (2013). *Singular Spectral Analysis for Time Series*. New York: Springer.

47 Golyandina, N. and Zhigljavsky, A. (2014). Basic singular spectral analysis and forecasting with R. *Comput. Stat. Data Anal.* 71: 1–40.

48 Candy, J. (2006). *Model Based Signal Processing*. Hoboken, NJ: Wiley/IEEE Press.

49 Thomson, D. (1982). Spectrum estimation and harmonic analysis. *Proc. IEEE* 70 (9): 1055–1096.

50 Percival, D. and Walden, A. (1993). *Spectral Analysis for Physical Applications*, 3e. Cambridge: Cambridge University Press.

51 Slepian, D. and Pollak, H. (1961). Prolate spheroidal wave functions, Fourier analysis and uncertainty—I. *Bell Syst. Tech. J.* 40 (1): 43–63.

52 Bronez, T. (1992). On the performance advantage of multitaper spectral analysis. *IEEE Trans. Signal Process.* 40 (12): 2941–2946.

53 Walden, A. (1995). Multitaper estimation of the innovation variance of a stationary time series. *IEEE Trans. Signal Process.* 43 (1): 181–187.

54 Haykin, S., Thomson, D., and Reed, J. (2009). Spectrum sensing for cognitive radio. *Proc. IEEE* 97 (5): 849–877.

55 Babadi, B. and Brown, E. (2014). A review of multitaper spectral analysis. *IEEE Trans. Biomed. Eng.* 61 (5): 1555–1564.

56 Pisarenko, V. (1973). The retrieval of harmonics from a covariance function. *Geophys. J. R. Astron. Soc.* 30: 347–356.

57 Schmidt, R. (1983). Multiple signal classification. *IEEE Antennas Propag. Mag.* 30: 276–280.

58 Johnson, D. and Dudgeon, D. (1993). *Array Signal Processing: Concepts and Techniques*. Englewood Cliffs, NJ: Prentice-Hall.

59 Candy, J. (2019). *Model Based Processing: An Applied Subspace Approach*. Hoboken, NJ: Wiley.

60 Hayes, M. (1996). *Statistical Digital Signal Processing and Modeling*. New York: Wiley.

61 Kumaresan, R. and Tufts, D. (1982). Estimating the parameters of exponentially damped sinusoids and pole-zero modeling in noise. *IEEE Trans. Acoust. Speech Signal Process.* ASSP-30 (6): 833–840.

62 Cadzow, J.A. and Wu, M.-M. (1987). Analysis of transient data in noise. *IEE Proceedings* 134 (1): 69–78.

63 Kung, S., Arun, K., and Rao, D. (1983). State-space and singular-value decomposition-based approximation methods for the harmonic retrieval problem. *J. Opt. Soc. Am.* 73 (12): 1799–1811.

64 Cadzow, J., Baseghi, B., and Hsu, T. (1983). Singular-value decomposition approach to time series modeling. *IEE Proceedings* 130 (3): 202–210.

65 Tufts, D. and Kumaresan, R. (1982). Estimation of frequencies of multiple sinusoids: making linear prediction perform like maximum likelihood. *Proc. IEEE* 70 (9): 975–989.

66 Rao, B. and Arun, K. (1992). Model-based processing of signals: a state-space approach. *Proc. IEEE* 80 (2): 283–309.
67 Orfanidis, S. (1985). *Optimum Signal Processing*. New York: Macmillan.
68 Haykin, S. (1996). *Adaptive Filtering Theory*. New Jersey: Prentice-Hall.
69 Robinson, E. and Treitel, S. (1980). *Geophysical Signal Analysis*. New Jersey: Prentice-Hall.
70 Silvia, M. and Robinson, E. (1979). *Deconvolution of Geophysical Time Series in the Exploration of Oil and Natural Gas*. New York: Elsevier.
71 Markel, J. and Gray, A. (1976). *Linear Prediction of Speech*. New York: Springer.
72 Friedlander, B. (1982). Lattice filters for adaptive processing. *Proc. IEEE* 70 (8): 829–867.
73 Kay, S. (1988). *Modern Spectral Estimation*. New Jersey: Prentice-Hall.
74 Kailath, T. (1981). *Lectures on Kalman and Wiener Filtering Theory*. New York: Springer.
75 Candy, J. (2022). Subsystem Estimation: A Modal Decomposition Approach. LLNL Report, *LLNL-TR12345*.
76 Moore, B. (1981). Principal component analysis in linear systems: controllability, observability and model reduction. *IEEE Trans. Autom. Control* AC-26: 17–32.
77 Katayama, T. (2005). *Subspace Methods for System Identification*. London: Springer.
78 Tretter, S. (1976). *Introduction to Discrete-Time Signal Processing*. New York: Wiley.
79 van Overschee, P. and De Moor, B. (1996). *Subspace Identification for Linear Systems: Theory, Implementation, Applications*. Boston, MA: Kluwer Academic Publishers.
80 Reyenders, E. (2012). System identification methods for modal analysis: review and comparison. *Arch. Comput. Meth. Eng.* 19 (1): 51–124.
81 Rainieri, C. and Fabbrocino, G. (2014). *Operational Modal Analysis of Civil Engineering Structures*. New York: Springer.

Problems

5.1 Given the data matrix, $\mathbf{X} \in \mathrm{R}^{n \times m}$, find an orthonormal transformation matrix, $\mathcal{T}$ such that $\mathbf{Y} = \mathcal{T} \times \mathbf{X}$ diagonalizes the associated covariance matrix $\mathbf{R}_{YY}$ yielding the principal components of $\mathbf{X}$ as row vectors $\mathbf{t}'$ of $\mathcal{T}$.

(a) Develop the solution as an eigen-decomposition of $\mathbf{R}_{YY}$.

(b) Show that the row vectors $\mathbf{t}_i'$ are eigenvectors of $\mathbf{XX}'$ with corresponding eigenvalues λ_i.
(c) Obtain the PCA of the SVD($\mathbf{XX}'$).
(d) How are the results of (a) and (c) related?

5.2 Suppose we have obtained a set of 2×2 covariance matrices from different measurements given by

$$\begin{bmatrix} 1 & 1/4 \\ 1/4 & 1 \end{bmatrix}, \begin{bmatrix} 1 & 1/2 \\ 1/2 & 1 \end{bmatrix}, \begin{bmatrix} 1 & 3/4 \\ 3/4 & 1 \end{bmatrix}, \begin{bmatrix} 1 & 1 \\ 1 & 1 \end{bmatrix}$$

corresponding to the variations of estimated eigenvectors, construct the set of image ellipses associated with each, and discuss their characteristics.

5.3 Suppose we perform the singular value decomposition of an $n \times m$ data matrix $\mathbf{X}$ for ($SVD(X) = U\Sigma V'$):
(a) Show that the left singular unit vectors, that is, the columns of U are orthogonal.
(b) Show that the right singular unit vectors, that is, the rows of V' are orthogonal.
(c) Show that the singular values of the rectangular diagonal matrix Σ are related to the eigenvalues of $\mathbf{X}$.
(d) Employing the $SVD(X)$ and specifying the product $V'X$, how are the resulting matrices, $\hat{U}$, $\hat{V}$, and $\hat{\Sigma}$ related?
(e) How are the columns of the "score" vector $S = XV$ related to the left singular vectors U?

5.4 We are given the linear transformation matrix,

$$\mathcal{T} = \begin{bmatrix} 6 & 2 \\ -7 & 6 \end{bmatrix}$$

(a) Construct an orthonormal basis $\{v_1, v_2\}$ for this transformation.
(b) Show that the image of the unit circle consisting of the vectors $x = \cos(t)v_1 + \sin(t)v_2$ with the image of the transformed vectors $\mathcal{T}x = \cos(t)\mathcal{T}(v_1) + \sin(t)\mathcal{T}(v_2)$ is an ellipse and specify the length of each axis (Hint: eigenvalues of $\mathcal{T}'\mathcal{T}$).
(c) Sketch the images.

5.5 Similar to the previous problem with the transformation now given by

$$\mathcal{T} = \begin{bmatrix} 0 & 1 & 1 \\ 1 & 1 & 0 \end{bmatrix}$$

(a) Find the singular values of $\mathcal{T}$.

(b) Find the set of orthonormal vectors $\{v_1, v_2, v_3\}$ such that $\{\mathcal{T}(v_1), \mathcal{T}(v_2), \mathcal{T}(v_3), \}$ are orthogonal.
(c) Sketch the images under $\mathcal{T}$.

5.6 Consider the singular value decomposition of a 2×2 matrix A. Progress with each step of the *SVD*(A) operation and sketch each of the relationships showing explicitly the scaling and rotations created by the left/right singular vectors.

5.7 Perform the SVD transformation of the data matrix given by

$$\mathcal{X} = 0.5 \times \begin{bmatrix} 2 & 0 & 0 & 0 & 4 \\ 0 & 0 & 6 & 0 & 0 \\ 0 & 0 & 0 & 0 & 0 \\ 0 & 4 & 0 & 0 & 0 \end{bmatrix}$$

(a) Find the left singular vectors from the columns of U. Show that they are orthonormal.
(b) Find the right singular vectors from the rows of V'. Show that they are orthonormal.
(c) What is the rank of X? What is the associated scaling matrix, Σ?
(d) Estimate the pseudo-inverse of X using both brute force and the *SVD*(X). How do the values compare?

5.8 The *ICA* approach consists of two steps: (i) pre-whitening the raw data matrix (PCA step) followed by (ii) extraction of the required rotation matrix. Suppose we are given the data matrix X that consists of correlated components.
(a) Define the whitening filter design approach to transform this data.
(b) How does this relate to the PCA approach? What makes ICA different?
(c) Develop the required transformation matrix.
(d) Show how using the pre-whitening procedure leads to the estimation of the required rotation matrix.

5.9 Synthesize a dataset of two sinusoids: one clipped (± 0.4 amp.) at 1 Hz and one at 2 Hz. The data are sampled at $F_s = 10$ Hz for 256 samples. The underlying (unknown) mixing matrix for simulation ($Y = M \times S$) is given by:

$$M = \begin{bmatrix} 1 & 0.5 \\ 0.25 & 1 \end{bmatrix}$$

(a) Show each of the synthesized signals and the mixed data.
(b) Show the estimated mixed data output power spectrum.

(c) Estimate the ICA decomposition and show each of the unmixed signals compared to the actual.

5.10 Simulate a random periodic ($P = 22$ samples) sinusoid of 200 samples with a 0 dB SNR. Develop a singular spectral decomposition of the dataset using 30 sample window section length.

(a) Display the synthesized signal and estimated covariance matrix (image) obtained by sectioning.

(b) Show the original and reconstructed sinusoidal signal for both full and first two components.

(c) Estimate the PCA decomposition for the sinusoidal signal and show each of its components.

(d) Estimate the SSA decomposition for the sinusoidal signal and show each of its components.

5.11 For a critical system component under test containing material properties that must be determined based on a subset of its inherent resonant frequencies. The set of resonant frequencies are given by $F_{res} :=$ $\{15, 16, 17, 20, 24\}$ KHz at amplitudes of $A_{res} := \{0.5, 0.6, 0.35, 0.7, 0.25\}$.

(a) Synthesize this *material response problem* using 100 KHz sampling frequency for 4000 samples and display and display both the true and noisy signals along with their raw power spectra.

(b) Apply the *MTM* and estimate:

1. The number of DPSS windows required.
2. The corresponding decomposed *MTM* power spectrum showing the set of resonances.
3. The spectral peaks extracted.

5.12 For the synthesized material response problem data (above) apply both the Pisarenko and *MUSIC* spectral decomposition methods to:

(a) Estimate and compare the decomposed resonance frequency responses.

(b) Extract the resonant peaks.

(c) Compare the results.

5.13 Again for the synthesized material response data apply the model-based decomposition methods: *DExp*, lattice, and state-space to:

(a) Estimate and compare the decomposed resonance frequency responses.

(b) Extract the resonant peaks.

(c) Compare the results.

6

Model-based Decomposition: Time Domain

6.1 Background: State-space Systems

This chapter focuses on the foundation of the system identification problem for state-space systems that lead to Subspace IDentification (SID) techniques for subsequent system/signal decomposition. The original basis has evolved from systems theory and the work of Kalman for control system design [1]. The fundamental problem is called the "realization problem"—the extraction of a linear state-space model from impulse response data or equivalently the transfer function representation. The basic paper by Ho and Kalman has been cited as the seminal publication showing how to extract the state-space model [2, 3]. Initially, we discuss this problem for two distinct data sets: impulse sequences of an infinite length (number of samples) data leading to the realization problem and input/output data leading to the SID problem. Subspace refers primarily to a vector (state) space in which its essential characteristics can be captured by a "smaller" space—the subspace. For example, an Nth order state-space dynamic system can be approximated by an Mth order ($M < N$) system capturing the majority of its properties. The subspace approach does *not* require any specific structure (black-box) of the state-space model only its order (number of states) that can be derived directly from the embedded algebraic decomposition.

SID is a technique to extract a model in generic state-space form from input/output data using robust, numerically stable, linear algebraic methods. It performs a black-box identification of a linear, time-invariant, and state-space model $\Sigma_{ABCD} := \{A, B, C, D\}$ from uncertain measurements. Here, we focus on the *deterministic* problem ignoring the uncertainties created by noisy measurements. We concentrate on the discrete linear time-invariant (LTI) system that is an integral component of the subspace decomposition techniques to follow. First, we develop a compact vector-matrix form of the system input/output relations that

Signal Processing: An Applied Decomposition Approach, First Edition. James Vincent Candy.

Companion website: www.wiley.com/go/CandySignProc

will prove crucial in extracting a generic model from noisy data and subsequently performing a coordinate transformation to "modal space" and then decomposing the transformed system.

Recall that the discrete *linear time-varying state-space representation* is given by the *system* or *process model* as

$$x(t) = A(t-1)x(t-1) + B(t-1)u(t-1) \tag{6.1}$$

with the corresponding discrete *output* or *measurement model* as

$$y(t) = C(t)x(t) + D(t)u(t) \tag{6.2}$$

where x, u, y are the respective N_x-state, N_u-input, N_y-output, and A, B, C, D are the $(N_x \times N_x)$-system, $(N_x \times N_u)$-input, $(N_y \times N_x)$-output, $(N_y \times N_u)$-feedthrough matrices.

Therefore, we have that the state-transition matrix for this system is given by

$$\Phi(t,k) = A(t-1) \cdot A(t-2) \cdots A(k) \overbrace{=}^{A(t)\to A} A^{t-k} \text{ for } t > k$$

Constraining the system to be time invariant, that is,

$$\begin{aligned} x(t) &= Ax(t-1) + Bu(t-1) \\ y(t) &= Cx(t) + Du(t) \end{aligned} \tag{6.3}$$

leads to the discrete *LTI* solution as

$$x(t) = A^t x(0) + \sum_{k=0}^{t-1} A^{t-k} Bu(k); \qquad t = 0, 1, \ldots, K \tag{6.4}$$

with the measurement or output system given by

$$y(t) = CA^t x(0) + \sum_{k=0}^{t-1} CA^{t-k} Bu(k) + D\delta(t) \tag{6.5}$$

Next, we discuss some of the pertinent system theoretic results that will enable us to comprehend much of the subspace realization algorithms to follow.

6.1.1 Discrete Systems Theory

In this section, we investigate the discrete state-space model from a systems theoretic viewpoint. There are certain properties that a dynamic system must possess in order to assure a consistent representation of the dynamics under investigation. For instance, it can be shown [4] that a necessary requirement of a measurement system is that it is *observable* providing enough information to reconstruct the internal variables or states.

Mathematically, a system is said to be *completely observable*, if for any initial state, say $x(0)$, in the state space, there exists a finite $t > 0$ such that knowledge of the input $u(t)$ and the output $y(t)$ is sufficient to specify $x(0)$ uniquely. Recall that the linear *state-space* representation of a discrete system is defined by the following set of equations:

$$\text{State Model:}\quad x(t) = A(t-1)x(t-1) + B(t-1)u(t-1)$$

with corresponding measurement system or output defined by

$$\text{Measurement Model:}\quad y(t) = C(t)x(t) + D(t)u(t)$$

Using this representation, the simplest example of an observable system is one in which each state is measured directly, therefore and the measurement matrix C is a $N_x \times N_x$ matrix. In order to reconstruct $x(t)$ from its measurements $y(t)$, then from the measurement system model, C must be invertible. In this case, the system is said to be completely observable; however, if C is not invertible, then the system is said to be *unobservable*.

The next level of complexity involves the solution to this same problem when C is a $N_y \times N_x$ matrix, then a pseudo-inverse must be performed instead [5, 6]. In the general case, the solution gets more involved because we are not just interested in reconstructing $x(t)$, but $x(t)$ over all finite values of t; therefore, we must include the state model, that is, the dynamics as well.

With this motivation in mind, we now formally define the concept of observability. The solution to the state representation is governed by the state-transition matrix, $\Phi(t,0)$, where recall that the solution of the state equation is [7]

$$x(t) = \Phi(t,0)x(0) + \sum_{k=0}^{t-1} \Phi(t,k)B(k)u(k)$$

Therefore, pre-multiplying by the measurement matrix, the output relations are

$$y(t) = C(t)\Phi(t,0)x(0) + \sum_{k=0}^{t-1} C(t)\Phi(t,k)B(k)u(k) \tag{6.6}$$

or rearranging, we define

$$\tilde{y}(t) := y(t) - \sum_{k=0}^{t-1} C(t)\Phi(t,k)B(k)u(k) = C(t)\Phi(t,0)x(0) \tag{6.7}$$

The problem is to solve this resulting equation for the initial state; therefore, multiplying both sides by $\Phi' C'$, we can infer the solution from the relation

$$\Phi'(t,0)C'(t)C(t)\Phi(t,0)x(0) = \Phi'(t,0)C(t)\tilde{y}(t)$$

Thus, the observability question now becomes under what conditions can this equation uniquely be solved for $x(0)$? Equivalently, we are asking if the null space of $C(t)\Phi(t,0)$ is $\mathbf{0} \in \mathcal{R}^{N_x \times 1}$. It has been shown [4, 5] that the following $N_x \times N_x$ *observability Gramian* has the identical null space, that is,

$$\mathcal{O}(0,t) := \sum_{k=0}^{t-1} \Phi'(k,0)C'(k)C(k)\Phi(k,0) \tag{6.8}$$

which is equivalent to determining that $\mathcal{O}(0,t)$ is nonsingular or rank N_x.

Further assuming that the system is *LTI*, then over a finite time interval for $t = 0, \ldots, K-1$ for $K > N_x$ leads to the $KN_y \times N_x$ *observability matrix* [4] defined by

$$\mathcal{O}_K := \begin{bmatrix} C \\ --- \\ \vdots \\ --- \\ CA^{K-1} \end{bmatrix} \tag{6.9}$$

A necessary and sufficient condition for a system to be completely observable is that the *rank*, $\rho\left(\mathcal{O}_K\right)$, must be N_x. Thus, for the *LTI* case, checking that all of the measurements contain the essential information to reconstruct the states reduces to checking the rank of the observability matrix. Although this is a useful mathematical concept, it is primarily used as a rule-of-thumb in the analysis of complicated measurement systems.

Analogous to the system theoretic property of observability is that of controllability, which is concerned with the effect of the input on the states of the dynamic system. A discrete system is said to be *completely controllable*, if for any $x(t), x(0) \in \mathcal{R}^{N_x \times 1}$, there exists an input sequence, $\{u(t)\}$ such that the solution to the state equations with initial condition $x(0)$ is $x(t)$ for some finite $t > 0$.

Following the same approach as that for observability, we obtain the *controllability Gramian* defined by

$$C(0,t) := \sum_{k=0}^{t-1} \Phi(0,k)B(k)B'(k)\Phi'(0,k) \tag{6.10}$$

and then require that it is nonsingular or $\rho\,[C(0,t)] = N_x$ for controllability.

Again for the *LTI* system, then over a finite time interval for $t = 0, \ldots, K-1$ for $K > N_x$, the $N_x \times KN_u$ *controllability matrix* defined by

$$C_K := \left[B\,|\,AB\,|\,\cdots\,|\,A^{K-1}B \right] \tag{6.11}$$

must satisfy the rank condition, $\rho\left(C_K\right) = N_x$ to be completely controllable [4].

6.1.2 Stable Linear Systems

Stability of a linear system can also be cast in conjunction with the properties of controllability and observability [4–8]. For the homogeneous discrete system, it has been shown that asymptotic stability follows directly from the *eigenvalues* of the system matrix expressed as $\lambda(A)$. These eigenvalues must lie within the unit circle or equivalently $\lambda_i(A) < 1 \; \forall \; i$.

Besides determining eigenvalues of A or equivalently factoring the characteristic equation to extract the system poles, another way to evaluate this condition follows directly from *Lyapunov stability* theory that incorporates observability [4, 8].

An observable system is *stable*, if an only if the *Lyapunov equation*

$$P = A'PA + C'C \qquad \text{[Lyapunov Equation]} \tag{6.12}$$

has a *unique positive definite* solution, $P > 0$. The proof is available in [4] or [8]. As we shall see, this is an important property that appears in SID theory, when extracting a so-called *balanced realization* [8].

6.1.3 Equivalent Linear Systems

Equivalent systems are based on the concept that there are an infinite number of state-space systems that possess "identical" input/output responses. In the state-space this property is termed a set of coordinates, that is, we can change the state vectors describing a system by a *similarity transformation* such that the system matrices $\Sigma_{ABCD} := \{A, B, C, D\}$ are transformed to a different set of coordinates by the transformation matrix T [4–8], that is,

$$\Sigma_{ABCD} \xrightarrow{\;T\;} \overline{\Sigma}_{\overline{ABCD}}$$

where we have

$$\{A, B, C, D\} \longrightarrow \left\{\overline{A}, \overline{B}, \overline{C}, \overline{D}\right\} := \left\{TAT^{-1}\,, TB,\; CT^{-1}, D\right\}$$

yields an "equivalent" system from an input/output perspective, that is, the *transfer functions* are identical

$$\begin{aligned} H(z) &= \overline{C}(zI - \overline{A})^{-1}\overline{B} + \overline{D} = CT^{-1} \times (zI - TAT^{-1})^{-1} \times TB + D \\ &= C(zI - A)^{-1}B + D \end{aligned}$$

as well as the corresponding impulse response matrices

$$\begin{aligned} H_t &= \overline{C}\,\overline{A}^{t-1}\,\overline{B} + \overline{D}\delta(t) = CT^{-1} \times TA^{t-1}T^{-1} \times TB + D\delta(t) \\ &= C\,A^{t-1}\,B + D\delta(t) \end{aligned}$$

There does exist unique representations of state-space systems termed "canonical forms" of Section 3.3 for single input/single output (*SISO*) systems as well as for multiple input/multiple output (*MIMO*) systems [9]. In fact, much of control theory is based on designing controllers in a *modal coordinate system* with a diagonal system or modal matrix A_M and then transforming the modal system back to the physical coordinates using the inverse transform $T \rightarrow T_M^{-1}$ [4].

6.1.4 Modal Systems

Perhaps one of the most expository representations of a system is its modal representation, where its internal structure and response to various excitations are exposed [4, 8]. The modal *decomposition* of a system can easily be determined from state-space systems by transforming its current coordinates to modal space which is accomplished through a *similarity transformation* constructed from an eigen-decomposition, that is, the system matrices $\Sigma := \{A, B, C, D\}$ are transformed to modal coordinates by the transformation matrix T_M constructed of the eigenvectors of the underlying system [5–7], that is,

$$\Sigma \xrightarrow{T_M} \Sigma_M$$

where we have

$$\{A, B, C, D\} \longrightarrow \{A_M, B_M, C_M, D_M\} := \{T_M A T_M^{-1}, T_M B, CT_M^{-1}, D\}$$

that yields an "equivalent" system from an input/output perspective, that is, the transfer functions are identical

$$\begin{aligned} H(s) &= C_M(sI - A_M)^{-1}B_M + D_M = CT_M^{-1} \times (sI - T_M A T_M^{-1})^{-1} \times T_M B \\ &= C(sI - A)^{-1}B + D \end{aligned}$$

as well as the corresponding impulse response matrices

$$H(\tau) = C_M e^{A_M(\tau)} B_M = CT_M^{-1} \times T_M e^{A(\tau)} T_M^{-1} \times T_M B = Ce^{A(\tau)}B$$

It is well-known from systems theory that the solution to the *continuous-time*, state-space system is given by[1] [5–8]

$$\mathbf{x}(\tau) = \underbrace{e^{A(\tau-\tau_0)}\mathbf{x}(\tau_0)}_{\text{homogeneous}} + \underbrace{\int_{\tau_0}^{\tau} e^{A(\tau-\alpha)}\mathbf{B}\mathbf{u}(\alpha)d\alpha}_{\text{forced}} \tag{6.13}$$

consisting of the homogeneous and forced solutions from which can define the *state-transition matrix* for the *LTI* system as

$$\mathbf{\Phi}(\tau, \tau_0) = e^{A(\tau-\tau_0)} \qquad \text{[State-Transition Matrix]} \tag{6.14}$$

Transforming this solution by performing an eigen-decomposition for distinct (independent) eigenvalues leads to a diagonal transition matrix, that is, selecting the modal similarity transformation T_M for

$$x(\tau) = T_M \times x_M(\tau)$$

1 Here τ is the time variable used for the continuous-time system.

satisfying the eigen-equations

$$A\boldsymbol{\varepsilon}_i = \lambda_i \boldsymbol{\varepsilon}_i \qquad \text{for} \qquad i = 1, 2, \ldots, N_x$$

where $\{\lambda_i\}$ are the distinct eigenvalues (real and/or complex) of A or equivalently the roots (modes) of the determinant $|\lambda I - A|$ (characteristic equation) and $\{\boldsymbol{\varepsilon}_i\}$ are the corresponding eigenvectors that lead to [4]

$$T_M = \left[\boldsymbol{\varepsilon}_1 \,|\, \boldsymbol{\varepsilon}_2 \,|\, \cdots \,|\, \boldsymbol{\varepsilon}_{N_x}\right]$$

the *modal matrix* consisting of independent columns such that

$$AT_M = T_M \Lambda$$

for $\Lambda = \text{diag}[\lambda_1, \lambda_2, \ldots, \lambda_{N_x}]$.

For this modal system, we have that the *modal* state-transition matrix $\boldsymbol{\Phi}_M$ follows from the eigen-decomposition as

$$\boldsymbol{\Phi}_M(\tau, \tau_0) = T_M e^{A(\tau-\tau_0)} T_M^{-1} = e^{T_M A(\tau-\tau_0) T_M^{-1}} = e^{\Lambda(\tau-\tau_0)} = \sum_{i=1}^{N_x} \varepsilon_i e^{\lambda_i(\tau-\tau_0)} \eta_i^T \tag{6.15}$$

where we have incorporated the *reciprocal eigenvectors* $\{\eta_i^T\}$; $i = 1, \ldots, N_x$ (rows of T_M^{-1}) to obtain [6]

$$\boldsymbol{\Phi}_M(\tau, \tau_0) = e^{\Lambda(\tau-\tau_0)} = \underbrace{\begin{bmatrix} e^{\lambda_1(\tau-\tau_0)} & & 0 \\ & \ddots & \\ 0 & & e^{\lambda_{N_x}(\tau-\tau_0)} \end{bmatrix}}_{\text{Modal Transition Matrix}} \tag{6.16}$$

In this coordinate system, the *modal* state solution is:

$$\mathbf{x}_M(\tau) = e^{\Lambda(\tau-\tau_0)} \mathbf{x}_M(\tau_0) + \int_{\tau_0}^{\tau} e^{\Lambda(\tau-\alpha)} B\mathbf{u}(\alpha) d\alpha \tag{6.17}$$

which can be written explicitly for the ith-mode x_i as

$$x_i(\tau) = e^{\lambda_i(\tau-\tau_0)} x_i(\tau_0) + \int_{\tau_0}^{\tau} e^{\lambda_i(\tau-\alpha)} b_{im} u_m(\alpha) d\alpha \text{ for } i = 1, 2, \ldots, N_x; \quad m = 1, \ldots, N_u \tag{6.18}$$

with b_{im} the (i, m)th component of the input transmission matrix B.

The corresponding measurement or output of the state-space system is easily found by multiplying this relation by the measurement matrix C, that is,

$$\mathbf{y}_M(\tau) = C_M e^{\Lambda(\tau-\tau_0)} \mathbf{x}(\tau_0) + \int_{\tau_0}^{\tau} C_M e^{\Lambda(\tau-\alpha)} B_M \mathbf{u}(\alpha) d\alpha \tag{6.19}$$

which can also be expressed by applying the modal transformation to obtain the mth-component of the output y_m as

$$y_n(\tau) = \sum_{i=1}^{N_x} c_{ni}^T e^{\lambda_i(\tau-\tau_0)}\, x_i(\tau_0) + \sum_{i=1}^{N_x} \int_{\tau_0}^{\tau} c_{ni} e^{\lambda_i(\tau-\alpha)}\, b_{im} u_m(\alpha) d\alpha \text{ for } n = 1, \ldots, N_y \tag{6.20}$$

with c_{ni} the (n, i)th component of the output transmission matrix C. This is defined as the *modal representation* of the system [6].

It is also interesting to note that if we *constrain* this solution to be *only* homogeneous, then the response to the initial (condition) state vector is simply

$$\mathbf{x}_M(\tau) = e^{\Lambda(\tau-\tau_0)}\mathbf{x}_M(\tau_0)$$

or component-wise, we have

$$\mathbf{x}_i(\tau) = e^{\lambda_i(\tau-\tau_0)}\mathbf{x}_i(\tau_0) \quad \text{for} \quad i = 1, 2, \ldots, N_x \tag{6.21}$$

where motion is expressed in terms of its eigenvalue–the natural frequency of the ith-principal *mode* along with its corresponding eigenvector.

Expanding this expression over i leads to the total solution or *modal decomposition* of the state-space as

$$\mathbf{x}_M(\tau) = e^{\Lambda(\tau-\tau_0)}\mathbf{x}_M(\tau_0) = \sum_{i=1}^{N_x} \left(\varepsilon_i e^{\lambda_i(\tau-\tau_0)} \eta_i^T\right) \mathbf{x}_i(\tau_0) \quad \text{[Modal Decomposition]} \tag{6.22}$$

that is, the homogeneous response is represented by a weighted superposition of system modes $\varepsilon_i e^{\lambda_i(\tau-\tau_0)}$ with the strength of each mode effected by $\mathbf{x}_i(\tau_0)$. Next, we discuss the special case of distinct eigenvalues that are complex.

With this in mind, we now extend the *modal state-space system* to the complex modal case which is quite common many applications [4, 6]. For a typical system, the eigenvalues are complex, but still distinct. In this case, the system matrix can be decomposed, as before, using the eigen-decomposition which now yields complex eigen-pairs along with the corresponding complex eigenvectors, that is,

$$\begin{aligned} &\lambda_i = -\sigma_i \pm j\omega_i \\ &\mathbf{t}_i = [\xi_i \mid \xi_i^*] \quad \text{for} \quad i = 1, 2, \ldots, N_d \text{ for } N_x = 2 \times N_d \end{aligned} \tag{6.23}$$

for N_d the number of the complex poles and $\mathbf{t}_i$, the ith-column vector of T_M with σ_i is the *damping coefficient*, ω_i the *damped natural frequency*, and ϕ_i the *phase* such that

$$e^{-\sigma_i\tau} \times \cos\left(\omega_i\tau + \phi_i\right) \qquad \text{[Damped Sinusoid]}$$

Note that the *homogeneous* state and output responses are *real*, since

$$\mathbf{x}_M(\tau) = \sum_{i=1}^{N_x} 2e^{-\sigma_i(\tau-\tau_0)} \operatorname{Re}\{\mathbf{x}_i(\tau_0)\} = \sum_{i=1}^{N_x} 2e^{-\sigma_i(\tau-\tau_0)} |\mathbf{x}_i(\tau_0)| \cos(\omega_i \tau + \phi_i) \tag{6.24}$$

We also have that the modal transformation matrix becomes

$$T_M = \left[\xi_1\ \xi_1^* \mid \xi_2\ \xi_2^* \mid \cdots \mid \xi_{N_x/2}\ \xi_{N_x/2}^* \mid\right] \tag{6.25}$$

and applying this transformation to the system matrix A, we obtain

$$\boldsymbol{\Lambda} := T_M \times A \times T_M^{-1} = \operatorname{diag}\left(\begin{bmatrix} \sigma_1 & \omega_1 \\ -\omega_1 & \sigma_1 \end{bmatrix}, \ldots, \begin{bmatrix} \sigma_{N_x} & \omega_{N_x} \\ -\omega_{N_x} & \sigma_{N_x} \end{bmatrix}\right) \tag{6.26}$$

with

$$B_M = T_M \times B; C_M = C \times T_M^{-1} \qquad \text{and} \qquad D_M = D$$

which leads to the modal state-transition matrix for the complex eigen system as

$$\Phi_M(\tau, \tau_0) = e^{\Lambda(\tau-\tau_0)} = \exp\left\{ \underbrace{\begin{bmatrix} \Lambda_1 & & 0 \\ & \ddots & \\ 0 & & \Lambda_{Nx} \end{bmatrix}}_{\Lambda} (\tau - \tau_0) \right\} \text{ for } \Lambda_i = \begin{bmatrix} \sigma_i & \omega_i \\ -\omega_i & \sigma_i \end{bmatrix} \tag{6.27}$$

Thus, the complex modal state-space system is given by

$$\begin{aligned} \dot{\mathbf{x}}(\tau) &= A_M \mathbf{x}(\tau) + B_M \mathbf{u}(\tau) \\ \mathbf{y}(\tau) &= C_M \mathbf{x}(\tau) + D_M \mathbf{u}(\tau) \end{aligned} \tag{6.28}$$

where the system matrices become

$$A_M = \begin{bmatrix} A_{M_1} & 0 & \cdots & 0 \\ 0 & A_{M_2} & \cdots & 0 \\ \vdots & \vdots & \ddots & 0 \\ 0 & 0 & \cdots & A_{M_{N_x}} \end{bmatrix}; B_M = \begin{bmatrix} B_{M_1} \\ \vdots \\ B_{M_{N_x}} \end{bmatrix}; \ C_M = \left[C_{M_1} \mid \cdots \mid C_{M_{N_x}}\right];$$

$$D_M = D$$

and $A_{M_i} = \Lambda_i$.

This completes the *MIMO modal state-space description*.

6.2 Realization Problem

In this section, we develop the fundamental realization problem that provides the foundation for the SID problems to follow. Here we review the basic realization (systems) theory (without proof) of the results developed by Kalman and show how they lead directly to the deterministic realization problem and its solution [1, 2].

The deterministic realization problem has evolved in a twofold manner: (i) the classical approach and (ii) the subspace approach as illustrated in Figure 6.1. The classical approach essentially evolved from system theoretic developments based on the Hankel matrix composed of impulse response matrices (Markov parameters) and its decomposition into the observability and controllability matrices, while the subspace approach is based on decomposing the data space employing projection theory (orthogonal and oblique) leading to the decomposition of a block input/output Hankel matrix. Both of the modern approaches use a singular value decomposition (SVD) of their respective Hankel matrix to estimate

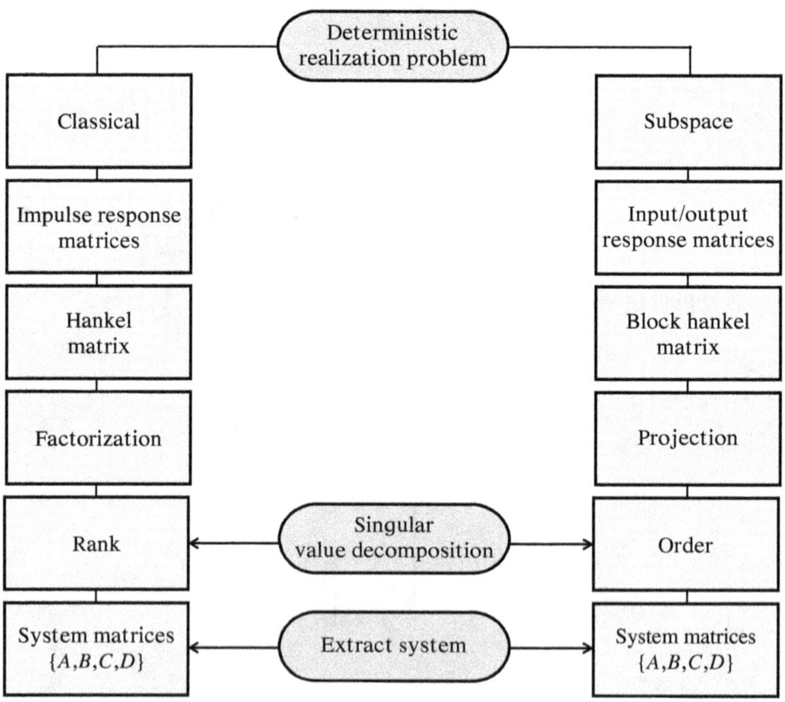

Figure 6.1 Deterministic realization problem: classical approach and subspace approach.

its rank or equivalently system order leading to the extraction of the underlying system Σ_{ABCD}. In fact, besides the critical system theoretical properties of the Hankel matrix, the *SVD* is the common thread or tool critical to a solution of the deterministic realization problem. The *SVD*, besides its superior numerical properties, enables a reliable and robust approach to extracting the underlying system.

We first develop the important system theoretic results essential for comprehending solutions to the deterministic realization and decomposition problem and then develop a set of classical numerical techniques based on the Ho–Kalman [3] approach in order to extract the system along with its subsystems. Although quite important, these algorithms are not numerically reliable. The evolution of the *SVD* has changed that fact; however, unfortunately, the underlying internal structure is difficult to unravel, since no particular coordinate system evolves—this is a drawback of both the *SVD* classical and subspace approaches leading to the idea of a black-box identification followed by a similarity transformation to the desired coordinate system—a necessary and reasonable tradeoff.

The subspace approach evolves from projection theory leading to a factorization of the block Hankel matrix, similar to the classical approach leading to its decomposition using the *SVD* and eventual extraction of the system model Σ_{ABCD}. This approach follows leading to the two most popular *SID* algorithms (*MOESP*, *N4SID*). Performance metrics are discussed and applied to illustrate their utility. Examples are then developed to demonstrate all of the algorithms, classical, and subspace along with a detailed structural vibration response case study.

6.2.1 Realization Theory

A *realization* is defined in terms of the system transfer function and/or a state-space model, that is, a matrix transfer function $H(z)$ is defined as "realizable," if there exists a finite-dimensional state-space description, $\Sigma_{ABCD} = \{A, B, C, D\}$, such that [4]

$$H(z) = C(zI - A)^{-1}B + D \tag{6.29}$$

where Σ_{ABCD} is called the *realization* of $H(z)$. The question of the existence of such a realization, termed "realizability," is a complex issue. We tacitly assume that $H(z)$ is a proper rational matrix (numerator polynomial degree less than denominator degree) to ensure such realizability exists (see [4] for details).

Realizability also follows from the system impulse response, since the discrete transfer function can be represented by the power series [4]

$$H(z) = \sum_{t=0}^{\infty} H_t z^{-t} \text{ for } H_t = CA^{t-1}B + D\delta(t) \tag{6.30}$$

and H_t is the $N_y \times N_u$ unit impulse response matrix with $H_0 = D$. Therefore, the infinite *Markov sequence* $\{H_t\}_0^\infty$ composed of the set of *Markov parameters* $(CA^{t-1}B)$ is *realizable*, if there exists a finite-dimensional Σ_{ABCD} that satisfies $H_t = CA^{t-1}B + D\delta(t); t = 0, \ldots, \infty$. Recall that the transfer function matrix $H(z)$ can be used to extract Markov parameter sequence $\{H_t\}$ by *long division* of each component transfer functions, that is, $H_{ij}(z) = N_{ij}(z)/D_{ij}(z)$.

With the property of system realizability established, we are next concerned with the concept of a "minimality." By definition, out of all the possible realizations Σ_{ABCD}, those that have the "smallest" order (number of states) are termed *minimal realizations*. This property is specified by the so-called *McMillan degree* of the matrix transfer function $H(z)$ that is equivalent to the dimension (order) of the system matrix, that is,

$$\dim A = \deg H(z) \tag{6.31}$$

Since the transfer function is the external description of the system from an input/output perspective, only the controllable and observable components are available (see [1] for details). With this in mind, it is clear that system minimality must be connected to this property. In fact, it is based on the fact proven in [2] that Σ_{ABCD} is a minimal realization of $H(z)$ or equivalently H_t, if and only if, the pairs (A, B) and (A, C) are *completely* controllable and observable, respectively, which is equivalent to Eq. (6.31). So we see that controllability and observability of Σ_{ABCD} is a necessary and sufficient condition for minimality!

This minimality concept leads to a powerful theoretical result proved by Kalman [3]. The Markov sequence $\{H_t\}$ has a *finite*-dimensional realization, if and only if, there exists an integer N and coefficients $\alpha_1, \ldots, \alpha_N$ such that

$$H_N + \sum_{t=1}^{N} \alpha_t H_{N-t} = \mathbf{0} \tag{6.32}$$

This result follows from the Cayley–Hamilton theorem of linear algebra that states that an N-dimensional matrix satisfies its own characteristic equation, that is, the Nth order characteristic equation $\alpha(A)$ is given by

$$A^N + \alpha_1 A^{N-1} + \alpha_2 A^{N-2} + \cdots + \alpha_{N-1} A + \alpha_N I = 0 \tag{6.33}$$

Pre- and post-multiplying this relation by the measurement and input transmission matrices C and B, respectively, it follows that the Markov parameters satisfy the recursion for the N-degree $\alpha(z)$

$$H_N = -\sum_{t=1}^{N} \alpha_t H_{N-t} = -\sum_{t=1}^{N} \alpha_t \left(CA^{N-t-1}B\right) \tag{6.34}$$

The fundamental problem of extracting the state-space system Σ_{ABCD} is called the realization problem. Simply stated, the deterministic *realization problem* is:

GIVEN a transfer function matrix $H(z)$ or equivalently a set of impulse response matrices $\{H_t\}$ with corresponding Markov parameters, $H_t = CA^{t-1}B + D\delta(t)$; $t = 0, \ldots, \infty$, FIND the underlying minimal state-space system, $\Sigma_{ABCD} := \{A, B, C, D\}$.

These properties evolve completely from the development of the Hankel matrix of systems theory [2] which plays an intrinsic role in the solution of the deterministic realization problem. Note that the Hankel matrix is to state-space identification as the Toeplitz matrix is to signal processing identification (e.g. spectral estimation [10]).

We define the $(K \times N_y N_u) \times (K \times N_y N_u)$ block *Hankel matrix* $(K \to \infty)$ by

$$\mathcal{H}_{K,K} := \begin{bmatrix} H_1 & H_2 & \cdots & H_K \\ H_2 & H_3 & \cdots & H_{K+1} \\ \vdots & \vdots & \cdots & \vdots \\ H_K & H_{K+1} & \cdots & H_{2K-1} \end{bmatrix} \tag{6.35}$$

If Σ_{ABCD} is a realization of $\{H_t\}$, then based on the Markov parameters, the Hankel matrix can be *factored* as the product of the controllability and observability matrices

$$\mathcal{H}_{K,K} = \begin{bmatrix} CB & CAB & \cdots & CA^{K-1}B \\ CAB & CA^2B & \cdots & CA^KB \\ \vdots & \vdots & \cdots & \vdots \\ CA^{K-1}B & CA^{K-2}B & \cdots & CA^{2K-2}B \end{bmatrix}$$

$$= \begin{bmatrix} C \\ CA \\ \vdots \\ CA^{K-1} \end{bmatrix} [B \,|\, AB \,|\, \cdots \,|\, A^{K-1}B] \tag{6.36}$$

or simply

$$\mathcal{H}_{K,K} = \mathcal{O}_K \times \mathcal{C}_K \tag{6.37}$$

Another interesting property of the Hankel matrix utilized by realization algorithms is its inherent *shift-invariant property*, that is, multiplication by the system matrix A shifts the Hankel matrix "up" ($\uparrow$) or "left" ($\leftarrow$) since

$$\mathcal{O}_K^{\uparrow} = \underbrace{\begin{bmatrix} C \\ CA \\ \vdots \\ CA^{K-1} \end{bmatrix}}_{\mathcal{O}_K} \times A = \underbrace{\begin{bmatrix} CA \\ CA^2 \\ \vdots \\ CA^K \end{bmatrix}}_{\mathcal{O}_K^{\uparrow}} \tag{6.38}$$

$$\mathcal{C}_K^{\leftarrow} = A \times \underbrace{[B \; AB \; \cdots \; A^{K-1}B]}_{\mathcal{C}_K} = \underbrace{[AB \; A^2B \; \cdots \; A^KB]}_{\mathcal{C}_K^{\leftarrow}} \tag{6.39}$$

therefore

$$\mathcal{H}^{\uparrow}_{K,K} := \mathcal{O}^{\uparrow}_{K} C_K = \underbrace{(\mathcal{O}_K \times A)}_{\mathcal{O}^{\uparrow}_{K}} \times C_K \iff \mathcal{H}^{\leftarrow}_{K,K} := \mathcal{O}_K C^{\leftarrow}_{K} = \mathcal{O}_K \times \underbrace{(A \times C_K)}_{C^{\leftarrow}_{K}} \tag{6.40}$$

Suppose the dimension of the minimal realization is given from a matrix transfer function as $N = N_x$, then the Hankel matrix could be constructed with $K = N_x$ using $2N_x$ impulse response matrices. Knowledge of the order N_x indicates the minimum number of terms required to exactly "match" the Markov sequence and extract the Markov parameters. It follows that the rank of the Hankel matrix must be N_x, that is, $\rho\left(\mathcal{H}_{K,K}\right) = N_x$, which is the dimension of the *minimal realization*. This leads to a very important property of the Hankel matrix. If we did *not* know the dimension of the system, then we would let (in theory) $K \to \infty$ and determine the rank of $\mathcal{H}_{\infty,\infty}$. Thus, the minimal dimension of an "unknown" system is given by the *rank* of the Hankel matrix.

This rank property of the Hankel matrix is fundamental to linear systems theory [2]. In order for a system to be *minimal*, it must be *completely controllable* and *completely observable*; therefore, it follows that the *rank condition*

$$\rho\left(\mathcal{H}_{K,K}\right) = \min\left[\rho\left(\mathcal{O}_K\right), \rho\left(C_K\right)\right] = N_x \qquad \text{[Rank Condition]} \tag{6.41}$$

Therefore, we see that the properties of controllability and observability are carefully woven into that of minimality. Testing the rank of the Hankel matrix yields the dimensionality of the underlying dynamic system. This fact will prove crucial when we must "identify" a system, $\Sigma_{ABCD} = \{A, B, C, D\}$, from noisy measurement data.

It is also important to realize that "shifting" the Hankel matrix preserves its rank N_x. If $H_t = CA^{t-1}B; t = 1, \ldots, 2N_x$, then it follows that the corresponding rank

$$\rho\left(\mathcal{H}_{N_x,N_x}\right) = \rho\left(\mathcal{H}_{N_x+1,N_x}\right) = \rho\left(\mathcal{H}_{N_x,N_x+1}\right) = N_x \tag{6.42}$$

is satisfied [8].

Finally, the realization problem based on an infinite Markov sequence $\{H_t\}_0^\infty$ leads to $\mathcal{H}_{\infty,\infty}$ with rank preserved

$$\rho\left(\mathcal{H}_{\infty,\infty}\right) = \sup_t \rho\left(\mathcal{H}_{K,K}\right)$$

that is satisfied by a minimal realization with $K = N_x$.

Another interesting system theoretic property that evolves from the observability, controllability, and Hankel matrices is the concept of a balanced realization. Simply stated a balanced realization is one in which "every state is as controllable as it is observable" [11]. It is a critical property in model order reduction characterizing certain features that uniquely capture the foundation of the underlying system.

We know that if a system is completely observable and controllable, then it is minimal and the Hankel matrix factors as in Eq. (6.37). Therefore, for a stable system, both observability and controllability Gramians must satisfy the following positivity constraints, that is, the Gramian matrices Σ_{AC}, Σ_{AB} are positive definite satisfying the *Lyapunov equations*

$$\begin{aligned} \Sigma_{AC} &= A'\Sigma_{AC}A + C'C \geq 0 && \text{[Observability Gramian]} \\ \Sigma_{AB} &= A\Sigma_{AB}A' + BB' \geq 0 && \text{[Controllability Gramian]} \end{aligned} \tag{6.43}$$

Alternatively, we can express these constraints in terms of the observability and controllability matrices as:

$$\begin{aligned} \Sigma_{AC}(K) &= \sum_{k=0}^{K} (A^k)'C'CA^k = \mathcal{O}_K'\mathcal{O}_K \geq 0 && \text{[Observability Gramian]} \\ \Sigma_{AB}(K) &= \sum_{k=0}^{K} A^kBB'(A^k)' = \mathcal{C}_K\mathcal{C}_K' \geq 0 && \text{[Controllability Gramian]} \end{aligned} \tag{6.44}$$

Performing a *SVD* of the observability and controllability matrices, we obtain

$$\mathcal{O}_K = U_{\mathcal{O}}(K)\Sigma_{\mathcal{O}}(K)V_{\mathcal{O}}'(K) \quad \text{and} \quad \mathcal{C}_K = U_{\mathcal{C}}(K)\Sigma_{\mathcal{C}}(K)V_{\mathcal{C}}'(K) \tag{6.45}$$

Note that applying the orthogonality properties of the *SVD* ($U'U = I$, $V'V = I$), we obtain the *Gramians* as

$$\begin{aligned} \Sigma_{AC}(K) &= \mathcal{O}_K'\mathcal{O}_K = V_{\mathcal{O}}(K)\Sigma_{\mathcal{O}}^2(K)V_{\mathcal{O}}'(K) \\ \Sigma_{AB}(K) &= \mathcal{C}_K\mathcal{C}_K' = U_{\mathcal{C}}(K)\Sigma_{\mathcal{C}}^2(K)U_{\mathcal{C}}'(K) \end{aligned} \tag{6.46}$$

From the factorization of the Hankel matrix and its SVD, we have

$$\begin{aligned} \mathcal{H}_{K,K} &= \mathcal{U}_K \times \Sigma_K \times \mathcal{V}_K' \\ &= \underbrace{\left(U_{\mathcal{O}}(K)\Sigma_{\mathcal{O}}(K)V_{\mathcal{O}}'(K)\right)}_{\mathcal{O}_K} \times \underbrace{\left(U_{\mathcal{C}}(K)\Sigma_{\mathcal{C}}(K)V_{\mathcal{C}}'(K)\right)}_{\mathcal{C}_K} \end{aligned} \tag{6.47}$$

establishing the relationship between the singular values of the Hankel matrix and those of the so-called second-order modes of the observability/controllability product values (see [11] or [12] for details).

Formally, the Hankel singular values are taken as the square root of the observability and controllability Gramian product values, that is, the *Hankel singular values* are given by

$$\sigma_{ABC} = \sqrt{\Sigma_{AC} \times \Sigma_{AB}}$$

It is important to realize that while the system eigenvalues are invariant under a similarity transformation those of the Gramians are *not*, but the eigenvalues of the product of the Gramians *are* similarity invariant [13]. It is these values that are the invariants of the system and are used in model order reduction [14].

6.2.2 Balanced Realizations

With these system theoretic properties in mind, we can now formally express the concept of a "balanced realization" in terms of the observability, controllability, and the process of simultaneously diagonalizing these matrices. Therefore, we define a *balanced realization* for a stable system as one in which its Gramians are equal and diagonal, that is, [12]

$$\Sigma_{AB}(K) = \Sigma_{AC}(K) = \Sigma_K \qquad \text{[Balanced Realization]}$$

where Σ_K is a diagonal matrix obtained from the *SVD* of the Hankel matrix of Eq. (6.47) such that

$$\Sigma_K = \text{diag}\left[\sigma_1, \sigma_2, \ldots, \sigma_K\right] \text{ for } \sigma_1 \geq \sigma_2 \geq \cdots \geq \sigma_K$$

and σ_i are the *Hankel singular values* of the system based on the product of the Gramian matrices. These values are *invariant* under similarity transformation (see [11–14] for details).

The maximum Hankel singular value defines the so-called *Hankel norm* of the system, a metric of system input/output energy as [12, 14]

$$\|\mathcal{H}_{K,K}\|_2 = \max\{\sigma_i\} \qquad \text{for} \qquad i = 1, \ldots, K \tag{6.48}$$

The resulting realization is said to be a *balanced realization* and is *scale invariant*, that is, the balanced system is robust to amplitude scaling [11–13]. It is also interesting that the balancing (similarity) transformation for any system is given by

$$T_{\text{bal}} = \Sigma_{AB}^{\frac{1}{2}} \times V_{\mathcal{P}} \times \Sigma_{\mathcal{P}}^{-\frac{1}{2}} \quad \text{and} \quad T_{\text{bal}}^{-1} = \Sigma_{\mathcal{P}}^{-\frac{1}{2}} \times U'_{\mathcal{P}} \times \Sigma_{AC}^{\frac{1}{2}} \tag{6.49}$$

where we define the product matrix, in terms of its Cholesky square-root decomposition as well as its corresponding *SVD* as

$$\mathcal{P} = \Sigma_{AC}^{\frac{1}{2}} \times \Sigma_{AB}^{\frac{1}{2}} = U_{\mathcal{P}} \times \Sigma_{\mathcal{P}} \times V'_{\mathcal{P}} \tag{6.50}$$

with $\Sigma_{\bullet}^{\frac{1}{2}}$ the respective square-root matrices of the observability and controllability Gramians.

6.2.3 Systems Theory Summary

We summarize these important system theoretic properties of Hankel matrices as:

- The system defined by Σ_{ABCD} is *minimal*, if and only if, it is *completely controllable* and *completely observable* [Minimal Order]
- For a minimal realization Σ_{ABCD}, the rank of the Hankel matrix satisfies:

$$\rho\left(\mathcal{H}_{K,K}\right) = \min\left[\rho\left(\mathcal{O}_K\right), \rho\left(\mathcal{C}_K\right)\right] = N_x \qquad \text{[Rank Property]}$$

- The constant rank under *shifted* Hankel matrix is:

$$\rho[\mathcal{H}_{K,K}] = \rho[\mathcal{H}_{K+1,K}] = \rho[\mathcal{H}_{K,K+1}] = N_x \qquad \text{[Shift Property]}$$

- The Hankel matrix is *shift invariant* such that

$$\mathcal{H}_{K,K}^{\uparrow} := \mathcal{O}_K^{\uparrow}\mathcal{C}_K \iff \mathcal{H}_{K,K}^{\leftarrow} := \mathcal{O}_K\mathcal{C}_K^{\leftarrow} \qquad \text{[Shift Invariant]}$$

- The eigenvalues of the system matrix A lie within the unit circle:

$$\lambda[A] \leq 1 \qquad \text{[Stability]}$$

  or if and only if the *Lyapunov equation*

$$P = A'PA + C'C \qquad \text{[Lyapunov Stability]}$$

  has a *unique positive definite* solution, $P > 0$.
- A *balanced realization* is defined by the equality of the diagonalized Gramians: $\Sigma_{AB}(K) = \Sigma_{AC}(K) = \Sigma_K$ such that

$$\Sigma_K = \text{diag}\left[\sigma_1, \sigma_2, \dots, \sigma_K\right] \text{ for } \sigma_1 \geq \sigma_2 \geq \cdots \sigma_K$$

  where Σ_K is given by the *SVD* of the Hankel matrix:

$$\mathcal{H}_{K,K} = \mathcal{U}_K\Sigma_K\mathcal{V}_K$$

- The resulting realization is said to be *scale invariant* [11], that is, the balanced system realized is robust to amplitude scaling.
- The *Hankel singular values* used in model order reduction are obtained as the root of the observability and controllability Gramian product, that is,

$$\sigma_{ABC} = \sqrt{\Sigma_{AC} \times \Sigma_{AB}}$$

  and are the underlying system similarity invariants.
- The *Hankel norm* is a measure of system input/output energy and it is given by the *maximum* Hankel singular value: $\|\mathcal{H}_{K,K}\|_2 = \max\{\sigma_i\}$ for $i = 1, \dots, K$
- The *balanced similarity transformation* is given in terms of the product matrix and its decompositions (Cholesky and *SVD*) as:

$$T_{\text{bal}} = \Sigma_{AB}^{\frac{1}{2}} \times V_{\mathcal{P}} \times \Sigma_{\mathcal{P}}^{-\frac{1}{2}} \quad \text{and} \quad T_{\text{bal}}^{-1} = \Sigma_{\mathcal{P}}^{-\frac{1}{2}} \times U_{\mathcal{P}}' \times \Sigma_{AC}^{\frac{1}{2}}$$

This completes the fundamental system theoretic properties of a realization Σ_{ABCD}. Next, we consider further properties that can lead to more robust realizations.

6.3 Realization Decomposition

The solution to the deterministic realization problem is based primarily on one fundamental property of the Hankel matrix—the *minimal order factorization* into the full-rank observability and controllability matrices. It has been shown that operations directly on the rows or columns of the Hankel matrix are identical to independently operating directly on the rows of the observability matrix or equivalently to operating directly on the columns of the controllability matrix [15, 16]. Therefore, minimality implies this unique decomposition of the Hankel matrix into its full-rank factors. Most realization algorithms assume this property (realizability) leading to a variety of operations. This fact leads to the development of not only solutions of the so-called canonical realization problem, but also to inherent techniques of similarity transformations to canonical forms [16]. We start with the basic classical algorithm developed by Ho and Kalman [3] and show its extension using the *SVD* of the Hankel matrix [17, 18].

6.3.1 Ho–Kalman Realization

The fundamental approach to solving the deterministic realization problem evolved from [3] and is called *Ho–Kalman algorithm*. Assuming the order is known, it proceeds in the following steps starting with performing transformations on the block Hankel matrix developing the elementary matrices,[2] $P \in \mathcal{R}^{KN_y \times KN_y}$ and $Q \in \mathcal{R}^{KN_u \times KN_u}$, that is,

$$\mathrm{P} \times \mathcal{H}_{K,K} \times \mathrm{Q} = \begin{bmatrix} I_{N_x} & | & 0 \\ - & - & - \\ 0 & | & 0 \end{bmatrix} \in \mathcal{R}^{KN_y \times KN_u} \tag{6.51}$$

After performing this decomposition, the realization Σ_{ABCD} is obtained by extracting the individual state-space matrices from:

$$\mathcal{P} := P \times \mathcal{H}_{K,K} = \left[\overbrace{\;B\;}^{N_u} |AB| \cdots |A^{K-1}B \right] \Longrightarrow B = \mathcal{P}(1 : N_x, 1 : N_u) \tag{6.52}$$

2 An elementary transformation is a nonsingular matrix, constructed by a sequence elementary operations of interchanging and scaling the rows and columns of the identity matrix such that $P = P_1 \times P_2 \times \cdots \times P_M$ [19].

$$\mathcal{Q} := \mathcal{H}_{K,K} \times Q = \begin{bmatrix} {}^{N_y}\{C \\ CA \\ \vdots \\ CA^{K-1} \end{bmatrix} \Longrightarrow C = \mathcal{Q}(1:N_y, 1:N_x) \tag{6.53}$$

$$\mathcal{A} := P \times \mathcal{H}^{\uparrow}_{K,K} \times Q \Longrightarrow A = \mathcal{A}(1:N_x, 1:N_x) \tag{6.54}$$

where recall that $\mathcal{H}^{\uparrow}_{K,K}$ is the shifted Hankel matrix of Eq. (6.40), N_u is the dimension of the input vector $\mathbf{u}$ and N_y that of the output or measurement vector $\mathbf{y}$ with the notation $1 : N \rightarrow 1, 2, \ldots, N$ or equivalently *select* the corresponding rows (columns) 1-to-N of the matrix.

Consider the following example to demonstrate this approach.

Example 6.1 We have an impulse response sequence given by

$$H(z) = \begin{bmatrix} 1 & 1 \\ 1 & 1 \end{bmatrix} z^{-1} + \begin{bmatrix} -2 & -4 \\ -3 & -1 \end{bmatrix} z^{-2} + \begin{bmatrix} 4 & 10 \\ 7 & 1 \end{bmatrix} z^{-3} + \begin{bmatrix} -8 & -22 \\ -15 & -1 \end{bmatrix} z^{-4} + \begin{bmatrix} 16 & 46 \\ 31 & 1 \end{bmatrix} z^{-5}$$

Constructing the corresponding Hankel and shifted Hankel matrices for this problem yields

$$H_{2,2} = \begin{bmatrix} 1 & 1 & | & -2 & -4 \\ 1 & 1 & | & -3 & -1 \\ - & - & - & - & - \\ -2 & -4 & | & 4 & 10 \\ -3 & -1 & | & 7 & 1 \end{bmatrix}; \quad H^{\uparrow}_{2,2} = \begin{bmatrix} -2 & -4 & | & 4 & 10 \\ -3 & -1 & | & 7 & 1 \\ - & - & - & - & - \\ 4 & 10 & | & -8 & -22 \\ 7 & 1 & | & -15 & -1 \end{bmatrix}$$

Performing elementary row and column operations on $\mathcal{H}_{2,2}$, we obtain

$$P = \begin{bmatrix} 1 & 0 & 0 & 0 \\ -1 & 0 & -1/2 & 0 \\ 1 & -1 & 0 & 0 \\ -2/3 & -1/6 & -1/6 & -1/6 \end{bmatrix}; \quad Q = \begin{bmatrix} 1 & -1 & 2 & 9 \\ 0 & 1 & 0 & 1 \\ 0 & 0 & 1 & 3 \\ 0 & 0 & 0 & 1 \end{bmatrix}$$

The input $B \Longrightarrow \mathcal{P}(1:4, 1:2)$ and output $C \Longrightarrow \mathcal{Q}(1:2, 1:4)$ matrices are found as:

$$P\mathcal{H}_{2,2} = B\begin{bmatrix} 1 & 1 & | & 2 & 9 \\ 0 & 1 & | & 0 & 1 \\ 0 & 0 & | & 1 & 3 \\ 0 & 0 & | & 0 & 1 \end{bmatrix} = [B\,|\,AB]; \quad \mathcal{H}_{2,2}Q = \overbrace{\begin{bmatrix} 1 & 0 & 0 & 0 \\ 1 & 0 & -1 & 0 \\ - & - & - & - \\ -2 & -2 & 0 & 0 \\ -3 & -2 & 1 & -5 \end{bmatrix}}^{C} = \begin{bmatrix} C \\ -- \\ CA \end{bmatrix}$$

Finally, the system matrix $A \Longrightarrow \mathcal{A}(1:4,1:4)$ is found from

$$PH_{2,2}^{\uparrow}Q = \begin{bmatrix} 1 & 0 & 0 & 0 \\ -1 & 0 & -1/2 & 0 \\ 1 & -1 & 0 & 0 \\ -2/3 & -1/6 & -1/6 & -1/6 \end{bmatrix} \times \begin{bmatrix} -2 & -4 & | & 4 & 10 \\ -3 & -1 & | & 7 & 1 \\ - & - & - & - & - \\ 4 & 10 & | & -8 & -22 \\ 7 & 1 & | & -15 & -1 \end{bmatrix}$$

$$\times \begin{bmatrix} 1 & -1 & 2 & 9 \\ 0 & 1 & 0 & 1 \\ 0 & 0 & 1 & 3 \\ 0 & 0 & 0 & 1 \end{bmatrix}$$

$$A = \begin{bmatrix} -2 & -2 & 0 & 0 \\ 0 & -1 & 0 & 0 \\ 1 & -4 & -1 & 6 \\ 0 & 1 & 0 & -2 \end{bmatrix}$$

completing the application of the Ho–Kalman algorithm to this *MIMO* problem. As a check of the realization, we reconstructed the impulse response sequence from the extracted model Σ_{ABCD} using $H_t = CA^{t-1}B; t = 1, \ldots, 5$ with a "perfect match," since the calculations were accomplished manually with perfect precision.

Note that if the minimal order was *not* known a priori as from the transfer function in this example, then a larger Hankel matrix would be processed until its rank was found using the elementary transformations to determine N_x as expressed in Eq. (6.51).

6.3.2 SVD Realization

A more efficient solution to the deterministic realization problem is obtained by performing the *SVD* of the Hankel matrix $\mathcal{H}_{m,K}$ where $m \geq N_x$ and $K \geq N_x$ with N_x the dimension of the minimum realization (system) or equivalently the underlying *true* number of states; therefore, [17, 18]

$$\mathcal{H}_{m,K} = \left[U_{N_x} \mid U_N\right] \begin{bmatrix} \Sigma_{N_x} & | & 0 \\ - & - & - \\ 0 & | & 0 \end{bmatrix} \begin{bmatrix} V'_{N_x} \\ -- \\ V'_N \end{bmatrix} = U_{N_x}\Sigma_{N_x}V'_{N_x} \tag{6.55}$$

for $\Sigma_{N_x} = \text{diag}\left[\sigma_1, \sigma_2, \ldots, \sigma_{N_x}\right]$.

From the factorization of the Hankel matrix and its *SVD*, we have that

$$\mathcal{H}_{m,K} = \mathcal{O}_m \times \mathcal{C}_K = \underbrace{\left(U_{N_x}\Sigma_{N_x}^{1/2}\right)}_{\mathcal{O}_{N_x}} \underbrace{\left((\Sigma'_{N_x})^{1/2}V'_{N_x}\right)}_{\mathcal{C}_{N_x}} \tag{6.56}$$

where $\Sigma_{N_x}^{1/2}$ is the matrix square root, that is, $R = S^{1/2} \times (S')^{1/2}$ obtained by a Cholesky decomposition [19].

Now, from shift-invariant property (see Eq. (6.40)) of the observability or controllability matrices, we can obtain the system matrices from either

$$\mathcal{O}_{N_x} \times A = \mathcal{O}_{N_x}^{\uparrow}$$
$$\text{or}$$
$$A \times C_{N_x} = C_{N_x}^{\leftarrow} \tag{6.57}$$

which yields the unique system matrices by applying the pseudo-inverse[3] $\left(Z^{\#} := (Z'Z)^{-1}Z'\right)$, since minimality guarantees full-rank observability and controllability matrices

$$A = \mathcal{O}_{N_x}^{\#} \times \mathcal{O}_{N_x}^{\uparrow} \qquad \text{or} \qquad A = C_{N_x}^{\leftarrow} \times C_{N_x}^{\#} \tag{6.58}$$

with the input and output matrices extracted directly from the controllability and observability matrices as

$$B = C(1 : N_x, 1 : N_u) \text{ and } C = \mathcal{O}(1 : N_y, 1 : N_x) \tag{6.59}$$

However, from the *SVD* of Eq. (6.56), we have

$$A = \mathcal{O}_{N_x}^{\#}\mathcal{O}_{N_x}^{\uparrow} = \left(\Sigma_{N_x}^{-1/2}U_{N_x}'\right) \times \mathcal{O}_{N_x}^{\uparrow} \text{ or } A = C_{N_x}^{\leftarrow}C_{N_x}^{\#} = C_{N_x}^{\leftarrow} \times \left(V_{N_x}(\Sigma_{N_x}^{-1/2})'\right) \tag{6.60}$$

which is a very efficient method of calculating the pseudo-inverse and extracting the model.

It is also interesting to note that with the *SVD* extraction of Σ_{ABCD}, both for the A and C-matrices all lower-order models are "nested" within these structures and can easily be extracted as sub-system matrices [20].

Example 6.2 Consider the previous example of Example 6.1 and assume that we do not know the order N_x. We would create an extended Hankel matrix in this case with $K = 5$ leading to

$$H_{3,3} = \begin{bmatrix} 1 & 1 & | & -2 & -4 & | & 4 & 10 \\ 1 & 1 & | & -3 & -1 & | & 7 & 1 \\ - & - & - & - & - & - & - & - \\ -2 & -4 & | & 4 & 10 & | & -8 & -22 \\ -3 & -1 & | & 7 & 1 & | & -15 & -1 \\ - & - & - & - & - & - & - & - \\ 4 & 10 & | & -8 & -22 & | & 16 & 46 \\ 7 & 1 & | & -15 & -1 & | & 31 & 1 \end{bmatrix}$$

3 Recall the *SVD* of $Z = U\Sigma V'$ efficiently provides $Z^{\#} = V\Sigma^{-1}U' = V(\Sigma^{-1/2})'\Sigma^{-1/2}U'$.

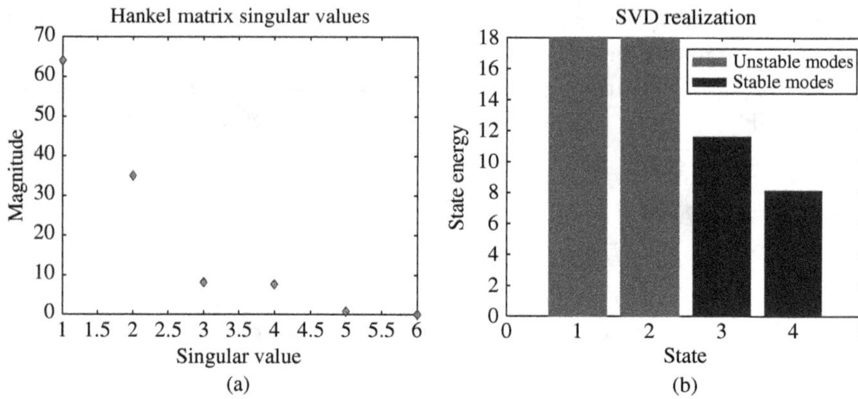

Figure 6.2 *SVD* realization. (a) estimated singular values of the Hankel matrix. (b) estimated Hankel singular values.

Investigating the Hankel matrix, we calculated its *SVD* to observe the rank in Figure 6.2 along with the corresponding *Hankel singular values* discussed previously. From the figure, it is clear that the order (rank) is four (4), since the first four singular values are large relative to the final two indicating a fourth-order system. The Hankel singular values indicate two unstable poles (eigenvalues); however; there should be four $(-1, -1, -2, -2)$, but inaccuracies in the identified model cause variations as indicated by the eigenvalues:

$$\lambda(A_{\text{true}}) = \{-1, -1, -2, -2\} \longrightarrow \lambda(\hat{A}) = \{-1, 0.474, -2, -1.606\}$$

indicating a nonperfect realization.

We applied the *SVD* approach to this set of impulse response matrices and extracted the following system Σ_{ABCD}:

$$\hat{A} = \begin{bmatrix} -2.095 & 0.376 & -0.448 & 0.004 \\ -0.057 & -1.884 & -0.303 & 0.064 \\ 0.260 & 0.756 & -0.745 & 0.344 \\ 0.140 & -0.147 & -0.418 & 5.92 \end{bmatrix}$$

$$\hat{B} = \begin{bmatrix} -0.873 & -0.943 \\ 0.940 & -0.235 \\ 0.0526 & -2.008 \\ -0.005 & 1.930 \end{bmatrix}; \hat{C} = \begin{bmatrix} -1.100 & -0.0248 & 1.934 & 1.989 \\ -0.657 & 0.9780 & -0.262 & 0.0463 \end{bmatrix}$$

As a check, we estimated the impulse response (as before) with $\hat{H}_t = \hat{C}\hat{A}^{t-1}\hat{B}$

$$H(z) = \begin{bmatrix} 1.08 & 1.00 \\ 1.68 & 1.03 \end{bmatrix} z^{-1} + \begin{bmatrix} -2.09 & 3.99 \\ -3.43 & -1.03 \end{bmatrix} z^{-2} + \begin{bmatrix} 3.97 & 9.96 \\ 6.98 & 0.76 \end{bmatrix} z^{-3} + \begin{bmatrix} -7.91 & -22.01 \\ -14.12 & -1.14 \end{bmatrix} z^{-4} + \begin{bmatrix} 15.83 & 46.18 \\ 28.73 & 1.85 \end{bmatrix} z^{-5}$$

Table 6.1 Ho–Kalman (SVD) realization algorithm.

Singular value decomposition

$$\mathcal{H}_{m,K} = \left[U_{N_x} \mid U_N\right]\begin{bmatrix} \Sigma_{N_x} & | & 0 \\ - & - & - \\ 0 & | & 0 \end{bmatrix}\begin{bmatrix} V'_{N_x} \\ -- \\ V'_N \end{bmatrix} = U_{N_x}\Sigma_{N_x}V'_{N_x} \quad \text{(rank matrix)}$$

Factorization

$$\mathcal{H}_{m,K} = \mathcal{O}_m \times \mathcal{C}_K = \underbrace{\left(U_{N_x}\Sigma_{N_x}^{1/2}\right)}_{\mathcal{O}_{N_x}}\underbrace{\left((\Sigma'_{N_x})^{1/2}V'_{N_x}\right)}_{\mathcal{C}_{N_x}} \quad \text{(observability/controllability matrices)}$$

System matrix

$$A = \mathcal{O}^{\#}_{N_x} \times \mathcal{O}^{\uparrow}_{N_x} = \left(\mathcal{O}'_{N_x}\mathcal{O}_{N_x}\right)^{-1}\mathcal{O}'_{N_x} \times \mathcal{O}^{\uparrow}_{N_x} = \left(\Sigma^{-1/2}_{N_x}U'_{N_x}\right) \times \mathcal{O}^{\uparrow}_{N_x} \quad \text{(system/process matrix)}$$

or

$$A = \mathcal{C}^{\leftarrow}_{N_x} \times \mathcal{C}^{\#}_{N_x} = \mathcal{C}^{\leftarrow}_{N_x} \times \mathcal{C}'_{N_x}\left(\mathcal{C}'_{N_x}\mathcal{C}_{N_x}\right)^{-1} = \mathcal{C}^{\leftarrow}_{N_x} \times \left(V_{N_x}(\Sigma^{-1/2}_{N_x})'\right)$$

Measurement (output) matrix

$$C = \mathcal{O}(1 : N_y, 1 : N_x) \quad \text{(output matrix)}$$

Input transmission matrices

$$B = \mathcal{C}(1 : N_x, 1 : N_u) \quad \text{(input matrix)}$$

Input–output transmission matrices

$$D = H_0 \quad \text{(input–output matrix)}$$

Examining the estimated impulse response, we see that errors have evolved but the results are still quite reasonable.

We summarize the Ho–Kalman/SVD algorithm in Table 6.1.

6.4 Subspace Decomposition: Orthogonal Projections

In this section, we develop the fundamental subspace realization approach to identifying the state-space representation from input–output data extending the realization from impulse response to input/output data—still assumed *deterministic*. In order to comprehend just how the deterministic realization problem can be solved when input/output sequences rather than impulse response sequences evolve, it is important to understand that most subspace realization techniques require the extraction of the extended observability matrix directly from acquired data first, followed by the system model extraction, second. The existing methods differ in the manner in which the observability matrix is estimated and its methodology for extracting the system model. To extract the state-space model, Σ_{ABCD}, it is only required to obtain either the extended column space of the observability matrix or the row space of the state sequence—both can easily be obtained using the *SVD*, that is, the *key* elements of SIDs are based on extracting:

$$\mathcal{O}_K = \underbrace{\begin{bmatrix} C \\ CA \\ \vdots \\ CA^{K-1} \end{bmatrix}}_{\text{column space}}; \qquad \mathcal{X}_k = \underbrace{[\mathbf{x}(k)\ \mathbf{x}(k+1)\ \cdots\ \mathbf{x}(K+k-1)]}_{\text{row space}}$$

Perhaps, the major "link" between classical realization theory employing impulse response sequences followed by the Hankel matrix factorization and subspace realization theory from input/output sequences lies within the inherent system theoretic "rank condition" defining the minimal dimension of the state-space. Subspace realization theory relies primarily on the concept of the persistent excitation of the input sequences as well as the minimal state-space dimension for solution [8]. That is, the observability matrix spans the N_x-dimensional state-space that contains that part of the output due to the states exclusive of the inputs. In order to see this, we first construct input/output data matrices. Recall that the discrete *LTI* solutions given by Eqs. (6.5) and (6.13). Then, expanding the latter measurement solution further over k-samples and collecting terms, that is,

$$y(t) = CA^t x(0) + \sum_{\ell=0}^{t-1} CA^{t-\ell-1} Bu(\ell) + D\delta(t)$$

$t = 0, 1, \ldots, k-1$, we obtain

$$\begin{bmatrix} y(0) \\ y(1) \\ \vdots \\ y(k-1) \end{bmatrix} = \underbrace{\begin{bmatrix} C \\ CA \\ \vdots \\ CA^{k-1} \end{bmatrix}}_{\mathcal{O}_k} x(0) + \underbrace{\begin{bmatrix} D & & \cdots & 0 \\ CB & D & \cdots & 0 \\ \vdots & \vdots & \vdots & \vdots \\ CA^{k-1}B & \cdots & CB & D \end{bmatrix}}_{\mathcal{T}_k} \begin{bmatrix} u(0) \\ u(1) \\ \vdots \\ u(k-1) \end{bmatrix} \tag{6.61}$$

where $\mathcal{O}_k \in \mathcal{R}^{kN_y \times N_x}$ is the *observability matrix* and $\mathcal{T}_k \in \mathcal{R}^{kN_y \times kN_u}$ is *impulse response matrix*—a Toeplitz matrix [21].

Shifting these relations in time $(0 \to t)$ yields

$$\begin{bmatrix} y(t) \\ y(t+1) \\ \vdots \\ y(t+k-1) \end{bmatrix} = \mathcal{O}_k x(t) + \mathcal{T}_k \begin{bmatrix} u(t) \\ u(t+1) \\ \vdots \\ u(t+k-1) \end{bmatrix} \tag{6.62}$$

leads to the *vector* input/output relation

$$\mathbf{y}_k(t) = \mathcal{O}_k \mathbf{x}(t) + \mathcal{T}_k \mathbf{u}_k(t) \tag{6.63}$$

where $\mathbf{y} \in \mathcal{R}^{kN_y \times 1}$, $\mathbf{x} \in \mathcal{R}^{N_x \times 1}$ and $\mathbf{u} \in \mathcal{R}^{kN_u \times 1}$.

We stack these vectors to create batch data (*block Hankel*) matrices to obtain the subsequent "data equation." That is, defining the block output data matrix ($\mathcal{Y}_{0|k-1}$) and block input data matrix ($\mathcal{U}_{0|k-1}$)[4] over K-samples are:

$$\mathcal{Y}_{0|k-1} := \begin{bmatrix} \mathbf{y}(0) & \mathbf{y}(1) & \cdots & \mathbf{y}(K-1) \\ \mathbf{y}(1) & \mathbf{y}(2) & \cdots & \mathbf{y}(K) \\ \vdots & \vdots & \vdots & \vdots \\ \mathbf{y}(k-1) & \mathbf{y}(k) & \cdots & \mathbf{y}(K+k-2) \end{bmatrix} \tag{6.64}$$

and

$$\mathcal{U}_{0|k-1} := \begin{bmatrix} \mathbf{u}(0) & \mathbf{u}(1) & \cdots & \mathbf{u}(K-1) \\ \mathbf{u}(1) & \mathbf{u}(2) & \cdots & \mathbf{u}(K) \\ \vdots & \vdots & \vdots & \vdots \\ \mathbf{u}(k-1) & \mathbf{u}(k) & \cdots & \mathbf{u}(K+k-2) \end{bmatrix} \tag{6.65}$$

or more compactly

$$\begin{aligned} \mathcal{Y}_{0|k-1} &= \left[\mathbf{y}_k(0) \ \cdots \ \mathbf{y}_k(K-1)\right] \ \in \mathcal{R}^{kN_y \times K} \\ \mathcal{U}_{0|k-1} &= \left[\mathbf{u}_k(t) \ \cdots \ \mathbf{u}_k(K-1)\right] \ \in \mathcal{R}^{kN_u \times K} \end{aligned} \tag{6.66}$$

with the corresponding state vector defined by

$$\mathcal{X}_0 := [\mathbf{x}(0) \ \mathbf{x}(1) \ \cdots \ \mathbf{x}(K-1)] \ \in \mathcal{R}^{N_x \times K} \tag{6.67}$$

Therefore, we have the *vector-matrix input/output equation* or the *data equation* that relates the system model to the data (input, state, and output matrices)

$$\mathcal{Y}_{0|k-1} = \mathcal{O}_k \mathcal{X}_0 + \mathcal{T}_k \mathcal{U}_{0|k-1} \tag{6.68}$$

The overall objective is to extract the observability matrix and subsequently the system matrices Σ_{ABCD} from the deterministic input/output data. That is,

---

4 The notation "$0|k-1$" defines the first and last row element of the K-column data matrix.

we seek a solution that possesses the column space (range) equivalent to $\mathcal{O}_K$ (e.g. range $(\mathcal{Y}_{0|k-1}) \equiv$ range $(\mathcal{O}_K)$). Clearly, if the impulse response matrix $\mathcal{T}_K$ is *known* besides the data and input matrices, then we can merely subtract the last term from the data equation and we would have

$$\mathcal{Y}_{0|k-1} = \mathcal{O}_K \mathcal{X}_0$$

providing the desired result enabling the extraction of Σ_{ABCD} after performing an *SVD* to obtain the equivalent column space, but unfortunately, $\mathcal{T}_K$ is *unknown*! Thus, we must somehow estimate the impulse response matrix, first, in order to remove it from the data equation.

6.4.1 Subspace Realization: Orthogonal Projections

One approach to extract the observability matrix is *least-squares* [22–26], that is, to formulate the problem as minimizing a cost function $\mathcal{J}$ with respect to $\mathcal{T}_k$

$$\mathcal{J} = \min_{\mathcal{T}_k} \|\mathcal{Y}_{0|k-1} - \mathcal{T}_k \mathcal{U}_{0|k-1}\| \tag{6.69}$$

in order to remove the effect of the input using the idea of developing a projection operator $\mathcal{P}$ that is orthogonal to the input space $\mathcal{U}$ such that $\langle \mathcal{U}, \mathcal{P}^{\perp} \rangle = \mathbf{0}$. Well-known methods exist to accomplish this estimation such as setting the gradient operator to zero $(\nabla_{\mathcal{T}} \mathcal{J} = \mathbf{0})$ and solving for $\hat{\mathcal{T}}_k$ or completing the square [23, 26]. Therefore, the least-squares estimate for this problem is obtained as

$$\hat{\mathcal{T}}_k := \mathcal{Y}_{0|k-1} \mathcal{U}'_{0|k-1} \left(\mathcal{U}_{0|k-1} \mathcal{U}'_{0|k-1} \right)^{-1} \tag{6.70}$$

Removing the input from the data equation gives

$$\mathcal{Y}_{0|k-1} - \hat{\mathcal{T}}_k \times \mathcal{U}_{0|k-1} = \mathcal{Y}_{0|k-1} - \mathcal{Y}_{0|k-1} \mathcal{U}'_{0|k-1} \left(\mathcal{U}_{0|k-1} \mathcal{U}'_{0|k-1} \right)^{-1} \times \mathcal{U}_{0|k-1}$$

and factoring we have

$$\begin{aligned} \mathcal{Y}_{0|k-1} - \hat{\mathcal{T}}_k \mathcal{U}_{0|k-1} &= \mathcal{Y}_{0|k-1} \underbrace{\left(\mathrm{I} - \mathcal{U}'_{0|k-1} \left(\mathcal{U}_{0|k-1} \mathcal{U}'_{0|k-1} \right)^{-1} \mathcal{U}_{0|k-1} \right)}_{\mathcal{P}^{\perp}_{\mathcal{U}'_{0|k-1}}} \\ &= \mathcal{Y}_{0|k-1} \mathcal{P}^{\perp}_{\mathcal{U}'_{0|k-1}} \end{aligned} \tag{6.71}$$

where

$$\mathcal{P}^{\perp}_{\mathcal{U}'_{0|k-1}} = \mathbf{I} - \mathcal{P}_{\mathcal{U}'_{0|k-1}} = \mathbf{I} - \mathcal{U}'_{0|k-1} \left(\mathcal{U}_{0|k-1} \mathcal{U}'_{0|k-1} \right)^{-1} \mathcal{U}_{0|k-1} \tag{6.72}$$

is the orthogonal projection $\langle \mathcal{U}_{0|k-1}, \mathcal{P}^{\perp}_{\mathcal{U}'_{0|k-1}} \rangle$ of the (column) space of $\mathcal{U}_{0|k-1}$ onto the null space of $\mathcal{U}$ such that

$$\mathcal{U}_{0|k-1} \times \mathcal{P}^{\perp}_{\mathcal{U}'_{0|k-1}} = \mathcal{U}_{0|k-1} - \left(\mathcal{U}_{0|k-1}\mathcal{U}'_{0|k-1}\right)\left(\mathcal{U}_{0|k-1}\mathcal{U}'_{0|k-1}\right)^{-1}\mathcal{U}_{0|k-1} = \mathbf{0} \tag{6.73}$$

Here the solution depends on assuming that $\mathcal{U}_{0|k-1}$ is "persistently exciting" or equivalently it is full rank, $\rho(\mathcal{U}) = kN_u$.

Applying the projection operator to the data equation of Eq. (6.68) projects the "past data" onto ("|") the null space of $\mathcal{U}$ defined by $\mathcal{P}^{\perp}_{\mathcal{Y}_{0|k-1}|\mathcal{U}_{0|k-1}}$ yielding the desired results

$$\begin{aligned}\mathcal{P}^{\perp}_{\mathcal{Y}_{0|k-1}|\mathcal{U}_{0|k-1}} := \mathcal{Y}_{0|k-1} \times \mathcal{P}^{\perp}_{\mathcal{U}'_{0|k-1}} &= \mathcal{O}_k\mathcal{X}_0\mathcal{P}^{\perp}_{\mathcal{U}'_{0|k-1}} + \mathcal{T}_k\underbrace{\mathcal{U}_{0|k-1}\,\mathcal{P}^{\perp}_{\mathcal{U}'_{0|k-1}}}_{\mathbf{0}} \\ &= \mathcal{O}_k\mathcal{X}_0\,\mathcal{P}^{\perp}_{\mathcal{U}'_{0|k-1}}\end{aligned} \tag{6.74}$$

of eliminating the input and extracting the observability matrix (see Figure 6.3).

Next, we must be assured that the column spaces of the data and the observability matrix are identical in order to extract Σ_{ABCD}. From the data equation of Eq. (6.68), we can construct the *augmented* (input–output)

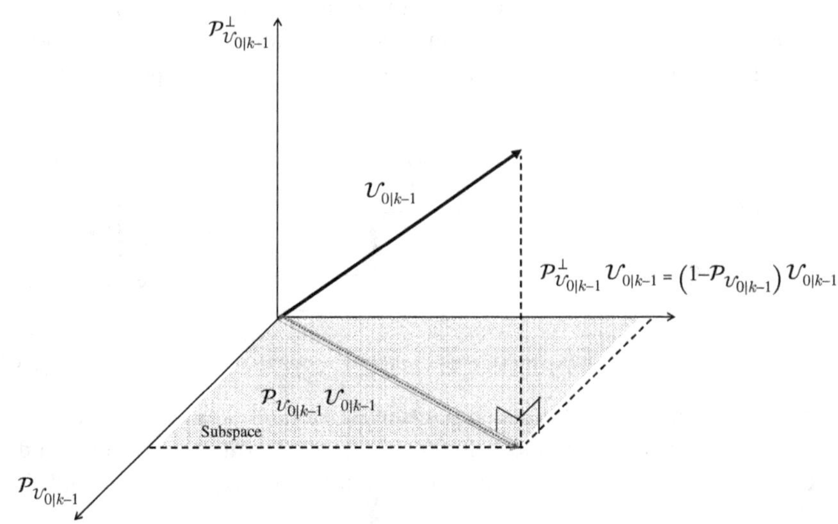

Figure 6.3 Orthogonal projection operator: orthogonal complement operation.

data matrix, $\mathcal{D} \in \mathcal{R}^{k(N_u+N_y)\times K}$

$$\mathcal{D}_{0|k-1} := \begin{bmatrix} \mathcal{U}_{0|k-1} \\ --- \\ \mathcal{Y}_{0|k-1} \end{bmatrix} = \begin{bmatrix} \mathbf{I}_{kN_u} & | & 0 \\ - & & - \\ \mathcal{T}_k & | & \mathcal{O}_k \end{bmatrix} \begin{bmatrix} \mathcal{U}_{0|k-1} \\ -- \\ \mathcal{X}_0 \end{bmatrix} \tag{6.75}$$

and

$$\rho\left(\begin{bmatrix} \mathcal{U}_{0|k-1} \\ -- \\ \mathcal{X}_0 \end{bmatrix}\right) = N_x + kN_u$$

leads to the *rank condition*

$$\rho\left(\mathcal{Y}_{0|k-1}\mathcal{P}^{\perp}_{\mathcal{U}'_{0|k-1}}\right) = \rho\left(\mathcal{O}_K\mathcal{X}_0\mathcal{P}^{\perp}_{\mathcal{U}'_{0|k-1}}\right) = N_x \qquad \text{[Rank Condition]}$$

therefore, it follows immediately that

$$\text{range}(\mathcal{Y}_{0|k-1}) \equiv \text{range}(\mathcal{O}_K)$$

the desired condition (see [8] or [23] for proof).

This "rank condition" provides a powerful result for the identification of the system model Σ_{ABCD} from deterministic input/output data just as the rank condition for impulse response data enabling the Hankel matrix factorization provides for the identification. It leads to pragmatic numerically efficient linear algebraic methods (to follow) providing a robust solution to solve the problem.

However, before we close this discussion, the system model must be extracted from the output data which can be accomplished using a truncated *SVD* decomposition—similar to the method of Section 6.3.2, that is, the *SVD* partitions the projected data, as

$$\begin{aligned} \mathcal{P}^{\perp}_{\mathcal{Y}_{0|k-1}|\mathcal{U}'_{0|k-1}} = \mathcal{Y}_{0|k-1}\mathcal{P}^{\perp}_{\mathcal{U}'_{0|k-1}} = \mathcal{O}_k\mathcal{X}_0\ \mathcal{P}^{\perp}_{\mathcal{U}_{0|k-1}} &= [U_S \mid U_N] \begin{bmatrix} \Sigma_S & | & \mathbf{0} \\ - & - & - \\ \mathbf{0} & | & \Sigma_N \end{bmatrix} \begin{bmatrix} V'_S \\ -- \\ V'_N \end{bmatrix} \\ &= \underbrace{U_S\Sigma_S V'_S}_{\text{Signal Subspace}} + \underbrace{U_N\Sigma_N V'_N}_{\text{Noise Subspace}} \end{aligned} \tag{6.76}$$

Note that an added benefit of this decomposition is that the system order N_x can be estimated directly from its singular values enabling the partition into signal and noise subspaces. There exists a variety of *order estimation* criteria to accomplish this choice [21].

For a *balanced realization* choice (see Section 6.2), the observability matrix can be extracted directly through this decomposition as

$$\mathcal{O}_k\mathcal{X}_0\ \mathcal{P}^{\perp}_{\mathcal{U}_{0|k-1}} \approx \underbrace{\left(U_S\Sigma_S^{1/2}\right)}_{\mathcal{O}_k} \times \left((\Sigma_S')^{1/2}V_S'\right) \tag{6.77}$$

assuming the noise subspace is negligible $\left(\Sigma_N \ll \Sigma_S\right)$.

Here unique system matrices can be obtained by applying the pseudo-inverse $\left(Z^{\#} = (Z'Z)^{-1}Z' \text{ or } Z^{\#} = Z'(ZZ')^{-1}\right)$, since the input/output rank condition guarantees full-rank of the observability matrix to extract[5]

$$A = \mathcal{O}^{\#}_{N_x} \times \mathcal{O}^{\uparrow}_{N_x} = \Sigma_S^{-1/2}U_S' \times \mathcal{O}^{\uparrow}_{N_x} \text{ and } C = \mathcal{O}_{N_x}(1 : N_y, 1 : N_x) \tag{6.78}$$

To obtain the input transmission matrix B and the input/output transmission matrix D, we can use the orthogonality of the left-singular matrix, U_N, from the *SVD* of Eq. (6.76) to isolate (extract) the impulse response matrix $\mathcal{T}_k$, that is, pre-multiplying the data equation by U_N gives [8, 27, 28]

$$U_N' \times \mathcal{Y}_{0|k-1} = \underbrace{U_N' \times \mathcal{O}_k\mathcal{X}_0}_{0} + U_N' \times \mathcal{T}_k\mathcal{U}_{0|k-1} = U_N' \times \mathcal{T}_k\mathcal{U}_{0|k-1} \tag{6.79}$$

since the first term is null, that is, from the *SVD* we have

$$\underbrace{U_N' \times}_{0} (\mathcal{O}_k\mathcal{X}_0\ \mathcal{P}^{\perp}_{\mathcal{U}_{0|k-1}}) = \underbrace{U_N'(U_S}_{0}\Sigma_S V_S') = 0$$

Post-multiplying by the pseudo-inverse

$$\mathcal{U}^{\#}_{0|k-1} = \mathcal{U}'_{0|k-1}\left(\mathcal{U}_{0|k-1}\mathcal{U}'_{0|k-1}\right)^{-1}$$

yields the relation

$$U_N'\mathcal{Y}_{0|k-1} \times \mathcal{U}^{\#}_{0|k-1} = U_N'\mathcal{T}_k\mathcal{U}_{0|k-1} \times \mathcal{U}^{\#}_{0|k-1} = U_N'\mathcal{T}_k \tag{6.80}$$

to isolate and extract the impulse response matrix.

Define the following matrices:

$$\begin{aligned} \mathcal{Z} := U_N'\mathcal{Y}_{0|k-1}\mathcal{U}^{\#}_{0|k-1} &:= \left[\mathbf{z}_1\ \mathbf{z}_1\ \cdots\ \mathbf{z}_k\right] \in \mathcal{R}^{(kN_y-N_x)\times kN_u} \\ U_N' &:= \left[\mathbf{u}_1^N\ \mathbf{u}_2^N\ \cdots\ \mathbf{u}_k^N\right] \in \mathcal{R}^{(kN_y-N_x)\times kN_y} \end{aligned} \tag{6.81}$$

for $\mathcal{T}_k \in \mathcal{R}^{kN_y\times kN_u}$ and $\mathbf{z}_i \in \mathcal{R}^{(kN_y-N_x)\times N_u}$, $\mathbf{u}_i \in \mathcal{R}^{(kN_y-N_x)\times N_y}$, $i = 1, \ldots, k$.

5 Recall the shift-invariant property of the observability matrix in Eq. (6.38).

Therefore, with these definitions, we have

$$\mathcal{Z} = U'_N \times \mathcal{T}_k \tag{6.82}$$

expanding gives

$$\mathcal{Z} = \begin{bmatrix}\mathbf{z}_1 & \mathbf{z}_1 & \cdots & \mathbf{z}_k\end{bmatrix} = \begin{bmatrix}\mathbf{u}_1^N & \mathbf{u}_2^N & \cdots & \mathbf{u}_k^N\end{bmatrix}\begin{bmatrix} D & & \cdots & 0 \\ CB & D & \cdots & 0 \\ \vdots & \vdots & \vdots & \vdots \\ CA^{k-1}B & \cdots & CB & D \end{bmatrix} \tag{6.83}$$

Performing the multiplications to expand, we obtain the following (vector) relations

$$\begin{aligned} \mathbf{z}_1 &= \mathbf{u}_1^N D + \mathbf{u}_2^N CB + \cdots + \mathbf{u}_k^N CA^{k-2}B \\ \mathbf{z}_2 &= \mathbf{u}_2^N D + \mathbf{u}_3^N B + \cdots + \mathbf{u}_k^N CA^{k-3}B \\ \vdots &= \quad \vdots \qquad\qquad \vdots \\ \mathbf{z}_{k-1} &= \mathbf{u}_{k-1}^N D + \mathbf{u}_k^N CB \\ \mathbf{z}_k &= \mathbf{u}_k^N D \end{aligned} \tag{6.84}$$

resulting in a stacking (reordering) of the $\mathbf{z}_i$ into a new structure defined by $\tilde{\mathcal{Z}} \in \mathcal{R}^{k(kN_y - N_x)\times N_u}$

$$\tilde{\mathcal{Z}} = \begin{bmatrix} \mathbf{z}_1 \\ \mathbf{z}_2 \\ \vdots \\ \mathbf{z}_{k-1} \\ \mathbf{z}_k \end{bmatrix} = \begin{bmatrix} \mathbf{u}_1^N & | & [\mathbf{u}_2^N \cdots \mathbf{u}_k^N] \times \mathcal{O}_{k-1} \\ \mathbf{u}_2^N & | & [\mathbf{u}_3^N \cdots \mathbf{u}_k^N] \times \mathcal{O}_{k-2} \\ \vdots & & \vdots \\ \mathbf{u}_{k-1}^N & | & [\ \mathbf{u}_k^N\ ] \times \mathcal{O}_1 \\ \mathbf{u}_k^N & | & [\,0\,] \end{bmatrix} \times \begin{bmatrix} D \\ -- \\ B \end{bmatrix}$$

or defining $\overline{\mathbf{u}}_i^N := [\mathbf{u}_i^N \ \mathbf{u}_{i+1}^N \ \cdots \mathbf{u}_k^N]$, we have

$$\tilde{\mathcal{Z}} = \underbrace{\begin{bmatrix} \mathbf{u}_1^N & | & \overline{\mathbf{u}}_2^N \times \mathcal{O}_{k-1} \\ \mathbf{u}_2^N & | & \overline{\mathbf{u}}_3^N \times \mathcal{O}_{k-2} \\ \vdots & & \vdots \\ \mathbf{u}_{k-1}^N & | & \overline{\mathbf{u}}_k^N \times \mathcal{O}_1 \\ \mathbf{u}_k^N & | & [\,0\,] \end{bmatrix}}_{\mathbf{U}} \times \begin{bmatrix} D \\ -- \\ B \end{bmatrix} := \mathbf{U} \times \begin{bmatrix} D \\ -- \\ B \end{bmatrix} \tag{6.85}$$

Since $\mathbf{U}$ is a full-rank matrix, we obtain a unique *least-squares* solution to extract the $\hat{B}$ and $\hat{D}$ input transmission matrices as

$$\begin{bmatrix} \hat{D} \\ -- \\ \hat{B} \end{bmatrix} = \mathbf{U}^{\#} \times \tilde{\mathcal{Z}} = \left(\mathbf{U}'\mathbf{U}\right)^{-1}\mathbf{U}' \times \tilde{\mathcal{Z}} \tag{6.86}$$

Even though this result is theoretically pleasing and provides insight into the motivation of the subsequent algorithms, it is *not* numerically efficient leading us to a more robust linear algebraic solution for this state-space identification problem.

6.4.2 Multivariable Output Error State-space (MOESP) Algorithm

A numerically efficient linear algebraic approach to solving these identification problems can be achieved by applying the *LQ*-decomposition[6] method to numerically calculate orthogonal projections enabling a reliable "separation" of data into subspaces in order to eliminate disturbances and noise. Therefore, to extract the state-space model, Σ_{ABCD}, it is required to obtain the extended column space of the observability matrix that can be obtained using the *SVD*, that is, one of the key elements of SID is based on extracting:

$$\mathcal{O}_K = \underbrace{\begin{bmatrix} C \\ CA \\ \vdots \\ CA^{K-1} \end{bmatrix}}_{\text{column space}}$$

The *MOESP*[7] technique is based on performing an orthogonal *LQ* decomposition on the data matrix (see Appendix C), that is, $\mathcal{D} = L \times Q$ with L a lower block triangular matrix, $L \in \mathcal{R}^{k(N_u+N_y)\times k(N_u+N_y)}$ and Q an orthogonal matrix, $Q \in \mathcal{R}^{k(N_u+N_y)\times K}$ such that

$$\mathcal{D}_{0|k-1} = \begin{bmatrix} I_{kN_u} & | & O_{kN_u \times N_x} \\ -- & | & -- \\ \mathcal{T}_k & | & \mathcal{O}_k \end{bmatrix} \begin{bmatrix} \mathcal{U}_{0|k-1} \\ --- \\ \mathcal{X}_0 \end{bmatrix} = \begin{bmatrix} L_{11} & | & O_{kN_u \times kN_y} \\ -- & | & -- \\ L_{21} & | & L_{22} \end{bmatrix} \begin{bmatrix} Q_1' \\ -- \\ Q_2' \end{bmatrix} \tag{6.87}$$

with the $L_{11} \in \mathcal{R}^{kN_u \times kN_u}$, $L_{21} \in \mathcal{R}^{kN_y \times kN_u}$, $L_{22} \in \mathcal{R}^{kN_y \times kN_y}$; $Q_1' \in \mathcal{R}^{kN_u \times K}$ and $Q_2' \in \mathcal{R}^{kN_y \times K}$.

This operation produces an efficient orthogonal decomposition of the augmented data matrix, $\mathcal{D}_{0|k-1}$, that is, the *LQ* decomposition is a numerical (matrix) version of the *Gram–Schmidt orthogonalization* procedure [19, 29, 30]. Recall that the *LQ* decomposition of the a matrix contains the set of orthonormal basis vectors in Q ($Q'Q = I$) with $\mathbf{q}_i \in \mathcal{R}^{K\times 1}$ and L contains the coefficients of linear dependence or equivalently the relationship the basis vectors of $\mathcal{D}$ to those of Q.

6 The *LQ*-decomposition is the *dual* of the well-known Gram–Schmidt orthogonalization method using the *QR*-decomposition of a matrix, that is, $LQ = (QR)' = R'Q'$ with L now lower triangular and replacing $Q' \rightarrow Q$ [8].

7 This method is called the *PI-MOESP* technique with instrument variable $\mathcal{U}_{0|k-1}$ [25].

In fact, from this viewpoint, we can think of partitioning $Q = [Q_{\mathcal{D}} \mid Q_{\mathcal{D}}^{\perp}]$ such that a rank-deficient data matrix $\mathcal{D}$ can (hypothetically) be decomposed as

$$\mathcal{D} = LQ = [L_{\mathcal{D}} \mid 0] \begin{bmatrix} Q_{\mathcal{D}} \\ -- \\ Q_{\mathcal{D}}^{\perp} \end{bmatrix} = L_{\mathcal{D}} \times Q_{\mathcal{D}}$$

The *projection matrix* associated with the decomposition is specified by [30]

$$\mathcal{P} = Q \times Q' \qquad \text{[Projection Matrix]}$$

From Eq. (6.75) (assuming full-rank matrices), we have that

$$\mathcal{D}_{0|k-1} = \begin{bmatrix} L_{11} & \mid & O \\ -- & \mid & -- \\ L_{21} & \mid & L_{22} \end{bmatrix} \begin{bmatrix} Q_1' \\ -- \\ Q_2' \end{bmatrix} = \begin{bmatrix} \mathcal{U}_{0|k-1} \\ --- \\ \mathcal{Y}_{0|k-1} \end{bmatrix} = \begin{bmatrix} L_{11} \times Q_1' \\ --- \\ L_{21}\, Q_1' + L_{22}\, Q_2' \end{bmatrix} \tag{6.88}$$

Orthogonally projecting $\mathcal{Y}_{0|k-1}$ onto $\mathcal{U}_{0|k-1}$ and $\mathcal{U}_{0|k-1}^{\perp}$ enables the decomposition by applying the projection matrices

$$\mathcal{P}_{\mathcal{Y}_{0|k-1} \mid \mathcal{U}_{0|k-1}} = \mathcal{Y}_{0|k-1} \underbrace{(Q_1 Q_1')}_{\mathcal{P}_{\mathcal{U}_{0|k-1}}} = (L_{21}Q_1' + L_{22}Q_2')\, Q_1 \times Q_1' = L_{21} \times Q_1'$$

$$\mathcal{P}_{\mathcal{Y}_{0|k-1} \mid \mathcal{U}_{0|k-1}^{\perp}} = \mathcal{Y}_{0|k-1} \underbrace{(Q_2 Q_2')}_{\mathcal{P}^{\perp}_{\mathcal{U}'_{0|k-1}}} = (L_{21}Q_1' + L_{22}Q_2')\, Q_2 \times Q_2' = L_{22} \times Q_2' \tag{6.89}$$

where we have applied the orthonormal properties of Q such that $Q_i'Q_i = I$ and $Q_i'Q_j = 0'$.

Returning to the original formulation of Eq. (6.75) and equating with these expressions, we see that

$$\mathcal{U}_{0|k-1} = L_{11}Q_1'$$
$$\mathcal{Y}_{0|k-1} = \mathcal{O}_k \mathcal{X}_0 + \mathcal{T}_k \mathcal{U}_{0|k-1} = \mathcal{O}_k \mathcal{X}_0 + \mathcal{T}_k \underbrace{(L_{11}Q_1')}_{U_{0|k-1}} = L_{21}Q_1' + L_{22}Q_2' \tag{6.90}$$

Post-multiplying Eq. (6.90) by Q_2 and imposing the *orthogonality conditions* of the *LQ*-decomposition as: $Q_2'\, Q_2 = I$ and $Q_1'\, Q_2 = 0$ enables the extraction of the observability matrix

$$\mathcal{Y}_{0|k-1} Q_2 = \mathcal{O}_k \mathcal{X}_0 Q_2 + \mathcal{T}_k L_{11} \underbrace{Q_1'Q_2}_{0} = L_{21} \underbrace{Q_1'Q_2}_{0} + L_{22} \underbrace{Q_2'Q_2}_{I} = L_{22}$$

or simply

$$\mathcal{O}_k \mathcal{X}_0 Q_2 = L_{22} \tag{6.91}$$

with $\rho(L_{22}) = N_x$ from the rank condition for input/output sequences.

Performing the *SVD* of L_{22} is given by

$$L_{22} = [U_S \mid U_N] \begin{bmatrix} \Sigma_S & | & 0 \\ - & - & - \\ 0 & | & \Sigma_N \end{bmatrix} \begin{bmatrix} V'_S \\ -- \\ V'_N \end{bmatrix} = \underbrace{U_S \Sigma_S V'_S}_{\text{Signal Subspace}} + \underbrace{U_N \Sigma_N V'_N}_{\text{Noise Subspace}} \tag{6.92}$$

Assuming the uncertainty or equivalently noise component is negligible compared to the signal component and selecting the balanced realization (as before) yields

$$\mathcal{O}_k \mathcal{X}_0 Q_2 \approx U_S \times \Sigma_S \times V'_S \approx \underbrace{\left(U_S \Sigma_S^{1/2}\right)}_{\mathcal{O}_k} \times \left((\Sigma'_S)^{1/2} V'_S\right) \tag{6.93}$$

extracting the observability matrix—the desired result.

The system matrices A, B, C, D can now be determined as before by:

$$A = \mathcal{O}_{N_x}^{\#} \times \mathcal{O}_{N_x}^{\uparrow} = \Sigma_S^{-1/2} U'_S; \qquad C = \mathcal{O}(1 : N_y, 1 : N_x) \tag{6.94}$$

with B and D obtained by solving a least-squares problem directly, since pre-multiplying Eq. (6.90) by U'_N gives

$$U'_N \mathcal{O}_k \mathcal{X}_0 + U'_N \mathcal{T}_k \mathcal{U}_{0|k-1} = U'_N L_{21} Q'_1 + U'_N L_{22} Q'_2$$

but after substituting for $\mathcal{U}_{0|k-1}$ and applying the orthogonality conditions $U'_N \mathcal{O}_k = 0$ and $U'_N L_{22} = 0$, we have

$$U'_N \mathcal{T}_k (L_{11} Q'_1) = U'_N L_{21} Q'_1$$

post-multiply by Q_1 and using its orthonormal property ($Q'_1 Q_1 = I$) gives

$$\mathrm{Z} := U'_N \mathcal{T}_k = U'_N L_{21} \times L_{11}^{-1} \tag{6.95}$$

analogous to Eq. (6.82) with $\mathrm{Z} \to \mathcal{Z}$. Stacking and reordering the vectors (as before) leads to an overdetermined set of linear equations and a least-squares solution with $\tilde{\mathrm{Z}} \to \tilde{\mathcal{Z}}$ to give

$$\begin{bmatrix} \hat{D} \\ -- \\ \hat{B} \end{bmatrix} = \mathcal{U}^{\#} \times \tilde{\mathrm{Z}} \tag{6.96}$$

where $\mathcal{U}$ is the block coefficient matrix equivalent to Eq. (6.86) ($\mathcal{U} \to \mathbf{U}$).

Therefore, we have the **M**ultivariable **O**utput **E**rror **S**tate-**SP**ace (*MOESP*[8]) algorithm given by:

- Compute the *LQ* decomposition of $\mathcal{D}$ of Eq. (6.87);
- Perform the *SVD* of L_{22} in Eq. (6.92) to extract $\mathcal{O}_k$;
- Obtain A and C from Eq. (6.94); and

8 Note that if the instrument variable is selected as both $\mathcal{U}_{0|k-1}$ and $\mathcal{Y}_{0|k-1}$, the past inputs and outputs, then the *PO-MOESP* technique evolves rather than the *PI-MOESP* [25].

Table 6.2 Multivariable output error state-space (MOESP) algorithm.

LQ decomposition

$$\mathcal{D}_{0|k-1} = \begin{bmatrix} I_{kN_u} & | & 0 \\ - & & - \\ \mathcal{T}_k & | & \mathcal{O}_k \end{bmatrix} \begin{bmatrix} \mathcal{U}_{0|k-1} \\ -- \\ \mathcal{X}_0 \end{bmatrix} = \begin{bmatrix} L_{11} & | & 0 \\ - & & - \\ L_{21} & | & L_{22} \end{bmatrix} \begin{bmatrix} Q_1' \\ -- \\ Q_2' \end{bmatrix} \quad \text{(data matrix)}$$

Singular value decomposition

$$L_{22} = [U_S \,|\, U_N] \begin{bmatrix} \Sigma_S & | & 0 \\ - & - & - \\ 0 & | & \Sigma_N \end{bmatrix} \begin{bmatrix} V_S' \\ -- \\ V_N' \end{bmatrix} \approx \underbrace{\left(U_S \Sigma_S^{1/2}\right)}_{\mathcal{O}_{N_x}} \times ((\Sigma_S')^{1/2} V_S') \quad \text{(observability matrix)}$$

System matrix

$$A = \mathcal{O}_{N_x}^{\#} \times \mathcal{O}_{N_x}^{\uparrow} = \left(\mathcal{O}_{N_x}' \mathcal{O}_{N_x}\right)^{-1} \mathcal{O}_{N_x}' \times \mathcal{O}_{N_x}^{\uparrow} = \Sigma_S^{-1/2} U_S' \times \mathcal{O}_{N_x}^{\uparrow} \quad \text{(system/process matrix)}$$

Measurement (output) matrix

$$C = \mathcal{O}(1 : N_y, 1 : N_x) \quad \text{(output matrix)}$$

Input/input–output transmission matrices

$$\begin{bmatrix} \hat{D} \\ -- \\ \hat{B} \end{bmatrix} = \mathcal{U}^{\#} \times \tilde{Z} \quad \text{(input matrices)}$$

for $\mathcal{U}^{\#} := (\mathcal{U}'\mathcal{U})^{-1}\mathcal{U}'$

- Solve the least-squares problem to obtain B and D from Eq. (6.96).

We summarize the deterministic *MOESP* algorithm in Table 6.2.

6.5 Subspace Decomposition: Oblique Projections

Finally, for the deterministic realization problem from input/output data, we develop an alternative technique, the *parallel* or *oblique* projection method based on both past and future data. This approach differs from the *MOESP* method in that it first estimates the state sequence directly from the data through a projection of the row space of block Hankel matrices. That is, in this approach the row space of the state sequence can be obtained using the *SVD*, here the *key* element of oblique SID is based on extracting:

$$\mathcal{X}_k = \underbrace{\left[\mathbf{x}(k)\ \mathbf{x}(k+1)\ \cdots\ \mathbf{x}(K+k-1)\right]}_{\text{State-Space Sequence}}$$

State estimation is then followed by a linear least-squares solution extracting the desired system matrices, Σ_{ABCD}. It is based on the concept that the projection of the "future outputs" depend solely on the past inputs/outputs as well as the "future" inputs that evolve directly from the available data. Since the state sequence is available from the input/output data, it is used to establish a solution to the least-squares problem enabling SID to be accomplished [8, 22–25].

We begin by defining the data matrices for samples ranging from k-to-$(2k-1)$—the *future* data as [8]

$$\mathcal{Y}_{k|2k-1} := \begin{bmatrix} \mathbf{y}(k) & \mathbf{y}(k+1) & \cdots & \mathbf{y}(K+k-1) \\ \mathbf{y}(k+1) & \mathbf{y}(k+2) & \cdots & \mathbf{y}(K+k) \\ \vdots & \vdots & \vdots & \vdots \\ \mathbf{y}(2k-1) & \mathbf{y}(2k) & \cdots & \mathbf{y}(K+2k-2) \end{bmatrix} \tag{6.97}$$

and

$$\mathcal{U}_{k|2k-1} := \begin{bmatrix} \mathbf{u}(k) & \mathbf{u}(k+1) & \cdots & \mathbf{u}(K+k-1) \\ \mathbf{u}(k+1) & \mathbf{u}(k+2) & \cdots & \mathbf{u}(K+k) \\ \vdots & \vdots & \vdots & \vdots \\ \mathbf{u}(2k-1) & \mathbf{u}(2k) & \cdots & \mathbf{u}(K+2k-2) \end{bmatrix} \tag{6.98}$$

Leading to the corresponding *data equation* given by

$$\mathcal{Y}_{k|2k-1} = \mathcal{O}_k \mathcal{X}_k + \mathcal{T}_k \mathcal{U}_{k|2k-1} \tag{6.99}$$

where the block (Hankel) matrices are defined to give

$$\begin{aligned} \mathcal{Y}_{k|2k-1} &= \left[\mathbf{y}_k(k)\ \mathbf{y}_k(k+1)\ \cdots\ \mathbf{y}_k(K+k-1)\ \right] \in \mathcal{R}^{kN_y \times K} \\ \mathcal{U}_{k|2k-1} &= \left[\mathbf{u}_k(k)\ \mathbf{u}_k(k+1)\ \cdots\ \mathbf{u}_k(K+k-1)\ \right] \in \mathcal{R}^{kN_u \times K} \\ \mathcal{X}_k &= \left[\mathbf{x}(k)\ \mathbf{x}(k+1)\ \cdots\ \mathbf{x}(K+k-1)\ \right] \in \mathcal{R}^{N_x \times K} \end{aligned}$$

Here, $\mathcal{U}_{0|k-1}$, $\mathcal{Y}_{0|k-1}$, $\mathcal{X}_0$ are defined as the *past* inputs, outputs, and states ($\mathcal{U}_p$, $\mathcal{Y}_p$, $\mathcal{X}_0$), while $\mathcal{U}_{k|2k-1}$, $\mathcal{Y}_{k|2k-1}$, and $\mathcal{X}_k$ are the *future* inputs, outputs, and states (U_f, Y_f, $\mathcal{X}_f$) which are all *block Hankel* matrices as before, that is,

$$\begin{aligned} &\mathcal{U}_p := \mathcal{U}_{0|k-1}; \qquad \mathcal{Y}_p := \mathcal{Y}_{0|k-1}; \qquad \mathcal{X}_p := \mathcal{X}_0 \\ &\mathcal{U}_f := \mathcal{U}_{k|2k-1}; \qquad \mathcal{Y}_f := \mathcal{Y}_{k|2k-1}; \qquad \mathcal{X}_f := \mathcal{X}_k \end{aligned} \tag{6.100}$$

Similar to the input/output relations developed previously in Eq. (6.68), the corresponding state equations can be developed by expanding the solution of the deterministic state-space model

$$\begin{aligned}\mathbf{x}(k+j) &= A^k\mathbf{x}(j) + \sum_{\ell=0}^{k-1} A^{k-\ell-1}Bu(\ell+j)\\ &= A^k\mathbf{x}(j) + A^{k-1}Bu(j) + A^{k-2}Bu(j+1) + \cdots + Bu(k+j-1)\end{aligned}$$

or

$$\mathbf{x}(k+j) = A^k\mathbf{x}(j) + \underbrace{\begin{bmatrix} A^{k-1}B & A^{k-2}B & \cdots & AB & B\end{bmatrix}}_{C_{AB}^{\leftarrow}} \underbrace{\begin{bmatrix} u(j) \\ u(j+1) \\ \vdots \\ u(j+k-1)\end{bmatrix}}_{\mathbf{u}}$$

simplifying and grouping, we have

$$\mathbf{x}(k+j) = A^k\mathbf{x}(j) + C_{AB}^{\leftarrow}(k)\ \mathbf{u}(j) \tag{6.101}$$

where the "reversed" (input) controllability matrix $C_{AB}^{\leftarrow} \in \mathcal{R}^{N_x \times kN_u}$ is defined by

$$C_{AB}^{\leftarrow}(k) := \begin{bmatrix} A^{k-1}B & A^{k-2}B & \cdots & AB & B\end{bmatrix} \text{ [reversed input controllability]} \tag{6.102}$$

Expanding this relation over K-columns with $j = 0$, then

$$\begin{aligned}\underbrace{[\mathbf{x}(k) \cdots \mathbf{x}(K+k-1)]}_{\mathcal{X}_k} &= A^k \underbrace{[\mathbf{x}(0) \cdots \mathbf{x}(K-1)]}_{\mathcal{X}_0}\\ &\quad + C_{AB}^{\leftarrow}(k) \underbrace{[\mathbf{u}_k(0) \cdots \mathbf{u}_k(K-1)]}_{\mathcal{U}_{0|k-1}}\end{aligned}$$

that is, we now have

$$\mathcal{X}_k = A^k\mathcal{X}_0 + C_{AB}^{\leftarrow}(k)\ \mathcal{U}_{0|k-1} \tag{6.103}$$

or in terms of our definitions (above), the set of future states in terms of past states $\mathcal{X}_p$, inputs $\mathcal{U}_p$ becomes

$$\mathcal{X}_f = A^k\mathcal{X}_p + C_{AB}^{\leftarrow}(k)\ \mathcal{U}_p \tag{6.104}$$

Therefore, the set of "deterministic" data equations for the past/future inputs/outputs/states are:

$$\begin{aligned}\mathcal{Y}_p &= \mathcal{O}_k\mathcal{X}_p + \mathcal{T}_k\mathcal{U}_p && \text{[Past Outputs]}\\ \mathcal{Y}_f &= \mathcal{O}_k\mathcal{X}_f + \mathcal{T}_k\mathcal{U}_f && \text{[Future Outputs]}\\ \mathcal{X}_f &= A^k\mathcal{X}_p + C_{AB}^{\leftarrow}(k)\mathcal{U}_p && \text{[Future States]}\end{aligned} \tag{6.105}$$

with the corresponding data matrices, past and future, defined by $\mathcal{D}_p := \mathcal{D}_{0|k-1}$ and $\mathcal{D}_f := \mathcal{D}_{k|2k-1}$ where

$$\mathcal{D}_p = \begin{bmatrix} \mathcal{U}_p \\ --- \\ \mathcal{Y}_p \end{bmatrix} \text{ and } \mathcal{D}_{0|k-1} = \begin{bmatrix} \mathcal{U}_{0|k-1} \\ --- \\ \mathcal{Y}_{0|k-1} \end{bmatrix}; \; \mathcal{D}_f = \begin{bmatrix} \mathcal{U}_f \\ --- \\ \mathcal{Y}_f \end{bmatrix} \text{ and}$$

$$\mathcal{D}_{k|2k-1} = \begin{bmatrix} \mathcal{U}_{k|2k-1} \\ --- \\ \mathcal{Y}_{k|2k-1} \end{bmatrix} \tag{6.106}$$

With these input/state/output relations available, we see the relationship between the future state vector with the corresponding past and future data as

$$\text{span}\{\mathcal{X}_f\} = \text{span}\{\mathcal{D}_p\} \cap \text{span}\{\mathcal{D}_f\} \tag{6.107}$$

under the assumption of a persistently excited, completely controllable, and observable system $\left(\rho(\mathcal{O}_k) = \rho(\mathcal{C}_k) = N_x\right)$ along with the constraint on the initial state matrix being independent of the past inputs $(\text{span}\{\mathcal{X}_0\} \cap \text{span}\{\mathcal{U}_{0|k-1}\} = \{0\})$.

Therefore, solving Eq. (6.105) for the past and future states employing the pseudo-inverse ($^{\#}$) of the full-rank observability matrix ($\mathcal{O}_k^{\#}$) provides

$$\mathcal{X}_p = \mathcal{O}_k^{\#} \times \left(\mathcal{Y}_p - \mathcal{T}_k\mathcal{U}_p\right) \;\text{ for } \mathcal{X}_p \in \text{span}\{\mathcal{D}_p\} \tag{6.108}$$

and

$$\mathcal{X}_f = \mathcal{O}_k^{\#} \times \left(\mathcal{Y}_f - \mathcal{T}_k\mathcal{U}_f\right) \;\text{ for } \mathcal{X}_f \in \text{span}\{\mathcal{D}_f\} \tag{6.109}$$

Substituting for $\mathcal{X}_p$ into Eq. (6.104) we have

$$\mathcal{X}_f = A^k \left(\mathcal{O}_k^{\#}\mathcal{Y}_p - \mathcal{O}_k^{\#}\mathcal{T}_k\mathcal{U}_p\right) + \mathcal{C}_{AB}^{\leftarrow}(k)\;\mathcal{U}_p \tag{6.110}$$

expanding this expression and grouping like-terms, we obtain a vector relation in terms of $\mathcal{D}_p$

$$\mathcal{X}_f = \underbrace{\left[\mathcal{C}_{AB}^{\leftarrow}(k) - A^k\mathcal{O}_k^{\#}\mathcal{T}_k \quad | \quad A^k\mathcal{O}_k^{\#}\right]}_{\Theta_p} \begin{bmatrix} \mathcal{U}_p \\ -- \\ \mathcal{Y}_p \end{bmatrix} = \Theta_p \times \mathcal{D}_p \tag{6.111}$$

demonstrating that the future states *depend only* on the past inputs and outputs and not the future inputs $\mathcal{U}_f$ indicating that Eq. (6.107) is satisfied, that is, the future states provide the set of basis vectors evolving from the intersection of the past and future data subspaces. Therefore, intuitively, $\mathcal{X}_f$, enables the exchange of information between the past and future.

In terms of Eq. (6.111), it also follows that the future output can be written as

$$\mathcal{Y}_f = \mathcal{O}_k \; \Theta_p \; \mathcal{D}_p + \mathcal{T}_k \; \mathcal{U}_f \tag{6.112}$$

With this information available, we can now establish the basis of *N4SID* approach.

6.5.1 Subspace Realization: Oblique Projections

Another approach to decomposing or separating the data spaces is to apply an *oblique projection operator* enabling the extraction of the state-space model and eliminating the past input/output data. The parallel (oblique) projection ("||") of the future output data $\mathcal{Y}_f$ projected ("|") *onto* the past data subspace $\mathcal{D}_p$ *along* ("o") the future $\mathcal{U}_f$ is depicted in Figure 6.4. As illustrated in the figure, an orthogonal projector can be constructed from the sum of two oblique projectors and one orthogonal complement projector. Here, the orthogonal projection is performed first enabling the projection onto the joint space ($\mathcal{D}_p \vee \mathcal{U}_f$) that is followed by the decomposition along either $\mathcal{D}_f$ or $\mathcal{U}_p$ [8]

$$\mathcal{P}_{\mathcal{Y}_f|\mathcal{D}_p \vee \mathcal{U}_f} = \mathcal{P}^{\|}_{\mathcal{Y}_f|\mathcal{D}_p \circ \mathcal{U}_f} + \mathcal{P}^{\|}_{\mathcal{Y}_f|\mathcal{U}_f \circ \mathcal{D}_p} + \mathcal{P}^{\perp}_{\mathcal{Y}_f|\mathcal{U}_f \vee \mathcal{D}_p} \tag{6.113}$$

Another way to express this idea from an algebraic rather than geometric perspective is that the "vectors in the row space of $\mathcal{Y}_f$ are the sum of linear combinations of vectors in the row space of the state $\mathcal{X}_f$ and with those in the row space of the future inputs $\mathcal{U}_f$."

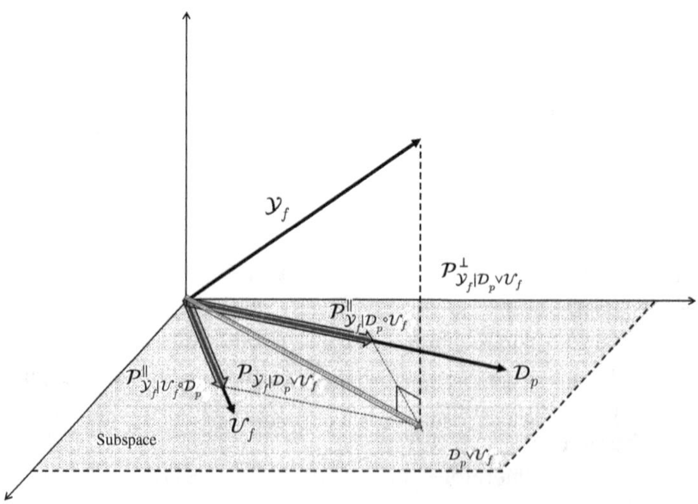

Figure 6.4 Oblique projections: projection $\left(\mathcal{P}^{\|}_{\mathcal{Y}_f|\mathcal{D}_p \circ \mathcal{U}_f}\right)$ of future outputs ($\mathcal{Y}_f$) onto past data ($\mathcal{D}_p$) along the future inputs ($\mathcal{U}_f$).

In contrast, the oblique projection matrix can also be defined in terms of orthogonal projections as well [22, 24]

$$\mathcal{P}^{\|}_{\mathcal{Y}_f|\mathcal{D}_p\circ\mathcal{U}_f} := \left[\mathcal{P}^{\perp}_{\mathcal{Y}_f|\mathcal{U}_f'}\right] \times \left[\mathcal{P}^{\perp}_{\mathcal{D}_p|\mathcal{U}_f'}\right]^{\#} \mathcal{D}_p = \left[\mathcal{Y}_f\mathcal{P}^{\perp}_{\mathcal{U}_f'}\right] \times \left[\mathcal{D}_p\mathcal{P}^{\perp}_{\mathcal{U}_f'}\right]^{\#} \mathcal{D}_p \tag{6.114}$$

Substituting Eq. (6.112) for the "future" output into this expression, we have

$$\begin{aligned}\mathcal{P}^{\|}_{\mathcal{Y}_f|\mathcal{D}_p\circ\mathcal{U}_f} &= \left[(\mathcal{O}_k\Theta_p\mathcal{D}_p + \underbrace{\mathcal{T}_k\mathcal{U}_f\ \mathcal{P}^{\perp}_{\mathcal{U}_f'}}_{0})\right]\left[\mathcal{D}_p\mathcal{P}^{\perp}_{\mathcal{U}_f'}\right]^{\#}\mathcal{D}_p \\ &= \mathcal{O}_k\Theta_p\underbrace{\left[\mathcal{D}_p\mathcal{P}^{\perp}_{\mathcal{U}_f'}\right]\left[\mathcal{D}_p\mathcal{P}^{\perp}_{\mathcal{U}_f'}\right]^{\#}}_{I}\mathcal{D}_p\end{aligned} \tag{6.115}$$

or simply from Eq. (6.111), we have

$$\mathcal{P}^{\|}_{\mathcal{Y}_f|\mathcal{D}_p\circ\mathcal{U}_f} = \mathcal{O}_k\Theta_p\mathcal{D}_p = \mathcal{O}_k\mathcal{X}_f \tag{6.116}$$

This is the underlying concept of the *N4SID* approach when applied to the deterministic realization problem. This relationship shows that obliquely projecting the future data nullifies the inputs while enabling the extraction of the observability and future states directly from the given data. Therefore, the *state vector* provides a basis for the *intersection* of the past and future subspaces and can be extracted from this projection.

Performing the *SVD* on Eq. (6.116), we obtain the partitioned signal and noise (uncertainty) relationship as

$$\mathcal{P}^{\|}_{\mathcal{Y}_f|\mathcal{D}_p\circ\mathcal{U}_f} = [U_S \mid U_N]\begin{bmatrix}\Sigma_S & | & 0 \\ - & - & - \\ 0 & | & \Sigma_N\end{bmatrix}\begin{bmatrix}V_S' \\ -- \\ V_N'\end{bmatrix} = \underbrace{U_S\Sigma_S V_S'}_{\text{Signal Subspace}} + \underbrace{U_N\Sigma_N V_N'}_{\text{Noise Subspace}} \tag{6.117}$$

Assuming the uncertainties are "negligible" ($\Sigma_N \ll \Sigma_S$) and selecting a *balanced deterministic realization* $\left(\Sigma_S = \Sigma_S^{1/2} \times (\Sigma_S')^{1/2}\right)$, we obtain the critical *N4SID* subspace relations:

$$\begin{aligned}\mathcal{P}^{\|}_{\mathcal{Y}_f|\mathcal{D}_p\circ\mathcal{U}_f} &\approx \underbrace{U_S\Sigma_S^{1/2}}_{\mathcal{O}_k}\underbrace{(\Sigma_S')^{1/2}V_S'}_{\mathcal{X}_f} \in \mathcal{R}^{kN_y\times K} && \text{[Projection]} \\ \mathcal{O}_k &= U_S\Sigma_S^{1/2} \in \mathcal{R}^{kN_y\times N_x} && \text{[Observability]} \\ \mathcal{X}_f &= \left(\Sigma_S'\right)^{1/2}V_S' \in \mathcal{R}^{N_x\times K} && \text{[States]}\end{aligned} \tag{6.118}$$

6.5.2 Numerical Algorithms for Subspace State-space System Identification (*N4SID*)

Again, a numerically efficient approach to extract the system Σ_{ABCD} from input/output data is to employ the LQ-decomposition of the block Hankel data matrices of the past $\mathcal{D}_p$ and future input/output data matrices $\{\mathcal{U}_f, \mathcal{Y}_f\}$ that will lead us to the oblique projection of $\mathcal{Y}_f$ onto $\mathcal{D}_p$ along $\mathcal{U}_f$ [8, 31]

$$\mathcal{D} := \begin{bmatrix} \mathcal{U}_f \\ -- \\ \mathcal{D}_p \\ -- \\ \mathcal{Y}_f \end{bmatrix} = \underbrace{\begin{bmatrix} L_{11} & | & 0 & | & 0 \\ -- & | & -- & | & -- \\ L_{21} & | & L_{22} & | & 0 \\ -- & | & -- & | & -- \\ L_{31} & | & L_{32} & | & L_{33} \end{bmatrix}}_{\text{L}} \times \underbrace{\begin{bmatrix} Q_1' \\ -- \\ Q_2' \\ -- \\ Q_3' \end{bmatrix}}_{\text{Q}} \tag{6.119}$$

with $L_{11} \in \mathcal{R}^{kN_u \times kN_u}$, $L_{21} \in \mathcal{R}^{k(N_u+N_y)\times kN_u}$, $L_{22} \in \mathcal{R}^{k(N_u+N_y)\times k(N_u+N_y)}$, $L_{31} \in \mathcal{R}^{kN_y\times kN_u}$, $L_{32} \in \mathcal{R}^{kN_y\times k(N_u+N_y)}$, $L_{33} \in \mathcal{R}^{kN_y\times kN_y}$; and $Q_1' \in \mathcal{R}^{kN_u\times K}$, $Q_2' \in \mathcal{R}^{k(N_u+N_y)\times K}$, $Q_3' \in \mathcal{R}^{kN_y\times K}$.

As illustrated in Figure 6.4, the oblique projection of the future data $\mathcal{Y}_f$ onto the "joint" space of future inputs $\mathcal{U}_f$ and past inputs/outputs data $\mathcal{D}_p$ can be expressed in terms of the summation of two oblique projections: one onto $\mathcal{U}_f$ along $\mathcal{D}_p$ ($\mathcal{P}^{\|}_{\mathcal{Y}_f\,|\,\mathcal{U}_f \circ \mathcal{D}_p}$) and one onto $\mathcal{D}_p$ along $\mathcal{U}_f$ ($\mathcal{P}^{\|}_{\mathcal{Y}_f\,|\,\mathcal{D}_p \circ \mathcal{U}_f}$) combined with complement of the orthogonal projection of $\mathcal{Y}_f$ onto the joint space of future inputs and past inputs/outputs ($\mathcal{P}^{\perp}_{\mathcal{Y}_f\,|\,\mathcal{U}_f \vee \mathcal{D}_p}$), that is,

$$\mathcal{P}^{\|}_{\mathcal{Y}_f\,|\,\mathcal{U}_f \vee \mathcal{D}_p} = \underbrace{\mathcal{P}^{\|}_{\mathcal{Y}_f\,|\,\mathcal{U}_f \circ \mathcal{D}_p} + \mathcal{P}^{\|}_{\mathcal{Y}_f\,|\,\mathcal{D}_p \circ \mathcal{U}_f}}_{\text{Oblique Projections}} + \underbrace{\mathcal{P}^{\perp}_{\mathcal{Y}_f\,|\,\mathcal{U}_f \vee \mathcal{D}_p}}_{\text{Orthogonal Complement Projection}} \tag{6.120}$$

The orthogonal projection of $\mathcal{Y}_f$ onto the joint space of $\mathcal{U}_f \vee \mathcal{D}_p$ is the subspace spanned by the orthonormal basis vectors of Q_1' and Q_2'; therefore, we have that

$$\mathcal{P}_{\mathcal{Y}_f\,|\,\mathcal{U}_f \vee \mathcal{D}_p} = \left[L_{31} \mid L_{32}\right] \begin{bmatrix} Q_1' \\ -- \\ Q_2' \end{bmatrix} \quad \text{[Orthogonal Projection]} \tag{6.121}$$

Alternatively, this projection can also be expressed in terms of *oblique projection operators* corresponding to the summation of the two oblique projections as in Eq. (6.120)

$$\mathcal{P}_{\mathcal{Y}_f|\mathcal{U}_f \vee \mathcal{D}_p} = \left[\mathcal{P}^{\|}_{\bullet|\mathcal{U}_f \circ \mathcal{D}_p} \mid \mathcal{P}^{\|}_{\bullet|\mathcal{D}_p \circ \mathcal{U}_f}\right] \times \begin{bmatrix} L_{11} & | & 0 \\ -- & & -- \\ L_{21} & | & L_{22} \end{bmatrix} \begin{bmatrix} Q_1' \\ -- \\ Q_2' \end{bmatrix} \quad \text{[Orthogonal Projection]} \tag{6.122}$$

since Q_1', Q_2' span the joint subspace of $\mathcal{U}_f \vee \mathcal{D}_p$.

Equating these relations for the "joint" orthogonal projection ($\mathcal{P}_{\mathcal{Y}_f|\mathcal{U}_f \vee \mathcal{D}_p}$) and solving in terms of the oblique projection operators gives

$$\left[\mathcal{P}^{\|}_{\bullet|\mathcal{U}_f \circ \mathcal{D}_p} \mid \mathcal{P}^{\|}_{\bullet|\mathcal{D}_p \circ \mathcal{U}_f}\right] = \left[L_{31} \mid L_{32}\right] \left[\begin{array}{c|c} L_{11} & 0 \\ \hline L_{21} & L_{22} \end{array}\right]^{-1}$$
$$= \left[L_{31} \mid L_{32}\right] \left[\begin{array}{c|c} L_{11}^{-1} & 0 \\ \hline -L_{22}^{-1} L_{21} L_{11}^{-1} & L_{22}^{-1} \end{array}\right] \tag{6.123}$$

or simply

$$\mathcal{P}^{\|}_{\bullet|\mathcal{U}_f \circ \mathcal{D}_p} = L_{31} L_{11}^{-1} - L_{32} L_{22}^{-1} L_{21} L_{11}^{-1}$$
$$\mathcal{P}^{\|}_{\bullet|\mathcal{D}_p \circ \mathcal{U}_f} = L_{32} L_{22}^{-1} \tag{6.124}$$

Obliquely projecting $\mathcal{Y}_f$ onto $\mathcal{D}_p$ along $\mathcal{U}_f$ and substituting for the past data from Eq. (6.119) as illustrated in Figure 6.5a, we obtain the *oblique projection* in terms of the reduced *LQ*-decomposition and perform the corresponding *SVD* as in Eq. (6.118) to obtain the set of relations:

$$\mathcal{P}^{\|}_{\mathcal{Y}_f|\mathcal{D}_p \circ \mathcal{U}_f} = \mathcal{O}_k \mathcal{X}_f = L_{32} L_{22}^{-1}\, \mathcal{D}_p \qquad \text{[Oblique Projection]}$$
$$\mathcal{O}_k = U_S \Sigma_S^{1/2} \qquad \text{[Observability]}$$
$$\mathcal{X}_f = \Sigma_S^{1/2} V_S' \qquad \text{[States]} \tag{6.125}$$

For the *N4SID* approach, both state sequences $\mathcal{X}_k$ and $\mathcal{X}_{k+1}$ will be required[9] subsequently to extract the model from the input/output data [22].

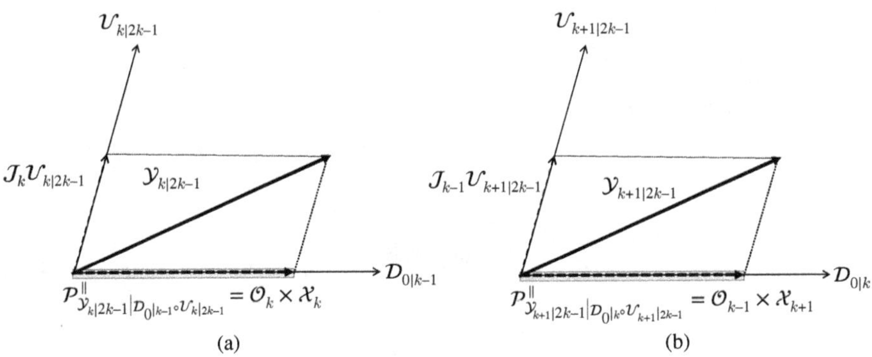

Figure 6.5 Extraction of future state and shifted state sequences: (a) oblique projection for future state sequence $\mathcal{X}_k$. (b) oblique projection for shifted state sequence $\mathcal{X}_{k+1}$.

9 Some approaches use the extracted states available from $\mathcal{X}_f \rightarrow \{\mathcal{X}_k\}; k \rightarrow k + K - 1$ for K large enabling a reasonable approximation for $\mathcal{X}_{k+1}$ [8].

Therefore, another oblique projection based on projecting the future (shifted) output $\mathcal{Y}_{k+1|2k-1}$ onto the (shifted) data $\mathcal{D}_{0|k}$ along the (shifted) future inputs $\mathcal{U}_{k+1|2k-1}$, that is, $\mathcal{P}^{\|}_{\mathcal{Y}_{k+1|2k-1}|\mathcal{D}_{0|k}\circ\mathcal{U}_{k+1|2k-1}}$ will be required (see Figure 6.5b).

These oblique projections are[10]:

$$
\begin{aligned}
\mathcal{P}^{\|}_{\mathcal{Y}_f|\mathcal{D}_p\circ\mathcal{U}_f} = \mathcal{P}^{\|}_{\mathcal{Y}_{k|2k-1}|\mathcal{D}_{0|k-1}\circ\mathcal{U}_{k|2k-1}} &= \mathcal{O}_k\mathcal{X}_k \quad \text{[Oblique Projection]}\\
\mathcal{P}^{\|}_{\mathcal{Y}_{k+1|2k-1}|\mathcal{D}_{0|k}\circ\mathcal{U}_{k+1|2k-1}} &= \mathcal{O}_{k-1}\mathcal{X}_{k+1} \quad \text{[Shifted Projection]}
\end{aligned}
\tag{6.126}
$$

where $\mathcal{O}_{k-1} \in \mathcal{O}_k$ is the extended observability matrix with the *last block row eliminated.*[11]

From these projections, we can estimate *both* required state sequences, $\overline{\mathcal{X}}_{k+1}$ and $\overline{\mathcal{X}}_k$ in order to solve a subsequent least-squares problem and extract the model Σ_{ABCD}. It is clear from the oblique projection of Eq. (6.125) that this can easily be accomplished from the *SVD* of Eq. (6.117) ($\overline{\mathcal{X}}_k \rightarrow \mathcal{X}_k \equiv \mathcal{X}_f$) by applying the pseudo-inverse of the full-rank observability matrix as

$$
\begin{aligned}
\overline{\mathcal{X}}_k &= \mathcal{O}_k^{\#} \times \mathcal{P}^{\|}_{\mathcal{Y}_{k|2k-1}|\mathcal{D}_{0|k-1}\circ\mathcal{U}_{k|2k-1}} = \Sigma_{S_k}^{-1/2}U'_{S_k} \times \mathcal{P}^{\|}_{\mathcal{Y}_{k|2k-1}|\mathcal{D}_{0|k-1}\circ\mathcal{U}_{k|2k-1}}\\
\overline{\mathcal{X}}_{k+1} &= \mathcal{O}_{k-1}^{\#} \times \mathcal{P}^{\|}_{\mathcal{Y}_{k+1|2k-1}|\mathcal{D}_{0|k}\circ\mathcal{U}_{k+1|2k-1}} = \Sigma_{S_{k-1}}^{-1/2}U'_{S_{k-1}} \times \mathcal{P}^{\|}_{\mathcal{Y}_{k+1|2k-1}|\mathcal{D}_{0|k}\circ\mathcal{U}_{k+1|2k-1}}
\end{aligned}
\tag{6.127}
$$

where $\Sigma_{S_{k-1}}^{-1/2} \in \Sigma_{S_k}^{-1/2}$ and $U_{S_{k-1}} \in U_{S_k}$ are the singular values and orthogonal matrices (respectively) with the last block row eliminated (as above for the observability matrix).

With these states now available, we can define the underlying "batch" state-space system $\overline{\Sigma}_{ABCD}$ as

$$
\begin{aligned}
\overline{\mathcal{X}}_{k+1} &= A\,\overline{\mathcal{X}}_k + B\,\overline{\mathcal{U}}_k\\
\overline{\mathcal{Y}}_k &= C\,\overline{\mathcal{X}}_k + D\,\overline{\mathcal{U}}_k
\end{aligned}
\tag{6.128}
$$

with

$$
\begin{aligned}
\overline{\mathcal{X}}_{k+1} &:= [x(k+1) \ \cdots \ x(k+K-1)]\\
\overline{\mathcal{X}}_k &:= [x(k) \ \cdots \ x(k+K-2)]\\
\overline{\mathcal{Y}}_k &:= [y(k) \ \cdots \ y(k+K-2)]\\
\overline{\mathcal{U}}_k &:= [u(k) \ \cdots \ u(k+K-2)]
\end{aligned}
$$

10 Recall that $\mathcal{X}_k \equiv \mathcal{X}_f$ and $\mathcal{X}_0 \equiv \mathcal{X}_p$.

11 The shifted projection relation follows from the same steps as Eqs. (6.112)–(6.116) replacing $\mathcal{Y}_f \rightarrow \mathcal{Y}_{k+1|2k-1}$, $\mathcal{U}_f \rightarrow \mathcal{U}_{k+1|2k-1}$, $\mathcal{O}_k \rightarrow \mathcal{O}_{k-1}$ and $\mathcal{X}_k \rightarrow \mathcal{X}_{k+1}$ in the derivation [22].

or more compactly

$$\underbrace{\begin{bmatrix} \overline{\mathcal{X}}_{k+1} \\ --- \\ \overline{\mathcal{Y}}_k \end{bmatrix}}_{\text{known}} = \begin{bmatrix} A & | & B \\ - & - & - \\ C & | & D \end{bmatrix} \underbrace{\begin{bmatrix} \overline{\mathcal{X}}_k \\ --- \\ \overline{\mathcal{U}}_k \end{bmatrix}}_{\text{known}} \tag{6.129}$$

which can be solved as a least-squares problem, that is,

$$\begin{bmatrix} \hat{A} & | & \hat{B} \\ - & - & - \\ \hat{C} & | & \hat{D} \end{bmatrix} = \min_{ABCD} \left\| \begin{bmatrix} \overline{\mathcal{X}}_{k+1} \\ --- \\ \overline{\mathcal{Y}}_k \end{bmatrix} - \begin{bmatrix} A & | & B \\ - & - & - \\ C & | & D \end{bmatrix} \begin{bmatrix} \overline{\mathcal{X}}_k \\ --- \\ \overline{\mathcal{U}}_k \end{bmatrix} \right\|^2 \tag{6.130}$$

where $\| \; \|$ is the Frobenius matrix norm.

Therefore, the *least-squares solution* follows as:

$$\begin{bmatrix} \hat{A} & | & \hat{B} \\ - & - & - \\ \hat{C} & | & \hat{D} \end{bmatrix} = \left(\begin{bmatrix} \overline{\mathcal{X}}_{k+1} \\ --- \\ \overline{\mathcal{Y}}_k \end{bmatrix} \begin{bmatrix} \overline{\mathcal{X}}_k \\ --- \\ \overline{\mathcal{U}}_k \end{bmatrix}' \right) \left(\begin{bmatrix} \overline{\mathcal{X}}_k \\ --- \\ \overline{\mathcal{U}}_k \end{bmatrix} \begin{bmatrix} \overline{\mathcal{X}}_k \\ --- \\ \overline{\mathcal{U}}_k \end{bmatrix}' \right)^{-1} \tag{6.131}$$

This *oblique* projection algorithm is termed **N**umerical algorithms **4** State-Space **S**ubspace **ID**entification (*N4SID*) [8] and can be summarized by the following steps:

- Compute the *LQ*-decomposition of the data matrix (augmented) $\mathcal{D}$ of Eq. (6.119);
- Perform the *SVD* of $\mathcal{P}^{\|}_{\mathcal{Y}_f|\mathcal{D}_p \circ \mathcal{U}_f}$ in Eq. (6.117) to extract $\mathcal{O}_k$ and $\mathcal{O}_{k-1}$;
- Compute $\overline{\mathcal{X}}_{k+1}$ and $\overline{\mathcal{X}}_k$ of Eq. (6.127);
- Construct the "batch" state-space system of Eq. (6.129); and
- Obtain A, B, C, D by solving the least-squares problem as in Eq. (6.131).

completing the algorithm.

We summarize the *N4SID* algorithm in Table 6.3. Next, we consider the important questions of order estimation and model validation techniques that are embedded in the SID process.

6.6 System Order Estimation and Validation

Once the SID technique is selected, then a sequence of questions arises that must be answered prior to application of the approach to measured data. The primary question is that of the overall *system order* required by the subspace identifier: how many states are needed to adequately capture the underlying data sets? Once the order or equivalently the number of states (N_x) is obtained or estimated based on

Table 6.3 Numerical algorithm for subspace state-space system identification (N4SID).

Oblique projection (*LQ* decomposition)

$$\mathcal{D} = \begin{bmatrix} \mathcal{U}_f \\ \mathcal{D}_p \\ \mathcal{Y}_f \end{bmatrix} = \begin{bmatrix} L_{11} & 0 & 0 \\ L_{21} & L_{22} & 0 \\ L_{31} & L_{32} & L_{33} \end{bmatrix} = \begin{bmatrix} \tilde{Q}'_1 \\ \tilde{Q}'_2 \\ \tilde{Q}'_3 \end{bmatrix} \qquad \text{(data matrix)}$$

Singular value decomposition

$$\mathcal{P}^{\|}_{\mathcal{Y}_f|\mathcal{D}_p \circ \mathcal{U}_f} = [U_S \; U_N] \begin{bmatrix} \Sigma_S & 0 \\ 0 & \Sigma_N \end{bmatrix} \begin{bmatrix} V'_S \\ V'_N \end{bmatrix} \approx U_S \Sigma_S V'_S \qquad \text{(projection matrix)}$$

Observability/state extraction

$$\mathcal{P}^{\|}_{\mathcal{Y}_{k|2k-1}|\mathcal{D}_{0|k-1} \circ \mathcal{U}_{k|2k-1}} = \mathcal{O}_k \mathcal{X}_k = L_{32} L_{22}^{-1} \mathcal{D}_{0|k-1} \qquad \text{(oblique projection)}$$

$$\mathcal{O}_k = U_S \Sigma_S^{1/2} \qquad \text{(observability matrix)}$$

$$\mathcal{X}_k = \Sigma_S^{1/2} V'_S \qquad \text{(state matrix)}$$

State matrices

$$\overline{\mathcal{X}}_k = \mathcal{O}_k^{\#} \times \mathcal{P}^{\|}_{\mathcal{Y}_{k|2k-1}|\mathcal{D}_{0|k-1} \circ \mathcal{U}_{k|2k-1}} = \Sigma_{S_k}^{-1/2} U'_{S_k} \times \mathcal{P}^{\|}_{\mathcal{Y}_{k|2k-1}|\mathcal{D}_{0|k-1} \circ \mathcal{U}_{k|2k-1}} \qquad \text{(state matrices)}$$

$$\overline{\mathcal{X}}_{k+1} = \mathcal{O}_{k-1}^{\#} \times \mathcal{P}^{\|}_{\mathcal{Y}_{k+1|2k-1}|\mathcal{D}_{0|k} \circ \mathcal{U}_{k+1|2k-1}} = \Sigma_{S_{k-1}}^{-1/2} U'_{S_{k-1}} \times \mathcal{P}^{\|}_{\mathcal{Y}_{k+1|2k-1}|\mathcal{D}_{0|k} \circ \mathcal{U}_{k+1|2k-1}}$$

System matrices (Σ_{ABCD})

$$\begin{bmatrix} \hat{A} & | & \hat{B} \\ - & - & - \\ \hat{C} & | & \hat{D} \end{bmatrix} = \left(\begin{bmatrix} \overline{\mathcal{X}}_{k+1} \\ --- \\ \overline{\mathcal{Y}}_k \end{bmatrix} \begin{bmatrix} \overline{\mathcal{X}}_k \\ --- \\ \overline{\mathcal{U}}_k \end{bmatrix}' \right) \left(\begin{bmatrix} \overline{\mathcal{X}}_k \\ --- \\ \overline{\mathcal{U}}_k \end{bmatrix} \begin{bmatrix} \overline{\mathcal{X}}_k \\ --- \\ \overline{\mathcal{U}}_k \end{bmatrix}' \right)^{-1}$$

with

$$\overline{\mathcal{X}}_{k+1} = [x(k+1) \; \cdots \; x(k+K-1)]$$

$$\overline{\mathcal{X}}_k = [x(k) \; \cdots \; x(k+K-2)]$$

$$\overline{\mathcal{Y}}_k = [y(k) \; \cdots \; y(k+K-2)]$$

$$\overline{\mathcal{U}}_k = [u(k) \; \cdots \; u(k+K-2)]$$

sound physical insight of the problem at hand, we select a data length K such that $K \gg N_x$ to ensure adequate coverage of the data space. In any case with this information available, we can perform some preliminary model "fits" and estimate the order based on some criteria (to follow) leading to the next pertinent question that must be answered—one of *validation*: how well does the model predict the data?

Validation is based on generating an ensemble of data gathered from a sophisticated simulation of the process under investigation or a controlled experiment. If an ensemble is not possible, then partitioning of the data is required to ensure validation from data that was *not* employed for the identification. Once the ensemble is generated, SID is performed on each realization to estimate the model $\hat{\Sigma}_{ABCD}$ and evaluate how well it predicts the data.

In any case, we must:

- *Estimate* the system order or equivalently the number of states such that $K \gg N_x$; and
- *Validate* the performance of the estimated models $\hat{\Sigma}_{ABCD}(n);\ n = 1, \ldots, N$ over the N-ensemble of data.

We start with the order estimation problem and then proceed to evaluate the ensemble performance of the identified models.

6.6.1 Order Estimation: SVD Approach

An important problem in SID schemes is that of estimating the order of the underlying system model. In fact, most schemes require the system order to be "known" prior to estimating the model parameters. Therefore, we must first determine the correct number of states to estimate from the data directly. Order estimation has evolved specifically from the identification problem [32] where the first step is to estimate or guess the system order and then attempt to validate error testing of the identified model. Much effort has been devoted to this problem [32, 33] with the most significant results evolving from the information-theoretical point of view [34, 35].

We consider some equivalent order estimation methods based on the SVD method applied to the data. The method is deterministic based on the Hankel singular values or the best rank approximation. These approaches are convenient for both the classical as well as SID, since the *SVD* has become an integral part of these algorithms and the singular values are available.

We know from "theoretical" systems (realization) theory that the key element is determining the order or number of states of the underlying system under investigation. Classically, order estimation is based on the rank condition that evolves from the Hankel matrix and its observability/controllability factors. However, the root of achieving a robust and reliable estimate lies in the SVD of the Hankel matrix, that is,

$$\mathcal{H}_{K,K'} = U\,\Sigma\,V' = \left[U_S \mid U_N\right]\begin{bmatrix} \Sigma_S & | & \mathbf{0} \\ - & - & - \\ \mathbf{0} & | & \Sigma_N \end{bmatrix}\begin{bmatrix} V_S' \\ -- \\ V_N' \end{bmatrix} = \underbrace{U_S\Sigma_S V_S'}_{\text{Signal}} + \underbrace{U_N\Sigma_N V_N'}_{\text{Noise}} \tag{6.132}$$

Ideally, $\Sigma_N \ll \Sigma_S$ and the number of nonzero singular values in Σ_S reveals the order or number of states of the underlying system. Therefore, in theory, it is a simple matter to perform the *SVD* of $\mathcal{H}_{KK'}$ and count the number of nonzero, singular values to obtain N_x, since $\rho(\mathcal{H}_{K,K'}) = \rho(\Sigma) = \rho(\Sigma_S) = N_x$. Unfortunately, zero singular values do not occur even in high-precision simulated data due to roundoff and truncation errors (e.g. see Example 6.32 of Sections 6.3 and 6.5).

With this in mind, we understand that the key *first step* in SID lies in the extraction of the singular values either employing the classical Hankel matrix of the *SVD*-based realization algorithm of Section 6.3 or the *SVD*-based projection algorithms of Section 6.4.

Next, let us consider properties of the Hankel matrix that aid in the order estimation problem. The Hankel matrix singular values enable us to estimate its rank and therefore the minimal number of states of the underlying system. However, when noise or computational error occurs, then it is necessary to apply judgment when evaluating these singular values. One approach is simply to search for a meaningful change in these values, referred to as a "gap" or "jump," selecting the order corresponding to this abrupt change as illustrated in Figure 6.2 for a 4th-order or 6th-order system.

Another approach is to employ the so-called *best rank approximation* that is calculated by the ratio of the summation of singular values up to the selected order over all the values, that is, β (percentage) based on the ratio of singular values, [29]

$$\beta(\hat{N}_x) := \left(\frac{\sum_{k=1}^{\hat{N}_x} \sigma_k}{\sum_{k=1}^{K} \sigma_k} \right) \times 100 \tag{6.133}$$

Here, β (%) is the percentage of the original matrix (Hankel) approximated by choosing $\hat{N}_x$. A threshold, τ_x (%) can be selected to search over k for that particular $\hat{N}_x$ that "best" approximates $\mathcal{H}_{K,K}$ up to the selected threshold, that is,

$$\beta_n(\hat{N}_x) \geq \tau_x(\%) \qquad \text{for } n = 1, \ldots, \hat{N}_x \tag{6.134}$$

Once the order is selected, then the error in this choice can be specified by the $\hat{N}_{x+1}$ singular value σ_{N_x+1} as a particular metric or the so-called truncation or *SVD-model error* metric defined by the summation over all of the remaining singular values as [11–13]

$$\mathcal{M}_\epsilon := 2 \sum_{k=N_x+1}^{K} \sigma_k \tag{6.135}$$

Other properties of these Hankel singular values are based on norms of the Hankel matrix and are used in intensity measures and model reduction [12, 14]. For instance, the *Hankel norm* is a measure of the energy stored in a system and is specified by the "largest" singular value of the Hankel matrix:

$$||\mathcal{H}_{\mathcal{K},\mathcal{K}}|| = \max\{\sigma_k\} = \sigma_1;\ k = 1, \dots, K$$

while the $||\mathcal{H}||_2$ norm is the *rms* sum of the impulse responses [14]. In any case, it is clear that the Hankel matrix is a critical entity in identifying a system from measured data.

Another popular *SVD* metric that is applied is the *model order estimator* (*MOE*)[12] [36] given by:

$$\hat{N}_x = \max\left\{N : \log\hat{\sigma}_N > \left(\log\hat{\sigma}_{\max} + \log\hat{\sigma}_{\min}\right)\right\} \tag{6.136}$$

is specified as the largest integer corresponding to the singular value (σ_{N_x}) *greater than* the geometric mean of the largest and smallest singular values extracted. Here, the idea is to formally search for the "gap" or "jump" in singular values. We summarize the metrics available from the *SVD* of the Hankel matrix as:

1. *Hankel order estimation (SVD, Best Rank)* of 6.133 $\Rightarrow \hat{N}_x$;
2. *Hankel singular values.* $\sigma_{ABC} = \sqrt{\Sigma_{AC} \times \Sigma_{AB}}$;
3. *Hankel norm (energy).* $||\mathcal{H}_{K,K'}|| = \max\{\sigma_i\};\ i = 1, \dots, N_x$;
4. *SVD-model error.* $\mathcal{M}_\epsilon = 2\sum_{k=\hat{N}_x+1}^{K} \sigma_k$; and.
5. *SVD-model order estimation.* $\hat{N}_x = \max\left\{N : \log\hat{\sigma}_N > \left(\log\hat{\sigma}_{\max} + \log\hat{\sigma}_{\min}\right)\right\}$.

This completes the discussion of order estimation for a deterministic system based on the *SVD* of the Hankel matrix.

6.6.2 Model Validation

Once the model order is selected, the SID completed and the "black-box" model Σ_{ABCD} identified, the question of its validity must be answered to decide just "how good" the model fits the data. For the deterministic SID case, we rely on the identified model response compared to the system response data to calculate the fitting error.

The response of the identified model $\Sigma_{\text{model}} := \{A_{\text{model}}, B_{\text{model}}, C_{\text{model}}, D_{\text{model}}\}$ is easily calculated using the available deterministic inputs that were used to extract it from the data. We define the *model response*, $\mathbf{y}_{\text{model}}(t)$ in state-space form as:

$$\begin{aligned} x(t) &= A_{\text{model}}x(t-1) + B_{\text{model}}u(t-1) && \text{[state response]} \\ y_{\text{model}}(t) &= C_{\text{model}}x(t) + D_{\text{model}}u(t) && \text{[measurement response]} \end{aligned} \tag{6.137}$$

12 This is the order estimator used in the identification toolbox of *MATLAB*®.

or equivalently as a *convolution response*

$$y_{\text{model}}(t) = C_{\text{model}}A^{t}_{\text{model}}x(0) + \sum_{k=0}^{t-1} C_{\text{model}}A^{t-k-1}_{\text{model}}B_{\text{model}}u(k-1) + D_{\text{model}}\delta(t) \tag{6.138}$$

or directly from the *impulse response*

$$y_{\text{model}}(t) = H_t = C_{\text{model}}A^{t-1}_{\text{model}}B_{\text{model}} + D_{\text{model}}\delta(t); \qquad t = 1, \ldots, K \tag{6.139}$$

We can now use any of these representations to estimate the *model (residual) error* or simulation error defined by [33] as

$$\epsilon(t) := \mathbf{y}(t) - \mathbf{y}_{\text{model}}(t) \qquad \text{[model residual error]} \tag{6.140}$$

and define the *model error metric* for the ith-measurement channel as

$$\varepsilon_i = \sqrt{\frac{\sum_{t=1}^{K} \epsilon_i^2(t)}{\sum_{t=1}^{K} \mathbf{y}_i^2(t)}}; \; i = 1, \ldots, N_y \qquad \text{[model error metric]} \tag{6.141}$$

to evaluate the performance of the model response $\{y_{\text{model}}(t)\}$ relative to the response data $\{y(t)\}$ [22, 23].

This metric can also be calculated over an N-ensemble of data for *each* measurement channel as well to provide statistics enabling a reasonable estimate of model quality. Here we have generated a data ensemble $\{y_n(t)\}; n = 1, \ldots, N$ with the corresponding *ensemble model error* for the nth realization defined by

$$\epsilon_n(t) := \mathbf{y}_n(t) - \mathbf{y}^n_{\text{model}}(t) \qquad \text{[ensemble model error]} \tag{6.142}$$

Ensemble statistics and confidence intervals can be estimated from the sample mean and variance estimates across the realizations $\{\epsilon_n(t)\}; n = 1, \ldots, N$ as

$$\begin{aligned} \mu_\epsilon(t) &:= \hat{E}\{\epsilon_n(t)\} = \frac{1}{N}\sum_{n=1}^{N} \epsilon_n(t) \\ \sigma_\epsilon(t) &:= \hat{E}\{\epsilon_n(t) - \mu_\epsilon(t)\} = \frac{1}{N}\sum_{n=1}^{N} \epsilon_n^2(t) - \mu_\epsilon^2(t) \end{aligned} \tag{6.143}$$

enabling a confidence interval estimate

$$I_\epsilon(t) = \mu_\epsilon(t) \pm 1.96\sigma_\epsilon(t) \qquad \text{for} \qquad N > 30$$

for *each* measurement channel realization providing another means of validating the identified model.

Another statistic that can be applied to validate the performance of the identified model is the *median absolute deviation* (*MAD*) statistic because of its inherent robustness property relative to the usual ensemble mean/standard deviation approach [37]

$$MAD(t) := \gamma \mathcal{M}_t \left(|\epsilon_n(t) - \mathcal{M}_i(\epsilon_i)| \right) \tag{6.144}$$

where $\epsilon_n(t) := \{\epsilon_n(0), \dots, \epsilon_n(t)\}$ is the nth realization of the set of model errors up to time t, $\mathcal{M}_t$ is the *median* of the data, and $\gamma = 1.4826$ is a constant based on the normalized (assumed Gaussian) data. A confidence limit that can be used for outlier detection can also be calculated as

$$\mathcal{M}_t - \alpha \times MAD(t) < \epsilon_n(t) < \mathcal{M}_t + \alpha \times MAD(t) \tag{6.145}$$

for α is a threshold equivalent to a confidence limit (e.g. $\alpha = 1, 2$).

If it is not possible to create a deterministic data ensemble through simulation or experiment, then the original measurement can be partitioned to create an ensemble or bootstrap data resampling techniques may be applied as well [38].

We summarize the steps in order estimation and model validation:

- *Generate* ensemble data: through experiments or simulation, if not possible, use partitioning or bootstrap techniques;
- *Create* Hankel impulse or input/output data matrices;
- *Perform* an *SVD* of the Hankel matrix extracting the observability matrix (classical) or extended observability matrix (subspace);
- *Estimate* the order from the extracted singular values and extract the corresponding statistics (Hankel order, Hankel singular values, Hankel norm, Hankel *SVD*-model error, *SVD*-MOE);
- *Identify* the model Σ_{ABCD} using classical realization or SID techniques;
- *Estimate* model (simulation) error(s) and its statistics over each realization of the ensemble; and
- *Validate* model performance using ensemble statistics and error metrics.

We now have the tools in place to perform a system identification and evaluate the identified model performance. Consider the following example.

6.7 Case Study: Multichannel Mechanical Systems

Mechanical systems are important in many applications, especially when considering vibrational responses of critical components such as turbine-generator pairs in nuclear systems on ships or even in local power plants as well as aircraft structures that transport people throughout the world. Here, we briefly present the generic multichannel mechanical system representation that is employed in examples and case studies to follow. Next, we incorporate synthesized data from the 3-mass structural system that has been discussed throughout the text to illustrate the performance of subspace techniques for comparative purposes [14].

6.7.1 Mechanical Systems

Linear, continuous time, time-invariant, *MIMO* mechanical systems are characterized by the vector-matrix differential equations that can be expressed in continuous time, t, as

$$M\ddot{\mathbf{d}}(t) + C_d\dot{\mathbf{d}}(t) + K\mathbf{d}(t) = B_p\mathbf{p}(t) \tag{6.146}$$

where $\mathbf{d}$ is the $N_d \times 1$ displacement vector, $\mathbf{p}$ is the $N_p \times 1$ excitation force, and M, C_d, K are the $N_d \times N_d$ lumped mass, damping, and spring constant matrices characterizing the vibrational process model, respectively. For series connections of the subsystems, these *MCK* matrices possess the following structural form

$$M = \begin{bmatrix} M_1 & 0 & 0 & 0 & 0 \\ 0 & M_2 & 0 & 0 & 0 \\ \vdots & \vdots & \ddots & \vdots & \vdots \\ 0 & 0 & 0 & M_{N_d-1} & 0 \\ 0 & 0 & 0 & 0 & M_{N_d} \end{bmatrix}, \qquad C_d = \left[C_{d_{ij}} \right] \tag{6.147}$$

and

$$K = \begin{bmatrix} (K_1 + K_2) & -K_2 & 0 & 0 & 0 \\ -K_2 & (K_2 + K_3) & \ddots & 0 & 0 \\ 0 & \ddots & \ddots & -K_{N_d-1} & 0 \\ 0 & 0 & -K_{N_d-1} & \left(K_{N_d-1} + K_{N_d}\right) & -K_{N_d} \\ 0 & 0 & 0 & -K_{N_d} & K_{N_d} \end{bmatrix} \tag{6.148}$$

If we define the $2N_d$-state vector as $\mathbf{x}(t) := \left[\ \mathbf{d}(t) \mid \dot{\mathbf{d}}(t)\right]'$, then the continuous-time state-space representation (see Section 3.3) of this process can be expressed as

$$\dot{\mathbf{x}}(t) = \underbrace{\begin{bmatrix} 0 & | & I \\ --- & | & --- \\ -M^{-1}K & | & -M^{-1}C_d \end{bmatrix}}_{A} \mathbf{x}(t) + \underbrace{\begin{bmatrix} 0 \\ --- \\ M^{-1}B_p \end{bmatrix}}_{B} \mathbf{p}(t) \tag{6.149}$$

The corresponding measurement or output vector relation can be characterized by

$$\mathbf{y}(t) = \mathbf{C_a}\ddot{\mathbf{d}}(t) + \mathbf{C_v}\dot{\mathbf{d}}(t) + \mathbf{C_{dp}}\mathbf{d}(t) \tag{6.150}$$

where the constant matrices: $\mathbf{C_a}$, $\mathbf{C_v}$, $\mathbf{C_{dp}}$ are the respective acceleration, velocity, and displacement weighting matrices of appropriate dimension.

In terms of the state vector relations of Eq. (6.146), we can express the acceleration vector as:

$$\ddot{\mathbf{d}}(t) = -M^{-1}K\mathbf{d}(t) - M^{-1}C_d\dot{\mathbf{d}}(t) + M^{-1}B_p\mathbf{p}(t) \tag{6.151}$$

Substituting for the acceleration term in Eq. (6.150), we have that

$$\mathbf{y}(t) = -\mathbf{C_a}M^{-1}\left[B_p\mathbf{p}(t) - C_d\dot{\mathbf{d}}(t) - K\mathbf{d}(t)\right] + \mathbf{C_v}\dot{\mathbf{d}}(t) + \mathbf{C_{dp}}\mathbf{d}(t)$$

or combining terms, we obtain

$$\mathbf{y}(t) = \underbrace{\left[C_{dp} - \mathbf{C_a}M^{-1}K \mid \mathbf{C_v} - \mathbf{C_a}M^{-1}\mathbf{C_{dp}}\right]}_{C}\begin{bmatrix}\mathbf{d}(t)\\ ---\\ \dot{\mathbf{d}}(t\end{bmatrix} + \underbrace{\mathbf{C_a}M^{-1}B_p}_{D}\mathbf{p}(t) \tag{6.152}$$

to yield the vibrational measurement system as:

$$\mathbf{y}(t) = C\mathbf{x}(t) + D\mathbf{p}(t) \tag{6.153}$$

where the output or measurement vector is $\mathbf{y} \in \mathcal{R}^{N_y\times 1}$ completing the *MIMO* mechanical system model description.

Next, we investigate a simple mechanical system developed previously in [14]. First, the classical realization algorithms are applied to synthesized deterministic data followed by the more sophisticated subspace techniques applied to uncertain but high *SNR* data.

6.7.2 Case Study: 3-mass Mechanical System

Consider the simple series *MCK*-structure represented by a 3-mass (spring-damper) mechanical system[13] illustrated in Figure 6.6 (as before) from [14] where $m_i = 1; k_i = 3; i = 1, \ldots, 3$ and $C_d = 0.01 \times K$. These parameters along

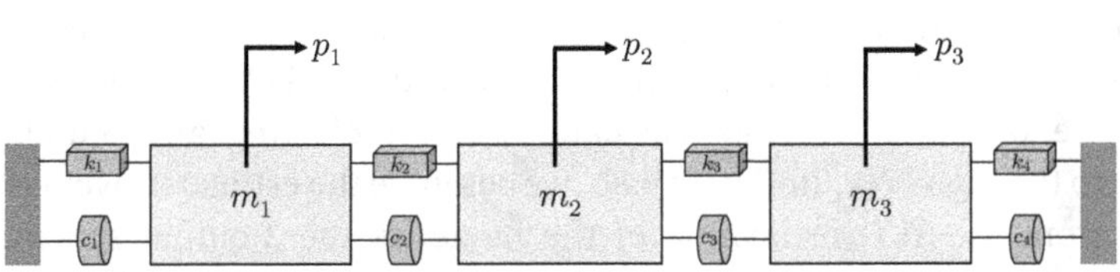

Figure 6.6 The 3-output mechanical system under investigation.

13 See [14] for a detailed analysis of this mechanical system.

with the input/output transmission matrices lead to the following state-space representation ($N_d = 3, N_x = 6$).

$$\dot{\mathbf{x}}(t) = A\mathbf{x}(t) + B\mathbf{p}(t)$$
$$\mathbf{y}(t) = C\mathbf{x}(t) + D\mathbf{p}(t)$$

$$A = \left[\begin{array}{ccc|ccc} 0 & 0 & 0 & 1 & 0 & 0 \\ 0 & 0 & 0 & 0 & 1 & 0 \\ 0 & 0 & 0 & 0 & 0 & 1 \\ - & - & - & - & - & - \\ -6 & 3 & 0 & -0.06 & 0.03 & 0 \\ 3 & -6 & 3 & 0.03 & -0.06 & 0.03 \\ 0 & 3 & -3 & 0 & 0.03 & -0.03 \end{array}\right]; \; B = \begin{bmatrix} 0 \\ 0 \\ 0 \\ 0 \\ 0 \\ 1 \end{bmatrix};$$

$$C = \left[\begin{array}{ccc|ccc} 1 & 0 & 0 & 0 & 0 & 0 \\ 0 & 0 & 0 & 1 & 0 & 0 \\ 0 & 0 & 0 & 0 & 0 & 1 \end{array}\right]$$

with $D = 0$.

Exciting this system with an impulse yields the 3-output response shown in Figure 6.7 along with its accompanying Fourier spectra. Discretizing the response with a sampling interval of $\Delta t = 0.01$ sec, creating the Hankel matrix and decomposing it as in Eq. (6.55) using the *SVD* reveal a 6th-order system ($N_d = 3, N_x = 6$). The log plot of the singular values is shown in Figure 6.8a. The modal frequencies of the system estimated from the identified "discrete" system matrix can be transformed from the discrete (z-domain) to the continuous (s-domain)—($\mathcal{Z} \rightarrow S$), that is,

$$s = \frac{1}{\Delta t} \ln z$$

yielding the modal frequencies at: $\{p_i = 0.497, 0.343, 0.123\ \text{Hz}\}$ that correspond to the peaks of the Fourier transform indicating a reasonable realization.

The *SVD*-realization technique is performed using the impulse response data at a high *SNR* of 50 dB. The rank of the Hankel matrix from the *SVD* decomposition indicates a 6th-order system (3 modes) as observed in Figure 6.8a. The realization results are shown in Figure 6.8b–d where we see the modal frequency estimates that match those of the system closely along with the estimated channel outputs overlaid onto the raw impulse response data—again indicating an accurate realization due to the high *SNR*. In Figure 6.8b, we compare the estimated average power spectra from the *SVD* realization of the model to the 3-output spectra and show the location of the extracted modal frequencies. Note how all of the peaks align indicating a valid realization of the underlying mechanical system. The pole-zero representation along with the extracted modal frequencies are shown in Figure 6.8c.

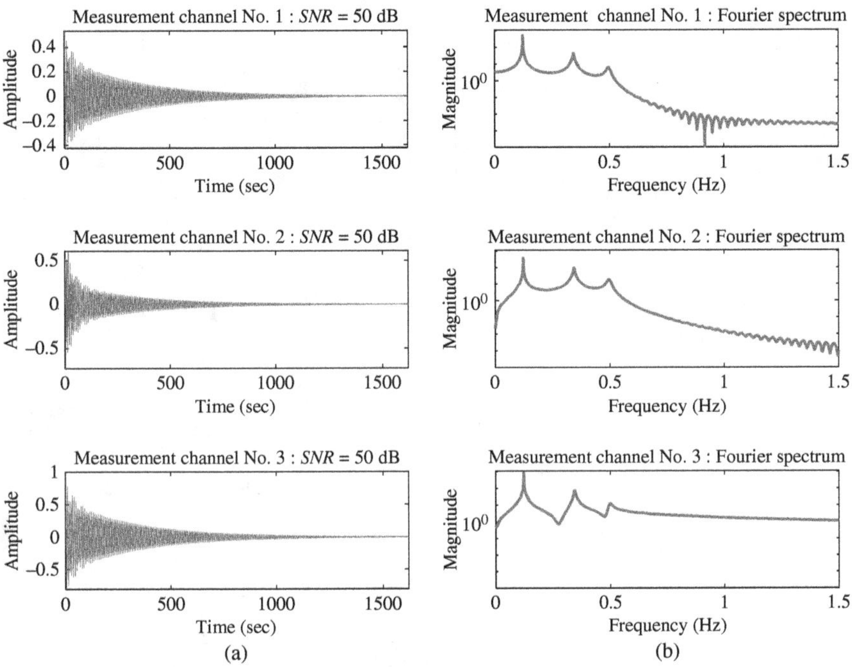

Figure 6.7 Deterministic impulse response and spectrum of 3-output mechanical system ($SNR = 50$ dB). (a) synthesized channel impulse responses. (b) Fourier transforms of channel responses with spectral peaks at: $0.497, 0.343, 0.123$ Hz.

Since the system is mechanical, it is common practice in operational modal analysis to excite the system "randomly" with the idea of a persistent excitation [39]. We chose a pseudo-random sequence to be applied as a single input. Exciting the system with this excitation yields the 3-output response shown in Figure 6.9 along with its accompanying Fourier spectra. The *SVD*-realization technique is performed as before at a high *SNR* of 50 dB. Discretizing the response with a sampling interval of $\Delta t = 0.01$ sec, creating the input/output Hankel matrices, and decomposing it with the *SVD* as in Eq. (6.55) using the *MOESP* or *N4SID* subspace realization technique reveals a 6th-order system as before. The log plot of the singular values is shown in Figure 6.10b. The corresponding Hankel metrics for the random excitation are:

$$
\begin{aligned}
\text{Hankel order estimation}: &\quad \hat{N}_x = 6\\
\text{Hankel singular values}: &\quad \sigma_i = \{56.0,\ 54.7,\ 7.5,\ 13.0,\ 2.5,\ 4.4\}\\
\text{Hankel norm (energy)}: &\quad \|\mathcal{H}_{K,K'}\| = 8.67 \times 10^{-4}\\
\text{SVD-model error}: &\quad \mathcal{M}_\epsilon = 8.3 \times 10^{-6}
\end{aligned}
$$

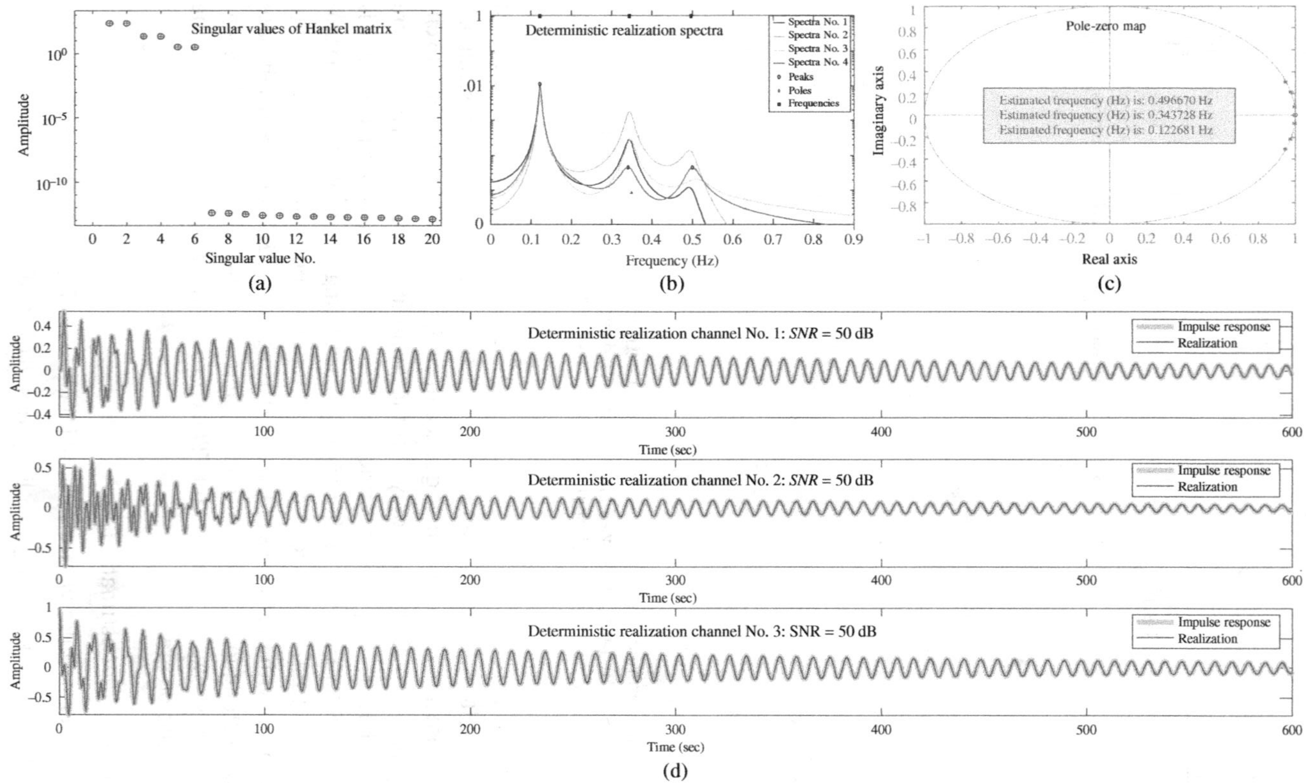

Figure 6.8 *SVD* realization of deterministic 3-output mechanical system. (a) singular values of the Hankel matrix. (b) power spectra: outputs/average with poles (boxes). (c) frequencies with poles/zeros. (d) realization output channels (data/realization) overlays.

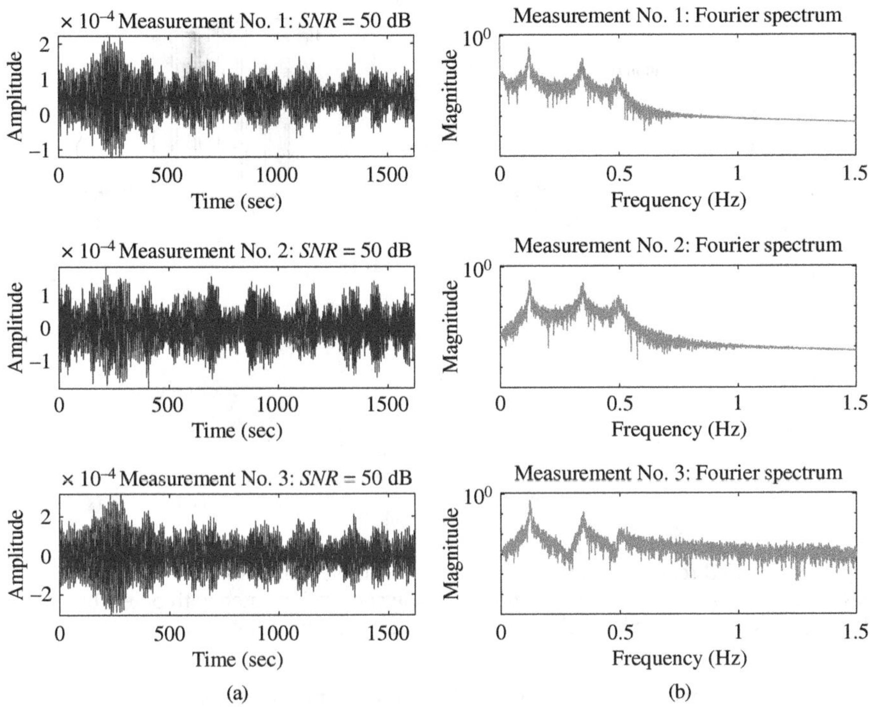

Figure 6.9 Pseudo-random response of 3-output mechanical system ($SNR = 50$ dB). (a) synthesized channel pseudo-random responses. (b) Fourier transforms of channel responses.

The corresponding output power spectra of each channel and the average power spectrum of the identified model are compared in Figure 6.10a indicating a reasonable realization of the data. Again the "identified" modal frequencies of the system estimated from the "discrete" identified system matrix and transformation from discrete ($\mathcal{Z}$-domain) to the continuous ($\mathcal{S}$-domain) given by ($z = \frac{1}{\triangle_t} \ln s$) yields $p_i = \{0.497, 0.343, 0.123\,\text{Hz}\}$ and are shown in Figure 6.10c. To confirm that the realization *matches* the modal frequency estimates of the system closely, the estimated channel outputs are overlaid onto the raw impulse response data as well as the random measurement data in Figure 6.11a,b—again indicating an accurate realization due to the high *SNR*.

Next, we calculate the various validation metrics to observe how well they predict/validate a reasonable realization of this data. The results are shown in

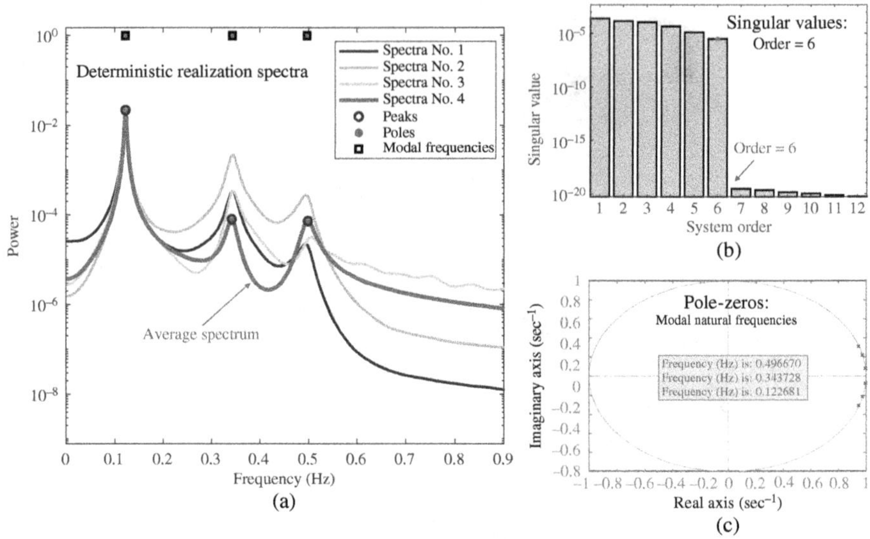

Figure 6.10 *SVD* realization of 3-output mechanical system for random excitation (*SNR* = 50 dB). (a) power spectra: outputs/average spectra indicated with poles (boxes). (b) singular values of the Hankel matrix. (c) frequencies with poles/zeros.

Table 6.4 Deterministic subspace realization: three 3-mass mechanical system.

		Model error/validation results		
Channel No.	**Residual Error**	**Average Model error**	**Deviation Model error**	**MAD Model error**
1	9.7×10^{-10}	5.4×10^{-4}	7.8×10^{-3}	5.9×10^{-3}
2	2.7×10^{-10}	4.8×10^{-4}	6.6×10^{-3}	5.2×10^{-3}
3	6.6×10^{-10}	5.4×10^{-4}	7.8×10^{-3}	5.9×10^{-3}
AVG	6.3×10^{-10}	5.2×10^{-3}	7.4×10^{-3}	5.6×10^{-3}

Table 6.4. From the table, we examine the average results of each of the statistics. Over an ensemble of 100-realizations and deterministic SIDs, the average residual error is 6.3×10^{-10} as expected, since the synthesized signal is essentially deterministic (50 dB *SNR*). The calculated *model errors* are reasonable for the corresponding average, standard deviation, and *MAD* statistics on the order of 10^{-3} corresponding to less than 1% error.

This completes the case study including order estimation and model validation techniques.

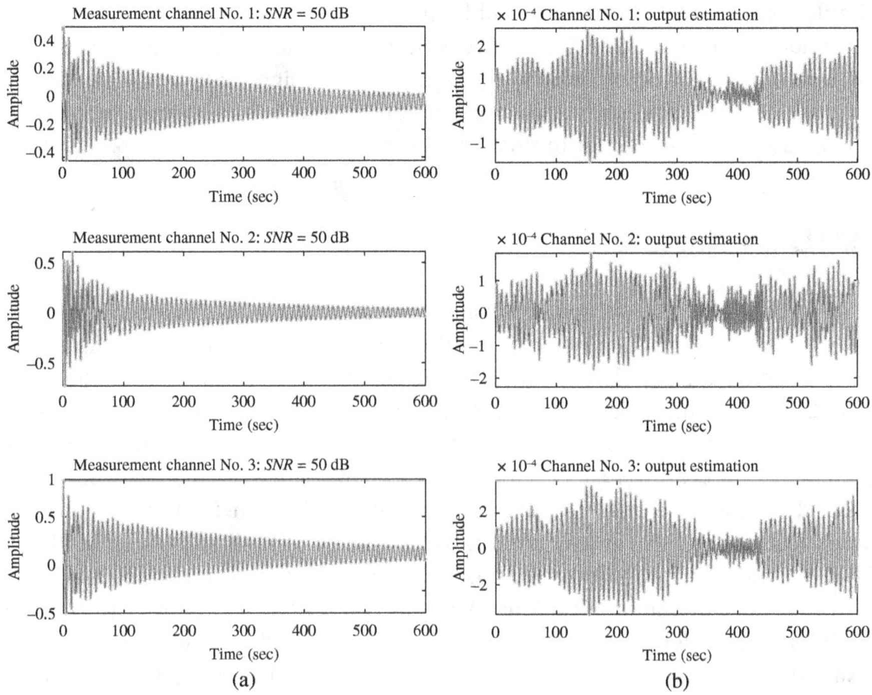

Figure 6.11 *SVD* realization of 3-output mechanical system for random excitation ($SNR = 50$ dB). (a) raw and estimated impulse response. (b) raw and estimated channel measurements.

6.8 Summary

In this chapter, we have discussed solutions to the deterministic realization problem ranging from the classical system theoretic approaches to the modern subspace projection approaches. We have shown that both *MOESP* and *N4SID* techniques rely on the SVD of the Hankel and block Hankel matrices each with underlying theoretical implications. Once the appropriate Hankel matrix is decomposed, the system order is estimated and the underlying system matrices Σ_{ABCD} identified. This is essentially a "black-box" approach using *MIMO*-embedded state-space models as the inherent model set. The philosophy is that once Σ_{ABCD} is extracted, it can be transformed into any coordinate system for analysis (e.g. canonical or modal coordinates). Using examples throughout, these approaches were developed from a pragmatic rather than theoretical perspective. It is demonstrated that the numerically superior *SVD* for both classical and subspace methods provides a robust and reliable solution to the deterministic realization problem. *SVD*-based order estimation is shown to reveal simple metrics (model error, ensemble model error) that can be used to extract the underlying order and state-space system. Validation techniques based on estimated model error evolve leading to pragmatic design solutions. A case study for both deterministic and random excitations of the 3-mass system problem is investigated to illustrate the approach and solution.

MATLAB® Notes

MATLAB® offers a wealth of linear algebraic algorithms that can be applied to the subspace realization problem including commands to create the **Hankel** and **Toeplitz** matrices and their decompositions using the singular value decomposition (**SVD**), the Gram–Schmidt or ***QR*** decomposition that is easily transformed to the lower triangular/orthogonal *LQ* transformation used throughout. Other techniques such as the **Cholesky** factorization for covariance matrices as well as the **LDU** decomposition methods extend these methods even further. The system identification toolbox (**ident**) offers a large variety of both subspace (**MOESP, N4SID**) and prediction-error methods for state-space models including *block-box* as well as *gray-box* representations. Analysis command enable the performance analysis of the identified models as well as well-founded validation methods. Using the embedded GUI allows the user to easily import the input/output data, perform preliminary pre-processing (e.g. mean and trend removal), partition the data for both identification and validation as well as enabling instantaneous analysis of results performing signal estimation plots, residual analysis, frequency domain plots, and transient response analysis. This is quite a powerful toolbox for this problem sets. Other implementations of subspace algorithms exist and can be extracted from the texts of [22] (disk) and [8].

References

1 Kalman, R. (1963). Mathematical description of linear dynamical systems. *SIAM J. Control* 82: 152–192.
2 Kalman, R., Falb, P., and Arbib, M. (1969). *Topics in Mathematical System Theory*. New York: McGraw-Hill.
3 Ho, B. and Kalman, R. (1966). Effective reconstruction of linear state variable models from input/output data. *Regelungstechnik* 14: 545–548.
4 Chen, C. (1984). *Introduction to Linear System Theory*. New York: Holt, Rhinehart, and Winston.
5 Kailath, T. (1980). *Linear Systems*. New Jersey: Prentice-Hall.
6 Reid, J. (1983). *Linear System Fundamentals: Continuous, and Discrete, Classic and Modern*. New York: McGraw-Hill.

7 DeCarlo, R. (1989). *Linear Systems: A State Variable Approach*. New Jersey: Prentice-Hall.

8 Katayama, T. (2005). *Subspace Methods for System Identification*. London: Springer. 545548.

9 Candy, J. (2019). *Model-Based Processing: An Applied Subspace Application*. Hoboken, NJ: Wiley/IEEE Press.

10 Rao, B. and Arun, K. (1992). Model based processing of signals: a state space approach. *Proc. IEEE* 80 (2): 283–309.

11 Moore, B. (1981). Principal component analysis in linear systems: controllability, observability and model reduction. *IEEE Trans. Autom. Control* AC-26: 17–32.

12 Glover, K. (1984). All optimal Hankel-norm approximations of linear multivariable systems and their L^{∞}-error bounds. *Int. J. Control* 39: 1115–1193.

13 Laub, A., Heath, M., Paige, C., and Ward, R. (1987). Computation of system balancing transformations and other applications of simultaneous diagonalization algorithms. *IEEE Trans. Autom. Control* AC-32: 115–122.

14 Gawronski, W. (2004). *Advanced Structural Dynamics and Active Control of Structures*. New York: Springer.

15 Rissanen, J. (1974). Basis of invariants and canonical forms for linear dynamic systems. *Automatica* 10: 175–182.

16 Candy, J., Warren, M., and Bullock, T. (1977). Realization of an invariant system description from Markov sequences. *IEEE Trans. Autom. Control* AC-23 (1): 93–96.

17 Zeiger, H. and McEwen, A. (1974). Approximate linear realization of given dimension via Ho's algorithm. *IEEE Trans. Autom. Control* AC-19 (2): 153.

18 Kung, S. (1978). A new identification and model reduction algorithm via the singular value decomposition. Proceedings of the 12th Annual Asilomar Conference on Circuits, Systems, and Computers, pp. 305–314.

19 Noble, B. and Daniel, J. (1977). *Applied Linear Algebra*, 2e. Englewood Cliffs, NJ: Prentice-Hall.

20 Aoki, M. (1990). *State Space Modeling of Time Series*, 2e. London: Springer.

21 Candy, J. (2006). *Model-Based Signal Processing*. Hoboken, NJ: Wiley/IEEE Press.

22 van Overschee, P. and De Moor, B. (1996). *Subspace Identification for Linear Systems: Theory, Implementation, Applications*. Boston, MA: Kluwer Academic Publishers.

23 Verhaegen, M. and Verdult, V. (2007). *Filtering and System Identification: A Least-Squares Approach*. Cambridge: Cambridge University Press.

24 Tangirala, A. (2015). *Principles of System Identification: Theory and Practice*. Boca Raton, FL: CRC Press.

25 Viberg, M. (1995). Subspace-based methods for the identification of linear time-invariant systems. *Automatica* 31 (12): 1835–1851.

26 Sharf, L. (1991). *Statistical Signal Processing: Detection, Estimation and Time Series Analysis*. Reading, MA: Addison-Wesley.

27 Swevers, J., Adams, M., and De Moor, B. (1987). A new direct time domain identification method for linear systems using singular value decomposition. Proceedings of the 12th International Seminar on Modal Analysis, 12 January-3 February, pp. 1–20.

28 De Moor, B., Vandewalle, J., Vandenberghe, L., and Mieghem, P. (1988). A geometrical strategy for the identification of state space models of linear multivariable systems with singular value decomposition. *Proc. IFAC* 88: 700–704.

29 Golub, G. and van Loan, C. (1996). *Matrix Computations*, 3e. Baltimore, MD: Johns Hopkins University Press.

30 Bretscher, O. (2015). *Linaer Algebra with Applications*, 5e. Upper Saddle River, NJ: Pearson Prentice-Hall.

31 Van Der Veen, A., Deprettere, E.F., and Swindlehurst, A.L. (1993). Subspace-based methods for the identification of linear time-invariant systems. *Proc. IEEE* 81 (9): 1277–1308.

32 Astrom, K. and Eykhoff, P. (1971). System identification—a survey. *Automatica* 7 (2): 123–162.

33 Ljung, L. (1987). *System Identification: Theory for the User*. Englewood Cliffs, NJ: Prentice-Hall.

34 Akaike, H. (1974). A new look at the statistical model identification. *IEEE Trans. Autom. Control* 19 (6): 716–723.

35 Sakamoto, Y., Ishiguro, M., and Kitagawa, G. (1986). *Akaike Information Criterion Statistics*. Boston, MA: D. Reidel/Kluwer Academic.

36 Bauer, D. (2001). Order estimation for subspace methods. *Automatica* 37: 1561–11573.

37 Pham-Gia, T. and Hung, T. (2001). The mean and median absolute deviations. *Math. Comput. Modell.* 34: 921–936.

38 Zoubir, A. and Iskander, D. (2004). *Bootstrap Techniques for Signal Processing*. Cambridge: Cambridge University Press.

39 Reynders, E. (2012). System identification methods for (operational) modal analysis: review and comparison. *Arch. Comput. Meth. Eng.* 19 (1): 51–124.

Problems

6.1 Classical realization theory is based on the factorization of the Hankel matrix of *MIMO* impulse response matrices into the full-rank, observability, and controllability matrices, $\mathcal{H}_{K,K'} = \mathcal{O}_K \times \mathcal{C}_{K'}$, showing that each row (column) of $\mathcal{H}_{K,K'}$ is dependent if and only if it is a dependent row (column) of $\mathcal{O}_K$ ($\mathcal{C}_{K'}$). (Proof of this result forms the foundation for canonical realization theory.)

6.2 Suppose we are given the following state-space description of a continuous-time system with [4]

$$\dot{x}(t) = \begin{bmatrix} -1 & -4/\theta \\ 4\theta & -2 \end{bmatrix} x(t) + \begin{bmatrix} 1 \\ \\ 2\theta \end{bmatrix} u(t)$$

$$y(t) = [-1 \quad 2/\theta]\, x(t)$$

The continuous-time (subscript "c") observability and controllability Gramians satisfy the following Lyapunov equations:

$$A_c'\mathcal{O}_c + \mathcal{O}_c A_c = -C_c' C_c \qquad \text{[Continuous Gramians]}$$
$$A_c \mathcal{C}_c + \mathcal{C}_c A_c' = -B_c B_c'$$

(a) Calculate the observability and controllability Gramians?
(b) Is this a balanced realization? Show results.
(c) If so, what values of θ are required?

6.3 Suppose we are given the transfer function

$$H(s) = \frac{4s^2 - 2s - 6}{2s^4 + 2s^3 + 2s^2 + 3s + 1}$$

(a) What order is this system? Verify your results.
(b) Find a balanced realization using the Ho–Kalman approach.

6.4 We are given the following continuous system

$$\dot{x}(t) = Ax(t)$$

with

$$A = \begin{bmatrix} -4 & -4 \\ 2 & -6 \end{bmatrix}$$

(a) Develop the solution to the corresponding *Lyapunov* equation:

$$A'P + PA = -Q \qquad \text{[Lyapunov Equation]}$$

with $Q = I$.
(b) Is this continuous system stable? Show result?
(c) Validate stability using the calculated eigenvalues.

6.5 Consider the following scalar example with impulse response, $H(k) = \{1, 1/2, 1/4, 1/8\ 1/16\}$ with $N = 5$. Using the *SVD* approach, we would like to extract the realization $\Sigma = (A, b, c)$. Develop the Hankel matrix and apply the *SVD*-Ho algorithm to extract the system.

6.6 Suppose we are given the data sequence:

$$\Lambda_{6,6} = \{6.5, 2.5, 0.833, 0.277, 0.0926, 0.0386\}$$

(a) Perform a deterministic realization of this sequence. What is your identified model, Σ_{abcd}?
(b) Verify that a particular realization is: $\Sigma_{abcd} = \{1/3, 5/4, 2, 13/4\}$.

6.7 In a simple radar system, a large-amplitude, narrow-width pulsed sinusoid is transmitted by an antenna. The pulse propagates through space at light speed ($c = 3 \times 10^8$ m/s) until it hits the object being tracked. The object reflects a portion of the energy. The radar antenna receives a portion of the reflected energy. By locating the time difference between the leading edge of the transmitted and received pulse $\triangle t$, the distance or range from the radar r can be determined, that is,

$$r = \frac{c\triangle t}{2}$$

The pulse is transmitted periodically, that is,

$$r(t) = \frac{c\triangle(t)}{2}$$

Assume that the object is traveling at a constant velocity ρ with random disturbance $w \sim N(0, 100\text{ m/s})$. Therefore, the range of the object at $t + 1$ is

$$r(t + 1) = r(t) + T\rho(t) \quad T = 1\text{ m/s}$$

where T is the sampling interval between received range values. The measured range is contaminated by noise

$$y(t) = r(t) + n(t) \quad \text{for } n \sim N(0, 10)$$

(a) Develop a model to simulate this radar system.
(b) Identify a realization from this synthesized dataset using the classical realization approach to obtain Σ_{ABCD}.

6.8 Suppose we are given the following deterministic system and asked to investigate its performance:

$$x(t) = Ax(t - 1) + Bu(t - 1)$$

$$y(t) = Cx(t) + Du(t)$$

with

$$A = \begin{bmatrix} 0 & 1 \\ -0.7 & 1.5 \end{bmatrix}; \; B = \begin{bmatrix} 0.25 \\ 0.625 \end{bmatrix}; \; C = [0.5 \;\; 0.5]; \; D = [1 \;\; 0]$$

(a) Generate impulse response data $\mathcal{H}_t$ with $\triangle t = 0.1$ sec and perform a deterministic identification.
(b) Add zero-mean random Gaussian noise with $\sigma_v = 0.001$ and let $u(t)$ be a unit pseudo-random noise sequence to synthesize noisy experimental data, $\mathcal{Y}_t$ and perform a deterministic identification.
(c) Calculate the impulse responses from each of the identified models. How do they compare to the set synthesized in (a)? Are they close?
(d) How do all of these realizations compare? (Hint: Compare the impulse responses from each of the identified models.)

6.9 Consider the design of an identifier and estimator for a series *RLC* circuit (second-order system) excited by a pulse train. Using Kirchhoff's voltage law, we can obtain the circuit equations with $i = C(de/dt)$:

$$\frac{d^2e}{dt^2} + \frac{R}{L}\frac{de}{dt} + \frac{1}{LC}e = \frac{1}{LC}e_{in}$$

where e_{in} is a unit pulse train. The dynamic equations can be placed in state-space form by choosing $x := [e \mid de/dt]'$ and $u = e_{in}$:

$$\frac{dx}{dt} = \begin{bmatrix} 0 & 1 \\ -\frac{1}{LC} & -\frac{R}{L} \end{bmatrix} x + \begin{bmatrix} 0 \\ -\frac{1}{LC} \end{bmatrix} u + \begin{bmatrix} 0 \\ -\frac{1}{LC} \end{bmatrix} w$$

where $w \sim N(0, R_{ww})$ is used to model component inaccuracies.

A high-impedance voltmeter is placed in the circuit to measure the capacitor voltage e. We assume that it is a digital (sampled) device contaminated with noise of variance R_{vv}, that is,

$$y(t) = e(t) + v(t)$$

where $v \sim N(0, R_{vv})$.

For this problem, we have the following parameters: $R = 5$ KΩ, $L = 2.5$ H, $C = 0.1$ μF, and $\triangle t = 0.1$ ms. We assume that the component inaccuracies can be modeled using $R_{ww} = 0.01$, characterizing a deviation of ± 0.1 V uncertainty in the circuit representation. Finally, we assume that the precision of the voltmeter measurements is ($e \pm 0.2$ V), the two standard deviation value, so that $R_{vv} = 0.01$ (V)^2. Summarizing the circuit model, we have the continuous-time representation

$$\frac{dx}{dt} = \begin{bmatrix} 0 & 1 \\ -4 & -2 \end{bmatrix} x + \begin{bmatrix} 0 \\ -4 \end{bmatrix} u + \begin{bmatrix} 0 \\ -4 \end{bmatrix} w$$

and the discrete-time measurements

$$y(t) = [1|0]\ x(t) + v(t)$$

where

$$R_{ww} = \frac{(0.1)^2}{\Delta t} = 0.1(\text{V})^2 \quad \text{and} \quad R_{vv} = 0.01(\text{V})^2$$

We first convert the system to a sampled-data (discrete) representation yielding the following discrete-time Gauss–Markov model:

$$x(t) = \begin{bmatrix} 0.98 & 0.09 \\ -0.36 & 0.801 \end{bmatrix} x(t-1) + \begin{bmatrix} -0.019 \\ -0.36 \end{bmatrix} u(t-1)$$
$$+ \begin{bmatrix} -0.019 \\ -0.36 \end{bmatrix} w(t-1)$$
$$y(t) = [1 \mid 0]x(t) + v(t)$$

where

$$R_{ww} = 0.1(\text{V})^2 \quad \text{and} \quad R_{vv} = 0.01(\text{V})^2$$

(a) Using the discrete model simulate the *RLC* circuit for a unit pulse train and generate the measured output voltage, $y(t)$ (measurement model).
(b) Using this synthesized data, identify a "black-box" state-space model from this data using a subspace method. Extract a realization Σ_{ABCD}.
(c) Repeat this identification using a pseudo-random sequence as the deterministic excitation? Compare these identifications.

6.10 We choose $K > N_x$, then perform an *SVD* of a given $\mathcal{H}_{K,K}$ and select a *balanced realization* such that

$$\mathcal{H}_{K,K} = U\Sigma V' = \left(U(\Sigma_{N_x}^{1/2})\right) \times \left(\left(\Sigma_{N_x}^{1/2}\right)' V'\right) = \mathcal{O}_K \times \mathcal{C}_K$$

where N_x is the rank of the Hankel matrix. Using the shift-invariance property of $\mathcal{H}_{K,K} \rightarrow \mathcal{H}_{K,K}^{\uparrow}$ show that if select the first N_1 of the N_x-singular values $(N_1 < N_x)$, then we can extract the N_1-dimensional realization $\Sigma_{A_1B_1C_1D_1}$ from the N_x th. Does this imply that we can extract *all* lower-order systems $(N < N_x)$ from a single N_x-realization as submatrices? Show this? Are these realizations "nested" $(A_1 \in A_{N_x})$?

6.11 Suppose we have the following discrete transfer function:

$$H(z) = \frac{z+3}{z^2 + 3z + 2}$$

(a) What is impulse response of this system? (Hint: Long Division.)
(b) Create the Hankel matrix from the impulse response.
(c) Using the original Ho–Kalman algorithm, extract the system Σ_{ABCD}.
(d) Using the *SVD* Ho–Kalman algorithm, also extract the system Σ_{ABCD}.
(e) Are the realizations equivalent? (Hint: Eigenvalues? Impulse Responses?)

6.12 We are given the following discrete system represented by the transfer function

$$H(z) = \begin{bmatrix} \frac{z-0.5}{z^2-1} & \frac{1}{z-1} \\ \frac{z^2}{z^2-1} & \frac{2z}{z-1} \end{bmatrix}$$

(a) Calculate the impulse response of this system.
(b) Create the Hankel matrix.
(c) Apply the original Ho–Kalman algorithm, extract the system Σ_{ABCD}.
(d) Apply the *SVD*-Ho–Kalman algorithm and also extract the system Σ_{ABCD}.
(e) Perform the canonical realization algorithm of Section 6.3 on the Hankel matrix, what are the observability and controllability invariants?
(f) Extract both row and column realizations Σ_R and Σ_C.
(g) Are the realizations equivalent? (Hint: Eigenvalues? Impulse Responses?)

6.13 Revisit the problem of Example 6.1 where we have a known $N_y \times N_u$ transfer function matrix with $N_y = N_u = 2$ as before. The impulse response sequence is given by

$$H(z) = \begin{bmatrix} 1 & 1 \\ 1 & 1 \end{bmatrix} z^{-1} + \begin{bmatrix} -2 & -4 \\ -3 & -1 \end{bmatrix} z^{-2} + \begin{bmatrix} 4 & 10 \\ 7 & 1 \end{bmatrix} z^{-3} + \begin{bmatrix} -8 & -22 \\ -15 & -1 \end{bmatrix} z^{-4} + \begin{bmatrix} 16 & 46 \\ 31 & 1 \end{bmatrix} z^{-5}$$

(a) Show the elementary row and column operations that can be performed, that is, develop the P (row) and Q (column) elementary matrices leading to a resulting realization Σ_{ABCD} using the Ho–Kalman algorithm. Verify the results with the example.
(b) Perform the *SVD*-Ho algorithm and compare with the Σ_{ABCD} of the example. Are they equivalent?
(c) Perform the canonical realization algorithm on this impulse response data and show the steps.

6.14 From the extracted model Σ_{ABCD} of the previous problem, determine the canonical similarity transformations T_R and T_C of Eq. (6.45) (see Ref. [9]). Transform the extracted system matrices $\{A, B, C, D\}$ to the Luenberger canonical forms $\{A_R, B_R, C_R, D_R\}$ and $\{A_C, B_C, C_C, D_C\}$. How do these realizations compare to those obtained above in (c)?

6.15 Suppose we are given the following *MIMO* transfer function matrix:

$$H(s) = \begin{bmatrix} \frac{s+1}{(s+2)(s+4)} & \frac{s+4}{s(s+2)} \\ \frac{s+1}{s^2(s+4)} & \frac{s+1}{(s+2)(s+4)} \end{bmatrix}$$

(a) What is impulse response of this system? (Hint: Long Division.)
(b) Create the Hankel matrix from the impulse response.
(c) Using the original *SVD*-Ho–Kalman algorithm, also extract the system Σ_{ABCD}.
(d) Is this system stable?
(e) What is the model error? (Compare Impulse Responses.)

6.16 Given the following transfer function matrix:

$$H(s) = \begin{bmatrix} \frac{4s-10}{2s+1} & \frac{3}{s+2} \\ \frac{1}{2s^2+5s+2} & \frac{1}{s^2+4s+4} \end{bmatrix}$$

(a) What is impulse response of this system?
(b) Create the Hankel matrix from the impulse response.
(c) Using the original *SVD* Ho–Kalman algorithm, also extract the system Σ_{ABCD}.
(d) Is this system stable?
(e) What is the model error? (Compare Impulse Responses)

6.17 We have a deterministic system characterized by sinusoids given by:

$$y(t) = \sum_{m=1}^{M} \alpha_m \sin(2\pi f_i t + \phi_i)$$

(a) Simulate this system for the set of frequencies, $f_i = \{1, 2, 5, 10, 20\}$ Hz with $dt = 0.01$ sec for $N = 5000$ samples with $\alpha_i = 1$ and $\phi_i = 0, \ \forall i$.
(b) Generate the required Hankel matrices and perform a subspace identification using the *MOESP* algorithm (MATLAB).
(c) What is the order of the system based on the *SVD* decomposition? Show the results.

(d) Repeat the order estimation using the *AIC* and *MDL* as well as the best rank approach. How do they compare to the *SVD* result above?
(e) Extract the underlying state-space model Σ_{ABCD}.
(f) Calculate the corresponding model error statistics for a single realization.
(g) Add a small amount of Gaussian noise ($SNR = 100$ dB) and generate an ensemble of 100-realizations. Calculate the corresponding error statistics as in (f) and compare the results. Are they close?
(h) Repeat the subspace identification for the *N4SID* algorithm. How do these results compare to that of the *MOESP* approach? Are the realizations equivalent?

6.18 Consider the following discrete Gauss–Markov system:

$$x(t) = 0.97x(t-1) + 0.03u(t-1) + w(t) \quad \text{for } w \sim \mathcal{N}(0, R_{ww}(i))$$
$$y(t) = 2.0x(t) + v(t) \qquad \text{for } v \sim \mathcal{N}(0, R_{vv})$$

with $x(0) = 2.5$, $R_{ww} = 1 \times 10^{-12}$ and $R_{vv} = R_{ww} = 1^1 0^{-10}$—essentially deterministic.
(a) Perform a simulation of this process with a sampling interval of $dt = 0.1$ sec and $u(t)$ a unit step function.
(b) Identify the system Σ_{ABCD} using the *MOESP* subspace technique and calculate the model error statistics for a single realization?
(c) Generate an ensemble from the system Σ_{ABCD} and again apply the *MOESP* subspace technique and calculate the model error statistics over the ensemble. How do the statistics compare to that of (b)?
(d) Repeat (b) and (c) identifying the system Σ_{ABCD} now using the *N4SID* subspace technique and calculate the model error statistics for both a single realization and an ensemble? How do the statistics compare? How do the statistics compare with those of the *MOESP* approach?
(e) For the system Σ_{ABCD} let $u(t)$ be a unit impulse and simulate the system response. Identify the model using the *SVD*-Ho algorithm and calculate the corresponding model error statistic for both single and ensemble realizations. How do these statistics compare?
(f) From the model error statistics calculated, summarize the identification results. Which of these techniques performed best?

6.19 An *RLC* circuit is characterized by the following continuous-time Gauss–Markov system:

$$\dot{x}(t) = \begin{bmatrix} 0 & 1 \\ -4 & -2 \end{bmatrix} x(t) + \begin{bmatrix} 0 \\ \\ -4\theta \end{bmatrix} u(t) + \begin{bmatrix} 0 \\ -4\theta \end{bmatrix} w(t)$$
$$y(t) = [1 \quad 0]\, x(t) + v(t)$$

with $x(0) = 0$, $R_{ww} = 1 \times 10^{-12}$ and $R_{vv} = R_{ww} = 1 \times 10^{-12}$—essentially deterministic.

(a) Perform a simulation of this process with a sampling interval of $dt = 0.1$ msec and $u(t)$ a unit pulse train with period 7 msec. (Hint: Convert to a sampled-data system.)

(b) Identify the system Σ_{ABCD} using the *MOESP* subspace technique and calculate the model error statistics for a single realization?

(c) Generate an ensemble of from the system Σ_{ABCD} and again apply the *MOESP* subspace technique and calculate the model error statistics over the ensemble. How do the statistics compare to that of (b)?

(d) Repeat (b) and (c) identifying the system Σ_{ABCD} now using the *N4SID* subspace technique and calculate the model error statistics for both a single realization and an ensemble? How do the statistics compare? How do the statistics compare with those of the *MOESP* approach.

(e) For the system Σ_{ABCD} let $u(t)$ be a unit impulse and simulate the system response. Identify the model using the *SVD*-Ho algorithm and calculate the corresponding model error statistic for both single and ensemble realizations? How do these statistics compare?

(f) From the model error statistics calculated, summarize the identification results. Which of these techniques performed best?

7

Model-Based Decomposition: Frequency Domain

7.1 Introduction

Resonances occur in a wide variety of applications enabling the unraveling of critical information about the underlying characteristics of the signal or system under investigation. Time-domain techniques usually dominate this area in practice with the availability of digital acquisition systems coupled with fast techniques capable of real-time application [1–3]. However, in some instances where more precise spectral information in a limited frequency band of interest is required, sweeping sinusoidal frequencies in tight bands can reveal characterizations of great importance. Technologies such as resonant ultrasound spectroscopy (*RUS*) or operational modal analysis (*OMA*) provide an answer to sometimes perplexing frequency responses [4]. The development of techniques to process frequency domain data directly has evolved from various technological areas ranging from geophysical applications, to speech and to nondestructive evaluation [5–7]. In this chapter, we discuss the evolution of techniques from the simple to highly sophisticated frequency domain identification techniques that provide spectral decomposition of the signal or system under investigation. Many of these approaches evolve directly from the structural dynamics area but recently have evolved from applications in nondestructive evaluation using *RUS* [8, 9].

Frequency domain identification of modal systems offers an alternative to the usual time-domain approach for modal (structural) analysis [10–12]. Once available in the spectral domain, a variety of techniques have evolved [1, 2]. For instance, the basic approach of extracting modal information is "peak picking" in which sharp resonant peaks are located and used as estimates for resonant spectral estimates. The "basic frequency domain" (*BFD*) method coupled with simple peak picking is used to estimate resonant peaks from least-squares estimates [10, 13]. Next, a more sophisticated spectral estimation method is employed based on the singular value decomposition (*SVD*) of the power spectrum [14]. An improvement follows based on the Blackman–Tukey spectral estimate of signal

Signal Processing: An Applied Decomposition Approach, First Edition. James Vincent Candy.

Companion website: www.wiley.com/go/CandySignProc

processing [15, 16]. This methodology is improved even further by incorporating the *SVD* of the multichannel spectral matrix and tracking the spectral eigenvalues at each frequency [16, 17]. This leads essentially to the frequency domain decomposition (*FDD*) method that is closely related to the complex mode indicator function (*CMIF*) formally decomposing the spectral-frequency response function (*FRF*) [18]. The PolyReference least-squares complex frequency (*pLSCF*) technique evolves using the polyreference decomposition method [12, 19]. Motivated by the multitude of resonances present in the data that must be screened and sorted for physical resonances, more sophisticated iterative methods evolved employing clustering techniques and numerical optimization methods leading to maximum likelihood frequency estimates (*ML-pLSCF*) [20].

7.1.1 Background

Here, we briefly develop the *multichannel* mechanical systems incorporated throughout this chapter. Most structures or equivalently vibrational systems are multiple input/multiple output (*MIMO*) systems. A linear, continuous-time, time-invariant, mechanical system can be expressed (parametrically) as a second-order vector-matrix, differential equation (as before) given by

$$M\ddot{\mathbf{d}}(t) + C_d\dot{\mathbf{d}}(t) + K\mathbf{d}(t) = B_p\mathbf{p}(t) \tag{7.1}$$

where $\mathbf{d}$ is the $N_d \times 1$ displacement vector, $\mathbf{p}$ is the $N_p \times 1$ excitation force, and M, C_d, K, are the $N_d \times N_d$ lumped mass, damping, and spring constant matrices characterizing the vibrational process model, respectively.

The corresponding measurement or output vector relation can be characterized by

$$\mathbf{y}(t) = \mathbf{C_v}\dot{\mathbf{d}}(t) + \mathbf{C_{dp}}\mathbf{d}(t) \tag{7.2}$$

where the constant matrices: $\mathbf{C_v}$ and $\mathbf{C_{dp}}$ are the respective velocity and displacement weighting matrices of appropriate dimension.

The mass and stiffness matrices can be transformed using the *modal matrix* transformation (Φ) resulting in diagonal modal matrices but not necessarily the damping matrix, that is,

$$\Phi := \underbrace{\begin{bmatrix} \boldsymbol{\phi}_1 & \boldsymbol{\phi}_2 & \cdots & \boldsymbol{\phi}_{N_d} \end{bmatrix}}_{\text{Mode Shapes}} \in R^{N_d \times N} \quad \text{for} \quad \boldsymbol{\phi}_k = \begin{bmatrix} \phi_{k1} \cdots \phi_{kN_d} \end{bmatrix}' \tag{7.3}$$

$$\{\mathbf{M}, \mathbf{C}, \mathbf{K}\} \quad \overset{\boldsymbol{\Phi}}{\rightarrow} \quad \{\mathbf{M}_m, \mathbf{C}_m, \mathbf{K}_m\}$$

such that,

$$\mathbf{M}_m = \mathbf{\Phi}'\mathbf{M}\mathbf{\Phi}; \quad \mathbf{K}_m = \mathbf{\Phi}'\mathbf{K}\mathbf{\Phi}; \quad \mathbf{C}_m = \mathbf{\Phi}'\mathbf{C}\mathbf{\Phi} \tag{7.4}$$

It is interesting to note that the modal transformation (matrix) simultaneously "decouples" or diagonalizes both the **M** and **K** matrices [21].

The modal matrix is essentially a *similarity transformation* that is used to transform the physical coordinates to modal coordinates $\mathbf{d}(t) = \mathbf{\Phi}\mathbf{q}(t)$ in continuous-time variable t so that the *MCK*-model becomes

$$\mathbf{M}_m\ddot{\mathbf{q}}(t) + \mathbf{C}_m\dot{\mathbf{q}}(t) + \mathbf{K}_m\mathbf{q}(t) = \mathbf{\Phi}'\mathbf{B}_p\mathbf{p}(t) \tag{7.5}$$

and the corresponding measurements become

$$\mathbf{y}(t) = \underbrace{\mathbf{C}_{vq}\mathbf{\Phi}}_{\mathbf{C}_m}\dot{\mathbf{q}}(t) + \underbrace{\mathbf{C}_{dq}\mathbf{\Phi}}_{\mathbf{C}_m}\dot{\mathbf{q}}(t) \tag{7.6}$$

Multiplying by the inverse mass matrix yields

$$\ddot{\mathbf{q}}(t) + \mathbf{M}_m^{-1}\mathbf{C}_m\dot{\mathbf{q}}(t) + \mathbf{M}_m^{-1}\mathbf{K}_m\mathbf{q}(t) = \underbrace{\mathbf{M}_m^{-1}\mathbf{\Phi}'B_p}_{\mathbf{B}_m}\mathbf{p}(t) \tag{7.7}$$

$$\mathbf{y}(t) = \underbrace{\mathbf{C}_{mq}\mathbf{\Phi}}_{\mathbf{C}_{m\dot{q}}}\dot{\mathbf{q}}(t) + \underbrace{\mathbf{C}_{dq}\mathbf{\Phi}}_{\mathbf{C}_m}\mathbf{q}(t) \tag{7.8}$$

or defining terms simplifies this *modal model* [21]

$$\ddot{\mathbf{q}}(t) + 2\mathbf{Z}\mathbf{\Omega}\dot{\mathbf{q}}(t) + \mathbf{\Omega}^2\mathbf{q}(t) = \mathbf{B}_m\mathbf{p}(t) \tag{7.9}$$

for $\mathbf{\Omega} := \text{diag}\left[\omega_1 \ \omega_2 \ \cdots \ \omega_N\right]$ and $\mathbf{Z} := \text{diag}\left[\zeta_1 \ \zeta_2 \ \cdots \ \zeta_N\right]$ with ζ_k the damping of the kth-mode.

One of the major features of high interest is that this modal decomposition "decouples" the modes, providing a set of individual modal equations for each, that is, [21]

$$\begin{aligned} &\ddot{\mathbf{q}}_k(t) + 2\zeta_k\omega_k\dot{\mathbf{q}}_k(t) + \omega_k^2\mathbf{q}_k(t) = \mathbf{b}_k p(t) \\ &y_k(t) = c_{mq}(k)\mathbf{q}_k(t) + c_{m\dot{q}}\dot{\mathbf{q}}_k(t) \qquad \text{for} \qquad k = 1, \ldots, N \end{aligned} \tag{7.10}$$

for the measurement the superposition of individual modal responses is

$$\mathbf{y}(t) = \sum_{k=1}^{N} y_k(t) \tag{7.11}$$

Corresponding to these decoupled modal representations are their particular FRFs leading to the composite *modal transfer function* that is parametrically given by

$$\mathbf{F}(\omega) = \left(\mathbf{C}_{mq} + j\omega\mathbf{C}_{m\dot{q}}\right)\left(\mathbf{\Omega}^2 - \omega^2\mathbf{I}_N + 2j\omega\mathbf{Z}\mathbf{\Omega}\right)^{-1}\mathbf{B}_m \tag{7.12}$$

with the individual modal *FRF*s due to the diagonal modal frequency and damping matrices specified by

$$F_{mk}(\omega) = \frac{\left(c_{mq}(k) + j\omega c_{m\dot{q}}(k)b_{mk}\right)}{\omega_k^2 - \omega^2 + 2j\zeta_k\omega_k\omega} \tag{7.13}$$

with the overall *FRF* given by the superposition

$$\mathbf{F}(\omega) = \sum_{k=1}^{N} \mathbf{F}_{mk}(\omega) \tag{7.14}$$

This completes the necessary background information on modal frequency domain modeling and analysis. Next, we consider the progression from simple to sophisticated techniques for modal information extraction from noisy data.

7.2 Frequency Response Functions (*FRF*)

The fundamental approach to frequency domain identification is data-based where a spectral input/output matrix of the frequency responses is constructed in which various techniques are applied to extract the modal frequencies [22]. Perhaps, most approaches in the frequency domain start with the *FRF* that is the transfer function evaluated along the $j\omega$-axis of the Laplace/Fourier domain. There are a variety of techniques developed to estimate the *FRF* from noisy data either in the time domain or frequency domain depending on the particular acquisition system employed. Techniques range from simple Fourier methods dividing input/output transforms to sophisticated parametric methods employing multiple channel state-space methods in the time domain to matrix fraction description (*MFD*) representations in the frequency domain [1, 2]. Here, we focus on multichannel frequency domain methods; however, estimating the *FRF* and its corresponding *MIMO* matrix must be accomplished first in order to create the input data required to estimate the modal system parameters.

We start with the N_o-output, N_i-input, N_ω-frequency (samples) *FRF*-matrix (*data cube*)

$$\mathbf{F}(\omega) := \begin{bmatrix} F_{11}(\omega) & F_{12}(\omega) & \cdots & F_{1N_i}(\omega) \\ F_{21}(\omega) & F_{22}(\omega) & \cdots & F_{2N_i}(\omega) \\ \vdots & \vdots & \ddots & \vdots \\ F_{N_o1}(\omega) & F_{N_o2}(\omega) & \cdots & F_{N_oN_i}(\omega) \end{bmatrix} \tag{7.15}$$

with $\mathbf{F}(\omega) \in \mathbf{C}^{N_o \times N_i \times N_\omega}$ and $F_{ij}(\omega)$ the ith-output, jth-input of N_ω frequency (samples) spectra.

With this in mind, we investigate various approaches of estimating the *FRF* from time domain as well as frequency domain data with the emphasis on the latter. We start with a well-known time-domain method based on Wiener filtering theory and then proceed to develop a variety of powerful frequency domain techniques–the primary objective of this chapter.

7.2.1 *FRF* Estimation: Impulse Response Method

Suppose we excite a modal system with an input, say $x(t)$, and we measure the output, $y(t)$. We would like to obtain a linear estimate of the embedded FRF $F(t)$ or equivalently impulse response function based on past excitation data, that is,

$$\hat{F}(t) = \hat{F}(t|X_{past})$$

Taking the parametric approach, we model the signal response by

$$\hat{s}(t) = \sum_{i=0}^{N_f} F(i)x(t-i) \tag{7.16}$$

If we represent the measurement $y(t)$ as

$$y(t) = s(t) + n(t) \tag{7.17}$$

where n is zero-mean, white noise of variance R_{nn}, then the optimal estimator is obtained by

$$\min_{\mathbf{F}} \mathcal{J}(t) = E\{e^2(t)\} \tag{7.18}$$

where $e(t) := y(t) - \hat{s}(t)$.

Following the standard minimization approach, we differentiate Eq. (7.18) with respect to $F(j)$, set the result to zero and obtain the *orthogonality condition*

$$\frac{\partial \mathcal{J}}{\partial F(j)} = 2E\left\{e(t)\frac{\partial e(t)}{\partial F(j)}\right\} = 0 \tag{7.19}$$

The error gradient is

$$\frac{\partial e(t)}{\partial F(j)} = \frac{\partial}{\partial F(j)}(y(t) - \hat{s}(t)) = \frac{\partial}{\partial F(j)}\left(y(t) - \sum_{i=0}^{N_f} F(i)x(t-i)\right) = x(t-j) \tag{7.20}$$

and therefore substituting Eq. (7.20) into Eq. (7.19), we obtain

$$E\{e(t)x(t-j)\} = E\left\{\left(y(t) - \sum_{i=0}^{N_f} F(i)x(t-i)\right)x(t-j)\right\} = 0; \quad j = 0, \dots, N_h \tag{7.21}$$

which yields the *normal equations* for the estimation problem as

$$\sum_{i=0}^{N_f} F(i)R_{xx}(i-j) = r_{yx}(j) \quad j = 0, \ldots, N_f \tag{7.22}$$

or expanding the equations over the indices gives

$$\begin{bmatrix} R_{xx}(0) & \cdots & R_{xx}(N_f) \\ \vdots & \ddots & \vdots \\ R_{xx}(N_f) & \cdots & R_{xx}(0) \end{bmatrix} \begin{bmatrix} F(0) \\ \vdots \\ F(N_f) \end{bmatrix} = \begin{bmatrix} r_{yx}(0) \\ \vdots \\ r_{yx}(N_f) \end{bmatrix} \tag{7.23}$$

This expression can be written in compact form as

$$\mathbf{R}_{xx}(N_f)\mathbf{F}(N_f) = \mathbf{r}_{yx}(N_f) \tag{7.24}$$

where $\mathbf{R}_{xx} \in \mathcal{R}^{(N_f+1)\times(N_f+1)}$ is a Toeplitz matrix. The optimal *FRF* estimate, $\hat{F}(N_h)$ is obtained by inverting $\mathbf{R}_{xx}$ using the "generalized Levinson" or equivalently the *LWR* algorithm [15]. This provides the *optimal Wiener solution* given by

$$\hat{\mathbf{F}}(N_f) = \mathbf{R}_{xx}^{-1}(N_f)\mathbf{r}_{yx}(N_f) \tag{7.25}$$

Note that the properties of the Toeplitz matrix decomposition can be applied to solve this problem in "batch" form. The backward prediction errors are developed by performing a *Gram–Schmidt orthogonalization* on the set of input data, $\{x(n)\}, n = t - N_f, \ldots, t-1, t$ as

$$\mathbf{e}_b(N_f) = \mathbf{L}(N_f)\mathbf{x}(N_f) \tag{7.26}$$

and therefore

$$R_{xx}(N_f) = E\{\mathbf{x}(N_f)\mathbf{x}'(N_f)\} = E\{(\mathbf{L}^{-1}(N_f)\mathbf{e}_b(N_f))(\mathbf{L}^{-1}(N_f)\mathbf{e}_b(N_f))'\}$$

or

$$R_{xx}(N_f) = \mathbf{L}^{-1}(N_f)E\{\mathbf{e}_b(N_f)\mathbf{e}_b'(N_f)\}\mathbf{L}^{-T}(N_f) = \mathbf{L}^{-1}(N_f)R_{e_be_b}(N_f)\mathbf{L}^{-T}(N_f) \tag{7.27}$$

and its corresponding inverse is

$$R_{xx}^{-1}(N_f) = \mathbf{L}'(N_f)R_{e_be_b}^{-1}(N_f)\mathbf{L}(N_f) \tag{7.28}$$

Similarly, the cross-correlation vector becomes

$$\mathbf{r}_{yx}(N_f) = E\{y(t)\mathbf{x}(N_f)\} = E\{y(t)(\mathbf{L}^{-1}(N_f)\mathbf{e}_b(N_f))\} = \mathbf{L}^{-1}(N_f)\mathbf{r}_{ye_b}(N_f) \tag{7.29}$$

Thus, the optimal Wiener solution of Eq. (7.25) becomes

$$\hat{\mathbf{F}}(N_f) = \left(\mathbf{L}'(N_f)R_{e_be_b}^{-1}(N_f)\mathbf{L}(N_f)\right)\left(\mathbf{L}^{-1}(N_f)\mathbf{r}_{ye_b}(N_f)\right) = \mathbf{L}'(N_f)R_{e_be_b}^{-1}(N_f)\mathbf{r}_{ye_b}(N_f) \tag{7.30}$$

which is equivalent to pre-whitening the input data. Taking the Fourier transform of the estimated response provides an estimate of the *FRF* in the frequency domain, that is, $\mathcal{F}\left[\hat{F}(t)\right] \rightarrow \mathbf{F}(\omega)$–the desired result.

7.2.2 *FRF* Spectral Estimation: Polynomial Models

Consider a single input/single output (*SISO*) linear continuous-time dynamic system having transfer function defined by

$$\mathcal{F}(s) = \frac{B(s)}{A(s)} \tag{7.31}$$

where the numerator and denominator polynomials are defined by $B(s) := b_0 + b_1 s + \cdots + b_{N_b} s^{N_b}$ and $a(s) := 1 + a_1 s + \cdots + a_{N_a} s^{N_a}$. Sweeping the system with sinusoidal excitations at frequencies $\{\omega_n\}; n = 1, \ldots, N$ generates the corresponding set of response data, $\{\mathcal{F}(j\omega_n)\}$.

To identify the system from the frequency response data, the *parameter vector*

$$\Theta := \left[a_1, a_2, \ldots, a_{N_a}, \quad b_0, b_1, \ldots, b_{N_b}\right]'$$

must be estimated from the measured responses.

Defining a frequency domain *error function* based on these parameters

$$\varepsilon(j\omega;\Theta) := A(j\omega;\Theta) \times \mathcal{F}(j\omega) - B(j\omega;\Theta) \tag{7.32}$$

leads to the error criterion function $\mathcal{J}$ that can be minimized over Θ, that is,

$$\mathcal{J}(\omega;\Theta) = (\mathcal{Y}(j\omega) - \mathcal{F}(j\omega)\Theta)^\dagger\,(\mathcal{Y}(j\omega) - \mathcal{F}(j\omega)\Theta) \tag{7.33}$$

where the data vector, $\mathcal{Y}(j\omega)$, and system response matrix, $\mathcal{F}(j\omega)$ are given by

$$\mathcal{Y}(j\omega) = \begin{bmatrix} F(j\omega_1) \\ \vdots \\ F(j\omega_N) \end{bmatrix}$$

$$\mathcal{F}(j\omega) = \begin{bmatrix} -j\omega_1 F(j\omega_1) & \cdots & (-j\omega_1)^{N_a} F(j\omega_1) & |1 & j\omega_1 & \cdots & (j\omega_1)^{N_b} \\ \vdots & \vdots & \vdots & \vdots & \vdots & \vdots & \vdots \\ -j\omega_N F(j\omega_N) & \cdots & (-j\omega_N)^{N_a} F(j\omega_N) & |1 & j\omega_N & \cdots & (j\omega_N)^{N_b} \end{bmatrix}$$

Minimizing the criterion function, that is, differentiating,[1] the criterion, setting to zero

$$\nabla_\Theta \mathcal{J}(\omega;\Theta) = -\mathcal{F}^*(j\omega)\,(\mathcal{Y}(j\omega) - \mathcal{F}(j\omega)\Theta) - \mathcal{F}'(j\omega)(\mathcal{Y}(j\omega) - \mathcal{F}(j\omega)\Theta)^\dagger = 0 \tag{7.34}$$

and solving for Θ leads to the parameter vector estimate, $\hat{\boldsymbol{\Theta}}$ [22]

$$\hat{\boldsymbol{\Theta}} = \text{Re}\left[\mathcal{F}^*(j\omega)\mathcal{F}(j\omega)\right]^{-1}\,\text{Re}\left[\mathcal{F}(j\omega)^*\mathcal{Y}(j\omega)\right] \tag{7.35}$$

1 The chain rule from vector calculus is defined by $\nabla_\Theta(\mathbf{a}'\mathbf{b}) = \left(\nabla_\Theta \mathbf{a}'\right)\mathbf{b} + \left(\nabla_\Theta \mathbf{b}'\right)\mathbf{a}$ for $\mathbf{a}, \mathbf{b} \in R^{N_\Theta \times 1}$.

Substituting the estimated parameters into the transfer function or equivalently *FRF* provides a solution to the problem, that is,

$$\hat{\mathcal{F}}(j\omega;\hat{\Theta}) = \frac{\hat{B}(j\omega)}{\hat{A}(j\omega)} \quad \text{for} \quad \hat{\Theta} = \left[\hat{a}_1,\ldots,\hat{a}_{N_a}|\hat{b}_0,\ldots,\hat{b}_{N_b}\right]' \tag{7.36}$$

7.2.3 *FRF*-Spectral Estimation: Power Spectra

A simple, yet effective method, to extract resonant frequencies, is to estimate the power spectral density (*PSD*) function from noisy time domain or frequency domain measurements [15]. The *BFD method* also known as the peak-picking method (*PPM*) estimates the *FRF* from spectral data using the division of the cross-*PSD*(input/output) by the auto-spectrum (input) with the true model frequencies confirmed by the coherence function.

Mathematically, the optimal *FRF* estimate can be obtained from the minimum mean-squared error (*MSE*) solution. In practice, one of the most popular methods of estimating the *FRF* is by exciting system with a unit variance, random input (white noise), $x(t)$ in the time domain and obtaining the Wiener filter solution given by the cross-power spectrum, that is, (generally) we have that [15]

$$\hat{\mathbf{F}}(\omega) = \frac{S_{yx}(\omega)}{S_{xx}(\omega)} \tag{7.37}$$

with corresponding squared *coherence function* estimated by

$$\gamma^2(\omega) = \frac{|S_{yx}(\omega)|^2}{S_{yy}(\omega)S_{xx}(\omega)} \quad \text{for} \quad 0 < \gamma^2(\omega) < 1 \tag{7.38}$$

Recall that when the *FRF* is near a resonance, the coherence function will be close to unity [2]. Combining the spectral information from the *PSD* and coherence function enables the extraction or separation of "close" modes in the frequency domain. Damping can also be extracted directly from the available half-power bandwidth at resonance [10]. *BFD* performs reasonably well, when the system is lightly damped with well-separated modes.

The *BFD* is considered a single degree of freedom (*SDOF*) method based on the premise that near a resonance one single dominant mode is present. The *SDOF* assumption is based on the multichannel partial fraction expansion of the *PSD*, that is, we have

$$\mathbf{S}_{yy}(\omega) = \sum_{n=1}^{N}\sum_{m=1}^{N}\left[\frac{\mathbf{R}_n}{j\omega-\lambda_n}+\frac{\mathbf{R}_n^*}{j\omega-\lambda_n^*}\right]\mathbf{C}\left[\frac{\mathbf{R}_m}{j\omega-\lambda_m}+\frac{\mathbf{R}_m^*}{j\omega-\lambda_m^*}\right] \tag{7.39}$$

assuming that $\mathbf{S}_{xx}(\omega) = \mathbf{C}$ is a purely random input (white noise) matrix. Therefore, near a resonance, the *SDOF* assumption can be approximated by a single dominant mode, say λ_k such that

$$S_{yy}(\omega) = \frac{R_k}{j\omega - \lambda_k} + \frac{R_k^*}{j\omega - \lambda_k^*} = \frac{\phi_k \ell_k}{j\omega - \lambda_k} + \frac{\phi_k^* \ell_k^*}{j\omega - \lambda_k^*} \tag{7.40}$$

with the residues related to the *mode shapes* ϕ_k and the so-called *participation* factors (of the excitation vectors), $\ell_k := Q_k \times \phi_k^*$, that can be extracted from the columns of the *PSD* matrix (input/output) [1, 2]. The *modal participation factors* indicate which modes respond most to the applied excitations.

Next, consider the following example as an illustration of the *BFD* method.

Example 7.1 Suppose we have a continuous time, *MCK* system given by 3-masses (spring and damper) with natural frequencies of $f = \{0.123, 0.345, 0.500\}$ Hz (as before) connected in series and excite it with an impulse (see [21] for details). The 3-mass (spring-damper) mechanical system where mass and stiffness M and K are shown below and damping is given by $C_d = 0.01 \times K$. These parameters along with the input/output transmission matrices leads to the "nodal model" of structural dynamics with $N_d = 3$ [21].

$$M\ddot{\mathbf{d}}(t) + C_d\dot{\mathbf{d}}(t) + K\mathbf{d}(t) = B_f\mathbf{f}(t)$$

$$\underbrace{\begin{bmatrix} 1 & 0 & 0 \\ 0 & 1 & 0 \\ 0 & 0 & 1 \end{bmatrix}}_{\mathbf{M}} \ddot{\mathbf{d}}(t) + \underbrace{\begin{bmatrix} -0.06 & 0.03 & 0 \\ 0.03 & -0.06 & 0.03 \\ 0 & 0.03 & -0.03 \end{bmatrix}}_{\mathbf{C}_d} \dot{\mathbf{d}}(t) + \underbrace{\begin{bmatrix} -6 & 3 & 0 \\ 3 & -6 & 3 \\ 0 & 3 & -3 \end{bmatrix}}_{\mathbf{K}} \mathbf{d}(t) = \underbrace{\begin{bmatrix} 0 \\ 0 \\ 1 \end{bmatrix}}_{\mathbf{B}_f} \delta(t)$$

with the independent variable t (continuous time) corresponding velocity output given by

$$y(t) = \underbrace{\begin{bmatrix} 1 \end{bmatrix}}_{\mathbf{C}_v} \dot{\mathbf{d}}(t)$$

Exciting this system with an impulse yields the 3-mass response shown in Figure 7.1a along with its accompanying Fourier spectra. The ideal impulse response is shown in Figure 7.1a and the *PSD* of the noisy data in Figure 7.1b. The estimated frequencies are shown as the annotated peaks in the figure. The corresponding coherence function is also shown in the Figure 7.1c indicating the high quality of the data with values near unity. The *BFD* can be quite effective when modes are well-separated in frequency even at low noise levels.

Finally, performing the modal decomposition of this 3-mode, *MCK* system with coordinate transformation (eigenvector matrix) from Example 1.8 gives [21]

$$\mathbf{T}_m = \begin{bmatrix} 0.5910 & 0.7370 & 0.3280 \\ -0.7370 & 0.3280 & 0.03 \\ 0.3280 & -0.5910 & 0.7370 \end{bmatrix}$$

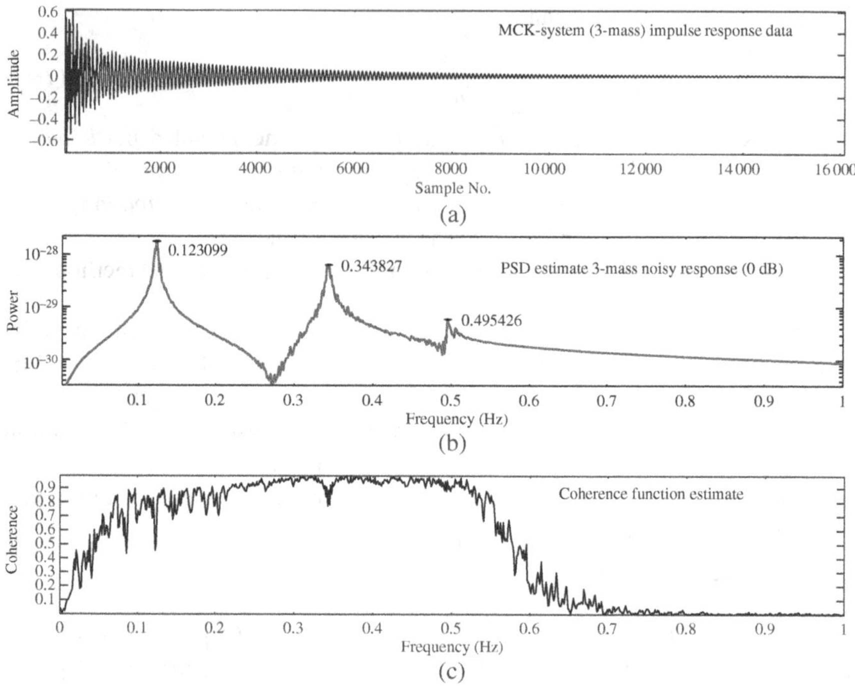

Figure 7.1 Basic frequency domain (*BFD*) method: 3-mass problem in noise ($SNR = 0$ dB): (a) deterministic impulse response (channel 2 output), (b) estimated power spectrum (correlation method), and (c) estimated coherence function.

yielding the resulting modal system of this *MCK* system given by

$$\ddot{\mathbf{q}}(t) + 2 \times \underbrace{\begin{bmatrix} 0.0156 & 0 & 0 \\ 0 & 0.0108 & 0 \\ 0 & 0 & 0.0039 \end{bmatrix}}_{\mathbf{Z}} \times \underbrace{\begin{bmatrix} 3.1210 & 0 & 0 \\ 0 & 2.1598 & 0 \\ 0 & 0 & 0.7708 \end{bmatrix}}_{\boldsymbol{\Omega}} \dot{\mathbf{q}}(t)$$

$$+ \underbrace{\begin{bmatrix} 3.1210 & 0 & 0 \\ 0 & 2.1598 & 0 \\ 0 & 0 & 0.7708 \end{bmatrix}^2}_{\boldsymbol{\Omega}^2} \mathbf{q}(t) = \underbrace{\begin{bmatrix} 0.3280 \\ -0.5910 \\ 0.7370 \end{bmatrix}}_{\mathbf{B}_m} \delta(t)$$

with the velocity output matrix given by

$$y_m(t) = \mathbf{c}_{mv,1}^T \mathbf{q}(t) + \mathbf{c}_{mv,2}^T \dot{\mathbf{q}}(t) = \underbrace{\begin{bmatrix} 0.7370 & 0 & 0 \end{bmatrix}}_{\mathbf{c}_{mv,1}^T} \mathbf{q}(t) + \underbrace{\begin{bmatrix} 0 & 0.7370 & -0.5910 \end{bmatrix}}_{\mathbf{c}_{mv,2}^T} \dot{\mathbf{q}}(t)$$

completing the example.

△△△

7.2.4 *FRF*-Spectral Estimation: Frequency Domain Decomposition (*FDD*) Method

An improvement of the *BFD* method evolves based on "decomposing" the *FRFs* by performing the numerically robust, SVD of the power spectral matrix.

7.2.4.1 Power Spectral Density Decomposition

The *multichannel PSD* is given by the N_o-output matrix

$$\mathbf{S}_{yy}(\omega) = \mathbf{F}^*(\omega) \times \mathbf{S}_{xx}(\omega) \times \mathbf{F}'(\omega) \tag{7.41}$$

for $\mathbf{S}_{yy} \in R^{N_o \times N_o}$, $\mathbf{S}_{xx} \in R^{N_i \times N_i}$, $\mathbf{F} \in C^{N_o \times N_i}$.

The multichannel *FRF*-matrix can be expressed in partial fraction form in terms of residues as well as modal parameters (frequencies, shapes, participation factors) as before in Eq. (7.40)

$$\mathbf{F}(\omega) = \sum_{k=1}^{K} \frac{\mathbf{R}_k}{j\omega - \lambda_k} + \frac{\mathbf{R}_k^*}{j\omega - \lambda_k^*} = \sum_{k=1}^{K} \frac{\mathbf{\Phi}_k \mathbf{L}_k}{j\omega - \lambda_k} + \frac{\mathbf{\Phi}_k^* \mathbf{L}_k^*}{j\omega - \lambda_k^*} \tag{7.42}$$

for the *residue matrix* at the kth-mode $\mathbf{R}_k \in C^{N_o \times N_i}$, the *rank-one* mode shape and participation factors, $\mathbf{\Phi}_k \in R^{N_o \times 1}$, $\mathbf{L}_k \in R^{1 \times N_i}$. K the number of poles defined by $\{\lambda_k\}$ with $\lambda_k = \sigma_k \pm j\omega_k$ and $(\{\sigma_k\}, \{\omega_k\})$ the respective damping coefficients and natural frequencies. The mode shapes $\mathbf{\Phi}_k$ embedded in the residue matrices can be extracted from the *PSD* matrix by performing an *SVD* [2, 18–20].

In principle, any response can be expressed directly in "modal coordinates" as

$$\mathbf{y}(t) = \mathbf{\Phi}\mathbf{q}(t) = \sum_{k=1}^{K} \mathbf{\Phi}_k \mathbf{q}_k(t) \tag{7.43}$$

with corresponding *covariance matrix*

$$\mathbf{R}_{yy}(\tau) = E\{\mathbf{y}(t+\tau)\mathbf{y}(t)\} = E\left\{\mathbf{\Phi}\mathbf{q}(t+\tau)\mathbf{q}'(t)\mathbf{\Phi}'\right\} = \mathbf{\Phi}\mathbf{R}_{qq}(\tau)\mathbf{\Phi}' \tag{7.44}$$

The power spectrum follows from the Wiener–Khintchine Fourier relationship ($\mathbf{S}_{yy}(\omega) = \mathcal{F}[\mathbf{R}_{yy}(\tau)]$) as [15]

$$\mathbf{S}_{yy}(\omega) = \mathbf{\Phi}\mathbf{S}_{qq}(\omega)\mathbf{\Phi}' \tag{7.45}$$

Decomposition of the *PSD* by applying the *SVD* technique is given by

$$\mathbf{S}_{yy}(\omega) = \mathbf{U}(\omega)\mathbf{\Sigma}(\omega)\mathbf{V}^\dagger(\omega) = \mathbf{U}(\omega) \times \mathbf{\Sigma}(\omega) \times \mathbf{U}^\dagger(\omega) \tag{7.46}$$

since $\mathbf{S}_{YY}(\omega) \geq \mathbf{0}$ and $\mathbf{U}(\omega) = [\mathbf{u}_1(\omega) \ \cdots \ \mathbf{u}_K(\omega)]^\dagger$ with

$$\mathbf{\Sigma}(\omega) = \text{diag}[\sigma_1(\omega) \ \cdots \ \sigma_K(\omega)] \ni \sigma_1 > \sigma_2 > \cdots > \sigma_K$$

Clearly, this decomposition can be used to relate the *SVD* to the modal representation of Eq. (7.45) [10, 13]. For instance, the singular vectors are directly related to the mode shapes and the singular values are related to the modal responses. Therefore, these parameters can be used to define equivalent *SDOF* systems. Near a resonance (ω_o), the first singular value corresponds to the dominant mode at that frequency. The rank of the output *PSD* matrix at a particular frequency indicates the number of modes contributing to the structural response (see Eq. (7.43)).

Assuming that a single dominant mode exists at ω, then the *SDOF* system in terms of the *SVD* parameters is

$$F(\omega) = \frac{\overbrace{R_o}^{\Phi_1 \ell_1}}{\underbrace{j\omega + \sigma_o - j\omega_o}_{\sigma_1}} + \frac{R_o^*}{j\omega + \sigma_o + j\omega_o} \tag{7.47}$$

where ω_n is the *undamped natural frequency* and ζ is the *damping ratio* of the system given by

$$\sigma_o = \zeta\omega_n$$
$$\omega_o = \omega_n\sqrt{1-\zeta^2} \tag{7.48}$$

In terms of our modal interpretation of Eq. (7.40) where the first mode corresponds to the first singular value at, say ω_k, then the corresponding mode shape is given by the first singular vector as

$$\mathbf{\Phi}_1(\omega_k) = \mathbf{u}_1(\omega_k)$$

Here, the first singular value belongs to the scalar *SDOF* spectrum corresponding to the mode of interest

$$S_{yy}(\omega_k) = \sigma_1 \mathbf{u}_1(\omega_k)\mathbf{u}_1^\dagger(\omega_k)$$

So we see how the decomposition of the *PSD* matrix leads directly to a SVD and how both singular values and singular vectors are related to the modal decomposition of the *MCK* system.

7.2.4.2 Complex Mode Indicator Function (*CMIF*)

Starting with the multichannel *FRF* matrix and applying the *SVD* technique enables an estimate of the *CMIF* to extract the desired modal information [17, 18]. The *CMIF* incorporates the ability to estimate the number of dominant modes, since it is based on detecting repeated poles by operating on the *PSD* directly [10, 13, 17, 18]. It evolves directly from the *FRF* or from its imaginary components or from its Hermitian product creating the *normal matrix* an estimate of the PSD matrix that can be decomposed using the *SVD* as in Eq. (7.46) above

$$\mathbf{S}_{FF}(\omega) \approx \mathbf{F}(\omega) \times \mathbf{F}^\dagger(\omega) \tag{7.49}$$

However, the *FRF* matrix (*data cube*) can be decomposed directly to create the *CMIF*, that is, applying the *SVD* to the *FRF* at *each* frequency provides a numerically robust, economical, and practical method leading directly to the *FRF* matrix [18]. Once the overall underlying *FRF* matrix is created, the dominant N_σ-singular values (modal frequencies) are determined with resulting partitions. The *FRF* matrix is approximated by choosing the particular columns and rows of the left and right singular vectors corresponding to the diagonal $N_\sigma \times N_\sigma$-matrix of singular values such that

$$\mathbf{F}(\omega) = \mathcal{U}(\omega) \times \Sigma(\omega) \times \mathcal{V}^{\dagger}(\omega) \tag{7.50}$$

with $\{\mathcal{U}, \mathcal{V}, \Sigma\}$, the respective left and right singular vectors with the corresponding matrix of ordered (largest-to-smallest) singular values such that

$$\hat{\mathbf{F}}(\omega) = \underbrace{\left[\mathbf{U}_{N_\sigma}(\omega) | \mathbf{U}_{N_o - N_o}(\omega)\right]}_{\mathcal{U}(\omega)} \underbrace{\begin{bmatrix} \mathbf{\Sigma}_{N_\sigma}(\omega) \mid & \mathbf{0} \\ --- & --- \\ \mathbf{0} & \mid \mathbf{\Sigma}_{N_o - N_\sigma}(\omega) \end{bmatrix}}_{\Sigma(\omega)} \underbrace{\begin{bmatrix} \mathbf{V}^T_{N_\sigma}(\omega) \\ --- \\ \mathbf{V}^T_{N_\sigma - N_\sigma}(\omega) \end{bmatrix}}_{\mathcal{V}(\omega)}$$

$$\approx \mathbf{U}_{N_\sigma}(\omega) \mathbf{\Sigma}_{N_\sigma}(\omega) \mathbf{V}^T_{N_\sigma}(\omega) \tag{7.51}$$

for

$$\mathbf{\Sigma}_{N_\sigma}(\omega) = \begin{bmatrix} \sigma_1(\omega) & 0 & \cdots & 0 \\ 0 & \sigma_2(\omega) & \cdots & 0 \\ \vdots & \vdots & \ddots & \vdots \\ 0 & 0 & \cdots & \sigma_{N_\sigma}(\omega) \end{bmatrix} \tag{7.52}$$

with the set of extracted singular values the root of the modal frequencies, $\{\sigma_i(\omega) = \{\sqrt{\lambda_i(\omega)}\}; i = 1, \ldots, N_\sigma$ and $\mathbf{U}_{N_\sigma}(\omega)$ containing the mode shapes and $\mathbf{V}_{N_\sigma}(\omega)$ the modal participation vectors—all functions of frequency. Note that from the properties of the *SVD*, the columns of $\mathbf{U}_{N_\sigma}$ and rows of $\mathbf{V}^{\dagger}_{N_\sigma}$ are unit vectors (unitary matrices) at *each* spectral line; therefore, as frequencies pass through a mode (pole) peaks occur since

$$\lim_{\omega \to \lambda_i} \left(\frac{1}{j\omega - \lambda_i} \right) \longrightarrow \infty \tag{7.53}$$

The *CMIF* is the set of dominant frequency-dependent singular value functions of the decomposed *FRF*-data cube providing $\mathbf{\Sigma}_{N_\sigma}(\omega)$–the diagonal matrix of Eq. (7.52) at each frequency. Mathematically, we define the complex modal indicator function, $C(\omega) \in C^{N_i \times N_\sigma \times N_\omega}$ (*data cube*) for the N_σ-dominant modes over the set of spectral frequency lines, $\{\omega_i\}; i = 1, \ldots, N_\omega$. Each member of the N_i-set of *CMIF*-functions is a result of an independent excitation of the system under investigation

$$C(\omega) := \mathbf{\Sigma}_{N_\sigma}(\omega) \tag{7.54}$$

The *CMIF* is based on the Expansion Theorem such that at every frequency the column vector, $\mathbf{f}_j(\omega)$ of the decomposed *FRF* matrix is the *summation* of all of the modal vectors. Since $C(\omega)$ is based on the set of diagonals of $\mathbf{\Sigma}_{N_\sigma}(\omega)$, then a N_σ-set of singular value functions, $\{\sigma_k(\omega)\}$, of frequencies results, such that,

$$\{C_k(\omega)\} := \{\sigma_k(\omega)\}; \quad k = 1, \ldots, N_\sigma \tag{7.55}$$

that is ordered from the largest (magnitudes) to smallest with the *primary* indicator curve (strongest) indicating a local maximum (or minimum) at each of the corresponding modal frequencies of the data.

The primary curve is followed by subsequent secondary curves indicating repeated poles that typically exist for axis-symmetric structural systems [17]. Thus, the *CMIF* provides a *decomposition* of the *FRF* or its functions that can be applied to identify dominant modal vectors participating at each frequency. It provides a measure of magnitude or strength of all modes at a particular frequency with the corresponding singular vectors providing shape and participation information [17, 18]. Thus, the peaks of the $C_k(\omega)$ provide estimates of modal frequencies with the total number of peaks yielding an estimate of the number of existing modes characterizing the corresponding system response.

Example 7.2 Reconsider the simple 3-mass (spring-damper) mechanical system of the previous example [21]. Exciting this system with an impulse yields the 3-mass, 3-output responses shown in Figure 7.2a. From the impulse responses, the *FRF*-matrix is estimated and the results are shown in Figure 7.2a where the *FRF*s, phase, and coherence functions (dotted line). The modal peaks (resonances) are clearly shown in each output channel along with their accompanying coherence functions indicating a quality response of unity in the frequency bands of interest. The estimated *CMIF* is illustrated in Figure 7.2b clearly discerning the 3-peak modal frequencies at 0.1225, 0.3432, and 0.4984 Hz corresponding to the *MCK*-system.

△△△

7.2.5 Stabilization Diagram (*SDIAG*)

The main objective of modal decomposition analysis techniques is the accurate and precise extraction of modal parameters describing the underlying structure. These techniques provide a mechanism to fit physical parameters to data and

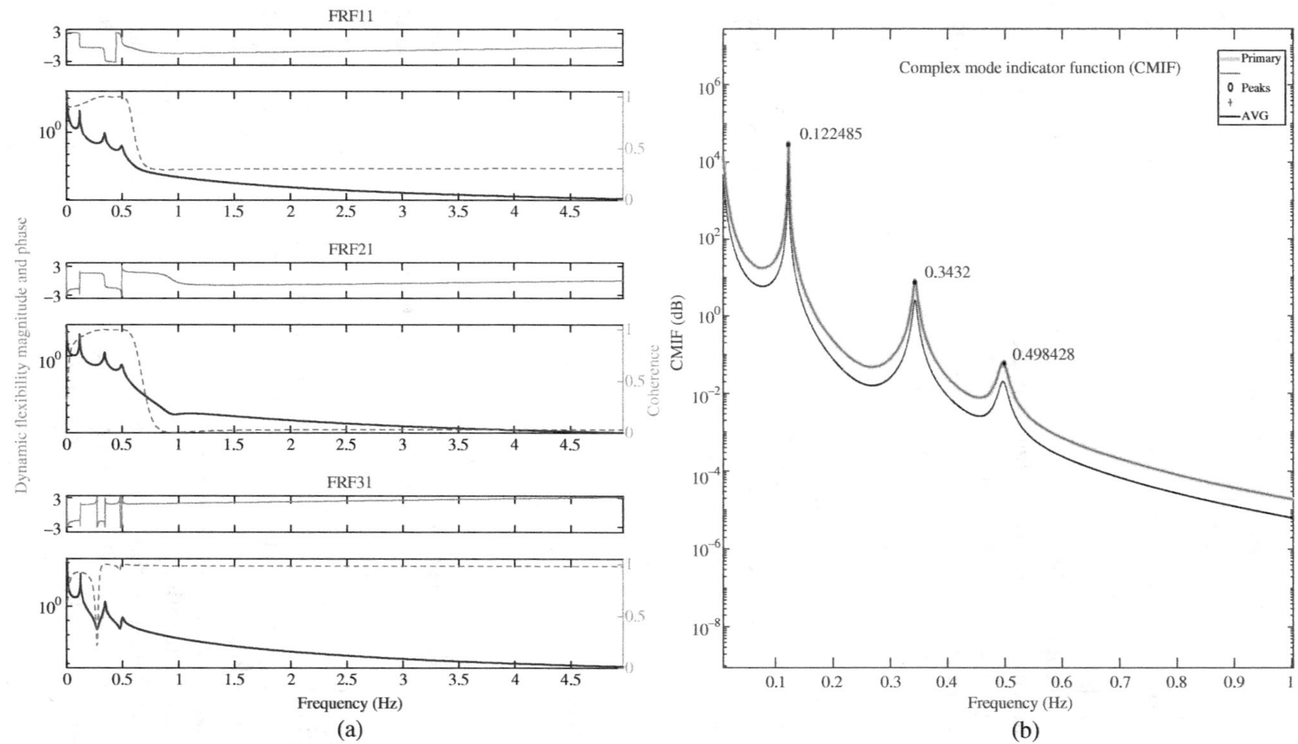

Figure 7.2 Complex mode indicator function (*CMIF*) for 3-mode/3-output *MCK*-system: (a) FRF (solid line, phase, coherence (dotted line) and (b) CMIF with identified modal peaks at 0.1225, 0.3432, and 0.4984 Hz.

validate their performance on measured data. Parametric-based methods must first decide on the proper *number of modes* leading to the correct model order and then extract corresponding modal parameters from the data.

Order estimation is a formidable task that must be resolved to achieve desired model performance and validation. The separation of the actual physical modes from estimated (mathematical) modes has motivated the development of the "stabilization diagram" to provide an answer to this nagging estimation problem [23–25].

The basic philosophy is that physical modes extracted from increasing orders of estimated mathematical models will *repeat* as the order is increased if the mode is truly a global characteristic of the underlying structure. Other artificial computational or mathematical modes will not maintain their consistency (repeatability) from model to model as the order is increased. This characteristic feature forms the basis of the *stabilization diagram* (*SDIAG*) that is employed to estimate the modes, but from parametric models in contrast to the usual peak picking of the *FRF*-matrix functions. Thus, the *SDIAG* provides a methodology to extract viable physical modes from data by classifying modal frequency and damping estimates as stable or unstable based on the closeness or small errors that physical (repeatable) modes exhibit. That is, following error constraints[2] are applied to determine modal closeness and consistency:

$$\mathcal{SD}(\omega,\zeta) \Rightarrow \begin{cases} \dfrac{\omega_k - \omega_{k-1}}{\omega_k} \leq \varepsilon_\omega \\[2ex] \dfrac{\zeta_k - \zeta_{k-1}}{\zeta_k} \leq \varepsilon_\zeta \end{cases}$$

where $\{\varepsilon_\omega, \varepsilon_\zeta\}$ are the corresponding error tolerance limits usually specified at 1% at each frequency/damping parameter estimate. A particular mode is classified as *stable*, if both the frequency and damping satisfy these constraints or if not, the modes are classified as *unstable* [23–25].

Thus, the *SDIAG* is simply a plot of model order versus the frequencies and damping estimated at each of these orders with the motivation that a "true" physical mode of the system should appear with *consistent* frequency, damping and mode shape at various orders, while spurious modes exhibit a more erratic appearance. This *SDIAG* calculation is usually initiated by selecting a high order for the *MCK* model and sequentially reducing the order, while estimating corresponding frequency and damping parameters to calculate the errors (ε). The $\mathcal{SD}(\omega,\zeta)$ plot of order of versus frequency with the order progressing from the largest-to-smallest characterizes the SDIAG is illustrated in Figure 7.3b) for the 3-mass problem.

2 Another constraint usually incorporated in *SDIAG* is the *modal assurance criterion* for mode shape comparisons ($MAC \leq \varepsilon_{\text{MAC}}$) that is discussed in Chapter 8 [23–25].

Next, we consider some more sophisticated parametric-based approaches to extract the modal properties directly from noisy frequency domain measurements.

7.3 Least-squares Complex Frequency (*LSCF*) Method

The estimation of modal frequencies is imprecise especially when attempting to extract them from uncertain frequency response spectra directly (peak picking) [2]. Although the *FDD* method offers significantly improved spectral response estimates, parametric-based estimation can provide a more reasonable alternative in either the time or frequency domain [1, 13]. Least-squares estimation of parametric *FRF* representations provides a foundation for frequency-domain techniques to accurately extract modal system parameters from uncertain measurements. In this section, we develop the least-squares complex frequency (*LSCF*) estimator based on a common denominator *FRF* model and apply it to synthesized data.

The *least-squares common denominator* (*LSCD*) characterization of the FRF is given by [11]

$$F_k(\omega_f) = \frac{N_k(\Omega_f, \Theta)}{d(\Omega_f, \Theta)} \quad \text{for} \quad k = 1, \ldots, N_o \times N_i; \quad f = 1, \ldots, N_f \tag{7.56}$$

for $N_k \in \mathcal{C}^{N_o \times N_i}$, $d \in \mathcal{C}^{N_m \times 1}$ and N_o, N_i, N_m the number of outputs, inputs and degrees of freedom (modes), Θ is the vector of unknown parameters $\{N_k^i\}, \{d_i\}$ or coefficients with the numerator and denominator polynomials defined by:

$$N_k(\Omega_f, \Theta) := \sum_{i=0}^{n} N_k^i \times \Omega_f^i; \text{ and } \quad d(\Omega_f, \Theta) := \sum_{i=0}^{n} d_i \times \Omega_f^i \tag{7.57}$$

and the *discrete* radian frequency at sampling interval $\triangle t$ defined by

$$\Omega_f^i := \exp\underbrace{(j\omega_f \triangle t)}_{z_f} \times i = z_f^i \tag{7.58}$$

In order to define an underlying "linear" (approximate) least-squares problem, we define the equation error (ϵ_k) after multiplying by the denominator polynomial and replacing the model of Eq. (7.56) with the measurement $\tilde{F}_k$ such that [11, 19]

$$\epsilon_k(\omega_f, \Theta) := N_k(\Omega_f, \Theta) - \tilde{F}_k(\omega_f) \times d(\Omega_f, \Theta) \approx \mathbf{0} \tag{7.59}$$

for the measured spectrum and the *k-output* channels at *frequency line number* f. Therefore, the least-squares cost function based on this *equation error* is

$$\mathcal{J}(\boldsymbol{\Theta}) = \sum_{k=1}^{N_o \times N_i} \sum_{f=1}^{N_f} |\epsilon_k(\omega_f, \boldsymbol{\Theta})|^2 \tag{7.60}$$

Expanding over these indices, we obtain the set of relations in terms of the Jacobian matrix ($\mathbf{J}$) and parameter vector ($\boldsymbol{\Theta}$) as

$$\epsilon = \mathbf{J} \times \boldsymbol{\Theta} = \begin{bmatrix} \Gamma_1 & 0 & \cdots & 0 & \Upsilon_1 \\ 0 & \Gamma_2 & \cdots & 0 & \Upsilon_2 \\ \vdots & \vdots & \ddots & \vdots & \vdots \\ 0 & 0 & \cdots & \Gamma_{N_oN_i} & \Upsilon_{N_oN_i} \end{bmatrix} \begin{bmatrix} \boldsymbol{\Theta}_{N_1} \\ \boldsymbol{\Theta}_{N_2} \\ \vdots \\ \boldsymbol{\Theta}_{N_oN_i} \\ --- \\ \boldsymbol{\Theta}_d \end{bmatrix} \approx \mathbf{0} \tag{7.61}$$

where

$$\epsilon = \begin{bmatrix} \epsilon_1 \\ \epsilon_2 \\ \vdots \\ \epsilon_{N_oN_i} \end{bmatrix}; \quad \text{for} \quad \epsilon_k = \begin{bmatrix} \epsilon_k(\omega_1) \\ \epsilon_k(\omega_2) \\ \vdots \\ \epsilon_k(\omega_{N_f}) \end{bmatrix}; \quad \boldsymbol{\Theta}_{N_k} = \begin{bmatrix} N_{k,0} \\ N_{k,1} \\ \vdots \\ N_{k,n} \end{bmatrix}; \quad \boldsymbol{\Theta}_d = \begin{bmatrix} d_0 \\ d_1 \\ \vdots \\ d_n \end{bmatrix}$$

and

$$\Gamma_k = \begin{bmatrix} \Gamma'_k(\omega_1) \\ \Gamma'_k(\omega_2) \\ \vdots \\ \Gamma'_k(\omega_{N_f}) \end{bmatrix} = \begin{bmatrix} z^0(\omega_1) & z^1(\omega_1) & \cdots & z^n(\omega_1) \\ z^0(\omega_2) & z^1(\omega_2) & \cdots & z^n(\omega_2) \\ \vdots & \vdots & \ddots & \vdots \\ z^0(\omega_{N_f}) & z^1(\omega_{N_f}) & \cdots & z^n(\omega_{N_f}) \end{bmatrix}$$

$$\Upsilon_k = \begin{bmatrix} \Upsilon'_k(\omega_1) \\ \Upsilon'_k(\omega_2) \\ \vdots \\ \Upsilon'_k(\omega_{N_f}) \end{bmatrix} = \begin{bmatrix} -\tilde{F}_k(\omega_1)z^0(\omega_1) & -\tilde{F}_k(\omega_1)z^1(\omega_1) & \cdots & -\tilde{F}_k(\omega_1)z^n(\omega_1) \\ -\tilde{F}_k(\omega_2)z^0(\omega_2) & -\tilde{F}_k(\omega_2)z^1(\omega_2) & \cdots & -\tilde{F}_k(\omega_2)z^n(\omega_2) \\ \vdots & \vdots & \ddots & \vdots \\ -\tilde{F}_k(\omega_{N_f})z^0(\omega_{N_f}) & -\tilde{F}_k(\omega_{N_f})z^1(\omega_{N_f}) & \cdots & -\tilde{F}_k(\omega_{N_f})z^n(\omega_{N_f}) \end{bmatrix}$$

A more computationally efficient approach is to formulate the *normal* matrix and solve such that

$$\underbrace{(\mathbf{J}^\dagger \mathbf{J})}_{\text{Normal Matrix}} \times \boldsymbol{\Theta} \approx \mathbf{0} \tag{7.62}$$

or expanding

$$(\mathbf{J}^{\dagger}\mathbf{J})\,\mathbf{\Theta} = \begin{bmatrix} \mathbf{R}_1 & 0 & \cdots & \mathbf{S}_1 \\ 0 & \mathbf{R}_2 & \cdots & \mathbf{S}_2 \\ \vdots & \vdots & \ddots & \vdots \\ \mathbf{S}_1^{\dagger} & \mathbf{S}_2^{\dagger} & \cdots & \sum_{k=1}^{N_oN_i}\mathbf{T}_k \end{bmatrix} \begin{bmatrix} \mathbf{\Theta}_{N_1} \\ \mathbf{\Theta}_{N_2} \\ \vdots \\ \mathbf{\Theta}_{N_oN_i} \\ \mathbf{\Theta}_d \end{bmatrix} \approx \mathbf{0} \tag{7.63}$$

where

$$\begin{aligned} \mathbf{R}_k &:= \Gamma_k^{\dagger}\,\Gamma_k \\ \mathbf{S}_k &:= \Gamma_k^{\dagger}\,\Upsilon_k \\ \mathbf{T}_k &:= \Upsilon_k^{\dagger}\,\Upsilon_k \end{aligned} \tag{7.64}$$

The denominator coefficient is constrained to unity ($d_n = 1$) for uniqueness of the estimation. Differentiating Eq. (7.62) and setting it to zero in terms of each of the sets of numerator and denominator coefficients enables the elimination of the numerator coefficients (see [19] for details). Therefore, the numerator coefficients can be eliminated by substituting the expression

$$\mathbf{\Theta}_k = -\mathbf{R}_k^{-1}\mathbf{S}_k\theta_d \tag{7.65}$$

into Eq. (7.62) to obtain the *reduced normal equations* as

$$\underbrace{\left[\sum_{k=1}^{N_oN_i}\mathbf{T}_k - \mathbf{S}_k^{\dagger}\mathbf{R}_k^{-1}\mathbf{S}_k\right]}_{\mathbf{M}}\theta_d = \mathbf{M}\theta_d \approx \mathbf{0} \tag{7.66}$$

with the $(n+1)\times(n+1)$ matrix $\mathbf{M}$ much smaller than that of $\mathbf{J}^{\dagger}\mathbf{J}$. Therefore, the least-squares estimate of the denominator coefficients can be obtained as ($d_n = 1$)

$$\hat{\theta}_d = \begin{bmatrix} -\overbrace{\mathbf{M}_{1:n,1:n}^{-1}}^{n\times n} \times \overbrace{\mathbf{M}_{1:n,n+1}}^{n\times 1} \\ \mathbf{1} \end{bmatrix} \tag{7.67}$$

for $\mathbf{M}_{m,\ell}$ a sub-matrix defined by m-rows and ℓ-columns.

With the estimated denominator coefficients now available, the corresponding natural frequency, damped modal frequency, and damping ratio can be estimated by applying the continuous-to-discrete (Laplace-to-Z) *impulse invariant transformation* given by

$$z_r = e^{\lambda_r \Delta t} \longrightarrow \lambda_r = \frac{\ln z_r}{\Delta t} \tag{7.68}$$

to obtain the modal parameters

$$f_r = \frac{|\lambda_r|}{2\pi}; \quad f_d = \frac{\text{Im } \lambda_r}{2\pi}; \quad \zeta_r = \frac{\text{Re } \lambda_r}{|\lambda_r|} \tag{7.69}$$

With the poles estimated and available, the mode shapes can now be extracted from the corresponding residue matrix using the so-called *least-squares frequency domain (LSFD)* estimator [23].

Recall that the *multichannel FRF* in *partial fraction form* for N_m-modes estimated is given by

$$\mathcal{F}_k(\omega) = \sum_{i=1}^{N_m} \frac{R_{k,i}}{(j\omega - \lambda_i)} + \frac{R^*_{k,i}}{(j\omega - \lambda^*_i)} = \sum_{i=1}^{N_m} \frac{\phi_{k,i}\ell_{k,i}}{(j\omega - \lambda_i)} + \frac{\phi^*_{k,i}\ell^*_{k,i}}{(j\omega - \lambda^*_i)} \tag{7.70}$$

for poles, $\lambda := -\zeta\omega_n \pm \omega_n\sqrt{1-\zeta^2}$, in terms of the damping ratio, natural frequency, and residues with their inherent *participation factor* defined by

$$\ell_{k,i} := Q_{k,i}\phi_{k,i}$$

as well as the modal scaling factor and mode shape with † the Hermitian conjugate.

In terms of this representation, the corresponding cost function for the *LSFD* approach is defined in terms of the residue matrices as

$$\mathcal{J}_k(R_{k,i}) = \sum_{k=1}^{N_f} \left[\tilde{F}_k(\omega_f) - \sum_{i=1}^{N_m} \frac{R_{k,i}}{(j\omega - \lambda_i)} + \frac{R^*_{k,i}}{(j\omega - \lambda^*_i)} \right] \quad \text{for} \quad k = 1, \ldots, N_o N_i \tag{7.71}$$

Differentiating $\mathcal{J}_k(R_{k,i})$ with respect to the residue terms and setting the results to zero yields (see [23] for more details)

$$\frac{\partial \mathcal{J}_k}{\partial R_{k,i}} = 2\sum_{k=1}^{N_f} \left[\tilde{F}_k(\omega_f) - \sum_{i=1}^{N_m} \frac{R_{k,i}}{(j\omega - \lambda_i)} + \frac{R^*_{k,i}}{(j\omega - \lambda^*_i)} \right] \left[\frac{-1}{(j\omega - \lambda_i)} \right] \approx 0 \tag{7.72}$$

leading to the following set of equations for mode shape estimation

$$\begin{aligned} \Lambda_{\Lambda R k} &= G_\Lambda \\ \Lambda_\Lambda &= \sum_{f=1}^{N_f} \Lambda(\omega_f)\Lambda'(\omega_f) \\ G_\Lambda &= \sum_{f=1}^{N_f} \tilde{F}_k(\omega_f)\Lambda'(\omega_f) \end{aligned} \tag{7.73}$$

with the vector arrays defined by

$$\Lambda(\omega_f) := \begin{bmatrix} \frac{1}{(j\omega - \lambda_1)} \\ \vdots \\ \frac{1}{(j\omega - \lambda_{N_m})} \\ --- \\ \frac{1}{(j\omega - \lambda_1^*)} \\ \vdots \\ \frac{1}{(j\omega - \lambda_{N_m}^*)} \end{bmatrix}; \quad R_k := \begin{bmatrix} R_{k,1} \\ \vdots \\ R_{k,N_m} \\ --- \\ R_{k,1}^* \\ \vdots \\ R_{k,N_m}^* \end{bmatrix} \tag{7.74}$$

Therefore from Eq. (7.73), we have that the residues are given by

$$R_k = \Lambda_\Lambda^{-1} \times G_\Lambda \quad \text{for} \quad k = 1, \ldots, N_o N_i \tag{7.75}$$

with the corresponding *mode shapes* obtained from Eq. (7.70) by performing the *SVD* of the kth-residue matrix at the ith-mode, that is,

$$R_{k,i} = \underbrace{U_{k,i}}_{\text{1st column}} \times \Sigma_{k,i} \times V_{k,i} \tag{7.76}$$

with the first column of $U_{k,i}$ an estimate of the kth-mode shape at the ith-mode, $\hat{\phi}_{k,i}$ completing the overall frequency domain parameter estimation method.

We summarize the steps in the *LSCF* approach:

***Least-Squares Complex Frequency Technique*:**

1. *Develop* the reduced normal equations of Eq. (7.66)
2. *Estimate* the *LSCD*-model denominator parameters applying Eq. (7.67)
3. *Calculate* the modal parameters as in Eq. (7.69)
4. *Estimate* the residue matrices from Eq. (7.75)
5. *Perform* the *SVD* of the residue matrices (SVD(R_k)) at the kth-mode
6. *Estimate* the set of k-mode shapes from the first column of the left singular matrix as in Eq. (7.76)
7. *Select* the set of physical modes based on the system order and corresponding *SDIAG*

completing the *LSCF* technique. Next, we investigate the application of the *LSCF* technique to out 3-mass modal system.

Example 7.3 We now apply this technique to our 3-mass modal system as before for comparison. The results are shown in Figure 7.3 where we see the estimated *FRF*s in Figure 7.3a with the peak frequencies clearly located and

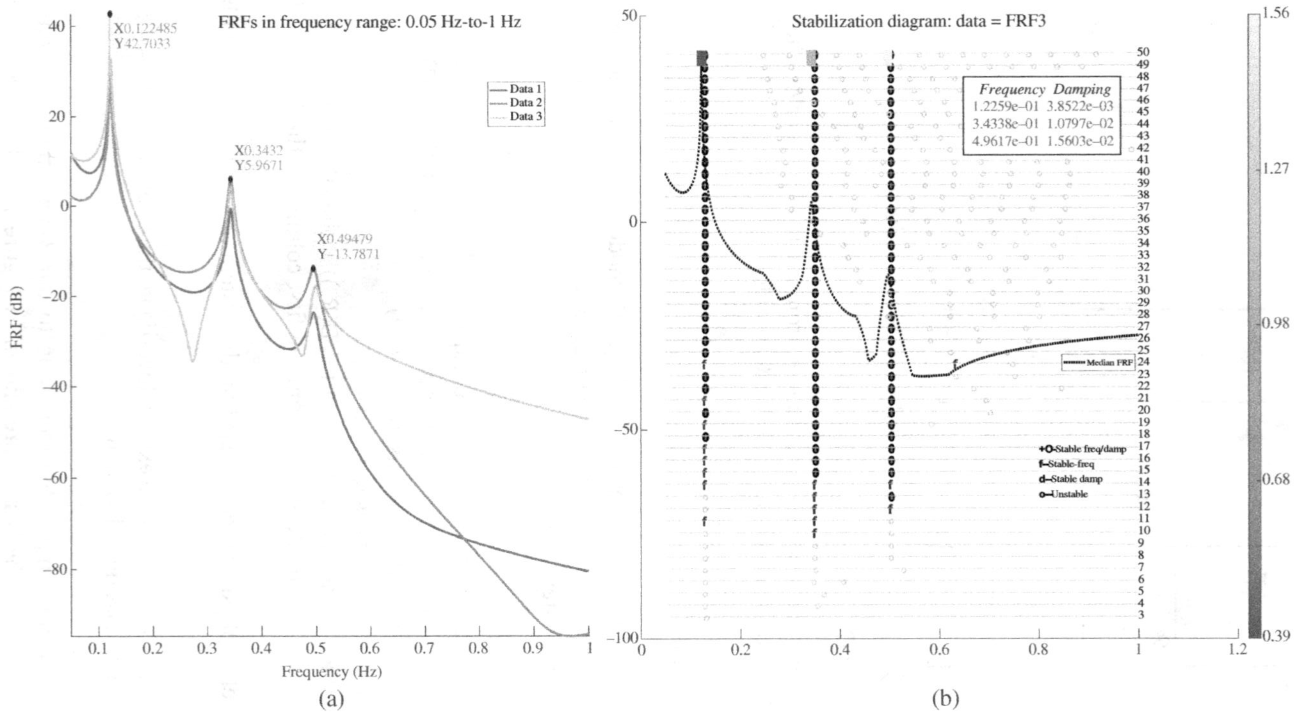

Figure 7.3 Least-square complex frequency domain (*LSCF*) method for the 3-mode *MCK*-system: (a) FRF data and (b) stabilization diagram with identified modal peaks at 0.1226, 0.3434, 0.4962 Hz, and associated damping coefficients of 0.0039, 0.011, 0.016.

the corresponding modal frequencies identified. In Figure 7.3b, we observe the SDIAG illustrating the "stable" modal frequencies and damping estimate for the 3 modes at an error of less than 1%. The estimated values are given by the modal frequencies: 0.1226, 0.3434, 0.4962 Hz and damping coefficients of 0.0039, 0.011, 0.016 shown in the figure completing the example.

△△△

Next, we develop PolyReference least-squares frequency domain estimator and apply it again to this 3-mass problem.

7.4 PolyReference Least-Squares Complex Frequency (*pLSCF*) Method

In this section, we discuss the development of the PolyReference least-squares complex frequency (*pLSCF*) estimator that evolves due to the weaknesses of the *LSCF* approach. It was found that this common denominator method (*LSCD*) could produce reasonable *FRF* spectral fits using its estimated model but failed to separate closely spaced modal resonances. Therefore, the corresponding *SDIAG* constructed using only pole information (eigenfrequencies and damping ratios) leads to a single pole instead of multiple resonances fits. It was also shown that due to the inaccuracy of the residue estimation (*LSFD*) technique employing the *SVD* decomposition of Eq. (7.75), the *LSCF* was lacking. That is, when converting the estimated model to a modal representation by using the *SVD* to obtain the rank-one residue matrix required by theory, the quality of the fit decreased [11–13]. For these reasons, the *pLSCF* approach evolved based on the right matrix fraction model description (*RMFD*). In this method, the participation factors become available enabling a much improved *SVD* used to separate physical from computational modal resonances producing a more reliable estimate of the system order or number of poles using *SDIAG*.

With this in mind, we start by defining the right matrix fraction model as

$$\mathbf{F}(\Omega_f, \mathbf{\Theta}) = \mathbf{B}(\Omega_f, \mathbf{\Theta}) \times \mathbf{A}^{-1}(\Omega_f, \mathbf{\Theta}) \tag{7.77}$$

where $\mathbf{F} \in C^{N_o \times N_i}$, $\mathbf{B} \in C^{N_o \times N_i}$, $\mathbf{A} \in C^{N_i \times N_i}$ with $\mathbf{\Theta}$ a parametric matrix to be estimated.

For each output channel, $o = 1, \ldots, N_o$, the corresponding *row vector* can be expressed by the $\mathbf{B}_o(\Omega_f, \mathbf{\Theta})$, the numerator matrix polynomial and $\mathbf{A}(\Omega_f, \mathbf{\Theta})$ the denominator matrix polynomial

$$\mathbf{F}_o(\Omega_f, \Theta) := \underbrace{\mathbf{B}_o(\Omega_f, \mathbf{\Theta})}_{1 \times N_i} \times \underbrace{\mathbf{A}^{-1}(\Omega_f, \mathbf{\Theta})}_{N_i \times N_i} \tag{7.78}$$

such that for n the polynomial order

$$\mathbf{B}_o(\Omega_f, \mathbf{\Theta}) = \sum_{i=1}^{n} \mathbf{B}_{o,i} \Omega_f^i \quad \text{and} \quad \mathbf{A}(\Omega_f, \mathbf{\Theta}) = \sum_{i=0}^{n} \mathbf{A}_i \Omega_f^i \tag{7.79}$$

The corresponding parametric vectors are defined by $\mathbf{\Theta} := [\underline{\beta}|\underline{\alpha}]'$

$$\underline{\beta} := \begin{bmatrix} \beta_1 \\ \vdots \\ \beta_n \end{bmatrix} \in R^{nN_o \times N_i}; \quad \text{and} \quad \underline{\alpha} := \begin{bmatrix} \mathbf{A}_0 \\ \vdots \\ \mathbf{A}_n \end{bmatrix} \in R^{(n+1)N_i \times N_i} \tag{7.80}$$

with the row vectors specified by

$$\underline{\beta}_o := \begin{bmatrix} \beta_{o,0} \\ \vdots \\ \beta_{o,n} \end{bmatrix} \in R^{(n+1) \times N_i}; \tag{7.81}$$

and (as before) with

$$\Omega_f^i := \underbrace{\exp\left(j\omega_f \triangle t\right)}_{z_f} \times i = z_f^i \tag{7.82}$$

Following the development of [2], the linear least-squares problem can be approximated by the equation error formulation in terms of the *MFD* model and *FRF* data $\tilde{\mathbf{F}}_o(\omega_f)$ by the cost function as in the *LSCF*-approach

$$\mathcal{J}(\mathbf{\Theta}) = \sum_{o=1}^{No} \sum_{f=1}^{N_f} \text{tr}\left(\varepsilon_o^\dagger(\omega_f, \mathbf{\Theta})\ \varepsilon_o(\omega_f, \mathbf{\Theta})\right) \tag{7.83}$$

where the error is defined as

$$\varepsilon_o(\omega_f, \mathbf{\Theta}) := \mathbf{B}_o(\omega_f, \mathbf{\Theta}) - \tilde{\mathbf{F}}_o(\omega_f)\mathbf{A}(\omega_f, \mathbf{\Theta}) \tag{7.84}$$

Minimizing Eq. (7.83) leads directly to the relation

$$\mathbf{J} \times \mathbf{\Theta} = \mathbf{0} \tag{7.85}$$

with the Jacobian matrix J given by

$$\mathbf{J} = \begin{bmatrix} \Gamma_1 & 0 & \cdots & 0 & \Upsilon_1 \\ 0 & \Gamma_2 & \cdots & 0 & \Upsilon_2 \\ \vdots & \vdots & \ddots & \vdots & \vdots \\ 0 & 0 & \cdots & \Gamma_{N_o} & \Upsilon_{N_o} \end{bmatrix} \in C^{N_f \times (n+1)(N_o+N_i)}; \quad \Gamma_o = \begin{bmatrix} \mathbf{z}_1' \\ \vdots \\ \mathbf{z}_{N_f}' \end{bmatrix} \in C^{N_f \times (n+1)} \tag{7.86}$$

for the row vector defined by $\mathbf{z}_f' := [1\ z_f\ z_f^2 \cdots z_f^n]$ with $f = 1, \ldots, N_f$.

$$\Upsilon_o = \begin{bmatrix} -\mathbf{z}_1' \otimes \tilde{F}_o(\omega_1) \\ -\mathbf{z}_2' \otimes \tilde{F}_o(\omega_2) \\ \vdots \\ -\mathbf{z}_{N_f}' \otimes \tilde{F}_o(\omega_{N_f}) \end{bmatrix} \in C^{N_f \times (n+1)N_o} \tag{7.87}$$

and $\otimes$ the Kronecker element-by-element product.

A more efficient numerical approach is to create the "normal matrix," $\mathbf{J}^\dagger\mathbf{J}$ and minimize

$$C(\boldsymbol{\Theta}) = \text{tr}\left(\boldsymbol{\Theta}^\dagger \mathbf{J}^\dagger \mathbf{J} \boldsymbol{\Theta}\right) \tag{7.88}$$

leading to the *reduced* set of normal equations as before specified by

$$\left(\mathbf{J}^\dagger\mathbf{J}\right) \times \boldsymbol{\Theta} \approx \mathbf{0} \tag{7.89}$$

for the normal matrix defined by

$$\mathbf{J}^\dagger\mathbf{J} := \begin{bmatrix} \mathbf{R}_1 & 0 & \cdots & 0 & \mathbf{S}_1 \\ 0 & \mathbf{R}_2 & \cdots & 0 & \mathbf{S}_2 \\ \vdots & \vdots & \ddots & 0 & \vdots \\ 0 & 0 & & \mathbf{R}_{N_o} & \mathbf{S}_{N_o} \\ \mathbf{S}_1^\dagger & \mathbf{S}_2^\dagger & \cdots & \mathbf{S}_{N_o}^\dagger & \sum_{k=1}^{N_o} \mathbf{T}_k \end{bmatrix} \tag{7.90}$$

where

$$\begin{aligned} \mathbf{R}_o &:= \Gamma_o^\dagger\, \Gamma_o \in C^{(n+1)\times(n+1)} \\ \mathbf{S}_o &:= \Gamma_o^\dagger\, \Upsilon_o \in C^{(n+1)\times(n+1)N_i} \\ \mathbf{T}_o &:= \Upsilon_o^\dagger\, \Upsilon_o \in C^{(n+1)N_i\times(n+1)N_i} \end{aligned} \tag{7.91}$$

If the denominator coefficients are constrained to the identity matrix ($\mathbf{A}_n = \mathbf{I}$) for uniqueness of the estimation, then differentiating Eq. (7.89) and setting to zero in terms of each of the sets of numerator and denominator coefficients enables the elimination of the numerator coefficients (see [19] for details). Therefore, the $\underline{\beta}_o$-coefficients can be eliminated, since

$$\underline{\beta}_o = -\mathbf{R}_o^{-1} \times \mathbf{S}_o \times \underline{\alpha} \tag{7.92}$$

Finally, the reduced normal equations are given by

$$\underbrace{\sum_{o=1}^{N_o} \left(\mathbf{T}_o - \mathbf{S}_o^\dagger \times \mathbf{R}_o^{-1} \times \mathbf{S}_o\right)}_{\mathbf{M}} \times \underline{\alpha} = \mathbf{M} \times \underline{\alpha} \approx \mathbf{0} \tag{7.93}$$

for $\mathbf{M} \in C^{(n+1)N_i\times(n+1)N_i}$.

For this approach, the highest-order coefficient (denominator) matrix is constrained to an N_i identity matrix, that is, $\mathbf{A}_n = \mathbf{I}_{N_i}$ for uniqueness with the resulting least-squares solution given by:

$$\underline{\hat{\alpha}} = \begin{bmatrix} -\mathbf{M}^{-1}_{1:nN_i,1:nN_i} \times \mathbf{M}_{1:nN_i,(nN_i+1):(n+1)N_i} \\ \mathbf{I}_{N_i} \end{bmatrix} \tag{7.94}$$

with the $\mathbf{M}_{1:nN_i,1:nN_i}$ sub-matrices consisting of the first $(n \times N_i)$ rows and columns and the sub-matrix $\mathbf{M}_{1:nN_i,(nN_i+1):(n+1)N_i}$ dimensioned by the first $(nN_i$ rows and last $((n+1)N_i)$ columns (see [2] for more details). After the $\underline{\alpha}$-coefficients have been estimated, the $\underline{\beta}$-coefficients can be determined from Eq. (7.92).

The *roots* of the denominator polynomial $(\mathbf{A}(\Omega_f; (\Theta))$ are the *eigenvalues* of the *companion matrix* [26]

$$\mathbf{A} = \begin{bmatrix} -\mathbf{A}_o^{-1}\mathbf{A}_n & -\mathbf{A}_o^{-1}\mathbf{A}_{n-1} & \cdots & -\mathbf{A}_o^{-1}\mathbf{A}_2 & \mathbf{A}_o^{-1}\mathbf{A}_1 \\ \mathbf{I}_{\mathbf{N}_i} & \mathbf{0} & \cdots & \mathbf{0} & \mathbf{0} \\ \vdots & \vdots & \ddots & \vdots & \vdots \\ \mathbf{0} & \mathbf{0} & \cdots & \mathbf{I}_{N_i} & \mathbf{0} \end{bmatrix} \in \mathcal{C}^{nN_i \times nN_i} \tag{7.95}$$

with the eigenvalues $\{z_r\}$ now available, they can be transformed to the $\mathcal{Z}$-domain (as before in Eq. (7.68)) enabling the estimation of the natural frequency, damped modal frequency and damping ratio of the rth-mode as f_r, f_d, ζ_r of Eq. (7.69).

Recall that the multichannel *FRF*-matrix can be expressed in partial fraction form in terms of residues as well as modal parameters (frequencies, shapes, participation factors) as before in Eq. (7.40)

$$\mathbf{F}(\omega) = \sum_{m}^{N} i = 1 \frac{\mathbf{\Phi}_i \mathbf{L}_i}{j\omega - \lambda_i} + \frac{\mathbf{\Phi}_i^* \mathbf{L}_i^*}{j\omega - \lambda_i^*} - \frac{LR}{\omega^2} + UR \tag{7.96}$$

for the *rank-one* respective mode shapes and participation factors, $\mathbf{\Phi}_i \in \mathcal{C}^{N_o \times 1}$, $\mathbf{L}_i \in \mathcal{C}^{1 \times N_i}$ with N_m the number of modes defined by $\{\lambda_i\}$. Here the LR and UR refer to "lower" and "upper" residual terms modeling influential out-of-band modes that are also included as part of the unknown parameters (see [12, 20] for details). These parameters can be estimated by solving a least-squares problem (*LSFD*) and minimizing the following cost function [2, 12, 18–20]

$$\mathcal{J}_k(\mathbf{R}_i) = \sum_{f=1}^{N_f} \left[\tilde{\mathbf{F}}(\omega_f) - \left(\overbrace{\frac{\mathbf{\Phi}_i \mathbf{L}_i}{j\omega - \lambda_i}}^{\mathbf{R}_i} + \frac{\mathbf{\Phi}_i^* \mathbf{L}_i^*}{j\omega - \lambda_i^*} - \frac{LR}{\omega^2} + UR \right) \right]^2 \tag{7.97}$$

Differentiating this expression with respect to the residue parameters leads to the following set of equations [2, 12]

$$\mathbf{\Lambda}_\Lambda \times \mathbf{R} = \mathbf{F}_\Lambda \tag{7.98}$$

where

$$\mathbf{\Lambda}_\Lambda = \sum_{f=1}^{N_f} \begin{bmatrix} \frac{\mathbf{I}_{N_o}}{(j\omega_f-\lambda_1)(j\omega_f-\lambda_1)} & \frac{\mathbf{I}_{N_o}}{(j\omega_f-\lambda_1^*)(j\omega_f-\lambda_1)} & \cdots & \frac{\mathbf{I}_{N_o}}{(j\omega_f-\lambda_{N_m})(j\omega_f-\lambda_1)} & \frac{\mathbf{I}_{N_o}}{(j\omega_f-\lambda_{N_m}^*)(j\omega_f-\lambda_1)} \\ \vdots & \vdots & \ddots & \vdots & \vdots \\ \frac{\mathbf{I}_{N_o}}{(j\omega_f-\lambda_1)(j\omega_f-\lambda_{N_m}^*)} & \frac{\mathbf{I}_{N_o}}{(j\omega_f-\lambda_1^*)(j\omega_f-\lambda_{N_m}^*)} & \cdots & \frac{\mathbf{I}_{N_o}}{(j\omega_f-\lambda_{N_m})(j\omega_f-\lambda_{N_m}^*)} & \frac{\mathbf{I}_{N_o}}{(j\omega_f-\lambda_{N_m}^*)(j\omega_f-\lambda_{N_m}^*)} \end{bmatrix}$$

for $\mathbf{\Lambda}_\Lambda \in \mathcal{C}^{nN_o \times nN_o}$ and

$$\mathbf{R} = \begin{bmatrix} R_1 \\ R_1^* \\ \vdots \\ R_{N_m} \\ R_{N_m}^* \end{bmatrix}$$

$$\mathbf{F}_\Lambda = \sum_{f=1}^{N_f} \begin{bmatrix} \frac{\tilde{\mathbf{F}}(\omega_f)}{j\omega_f - \lambda_1} \\ \frac{\tilde{\mathbf{F}}(\omega_f)}{j\omega_f - \lambda_1^*} \\ \vdots \\ \frac{\tilde{\mathbf{F}}(\omega_f)}{j\omega_f - \lambda_{N_m}} \\ \frac{\tilde{\mathbf{F}}(\omega_f)}{j\omega_f - \lambda_{N_m}^*} \end{bmatrix}$$

The mode shape of the ith-mode can then be estimated using the *LSFD* method of Eq. (7.70) as before applying the *SVD* method, but with the required rank one properties for shapes and participation factors [2, 12, 18–20].

We summarize the steps in the *pLSCF* approach:

***PolyReference Least-Squares Complex Frequency Technique*:**

1. *Develop* the reduced normal equations of Eq. (7.93)
2. *Estimate* the *pLSCF*-model denominator parameters from the companion matrix applying Eq. (7.95)
3. *Calculate* the modal parameters from the eigenvalues of Eq. (7.95)
4. *Extract* the model numerator parameters by applying Eq. (7.92)
5. *Estimate* the residue matrices from the inversion of Eq. (7.98)

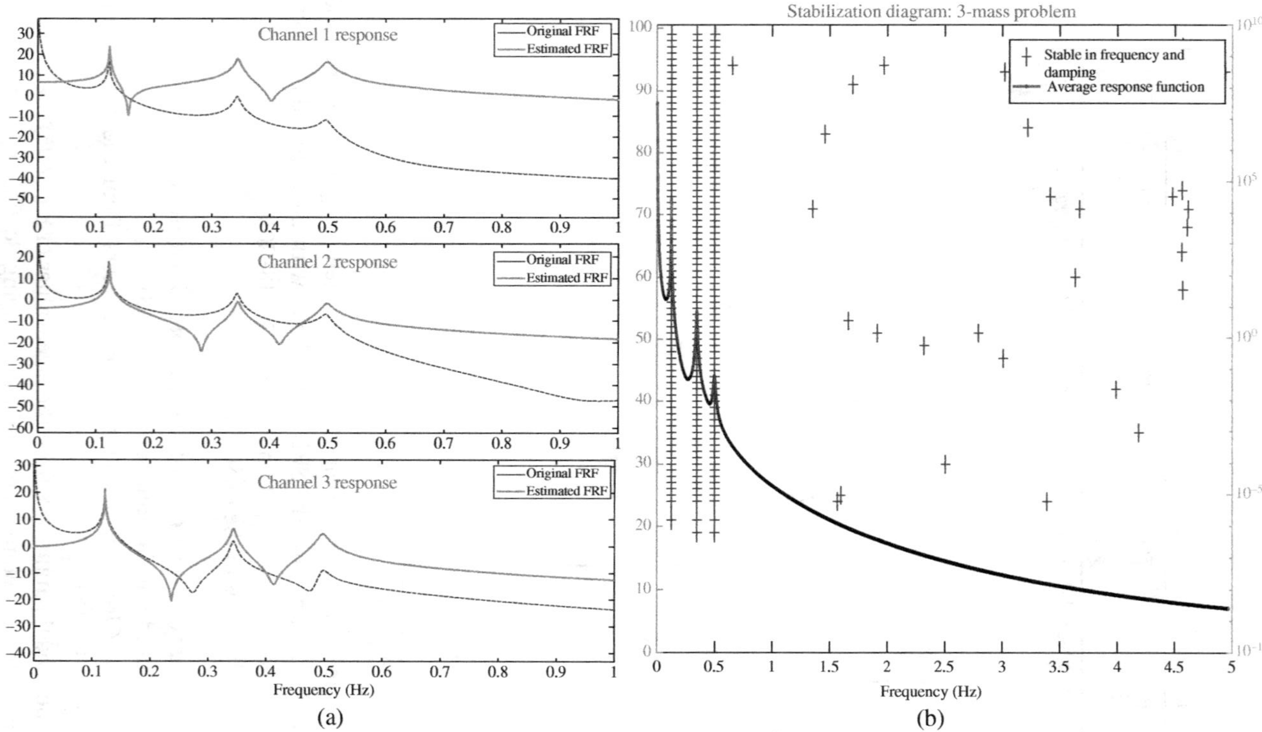

Figure 7.4 PolyReference least-square complex frequency domain method for the 3-mode MCK system: (a) estimated FRFs and (b) stabilization diagram with identified modal peaks at 0.1226, 0.3434, 0.4962 Hz and associated damping coefficients of 0.0039, 0.011, 0.016.

6. *Perform* the *SVD* of the residue matrices (SVD(R_k)) at the kth-mode
7. *Estimate* the set of k-mode shapes from the left singular (rank one) vector
8. *Select* the set of physical modes based on the system order and corresponding *SDIAG*

completing the *pLSCF* method. Next, we investigate the application of the *pLSCF*-technique to our 3-mass problem.

Example 7.4 Again we choose our 3-mass *MCK*-system data and apply the *pLSCF* method to the impulse response data with the results shown in Figure 7.4. Here, we observe the estimated *FRF* spectra using the extracted modal parameters and the corresponding impulse response spectra indicating a reasonable comparison in Figure 7.4a. In Figure 7.4b, we show the corresponding SDIAG with each of the "physical" modal frequencies are consistent and repeatable over the order estimates (50 maximum). Both modal frequency estimates: 0.1226, 0.3434, 0.4962 Hz, and associated damping coefficients of 0.0039, 0.011, 0.016 are reasonable for this frequency domain data completing the example.

△△△

Next, we develop a more sophisticated approach coupling the PolyReference estimator to nonlinear optimization techniques to improve the modal parameter estimates.

7.5 Maximum Likelihood PolyReference Frequency Domain Estimation (*ML-pLSCF*)

A more sophisticated approach to frequency domain (parametric) estimation evolves based on fundamental techniques coupled with the order estimation applying the SDIAG [23–27]. The basic concept is to employ the *pLSCF*-estimator to obtain the *initial* modal parameter estimates in order to develop the more sophisticated maximum likelihood estimator [27]. This approach enables an improved SDIAG for order estimation. The *SDIAG* coupled with a set of order estimators and corresponding modal parameters is processed using *clustering* techniques to extract the "best" modal model. This model is then refined even further using an iterative (numerical) maximum likelihood parameter estimator employing a nonlinear Gauss–Newton and/or Levenberg–Marquardt optimizer to extract the parametric model [28]. The final estimates are validated by reconstruction of the original *FRF* data [23–25]. Although more computationally intensive, the *ML-pLSCF* method provides improved and reliable estimates of the underlying modal system even with closely spaced resonances. We start with the basic maximum likelihood estimation technique and then apply this to the modal estimation problem.

The Bayesian representation of the posterior distribution of the multichannel *FRF* is given by

$$\Pr(\Theta|F(\omega)) = \Pr(F(\omega)|\Theta)\frac{\Pr(\Theta)}{\Pr(F(\omega))} \tag{7.99}$$

where we are estimating the modal parameters, Θ, from *FRF* data, $F(\omega)$. The associated conditional density $\Pr(\Theta|F(\omega))$ is the *posterior density*, because the estimate is conditioned "after" (*post*) the measurements have been acquired while $\Pr(\Theta)$ is called the *prior density* (before measurement) and $\Pr(F(\omega)|\Theta)$, the likelihood (more likely to be true) with $\Pr(F(\omega))$ is the *evidence* (scales the posterior to assure its integral is unity) as discussed in Section 4.8.

The likelihood method views the parameter as deterministic, but *unknown*. It produces the "best" estimate as the value that maximizes the probability of the measurements given that the parameter value is "most likely" true. In the estimation problem, the measurement data are given along with the underlying structure of the probability density function, but the parameters of the density are unknown and must be determined from the measurements. Therefore, the maximum likelihood estimate can be considered heuristically as that value of the parameter that best "explains" the measured data giving the most likely estimation.

Let $\boldsymbol{\Theta}$ be a vector of unknown *modal parameters*, $\boldsymbol{\Theta} \in C^{N_\theta \times 1}$ along with the corresponding set of N-independent (multichannel) frequency response measurements, $F(\omega) := \{\mathbf{F}(\omega_1) \cdots \mathbf{F}(\omega_N)\}$. The *likelihood* of $\boldsymbol{\Theta}$, given the measurements is defined to be proportional to the value of the probability density of the measurements given the parameters, that is,

$$\mathcal{L}(F(\omega);\Theta) \propto \Pr(F(\omega|\Theta) = \Pr(\mathbf{F}(\omega_1)\cdots\mathbf{F}(\omega_N)|\Theta) = \Pi_{i=1}^{N}\ \Pr(\mathbf{F}(\omega_i)|\Theta) \tag{7.100}$$

where $\mathcal{L}$ is the likelihood function and $\Pr(F(\omega)|\Theta)$ is the joint probability density functions of the measurements given the unknown parameters. Once we have the measurements (given) and the likelihood function, then we would like to find the best estimate of the (modal) parameters. Searching over all Θ and selecting the value of Θ that is maximum (most probable) leads to the maximum likelihood estimate (*ML*) given by

$$\hat{\Theta}_{ML}(F(\omega)) = \arg\max_{\Theta}\ \Pr(F(\omega)|\Theta) \tag{7.101}$$

We obtain the maximum likelihood estimate by solving the *likelihood equation* and checking for the existence of a maximum, that is, differentiating and setting to zero

$$\nabla_{\Theta}\Pr(F(\omega)|\Theta)|_{\Theta=\hat{\Theta}_{ML}} = 0 \tag{7.102}$$

Of course, to check for a maximum, we have that $\nabla_{\Theta}(\nabla_{\Theta}\Pr(F(\omega)|\Theta)') < 0$.

Applying Bayes' rule, we have [27]

$$\nabla_{\Theta}\Pr(\Theta|F(\omega)) = \nabla_{\Theta}\Pr(F(\omega)|\Theta)\big|_{\Theta=\hat{\Theta}_{ML}} = 0. \tag{7.103}$$

With this background in mind, we now develop the *ML*-estimator for the modal estimation problem.

Assuming that the *FRF*s are Gaussian, that is, $\mathbf{F}(\omega) \sim \mathcal{N}(\mathbf{m}_F(\omega), \mathbf{R}_{FF}(\omega))$, then

$$\begin{aligned}\mathbf{F}(\omega) &\sim \mathcal{N}(\mathbf{m}_F(\omega), \mathbf{R}_{FF}(\omega))\\ &= \kappa \exp\left(-1/2\left(\mathbf{F}(\omega) - \mathbf{m}_F(\omega)\right)\mathbf{R}_{FF}^{-1}(\omega)\left(\mathbf{F}(\omega) - \mathbf{m}_F(\omega)\right)^{\dagger}\right)\end{aligned} \tag{7.104}$$

and the *FRF*s are assumed uncorrelated for κ a constant scale factor

$$\mathbf{R}_{FF}(\omega) := \mathrm{var}(\mathbf{F}(\omega)) = \mathrm{diag}[\sigma_1^2(F(\omega_1))\ \sigma_2^2(F(\omega_2))\ \cdots\ \sigma_N^2(F(\omega_N))]$$

Thus, the likelihood function can be expressed in terms of the *output errors* (over N_i-inputs) corresponding to the oth degree of freedom (*DOF* system) such that [20, 28]

$$\mathcal{E}_o(\omega_f;\Theta) := \left[\frac{F_{o1}(\omega_f) - \hat{F}_{o1}(\omega_f,\Theta)}{\sigma(F_{o1}(\omega_f))} \cdots \frac{F_{oN_i}(\omega_f) - \hat{F}_{oN_i}(\omega_f,\Theta)}{\sigma(F_{oN_i}(\omega_f))}\right] \in \mathcal{C}^{1\times N_i} \tag{7.105}$$

for $F_o(\omega_f); = [F_{o1}(\omega_f)\cdots F_{oN_i}(\omega_f)]$ the oth-complex row vector of the *FRF* matrix over N_i-inputs, $\sigma(F_{o1}(\omega_f))$ the standard deviation of the oth-row with $\hat{F}_o(\omega_f,\Theta)$–the maximum likelihood estimate.

The estimated (reconstructed) *FRF* leading to the corresponding likelihood function to be maximized is

$$\mathcal{L}(F(\omega_f);\Theta) = \sum_{o=1}^{N_o}\sum_{f=1}^{N_f}\mathcal{E}_o(\omega_f;\Theta)\times\mathcal{E}_o^{\dagger}(\omega_f;\Theta) \tag{7.106}$$

for N_f the number of frequencies, N_o the number of output responses, ω_f the fth radian frequency, and Θ the modal parameters or equivalently in terms of the underlying vector *FRF*s as

$$\mathcal{L}(F(\omega_N);\Theta) = \sum_{k=1}^{N_o}\sum_{f=1}^{N_f}\frac{|F_k(\omega_f) - \hat{F}_k(\omega_f,\Theta)|^2}{\sigma_F(F_k(\omega_f))} \tag{7.107}$$

where $F_k(\omega_f)$ and $\sigma_F(F_k(\omega_f))$ are the respective measured *FRF* vectors and their corresponding standard deviation at a frequency ω_f.

Typically, the iterative optimization is performed over two stages: the *first stage* encompasses a *Gauss–Newton* optimizing over poles/participation factors and the *second stage* updates the mode shapes and corresponding residuals as implicit functions [29–31]. That is, the *FRF*s are parameterized by the *FRF* matrix

$$\mathcal{F}(\omega_f;\Theta) = \left(\sum_{m=1}^{N_m} \frac{\Phi_m L_m}{j\omega_f - \lambda_m} + \frac{\Phi_m^* L_m^*}{j\omega_f - \lambda_m^*}\right) + \frac{[LR]}{\omega_f^2} + [\mathrm{UR}] \tag{7.108}$$

where $\mathcal{F}(\omega_f;\Theta) \in \mathbf{C}^{N_o \times N_i}$, the *FRF* matrix, N_m-number of modes, Φ_m, L_m are the N_o-mode shape and N_i-participation vectors with $[LR]$ and $[UR]$ the lower and upper compensated out-of-band *modal residual terms*. Note that the product of $\mathbf{R}_m = \Phi_m \times L_m$ is the corresponding residue matrix of the mth-mode in the partial fraction expansion [20, 28]. Some of the more pragmatic approaches incorporate constrained techniques such as reciprocity (symmetry) of the residue matrix as well as proportional damping (real mode shapes) [28].

The *Gauss–Newton iterator* for the pth-iteration is defined by [29]

$$\Theta_{p+1} = \Theta_p + \triangle\Theta_p \quad \text{for} \quad \triangle\Theta_p = \underbrace{-\left[\nabla_{\Theta\Theta}\mathcal{E}(\omega,\Theta)\right]^{-1}}_{\text{Hessian}} \times \underbrace{\nabla_{\Theta}\mathcal{E}(\omega,\Theta)}_{\text{Gradient}} \tag{7.109}$$

or in terms of the Jacobian matrix, $\mathbf{J}_p(\Theta)$

$$\triangle\Theta_p = -\left[\mathbf{J}_p^\dagger(\Theta)\mathbf{J}_p(\Theta)\right]^{-1} \mathbf{J}_p^\dagger(\Theta) \times \mathcal{E}_p(\omega,\Theta) \tag{7.110}$$

where the Jacobian matrix is defined in terms of the output error derivatives, that is,

$$\mathbf{J}_p(\Theta) := \frac{\partial \mathcal{E}_p(\omega,\Theta)}{\partial \Theta} \in \mathbf{C}^{N_o N_i N_f \times 2N_m(N_i+1)}$$

for the poles and participation factors over all the modes such that $\triangle\Theta_p = [\triangle L_p \;\; \triangle\lambda_p]$ where

$$\mathbf{J} = \begin{bmatrix} \Gamma_{L_1} & 0 & \cdots & 0 & \Upsilon_{L_1}^{\lambda} \\ 0 & \Gamma_{L_2} & \cdots & 0 & \Upsilon_{L_2}^{\lambda} \\ \vdots & \vdots & \ddots & \vdots & \vdots \\ 0 & 0 & \cdots & \Gamma_{L_{N_i}} & \Upsilon_{L_{N_i}}^{\lambda} \end{bmatrix} \in \mathbf{C}^{N_o N_i N_f \times 2N_m(N_i+1)} \tag{7.111}$$

and with $i = 1, \ldots, N_i$

$$\Gamma_{L_i} := \frac{\partial \mathcal{E}_p(\omega,\Theta)}{\partial L_i} \in \mathbf{C}^{N_o N_f \times 2N_m} \quad \text{and} \quad \Upsilon_{L_i}^{\lambda} := \frac{\partial \mathcal{E}_p(\omega,\Theta)}{\partial \lambda_i} \in \mathbf{C}^{N_o N_f \times 2N_m}$$

From the Jacobian matrix structure of Eq. (7.111) and substitution in Eq. (7.110), we have the corresponding Gauss–Newton iteration step as

$$\triangle\Theta_p = -\begin{bmatrix} \mathbf{R}_{L_1} & 0 & \cdots & 0 & \mathbf{S}_1^{L\lambda} \\ 0 & \mathbf{R}_{L_2} & \cdots & 0 & \mathbf{S}_2^{L\lambda} \\ 0 & 0 & \ddots & \vdots & \vdots \\ 0 & 0 & \ddots & \mathbf{R}_{L_{N_i}} & \mathbf{S}_{N_i}^{L\lambda} \\ \mathbf{S}_1^{L\lambda^\dagger} & \mathbf{S}_2^{L\lambda^\dagger} & \cdots & \mathbf{S}_{N_i}^{L\lambda^\dagger} & \sum_{i=1}^{N_i} \mathbf{T}_{L_i}^{\lambda} \end{bmatrix}^{-1} \begin{bmatrix} \mathbf{F}_1 \\ \mathbf{F}_2 \\ \vdots \\ \mathbf{F}_{N_i} \\ \sum_{i=1}^{N_i} \mathbf{V}_i \end{bmatrix} \tag{7.112}$$

with

$$\begin{aligned}
\mathbf{R}_{L_i} &= \mathrm{Re}[\Gamma_{L_i}^{\dagger}\Gamma_{L_i}] \in \mathbf{R}^{2N_m\times 2N_m}\\
\mathbf{S}_i^{L\lambda} &= \mathrm{Re}[\Gamma_{L_i}^{\dagger}\Upsilon_{L_i}^{\lambda}] \in \mathbf{R}^{2N_m\times 2N_m}\\
\mathbf{T}_{L_i}^{\lambda} &= \mathrm{Re}[\Upsilon_{L_i}^{\lambda^\dagger}\Upsilon_{L_i}^{\lambda}] \in \mathbf{R}^{2N_m\times 2N_m}\\
\mathbf{F}_i &= \mathrm{Re}[\Gamma_{L_i}^{\dagger}\underline{\mathcal{E}}_i(\omega,\Theta)]\\
\mathbf{V}_i &= \mathrm{Re}[\Upsilon_{L_i}^{\lambda^\dagger}\mathcal{E}_i(\omega,\Theta)]
\end{aligned}$$

for $\underline{\mathcal{E}}_i(\omega,\Theta) \in \mathbf{C}^{N_oN_f\times 1}$ corresponding to the input location i and all outputs at all frequencies ω.

It can be shown that by applying the matrix inversion lemma [29], the perturbations of the poles and participation vectors at the pth-iteration are:

$$\begin{aligned}
\triangle_{\lambda_p} &= \left[\sum_{i=1}^{N_i}\mathbf{T}_{Li}^{\lambda} - \mathbf{S}_i^{L\lambda^\dagger}\mathbf{R}_{L_i}^{-1}\mathbf{S}_i^{L\lambda}\right]^{-1}\left[\sum_{i=1}^{N_i}\mathbf{S}_i^{L\lambda^\dagger}\mathbf{R}_{L_i}^{-1}\mathbf{F}_i - \mathbf{V}_i\right] \in R^{2N_m\times 1}\\
\triangle_{L_p} &= -\mathbf{R}_{L_i}^{-1}\left[\mathbf{F}_i + \mathbf{S}_i^{L\lambda}\triangle_{\lambda_p}\right] \in R^{2N_m\times 1}
\end{aligned} \tag{7.113}$$

and therefore, the parametric iterations are

$$\begin{aligned}
\lambda_{p+1} &= \lambda_p + \triangle_{\lambda_p}\\
L_{p+1} &= L_p + \triangle_{L_p}
\end{aligned} \tag{7.114}$$

The initial parameter estimates Θ_0 to construct Eq. (7.111) are obtained from the *pLSCF* estimator. However, it should be noted that the Gauss–Newton approach improves the accuracy of the parameter estimates while using a Levenberg–Marquardt technique can force the method to converge by regularizing the correction as

$$\triangle\Theta_p = -\left[\mathbf{J}_p^{\dagger}(\Theta)\mathbf{J}_p(\Theta) + \kappa_p\,\mathrm{diag}\left(\mathbf{J}_p^{\dagger}(\Theta)\mathbf{J}_p(\Theta)\right)\right]^{-1}\mathbf{J}_p^{\dagger}(\Theta)\times\mathcal{E}_p(\omega,\Theta) \tag{7.115}$$

and increasing the factor κ_p at each iteration.

After updating, $\{\lambda_{p+1}, L_{p+1}\}$ the $[LR]$ and $[UR]$ are estimated using a least-squares method (see [28] for details) as well as the corresponding mode shapes. Once the shapes and residuals are estimated, the parameter vector $\boldsymbol{\Theta}_p = \{\lambda, L, \Phi, LR, UR\}$ is used in the next iterations. Note also that the parametric confidence limits are available from the normal Jacobian matrix by inversion leading to the well-known Fisher Information matrix, that is,

$$\text{Cov}\,(\lambda, L) = \text{Re}\,\left[\mathbf{J}^{\dagger}(\Theta)\mathbf{J}(\Theta)\right]^{-1} \tag{7.116}$$

The individual confidence limits can be obtained by using the structure of $\mathbf{J}(\Theta)$ coupled with the matrix inversion lemma [29] to obtain

$$\text{Cov}(\lambda) = \left[\sum_{i=1}^{N_i} \mathbf{T}_{Li}^{\lambda} - \mathbf{S}_i^{L\lambda^{\dagger}} \mathbf{R}_{L_i}^{-1} \mathbf{S}_i^{L\lambda}\right]^{-1}$$

$$\text{Cov}(L_i) = \mathbf{R}_{L_i}^{-1} \mathbf{S}_i^{L\lambda} \text{Cov}(\lambda) \mathbf{S}_i^{L\lambda^{\dagger}} \mathbf{R}_{L_i}^{-1} + \mathbf{R}_{L_i}^{-1} \tag{7.117}$$

We summarize the various major steps evolving from the maximum likelihood algorithm using the *pLSCF* method for the initial modal parameter values as illustrated in Figure 7.5 for the 3-mass problem. After loading the *FRF*-data, the *pLSCF*-estimator is applied to extract the modal parameters where the optimal order is determined by clustering modal poles. Maximum likelihood iterations are then performed using either the Gauss–Newton or Levenberg–Marquardt optimization techniques to improve the modal parameter estimates providing a reliable and precise set of estimates. This approach is validated by comparing the estimated *FRF* estimates with the raw measurement data minimizing the model error.

This completes the *ML-pLSCF*-estimator development, next we apply this approach to a large-scale structural problem.

7.6 Case Study: 15-DOF Structure

We apply the frequency domain methods to a well-known analytical model of 15-*DOF* structural system shown in Figure 7.6 [18]. The frequency range is 0–512 Hz with a sampling frequency of 1024 Hz for 2048 samples providing a frequency resolution of $\Delta f = 0.5$ Hz. The system is lightly damped with true modal frequencies ranging from 15 to 160 Hz and damping 1–5% as specified in Table 7.1. Four output channels of accelerometer measurements are available at mass locations of $\{1, 5, 6, 10\}$ as illustrated in the figures.

Simulations are performed by applying a deterministic impulse function exciting the system with the results shown in Figure 7.7. Both time domain and frequency domain data are used in this study with the temporal data provided

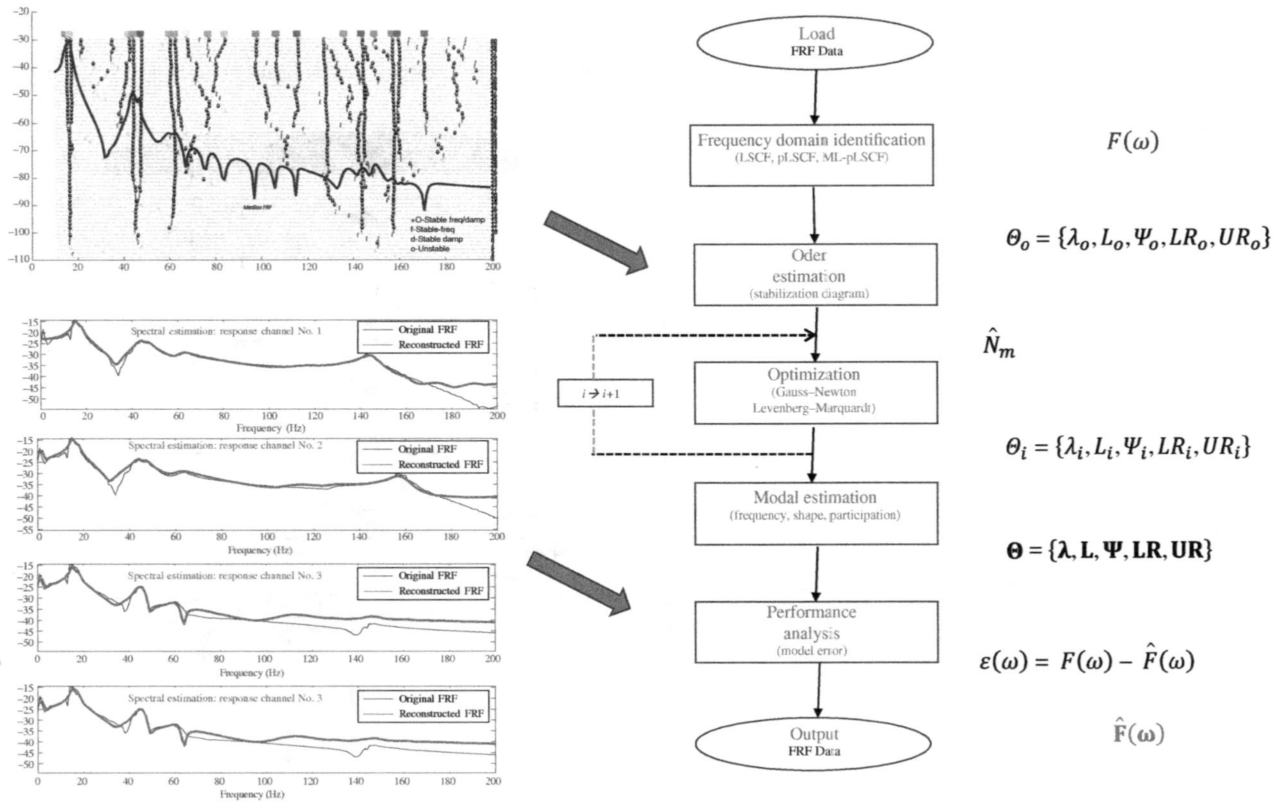

Figure 7.5 Maximum likelihood with PolyReference least-squares complex frequency estimation (*ML-pLSCF*): stabilization diagram and model reconstruction estimation results for the 3-mass *MCK* problem.

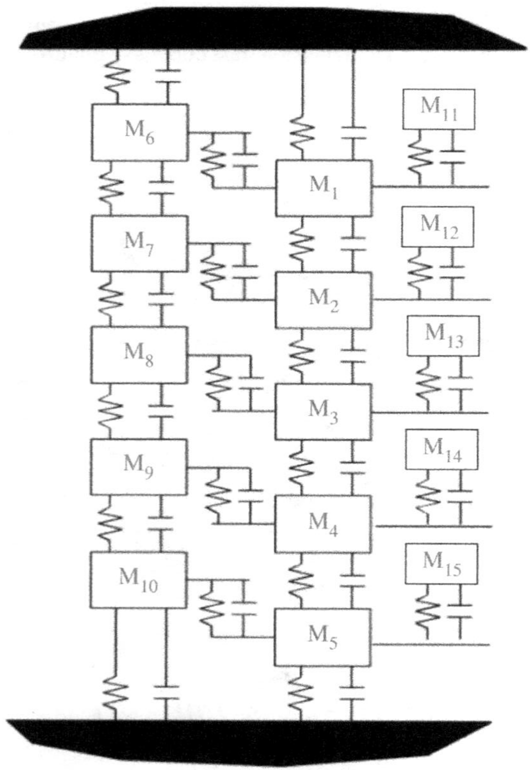

$$M = \begin{bmatrix}
0.0259 & 0 & 0 & 0 & 0 & 0 & 0 & 0 & 0 & 0 & 0 & 0 & 0 & 0 & 0 \\
0 & 0.0259 & 0 & 0 & 0 & 0 & 0 & 0 & 0 & 0 & 0 & 0 & 0 & 0 & 0 \\
0 & 0 & 0.0259 & 0 & 0 & 0 & 0 & 0 & 0 & 0 & 0 & 0 & 0 & 0 & 0 \\
0 & 0 & 0 & 0.0259 & 0 & 0 & 0 & 0 & 0 & 0 & 0 & 0 & 0 & 0 & 0 \\
0 & 0 & 0 & 0 & 0.0259 & 0 & 0 & 0 & 0 & 0 & 0 & 0 & 0 & 0 & 0 \\
0 & 0 & 0 & 0 & 0 & 0.0259 & 0 & 0 & 0 & 0 & 0 & 0 & 0 & 0 & 0 \\
0 & 0 & 0 & 0 & 0 & 0 & 0.0259 & 0 & 0 & 0 & 0 & 0 & 0 & 0 & 0 \\
0 & 0 & 0 & 0 & 0 & 0 & 0 & 0.0259 & 0 & 0 & 0 & 0 & 0 & 0 & 0 \\
0 & 0 & 0 & 0 & 0 & 0 & 0 & 0 & 0.0259 & 0 & 0 & 0 & 0 & 0 & 0 \\
0 & 0 & 0 & 0 & 0 & 0 & 0 & 0 & 0 & 0.0013 & 0 & 0 & 0 & 0 & 0 \\
0 & 0 & 0 & 0 & 0 & 0 & 0 & 0 & 0 & 0 & 0.0013 & 0 & 0 & 0 & 0 \\
0 & 0 & 0 & 0 & 0 & 0 & 0 & 0 & 0 & 0 & 0 & 0.0013 & 0 & 0 & 0 \\
0 & 0 & 0 & 0 & 0 & 0 & 0 & 0 & 0 & 0 & 0 & 0 & 0.0013 & 0 & 0 \\
0 & 0 & 0 & 0 & 0 & 0 & 0 & 0 & 0 & 0 & 0 & 0 & 0 & 0 & 0.0013
\end{bmatrix}$$

$$K = \begin{bmatrix}
4000 & -1000 & 0 & 0 & 0 & -1000 & 0 & 0 & 0 & 0 & -1000 & 0 & 0 & 0 & 0 \\
-1000 & 3900 & -1000 & 0 & 0 & 0 & -1000 & 0 & 0 & 0 & 0 & -900 & 0 & 0 & 0 \\
0 & -1000 & 4100 & -1000 & 0 & 0 & 0 & -1000 & 0 & 0 & 0 & 0 & -1100 & 0 & 0 \\
0 & 0 & -1000 & 3800 & -1000 & 0 & 0 & 0 & -1000 & 0 & 0 & 0 & 0 & -800 & 0 \\
0 & 0 & 0 & -1000 & 4200 & 0 & 0 & 0 & 0 & -1000 & 0 & 0 & 0 & 0 & -1200 \\
-1000 & 0 & 0 & 0 & 0 & 3000 & -1000 & 0 & 0 & 0 & 0 & 0 & 0 & 0 & 0 \\
0 & -1000 & 0 & 0 & 0 & -1000 & 3000 & -1000 & 0 & 0 & 0 & 0 & 0 & 0 & 0 \\
0 & 0 & -1000 & 0 & 0 & 0 & -1000 & 3000 & -1000 & 0 & 0 & 0 & 0 & 0 & 0 \\
0 & 0 & 0 & -1000 & 0 & 0 & 0 & -1000 & 3000 & -1000 & 0 & 0 & 0 & 0 & 0 \\
0 & 0 & 0 & 0 & -1000 & 0 & 0 & 0 & -1000 & 3000 & 0 & 0 & 0 & 0 & 0 \\
-1000 & 0 & 0 & 0 & 0 & 0 & 0 & 0 & 0 & 0 & 1000 & 0 & 0 & 0 & 0 \\
0 & -900 & 0 & 0 & 0 & 0 & 0 & 0 & 0 & 0 & 0 & 900 & 0 & 0 & 0 \\
0 & 0 & -1100 & 0 & 0 & 0 & 0 & 0 & 0 & 0 & 0 & 0 & 1100 & 0 & 0 \\
0 & 0 & 0 & -800 & 0 & 0 & 0 & 0 & 0 & 0 & 0 & 0 & 0 & 800 & 0 \\
0 & 0 & 0 & 0 & -1200 & 0 & 0 & 0 & 0 & 0 & 0 & 0 & 0 & 0 & 1200
\end{bmatrix}$$

Figure 7.6 Block diagram of large-scale structure (15-DOF) interconnections.

Table 7.1 Large-scale structure (15-DOF, IR data): modal frequency estimation.

	MODAL Frequency Estimates (Error Deviation/Relative Error ($\pm\sigma(\varepsilon)$))				
Frequency	**PSD**	**CMIF**	**LSCF**	**pLSCF-ML**	**pLSCF**
15.99 Hz	16.01(0.13)	15.00(6.19)	16.36(2.31)	16.79(5.00)	16.99(6.25)
30.86 Hz	–	–	–	–	–
43.60 Hz	44.02(0.96)	44.00(0.92)	43.71(0.25)	-	44.54(2.16)
46.44 Hz	–	–	47.22(1.67)	47.62(2.54)	48.03(3.42)
53.32 Hz	50.53(5.23)	–	–	–	–
53.39 Hz	–	–	–	–	–
59.41 Hz	–	–	59.64(0.39)	–	–
61.62 Hz	63.31(2.74)	63.00(2.24)	61.86(0.39)	62.11(0.80)	62.55(1.51)
68.81 Hz	65.03(5.49)	–	67.09(2.50)	–	–
73.63 Hz	–	–	76.65(4.10)	–	76.32(3.65)
128.84 Hz	128.06(0.61)	–	–	–	127.52(1.03)
136.55 Hz	137.57(0.75)	–	135.87(0.50)	–	135.47(0.79)
143.88 Hz	143.07(0.56)	143.00(0.61)	143.45(0.30)	–	144.67(0.55)
150.83 Hz	155.08(2.82)	145.00(3.87)	149.04(1.12)	144.54(4.17)	153.73(1.92)
157.47 Hz	157.58(0.07)	157.00(0.30)	156.65(0.52)	158.46(0.63)	158.39(0.58)
Avg (% ε)	1.94%	2.36%	1.26%	2.63%	2.19%

as inputs for the *PSD* estimators producing the multichannel time-domain responses. These responses are transformed to the frequency domain by "fitting" (*Prony method*) a set of complex exponential models of order 60 with the results shown in Figure 7.8. Here, each of the response channel *FRF*s is shown (solid line) with the corresponding *coherence functions* displayed (dashed line) as well. Recall that the coherence function is used to characterize a linear relationship between input and output channel responses providing a quantitative metric indicating the quality of the data. The coherence function spans the frequency range with values of unity for perfect coherence with reasonable results greater than 0.8 and poor data less than 0.2. It is clearly seen that the data between 10 and 100 Hz is of high quality, while the coherence decreases significantly for frequencies greater than 100 Hz indicating noisy, uncertain data of poor quality. Returning to the simulations of Figure 7.7, we observe that all of the modes are not excited as illustrated by the accompanying power spectra in both plots.

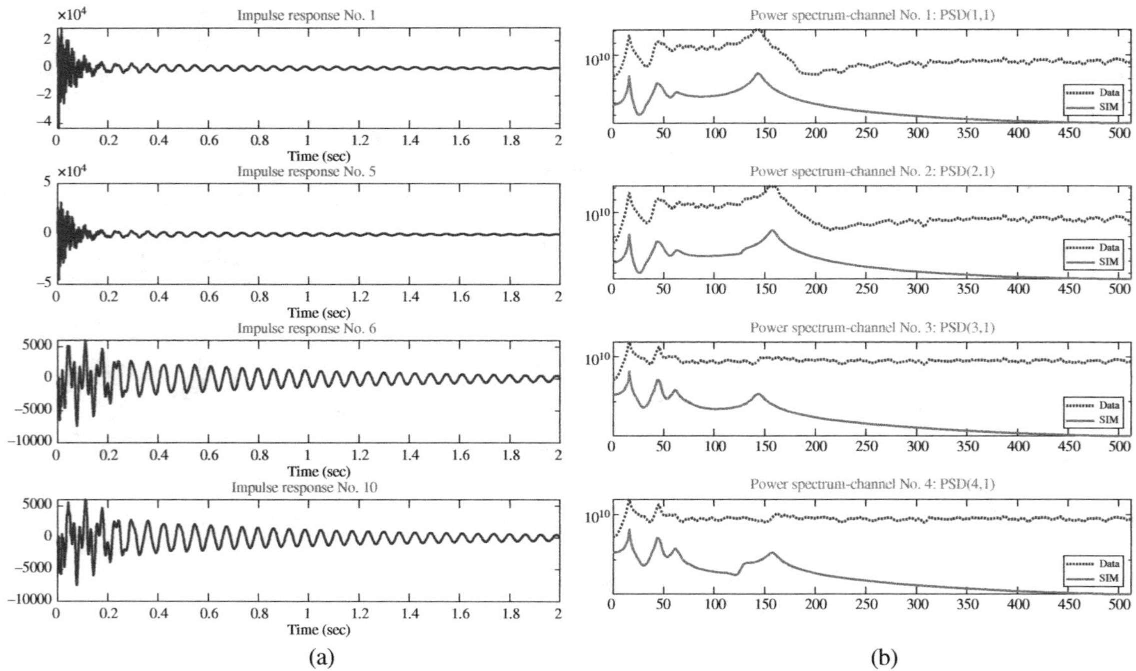

Figure 7.7 Impulse response of large-scale structure (15-DOF): (a) deterministic response (channels 1, 5, 6, 10) and (b) power spectral density.

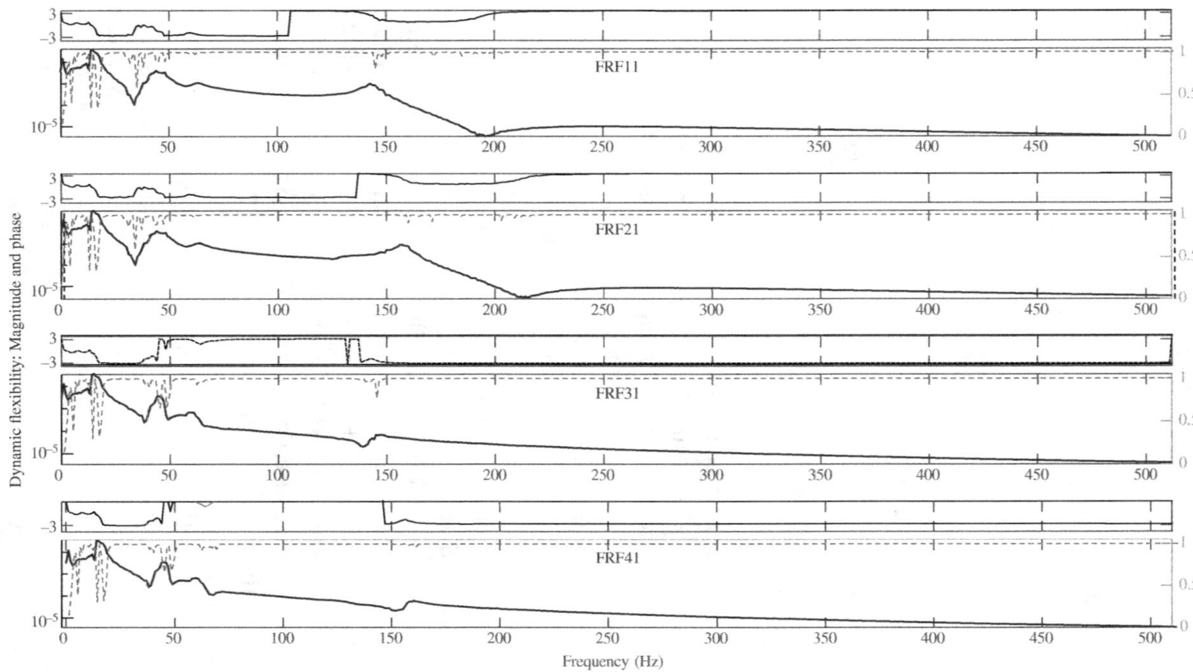

Figure 7.8 FRF estimation for large-scale structure (15-DOF): phase, FRF, coherence function.

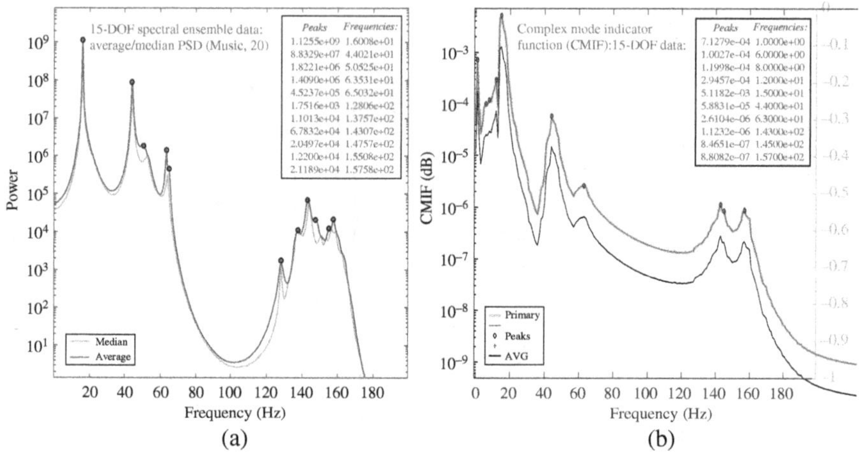

Figure 7.9 Spectral estimation for large-scale structure (15-DOF): (a) SVD PSD (Music, 20) and (b) CMIF.

An *SVD* based an ensemble power spectrum of the impulse response data is estimated (MUSIC, 20 modes) using the 4-output response measurements with the results shown in Figure 7.9a along with the *SVD*-based *CMIF* in Figure 7.9b. It is clear from the figures that all modes are not excited with the *PSD* (average) providing a rich spectrum and the *CMIF* extracting only the strongest modes with both estimators limited to the frequency range of 0–200 Hz. Table 7.1 indicates the extracted modal frequencies and relative errors based on the "true" 15-DOF model frequencies using the impulse response data with the resulting average relative errors of 1.94% for the *PSD* and 2.36% for the *CMIF* estimators.

Next, using the estimated *FRF*s as inputs the least-squares complex frequency domain (*LSCF*) estimator is applied to the impulse response data with the results shown in the *SDIAG* of Figure 7.10. Here a maximum order of 30-modes (60-poles) was selected in the accompanying *SDIAG* waterfall plot shown in the figure. The extracted modal frequencies are estimated using the 15-mode model and provided by the inset table with the results summarized in Table 7.1 indicating an overall relative error of 1.26%.

The PolyReference-based frequency domain processor (*pLSCF*) and maximum likelihood (ML-*pLSCF*) techniques are applied to the impulse response data with the annotated SDIAG shown in Figure 7.11. Recall that the ML-*pLSCF* estimates the modal parameters using the basic *pLSCF* estimator for the 60th-order model sequentially decreasing to the 10th-order model shown in the figure, performs cluster analysis over the set of order estimates. The cluster centers are used to extract the modal parameters that provide the initial starting parameters for the maximum likelihood parameter estimation algorithm eventually extracting

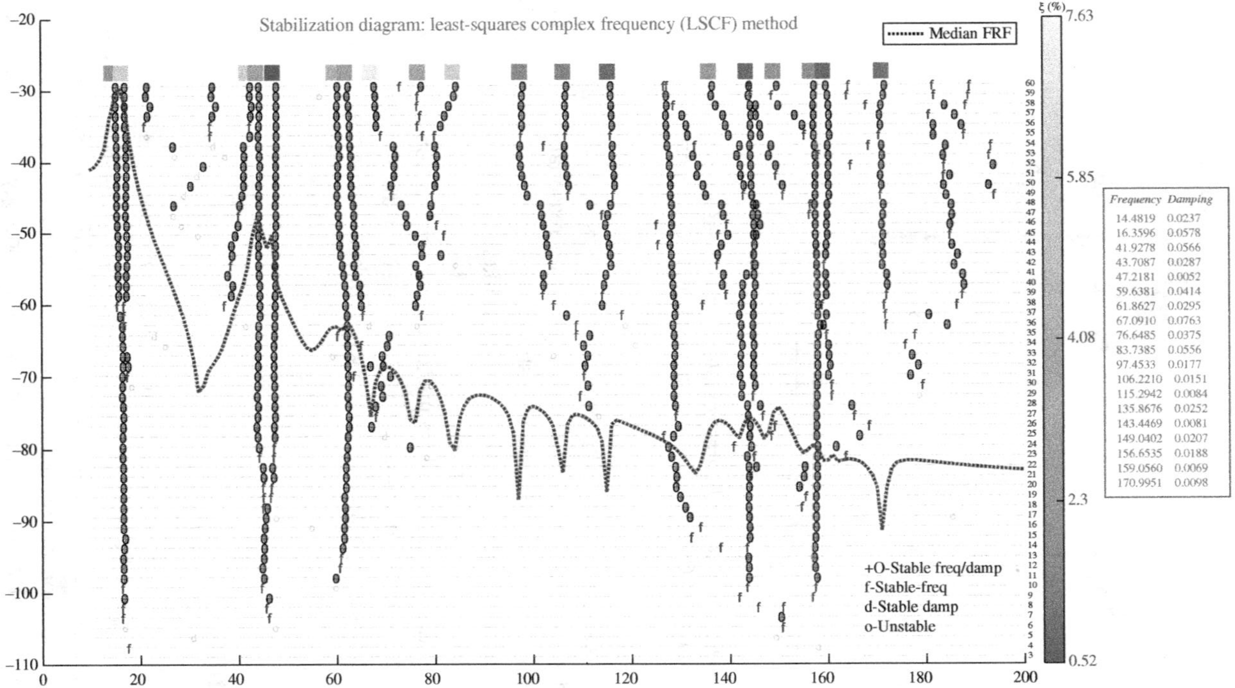

Figure 7.10 Least-squares complex frequency estimation (*LSCF*) for large-scale structure (15-DOF): stabilization diagram with frequency/damping estimates.

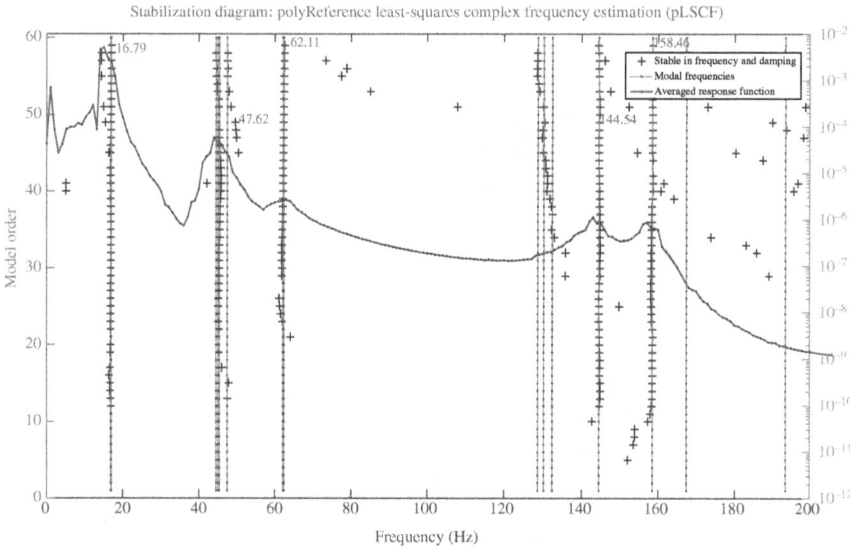

Figure 7.11 PolyReference least-squares complex frequency estimation (*pLSCF*) for large-scale structure (15-DOF): stabilization diagram with frequency estimates (annotated).

the final modal estimates. With the model parameters in hand, the *FRF*s are estimated and shown in Figure 7.12 again indicating reasonable results in the range of 0–100 Hz. These results are also shown explicitly in Table 7.1 for both methods: polyReference and with the maximum likelihood resulting in relative errors of 2.19% and 2.63%, respectively.

This completes the application of frequency domain estimation techniques to a large-scale structure for both noise-free and noisy data (0 dB SNR). Interestingly enough the processors performed reasonably well in both environments. Next, we summarize the results.

7.7 Summary

This chapter is based on the development and application of frequency domain techniques developed to extract modal parametric estimates from uncertain vibrational response data. Starting with the spectral estimation techniques based upon the decomposition of the multichannel frequency response matrix (*FRF* matrix) or the normal matrix using the SVD methodology, both the multiple signal classification method (MUSIC) of Section 5.3 and the *CMIF* are discussed. Next, the more sophisticated, parametric, multichannel, least-squares techniques

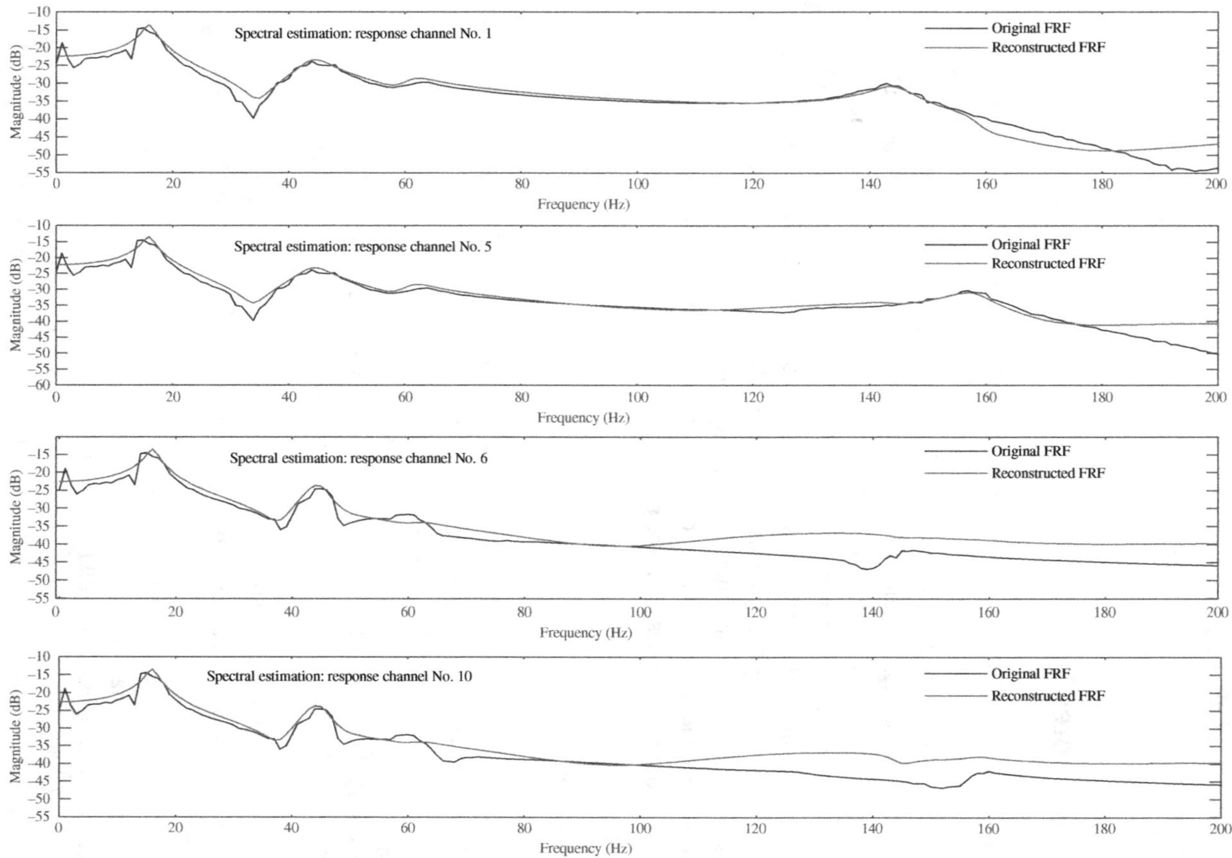

Figure 7.12 Maximum likelihood with PolyReference least-squares complex frequency estimation (*pLSCF*) for large-scale structure (15-DOF): estimated FRFs (model order = 60).

(*LSCF*, *pLSCF*) are developed and applied leading to the maximum likelihood parameter estimation approach (*ML-pLSCF*). Besides simple structural examples, a case study based on a well-known large-scale structure is simulated with the various algorithms applied to demonstrate their inherent capabilities.

MATLAB® Notes

MATLAB® provides a wealth of identification and analysis techniques for frequency domain data. Starting with the *System Identification Toolbox* frequency domain data can be input directly into a variety of algorithms capable of fitting parametric models. Next, included in the *Signal Processing Toolbox* are of wealth of tools to analyze and extract Modal *MCK*-systems. For instance, the **modalfit** command extracts modal parameters directly from *FRF*s, while the **modalfrf** command estimates *FRF*s from time-domain data sets. *MATLAB* provides a SDIAG estimator using the **modalsd** command. Besides these tools, *MATLAB* enables analytic calculations including: **envspectrum** envelope spectrum estimation, **order spectrum** average order spectrum, **ordertrack** tracking, and extracting order magnitudes, etc. (see **doc** Vibration Analysis for more details).

References

1 Reyenders, E. (2012). System identification methods for modal analysis: review and comparison. *Arch. Comput. Meth. Eng.* 19 (1): 51–124.

2 Rainieri, C. and Fabbrocino, G. (2014). *Operational Modal Analysis of Civil Engineering Structures*. New York: Springer.

3 Peeters, B. and De Roeck, G. (2001). Stochastic system identification for operational modal analysis: a review. *J. Dyn. Syst. Meas. Control* 123 (4): 659–667.

4 Migliori, A. and Sarrao, J. (1997). *Resonant Ultrasound Spectroscopy: Applications to Physics, Materials Measurements, Nondestructive Evaluation*. New York: Wiley.

5 Remillieux, M., Ulrich, T., Payan, C. et al. (2015). Resonant ultrasound spectroscopy for materials with high damping and samples of arbitrary geometry. *J. Geophys. Res.: Solid Earth* 1250: 4898–4916.

6 Vary, A. (1980). Ultrasonic measurements of material properties. In: *Research Techniques in Nondestructive Testing*, vol. 4, 159–204. London: Academic Press.

7 Zadler, B., Le Rousseau, J., Scales, J., and Smith, M. (2004). Resonant ultrasound spectroscopy: theory and application. *Geophys. J. Int.* 156: 154–169.

8 Fisher, K. (2020). Estimation of elastic properties of an additively manufactured lattice using resonant ultrasound spectroscopy. *J. Acoust. Soc. Am.* 148 (6): 4025–4036.

9 Ulrich, T. and McCall, K. (2002). Determination of elastic moduli of rock samples using resonant ultrasound spectroscopy. *J. Acoust. Soc. Am.* 111 (4): 1667–1674.

10 Brinker, R., Zhang, L., and Andersen, P. (2000). Modal identification from ambient responses using frequency domain decomposition. In: *Proceedings of the 18th International Modal Analysis Conference (IMAC)*, 625–630.

11 Van der Auweraer, H., Guillaume, P., Verboven, P., and Vanlanduit, S. (2001). Application of fast-stabilizing frequency domain parameter estimation method. *J. Dyn. Syst. Meas. Control* 123: 651–658.

12 Peeters, B., Van der Auweraer, H., Guillaume, P., and Leuridan, J. (2001). The PolyMAX frequency-domain method: a new standard for modal parameter estimation? *Shock Vib.* 11: 395–409.

13 Brinker, R., Andersen, P., and Zhang, L. (2001). Modal identification of output-only systems using frequency domain decomposition. *Smart Mater. Struct.* 10: 441–445.

14 Brinker, R., Andersen, P., and Jacobsen, N. (2007). Automated frequency domain decomposition for operational modal analysis. Proceedings of IMAC XXIV.

15 Candy, J. (2019). *Model-Based Processing: An Applied Subspace Identification Approach*. New Jersey: Wiley.

16 Candy, J., Fisher, K., Case, J., and Goodrich, T. (2020). Multichannel spectral estimation in acoustics: a state-space approach. *J. Acoust. Soc. Am.* 148 (2): 759–779.

17 Shih, C., Tsuei, Y., Allemang, R., and Brown, D. (1988). Complex mode indication function and its application to spatial domain parameter estimation. *Mech. Syst. Signal Process.* 2 (4): 367–377.

18 Allemang, R. and Brown, D. (2006). A complete review of the complex mode indicator function (CMIF) with applications. Proceedings of ISMA Conference, Leuven, Belgium.

19 Guillaume, P., Verboven, P., Vanlanduit, S. et al. (2003). A poly-reference implementation of the least-squares complex frequency-domain estimator. In: *Proceedings of the 21st International Modal Analysis Conference*, 183–192.

20 El-Kafafy, M., Guillaume, P., and Peeters, B. (2013). Modal parameter estimation by combining stochastic and deterministic frequency-domain approaches. *Mech. Syst. Signal Process.* 35: 52–68.

21 Gawronski, W. (2004). *Advanced Structural Dynamics and Active Control of Structures*. London: Springer.

22 Goodwin, G. and Payne, R. (1977). *Dynamic System Identification: Experiment Design and data Analysis*. New York: Academic Press.

23 Van der Auweraer, H. and Peeters, B. (2004). Discriminating physical poles from mathematical poles in high order systems: use and automation of the stabilization diagram. In: *Proceedings of the Instrumentation and Measurement Technology Conference (IMTC) 2004*, 2193–2198.

24 Verboven, P., Guillaume, P., Cauberghe, B. et al. (2003). Stabilization charts and uncertainty bounds for frequency-domain linear least squares estimators. PhD thesis. Vrije Universiteit Brussel.

25 Wu, C., Liu, H., Qin, X., and Wang, J. (2017). Stabilization diagrams to distinguish physical modes and spurious modes for structural parameter identification. *J. Vibroeng.* 19 (4): 2777–2794.

26 Golub, G. and Van Loan, C.F. (1989). *Matrix Computations*, 2e. Baltimore, MD: The Johns Hopkins University Press.

27 Candy, J. (2016). *Bayesian Signal Processing: Classical, Modern, and Particle Filtering Methods*, 2e. Wiley/IEEE Press.

28 El-Kafafy, M., Peeters, B., Guillaume, P., and De Troyer, T. (2016). Constrained maximum likelihood modal parameter identification applied to structural dynamics. *Mech. Syst. Signal Process.* 72-73 (4): 567–589.

29 Candy, J. (2006). *Model-Based Signal Processing*. Hoboken, NJ: Wiley/IEEE Press.

30 Ljung, L. (1999). *System Identification: Theory for the User*. Englewood Cliffs, NJ: Prentice-Hall.

31 Schoukens, J. and Pintelon, R. (1993). *Identification of Linear Systems: A Practical Guide to Accurate Modeling*. Elmsford, NY: Pergamon Press.

Problems

7.1 We are interested in characterizing the material properties of a critical system component that can be determined from a subset of frequencies. The set of resonant frequencies are given by $F_{res} := \{15, 16, 17, 20, 24\}$ KHz at amplitudes of $A_{res} := \{0.5, 0.6, 0.35, 0.7, 0.25\}$

(a) Synthesize this *material response problem* using 100 KHz sampling frequency for 4000 samples and display and display both the true and noisy signals along with their raw power spectra.

(b) Estimate the corresponding frequency response function (FRFs) and compare the results in the spectral domain using:

1. Impulse response method.
2. Power spectral method.
3. Damped exponential method (Section 5.4).
4. Frequency domain decomposition method (CMIF estimation).

7.2 For the material estimation problem specified by the set of resonant frequencies (above).

(a) Apply the *LSCF* method to extract the resonant frequencies from the FRF.

(b) Apply the *pSCF* method to extract the resonant frequencies from the FRF.

(c) Apply the *ML-pLSCF* method to extract the resonant frequencies from the FRF.

(d) Apply the *N4SID* state-space method to extract the resonant frequencies from the synthesized time domain data.

Compare all of the extracted power spectra of each and show the corresponding stabilization diagrams where applicable.

7.3 Consider the 4th order system characterized by 2 dominant frequencies provided in *MATLAB*® (load modal data). The system is excited by a random input (Xrand) producing a 5000-sample sequence at sampling frequency $F_{samp} = 4$ KHz with 2-outputs (Yrand) available.

(a) Estimate the corresponding 2×2 FRF matrix. What are the resonant peak frequencies estimated?

(b) Using the estimated the FRF matrix, estimate the corresponding stabilization diagram and discuss its predicted results.

(c) Transform the FRF matrix to an estimated the IRF matrix, how do the average spectra of each compare?

(d) Apply the *LSCF* method to extract the resonant frequencies from the FRF. Show annotated stabilization diagram.

(e) Apply the *ML-pLSCF* method to extract the resonant frequencies from the FRF. Show annotated stabilization diagram.

(f) Apply the *N4SID* state-space method to extract the resonant frequencies from the synthesized time domain data.

Compare all of the extracted power spectra of each and show the corresponding stabilization diagrams where applicable.

7.4 Reconsider the 2-body oscillator system provided in *MATLAB*® (doc modalfrf). Estimate the frequency-response function and modal parameters of a simple multi-input/multi-output system.

This problem is based on an ideal one-dimensional oscillating system consists of two masses ($M_1 = 1; M_2 = 2$) attached to the nearest wall by a spring with an elastic constant. Three dampers impede the motion of the masses by exerting on them forces proportional to speed (see *MATLAB* documents for details). The system is excited by 2-random input producing a 30 000-sample sequence at sampling frequency $F_{samp} = 50$ KHz with 2-outputs.

(a) Estimate the corresponding 2×2 FRF matrix. What are the resonant peak frequencies estimated?
(b) Using the estimated the FRF matrix, estimate the corresponding stabilization diagram and discuss its predicted results.
(c) Transform the FRF matrix to an estimated the IRF matrix, how do the average spectra of each compare?
(d) Perform FDD and estimate the CMIF, what are the estimate resonance peaks?
(e) Apply the *LSCF* method to extract the resonant frequencies from the FRF. Show annotated stabilization diagram.
(f) Apply the *ML-pLSCF* method to extract the resonant frequencies from the FRF. Show annotated stabilization diagram.
(g) Apply the *N4SID* state-space method to extract the resonant frequencies from the synthesized time domain data. Compare all of the extracted power spectra of each and show the corresponding stabilization diagrams where applicable.

7.5 For the *LSCF* method of Section 7.3, a critical step is to approximate the estimator by simplifying the nonlinear least-squares cost function in terms of the *LSCD*-parameters by introducing the equation error and then minimizing the cost.
Show by direct minimization applying the parameter constraint that ($d_n = 1$), the numerator coefficients can be removed from the normal equations by substituting the parametric relation of Eq. (7.65).

7.6 Given the following multichannel transfer function matrix

$$T(s) = \begin{bmatrix} s/s+1 & 1/s+2 \\ s+3/s+2 & 2s/s+2 \end{bmatrix}$$

(a) What is the least common denominator of this transfer function?
(b) What is its left matrix fraction form $T(s) = D^{-1}(s)N(s)$?

7.7 For the following multichannel transfer function matrix

$$T(s) = \begin{bmatrix} s/(s+1)^2 & s/(s+2)^2 \\ -s/(s+2)^2 & -s/(s+2)^2 \end{bmatrix}$$

(a) What is the right matrix fraction description of this multichannel transfer function matrix?
(b) What is its left matrix fraction?

7.8 For the following multichannel transfer function matrices, determine if they are nonsingular ($T_i(s) \neq 0$)

$$T_1(s) = \begin{bmatrix} \dfrac{(s+1)}{s^2+3s+2} & \dfrac{(s+3)}{s^2+5s+4} \end{bmatrix}$$

and

$$T_2(s) = \begin{bmatrix} \dfrac{(s+1)}{s^2+3s+2} & \dfrac{(s+3)}{s^2+5s+6} \end{bmatrix}$$

Show your results.

7.9 Given the following multichannel transfer function matrix

$$T(s) = \begin{bmatrix} s/s^2+s+1 & 2s+1/s^2+s+1 & s-1/s^2+s+1 \\ s-1/s^2+s+1 & s/s^2+s+1 & s+1/s^2+s+1 \end{bmatrix}$$

What is its right matrix fraction form $T(s) = N(s)D^{-1}(s)$?

8

Performance Analysis

8.1 Statistical Performance Methods

Statistical metrics provide a mechanism to evaluate the performance of a variety of processing techniques especially aimed at random data. In what might be called "sanity testing," classical methods of signal estimation employ the basic statistical philosophy that "if all of the signal information has been removed (explained) from measurement data, then the error residuals or prediction errors should be uncorrelated." That is, if the estimated signal "fits" the data then all that remains is an uncorrelated (white) error residual or equivalently innovations sequence. Therefore, performing statistical tests to ensure that only uncorrelated residual errors remain after processing leads to the following (simple) tests: zero-mean (*Z-M*), whiteness (*W-T*), and weighted sum-squared residual (*WSSR*) testing as well as checking the normalized mean-squared error (*NMSE*) of the error residual/signal.

8.1.1 Zero-Mean Test

A necessary and sufficient condition for a processor to be optimal is that the innovations or equivalently prediction error sequence is zero-mean and white (see Ref. [1] for the proof). These are the first properties that must be evaluated to ensure that a processor is operating optimally (see Section 4.9). If we assume further that the residual (innovation) sequence is ergodic and Gaussian, then we can use the sample mean as the test statistic to estimate, m_e, the population mean.

The sample mean for the ith component of e_i is given by

$$\hat{m}_e(i) = \frac{1}{N}\sum_{t=1}^{N} e_i(t) \quad \text{for} \quad i = 1, \ldots, N_y \tag{8.1}$$

where $\hat{m}_e(i) \sim \mathcal{N}(m_e, R_{ee}(i)/N)$ and N is the number of data samples. We perform a statistical hypothesis test to "decide" if the residual (innovation) mean

Signal Processing: An Applied Decomposition Approach, First Edition. James Vincent Candy.

Companion website: www.wiley.com/go/CandySignProc

is zero [2]. We test that the mean of the ith component of the innovation vector $e_i(t)$ is

$$H_o : m_e(i) = 0$$
$$H_1 : m_e(i) \neq 0$$

As our test statistic, we use the sample mean. At the α-significance level, the probability of rejecting the null hypothesis H_o is given by

$$\Pr\left(\left|\frac{\hat{m}_e(i) - m_e(i)}{\sqrt{R_{ee}(i)/N}} > \frac{\tau_i - m_e(i)}{\sqrt{R_{ee}(i)/N}}\right|\right) = \alpha \tag{8.2}$$

Therefore, the *zero-mean test* (*Z-M*) [2] on each component residual (innovation) e_i is given by

$$\hat{m}_e(t) \underset{\text{Accept } H_o}{\overset{\text{Reject } H_o}{\gtrless}} \tau_i \tag{8.3}$$

Under the null hypothesis H_o, each $m_e(i)$ is zero. Therefore, at the 5% significance level ($\alpha = 0.05$), we have that the threshold is

$$\tau_i = 1.96\sqrt{\frac{\hat{R}_{ee}(i)}{N}} \tag{8.4}$$

where $\hat{R}_{ee}(i)$ is the *sample variance* (assuming ergodicity) estimated by

$$\hat{R}_{ee}(i) = \frac{1}{N}\sum_{t=1}^{N} e_i^2(t) \tag{8.5}$$

8.1.2 Whiteness Test

Under the same assumptions, we can perform a *whiteness test* (*W-T*) [2], that is, check statistically that the innovations covariance corresponds to that of an uncorrelated (white) sequence. Again assuming ergodicity of the innovations sequence, we use the sample covariance function as our test statistic with the ith component covariance given by

$$\hat{R}_{ee}(i,k) = \frac{1}{N}\sum_{t=k+1}^{N} \left(e_i(t) - \hat{m}_e(i)\right)\left(e_i(t+k) - \hat{m}_e(i)\right) \tag{8.6}$$

We use the *normalized covariance* test statistic

$$\hat{\rho}_{ee}(i,k) = \frac{\hat{R}_{ee}(i,k)}{\hat{R}_{ee}(i)} \tag{8.7}$$

Asymptotically for large N, it can be shown that [2, 3]

$$\hat{\rho}_{ee}(i,k) \sim \mathcal{N}(0, 1/N)$$

Therefore, the 95% confidence interval estimate is

$$I_{\rho_{ee}} = \hat{\rho}_{ee}(i,k) \pm \frac{1.96}{\sqrt{N}} \quad \text{for} \quad N > 30 \tag{8.8}$$

Hence, under the null hypothesis, 95% of the $\hat{\rho}_{ee}(i,k)$ residual values must lie within this confidence interval, that is, for each *component* residual (innovation) sequence to be considered statistically white. Similar tests can be constructed for the *cross-covariance* properties of the innovations [4] as well, that is,

$$\text{Cov}\,(e(t), e(k)) = 0 \text{ and } \text{Cov}\,(e(t), u(t-1)) = 0$$

The whiteness test of Eq. (8.8) is very useful for detecting signal inaccuracies from *individual component* residual (innovations). However, for complex systems with a large number of measurement channels, it becomes computationally burdensome to investigate each residual (innovation) component-wise.

8.1.3 Weighted Sum-Squared Residual Test

A statistic capturing *all* of the residual (innovation) information is the *WSSR* [4–7]. It aggregates all of the residual (innovation) *vector* information over some finite window of length N. It can be shown that the *WSSR* is related to a maximum-likelihood estimate of the normalized innovations variance [2, 3].

The *WSSR* test statistic is given by

$$\hat{\rho}(\ell) := \sum_{k=\ell-N+1}^{\ell} e'(k)R_{ee}^{-1}(k)e(k) \text{ for } \ell \geq N \tag{8.9}$$

and is based on the hypothesis test

$$\begin{aligned} H_o &: \quad \{e(t)\} \ \textit{is} \ \text{white} \\ H_1 &: \quad \{e(t)\} \ \textit{is not} \ \text{white} \end{aligned}$$

with *WSSR* test statistic

$$\hat{\rho}(\ell) \underset{\text{Accept } H_o}{\overset{\text{Reject } H_o}{\gtrless}} \tau \tag{8.10}$$

Under the null hypothesis, the *WSSR* is chi-squared distributed, $\rho(\ell) \sim \chi^2(N_yN)$. However, for $N_yN > 30$, $\rho(\ell)$ is approximately Gaussian $\mathcal{N}(N_yN, 2N_yN)$ (see [8] for more details). At the α-significance level, the probability of rejecting the null hypothesis is given by

$$\Pr\left(\left|\frac{\rho(\ell) - N_yN}{\sqrt{2N_yN}} > \frac{\tau - N_yN}{\sqrt{2N_yN}}\right|\right) = \alpha \tag{8.11}$$

For a level of significance of $\alpha = 0.05$, we have

$$\tau = N_y N + 1.96\sqrt{2N_y N} \tag{8.12}$$

Thus, the *WSSR* can be considered a "whiteness test" of the innovations *vector* over a finite window of length N. Note that the adjustable parameter of the *WSSR* statistic is the window length N, which essentially controls the width of the window sliding through the residual (innovation) sequence.

8.1.4 Standard Error Test

Standard statistical error estimates

$$\epsilon(t) = y(t) - \hat{s}(t)$$

are employed to provide quantitative assessments of signal/parameter estimates consisting of the standard error, bias error, and *RMS* error given by [9]

$$\begin{aligned} \sigma_\epsilon &= \sqrt{E\{\epsilon^2(t)\} - E^2\{\epsilon(t)\}} \\ \beta_\epsilon &= E\{\epsilon(t)\} - 0 \\ RMS_\epsilon &= \sqrt{\sigma_\epsilon^2 + \beta_\epsilon^2} \end{aligned} \tag{8.13}$$

Thus, these are the basic steps that can be undertaken to ensure that signals/residuals have been implemented properly.

8.1.5 Correlation Coefficient Function Test

Perhaps, one of the most frequently applied statistical tests when comparing random variables or stochastic processes is the *correlation coefficient test* (*CC-T*) that essentially investigates, if the processes under scrutiny are linearly related or coherent (to follow next).

Fundamentally, the correlation coefficient is defined in terms of the relationship of processes $x(t)$ and $y(t)$ as [10]

$$R_{xy}(k) \leq R_{xx}(0) \times R_{yy}(0) \tag{8.14}$$

with the *correlation coefficient function* defined by

$$\rho_{xy}(k) := \frac{R_{xy}(k)}{R_{xx}(0) \times R_{yy}(0)} = \frac{R_{xy}(k)}{\sigma_x \times \sigma_y} \tag{8.15}$$

and satisfies (test)

$$-1 \leq \rho_{xy}(k) \leq +1 \qquad \forall k \tag{8.16}$$

If the processes are not zero-mean, then the covariance functions satisfy the same properties (see [9, 10] for proof).

8.1.6 Coherence Function Test

For engineers and scientists, the frequency domain transformation of data typically provides extremely meaningful information about the signal/system under investigation. When measurements or simulation data is available, then the frequency domain equivalent of the correlation coefficient for power spectra is the coherence function that can be applied as a tool to investigate the quality of data as well as the linearity of the system under investigation (see [9, 10] for more details). As before in Section 2.5, we recall the definition of the ordinary *squared-coherence function* in terms of input/output power spectral density functions as

$$\gamma_{xy}^2(\Omega) := \frac{|S_{xy}(\Omega)|^2}{S_{xx}(\Omega)S_{yy}(\Omega)} \tag{8.17}$$

where x is the input and y is the measurement or output. It follows immediately from the properties of the *PSD* that the coherence function satisfies

$$0 \leq \gamma_{xy}^2(\Omega) \leq 1 \tag{8.18}$$

It also follows that if the system under investigation is linear, then from the following spectral properties

$$S_{xy}(\Omega) = H(\Omega)S_{xx}(\Omega) \text{ and } S_{yy}(\Omega) = |H(\Omega)|^2 S_{xx}(\Omega) \tag{8.19}$$

the coherence function is

$$\gamma_{xy}^2(\Omega) = \frac{|H(\Omega)S_{xx}(\Omega)|^2}{S_{xx}(\Omega)(|H(\Omega)|^2 S_{xx}(\Omega))} = 1 \quad \forall \quad \Omega \tag{8.20}$$

Another way of examining this function follows directly from the first relation, that is,

$$S_{yx}(\Omega) = H^{-1}(\Omega)S_{yy}(\Omega) = S_{xy}^*(\Omega) \tag{8.21}$$

or

$$H^{-1}(\Omega) = \frac{S_{xy}^*(\Omega)}{S_{yy}(\Omega)} \tag{8.22}$$

Clearly, we see that

$$\gamma_{xy}^2(\Omega) = \left[\frac{S_{xy}(\Omega)}{S_{xx}(\Omega)}\right]\left[\frac{S_{xy}^*(\Omega)}{S_{yy}(\Omega)}\right] = H(\Omega)H^{-1}(\Omega) \tag{8.23}$$

and in the ideal case, these constraints must be satisfied. From this relation, we see that the coherence function can be used to determine the frequency bandover which data or a system can be considered linear. If it is perfectly linear, then the transfer function times its inverse leads to a perfectly *coherent* ($\gamma_{xy}^2(\Omega) = 1$) system.

On the other hand, if x and y are uncorrelated then $S_{xy}(\Omega) = 0$ and, therefore, completely *incoherent* ($\gamma_{xy}^2(\Omega) = 0$). We summarize these properties as:

$$\gamma_{xy}^2(\Omega) = \begin{cases} 0 & \text{Incoherent} \quad S_{xy}(\Omega) = 0 \\ 1 & \text{Coherent} \end{cases} \tag{8.24}$$

In practice, the coherence function ranges between 0 and 1 under the following conditions: (i) measurement noise, (ii) nonlinearities in the system, (iii) biased spectral estimates, and (iv) output is due to other inputs.

8.1.7 Ensemble Tests

Perhaps, another one of the most important aspects of signal estimation is to recall that processing results *only* in a *single realization* of its possible output. For instance, the conditional mean is just a statistic, but still a *random process*; therefore, if direct measurements are not available, it is best to generate an *ensemble* of realizations and extract the resulting ensemble statistics.

Simple ensemble statistics can be employed to provide an "average" solution enabling us to generate "what we would expect to observe," when the processor is applied to the actual problem. Once tuned, the processor is then applied to the raw measurement data and ensemble statistics are estimated to provide a thorough analysis of the processor performance.

Statistics are calculated over an ensemble of processor runs. The *updated* signal estimate is obtained directly from a K-ensemble, that is,

$$\hat{\hat{\mathbf{s}}}(t) \longrightarrow \{\hat{\mathbf{s}}_k(t)\}\ ;\ k = 1, \ldots, K$$

and therefore, the updated signal ensemble estimate is

$$\hat{\hat{\mathbf{s}}}(t) = \frac{1}{K}\sum_{k=1}^{K} \hat{\mathbf{s}}_k(t) \tag{8.25}$$

The *predicted* measurement estimate is also obtained directly from the ensemble, that is,

$$\hat{\hat{\mathbf{y}}}(t) \longrightarrow \{\hat{\mathbf{y}}_k(t)\}\ ;\ k = 1, \ldots, K$$

under the assumption that the measurement is a function of the signal estimate ($f(\cdot)$)

$$\hat{\mathbf{y}}_k(t) = f\left[\hat{\mathbf{s}}_k(t)\right] \tag{8.26}$$

Thus, the predicted measurement ensemble estimate is

$$\hat{\hat{\mathbf{y}}}(t) = \frac{1}{K}\sum_{k=1}^{K} \hat{\mathbf{y}}_k(t) \tag{8.27}$$

Finally, the corresponding predicted residual (innovations) ensemble estimate is given by

$$\mathbf{e}_k(t) = \mathbf{y}(t) - \hat{\mathbf{y}}_k(t); \; k = 1, \ldots, K \tag{8.28}$$

$$\bar{\mathbf{e}}(t) \longrightarrow \left\{\mathbf{e}_k(t)\right\}; \; k = 1, \ldots, K$$

and therefore, the predicted residual (innovations) ensemble estimate is

$$\bar{\mathbf{e}}(t) = \frac{1}{K}\sum_{k=1}^{K}\mathbf{e}_k(t) \tag{8.29}$$

The ensemble residual or innovations sequence can be "sanity tested" for zero-mean (*Z-M*), whiteness (*W-T*), and the *NMSE* over the ensemble to ensure that the processor has been tuned properly. Therefore, the ensemble statistics can be estimated and provide a better glimpse of the potential performance of the processor for an actual problem.

8.1.8 Statistical Order Estimation

Here, we consider statistical methods of estimating the signal/system order besides the model-based methods of Section 6.6 that applied the singular values from a Hankel or Toeplitz covariance matrix. These subsequent metrics evolve from information theory using the Kullback–Leibler (*KL*) approach (see [11] for more details).

The *information* (self) contained in the occurrence of the statistical event ω_i such that $Z(\omega_i) = z_i$, is

$$\mathcal{I}(z_i) = -\ln \Pr\left(Z(\omega_i) = z_i\right) = -\ln \Pr{}_Z(z_i) = -\ln \Pr(z_i) \tag{8.30}$$

The *entropy* or *average information* is defined by

$$\mathcal{H}(z_i) := E_Z\{I(z_i)\} = \sum_{i=1}^{K}\mathcal{I}(z_i)\Pr(z_i) = -\sum_{i=1}^{K}\Pr(z_i)\ln\Pr(z_i) \tag{8.31}$$

after substitution for $\mathcal{I}$.

Mutual information is defined in terms of the information available in the occurrence of the event $Y(\omega_j) = y_j$ about the event $Z(\omega_i) = z_i$ or

$$\mathcal{I}(z_i; y_j) = \ln\frac{\Pr(z_i|y_j)}{\Pr(z_i)} = \ln\Pr(z_i|y_j) - \ln\Pr(z_i) \tag{8.32}$$

Using these concepts, we take the information-theoretic approach to model order estimation following [12, 13]. Since many models are expressed in terms of their probability distributions, quality or "goodness" can be evaluated by their *similarity* to the true underlying probability distribution generating the measured data.

Suppose $\Pr(z_i)$ is the *true* discrete probability distribution and $\Pr(m_i)$ is that of the signal/system model, then the *Kullback–Leibler Information (KLI)* quantity of the true distribution relative to the model is defined by using

$$\mathcal{I}_{KL}\left(\Pr(z_i); \Pr(m_i)\right) := E_Z\left\{\ln \frac{\Pr(z_i)}{\Pr(m_i)}\right\} = \sum_{i=1}^{K} \Pr(z_i) \ln \frac{\Pr(z_i)}{\Pr(m_i)}$$
$$= \sum_{i=1}^{K} \Pr(z_i) \ln \Pr(z_i) - \sum_{i=1}^{K} \Pr(z_i) \ln \Pr(m_i) \tag{8.33}$$

The *KLI* possesses some very interesting properties that we state without proof (see [13] for details) such as

1) $\mathcal{I}_{KL}\left(\Pr(z_i); \Pr(m_i)\right) \geq 0$
2) $\mathcal{I}_{KL}\left(\Pr(z_i); \Pr(m_i)\right) = 0 \Leftrightarrow \Pr(z_i) = \Pr(m_i) \quad \forall i$
3) The negative of the *KLI* is the *entropy*, $\mathcal{H}_{KL}\left(\Pr(z_i); \Pr(m_i)\right)$

The second property implies that as the *model* approaches the *true* distribution, then the value of the *KLI* approaches *zero* (minimum). Thus, investigating Eq. (8.33), we see that the first term is a constant specified by the true distribution; therefore, we only need to estimate the *average* value of the model relative to the true distribution, that is,

$$\mathcal{L}(m_i) := E_Z\{\ln \Pr(m_i)\} = \sum_{i=1}^{K} \Pr(z_i) \ln \Pr(m_i) \tag{8.34}$$

where $\mathcal{L}(m_i)$ is defined as the *average log-likelihood* of the random variable of value $\ln \Pr(m_i)$. Clearly, the *larger* the average log-likelihood, the *smaller* the *KLI* implying a *better* model.

The third property, *entropy*, is approximately equal to $\frac{1}{K}$ times the probability that the relative frequency distribution of K measurements obtained from the assumed model equals the true distribution.

From the *law of large numbers*, it follows that [13]

$$\mathcal{L}(m) := E_Y\{\ln \Pr(m)\} \longrightarrow \frac{1}{K}\sum_{k=1}^{K} \ln \Pr\left(m(y_k)\right) \quad \text{for } K \to \infty \tag{8.35}$$

That is, if the event ω_i occurs at the kth-measurement, y_k, then $\Pr\left(m(y_k)\right) = \Pr\left(m_i\right)$ converges to the average log-likelihood. Further, if k_i is the number of times $\ln \Pr(m)$ takes on the value $\ln \Pr(m_i)$, then Eq. (8.35) becomes the log-likelihood of the model, that is,

$$\mathcal{L}(m_i) = \frac{1}{K}\sum_{i=1}^{K} k_i \ln \Pr\left(m_i\right) \tag{8.36}$$

and Eqs. (8.35) and (8.36) are identical in the limit. Thus, the conclusion from this discussion is that the *larger* $\mathcal{L}(m_i)$, the *better* the model. With this background information defined, let us develop the Akaike Information Criterion (*AIC*).

The basic concept of the *AIC* is that the average log-likelihood of the model, $\mathcal{L}(\cdot)$, where the average is with respect to the measurement data is used for the metric of goodness-of-fit as developed above in the general case. A more natural estimate seems to be the maximum log-likelihood; however, it is a *biased* estimator for this problem which tends to overestimate the statistic [13]. This tendency has been observed especially when the number of "free parameters" (or states) in the model, $\Theta = \{\theta_1, \ldots, \theta_{N_\Theta}\}$, is large. Akaike [12] observed that the bias is closely related to the number of free parameters and developed an asymptotically unbiased estimator of $\mathcal{L}(\cdot)$ defined verbally by

$$
\begin{aligned}
AIC = {} & -2 \times \left(\text{maximum log-likelihood of model}\right) \\
& +2 \times (\text{number of free parameters})
\end{aligned}
$$

where historical reasons account for the "2" [12, 13]. A model that minimizes the *AIC* is defined as the minimum *AIC* estimate or *MAICE*. This definition implies that if there are several models that have approximately the same *AIC*, then the one with the "fewest" free parameters (or states) should be selected. This choice is called *parsimony* in the identification literature [14–16].

From the model-based perspective, we restrict the distributions of truth and model to the same inherent structure (e.g. *AR*, *MA*, *ARMA*, state-space, etc.). We assume that the *true* model parameters (or states) are represented by Θ_{true}, while those of the estimated model are Θ. In terms of these parameters, the corresponding distributions are respectively, $\Pr\left(y(t)|\Theta_{\text{true}}\right)$ and $\Pr\left(y(t)|\Theta\right)$ with $y(t)$ the discrete measurement data generated by the true model distribution. Therefore, if we define the *AIC* in terms of log-likelihood and the N_Θ-parameter set, Θ, then

$$
AIC(N_\Theta) = -2 \times \mathcal{L}(\Theta) + 2 \times N_\Theta \tag{8.37}
$$

where $\mathcal{L}(\Theta)$ is the *average log-likelihood.*

Next, let us investigate the set of prediction errors or residuals (innovations). If the innovations are given by $\{e(t)\}$, then the log-likelihood function is specified by the following relations, since the $\{e(t)\}$ are independent. From Eq. (8.35), we have the average log-likelihood

$$
\mathcal{L}(K, \Theta) = E_e\{\ln \Pr(e(t)|\Theta)\} = \frac{1}{K}\sum_{t=1}^{K} \ln \Pr(e(t)|\Theta) \tag{8.38}
$$

which in the limit approaches the ensemble *mean log-likelihood* function

$$
\lim_{K\to\infty} \mathcal{L}(K, \Theta) \to E_e\{\ln \Pr(e(t)|\Theta)\} \tag{8.39}
$$

For the prediction error (innovations) models, the *AIC* is given by

$$AIC(N_\Theta) = -2 \times \mathcal{L}(\Pr\,(e(t)|\Theta)) + 2 \times N_\Theta \tag{8.40}$$

but from Eq. (8.35), we have that

$$E_e\{\ln \Pr\,(e(t)|\Theta)\} \approx \frac{1}{K}\sum_{t=1}^{K} \ln \Pr\,(e(t)|\Theta) \quad \text{for large } K \tag{8.41}$$

Now, in vector notation for ease of development, $\mathbf{e} = [e(1) \cdots e(K)]'$ and following [17], then for $AR(N_a)$ model, the criterion with $N_\Theta = N_a$ becomes

$$AIC(N_\Theta) = -\ln\left(\frac{1}{K}\mathbf{e}'\mathbf{e}\right) + 2\frac{N_\Theta}{K} \tag{8.42}$$

where N_Θ is the number of free parameters, K the number of data samples, and $\left(\frac{1}{K}\mathbf{e}'\mathbf{e}\right)$ is the (estimated) prediction error variance. Model order is determined by selecting the order corresponding to the minimum value of the criteria (*MAICE*) as before.

In the estimation problem, several families of $\Pr\,(e(t)|\Theta)$ with different forms of $\Pr\,(e(t)|\Theta)$ are given and we are required to select the best "fit." The *AIC* still provides a useful estimate of the entropy function in this case. Here, the *AIC* is calculated for various model orders, and the value with the minimum *AIC* is selected, that is, the *MAICE*.

Similar to the *AIC* is the minimum data length (*MDL*) order estimator [18, 19]. For the $AR(N_\Theta)$ model, we have

$$MDL(N_\theta) = -\ln\left(\frac{1}{K}\mathbf{e}'\mathbf{e}\right) + \frac{N_\Theta}{2}\ln K \tag{8.43}$$

The difference between the *AIC* and *MDL* is in the final free parameter penalty term for overestimating the order. The *AIC* tends to overestimate the order, while the *MDL* parameter penalty term increases with data length K and order N_Θ. It has also been shown that the *MDL* is a more consistent order estimator, since it converges to the true order as the number of measurements increase [18].

There are quite a wide variety of order estimators. For instance, it can also be shown that the Final Prediction Error (*FPE*) criterion given by [12]

$$FPE(N_\Theta) = \left(\frac{K+1+N_\Theta}{K-1-N_\Theta}\right)\left(\frac{1}{K}\mathbf{e}'\mathbf{e}\right) \tag{8.44}$$

is asymptotically equivalent to the $AIC(N_\Theta)$ for large K (data samples).

Other tests are also utilized to aid in the selection process. The *Signal Error Test* (*SErr-T*) of Table 8.1 consists of identifying models for a sequence of increasing orders, exciting each model with the exogenous input to produce an estimated response, $\hat{y}(t)$, and overlaying the estimated and the actual response ($y(t)$ versus $\hat{y}(t)$) to select a "reasonable" fit to the data. That is, these tests are performed by producing the estimated filtered or smoothed ($\hat{y}(t) = \hat{s}(t)$) response,

Table 8.1 Order estimation methods: statistical tests.

Test	Criterion	Remarks
Best rank:	$\beta(\hat{N}_x) := \left(\frac{\sum_{k=1}^{\hat{N}_x} \sigma_k}{\sum_{k=1}^{K} \sigma_k}\right) \times 100$	Choose $\hat{N}_x$ value
Model order estimate:	$\hat{N}_x = \max \{N : \log \hat{\sigma}_N > (\log \hat{\sigma}_{\max} + \log \hat{\sigma}_{\min})\}$	Select $\hat{N}_x$ value (gap)
Model error fit:	$\varepsilon_i = \sqrt{\frac{\sum_{t=1}^{K} \epsilon_i^2(t)}{\sum_{t=1}^{K} \mathbf{y}_i^2(t)}}$; $i = 1, \ldots, N_y$	Locate knee of curve
Model error:	$\mathcal{M}_\epsilon = 2 \sum_{k=\hat{N}_x+1}^{K} \sigma_k$	Calculate error/compare
Mean-squared prediction error:	$MSE = \frac{1}{K}\mathbf{e}'\mathbf{e}$	Calculate MSE/compare
Zero-mean:	$\hat{\mu}_\epsilon(i) \underset{\text{Accept}}{\overset{\text{Reject}}{\gtrless}} 1.96\sqrt{\frac{\hat{R}_{\epsilon\epsilon}(i)}{K}}$ for $i = 1, \ldots, N_y$	Check against threshold
Whiteness:	$I_{\rho_{\epsilon\epsilon}} = \hat{\rho}_{\epsilon\epsilon}(i, k) \pm \frac{1.96}{\sqrt{K}}$ for $K > 30$	Check 95% of covariances lie within bounds
Weighted sum-squared residual:	$\rho_\epsilon(\ell) \underset{\mathcal{H}_0}{\overset{\mathcal{H}_1}{\gtrless}} N_\epsilon K + 1.96\sqrt{2N_\epsilon K}$	Check threshold (below for white)
Signal error test:	$\hat{y}(t) = \hat{s}(t)$	Visual check Overlay (y versus $\hat{y}$) for fit
Final prediction error:	$FPE(N_\Theta) = \left(\frac{K+1+N_\Theta}{K-1-N_\Theta}\right)\left(\frac{1}{K}\mathbf{e}'\mathbf{e}\right)$	Penalties for overparameterization reduced for large K
Akaike information criteria:	$AIC(N_\Theta) = -\ln\left(\frac{1}{K}\mathbf{e}'\mathbf{e}\right) + 2\frac{N_\Theta}{K}$	Penalties for overparameterization biased—vary N_Θ knee (choose minimum)
Minimum description length:	$MDL(N_\Theta) = -\ln\left(\frac{1}{K}\mathbf{e}'\mathbf{e}\right) + \frac{N_\Theta}{2}\ln K$	Penalties for overparameterization consistent—vary N_Θ knee (choose minimum)

We list some of the more popular order tests in Table 8.1 and refer the interested reader to [20] for more details. Next, we consider validating the identified deterministic model.

8.1.9 Signal (Model) Validation

Once the order is selected the question of validity must be answered to decide just "how good" the estimated signal fits the data. For the deterministic estimation case, we rely on the estimated response comparison with the system response data to calculate the fitting error.

We can now use any of the signal representations to estimate the *model (residual) error* or simulation error defined by [14] as

$$\epsilon(t) := \mathbf{y}(t) - \mathbf{y}_{\text{model}}(t) \qquad \text{[model residual error]} \tag{8.45}$$

and define the *model error metric* for the *i*th-measurement channel as

$$\varepsilon_i = \sqrt{\frac{\sum_{t=1}^{K} \epsilon_i^2(t)}{\sum_{t=1}^{K} \mathbf{y}_i^2(t)}};\ i = 1, \ldots, N_y \qquad \text{[model error metric]} \tag{8.46}$$

to evaluate the performance of the system response $\{y_{\text{model}}(t)\}$ relative to the response data $\{y(t)\}$ [21–23].

This metric can also be calculated over a K ensemble of data for *each* measurement channel as well to provide statistics enabling a reasonable estimate of model quality. Here we have generated a data ensemble $\{y_k(t)\}; k = 1, \ldots, K$ with the corresponding *ensemble model error* for the kth realization defined by

$$\epsilon_k(t) := \mathbf{y}_k(t) - \mathbf{y}^k_{\text{model}}(t) \qquad \text{[ensemble model error]} \tag{8.47}$$

Ensemble statistics and confidence intervals can be estimated from the sample mean and variance estimates across the realizations $\{\epsilon_k(t)\}; k = 1, \ldots, K$ as

$$\begin{aligned}
\mu_\epsilon(t) &:= \hat{E}\{\epsilon_k(t)\} = \frac{1}{K}\sum_{k=1}^{K} \epsilon_k(t) \\
\sigma_\epsilon(t) &:= \hat{E}\{\epsilon_k(t) - \mu_\epsilon(t)\} = \frac{1}{K}\sum_{k=1}^{K} \epsilon_k^2(t) - \mu_\epsilon^2(t)
\end{aligned} \tag{8.48}$$

enabling a confidence interval estimate

$$I_\epsilon(t) = \mu_\epsilon(t) \pm 1.96\sigma_\epsilon(t) \qquad \text{for} \qquad K > 30$$

for *each* measurement channel realization providing another means of validating the identified model.

8.1.10 *MAD* Signal Validation

Another statistic that can be applied to validate the performance of the identified model is the *median absolute deviation* (*MAD*) statistic because of its inherent robustness property relative to the usual ensemble mean/standard deviation approach [24]

$$MAD(t) := \gamma \mathcal{M}_t \left(|\epsilon_k(t) - \mathcal{M}_k(\epsilon_k)| \right) \tag{8.49}$$

where $\epsilon_k(t) := \{\epsilon_k(0) \cdots \epsilon_k(t)\}$ is the kth realization of the set of model errors up to time t, $\mathcal{M}_k$ is the median of the data, and $\gamma = 1.4826$ is a constant based on the normalized (assumed Gaussian) data. A confidence limit that can be used for outlier detection can also be calculated as

$$\mathcal{M}_t - \beta \times MAD(t) < \epsilon_k(t) < \mathcal{M}_t + \beta \times MAD(t) \tag{8.50}$$

for β is a threshold equivalent to a confidence limit (e.g. $\beta = 1, 2$).

If it is not possible to create a deterministic data ensemble through simulation or experiment, then the original measurement can be partitioned to create an ensemble or bootstrap data resampling techniques may be applied as well [25].

Next, consider the application of these statistical methods to an electromagnetic signal processing problem of characterizing a test facility [26].

Example 8.1 An Electromagnetic Test Facility (*EMTF*) is developed primarily to determine the transient response characteristics of scale models subjected to various excitations. The facility consists of three major subsystems: (i) incident field simulator, (ii) test object, and (iii) measurement system. The system diagram is depicted in Figure 8.1 (see [26–28] for details).

The incident pulse simulator subsystem is modeled as pulse generator and antenna transfer functions and a propagation equation. The incident field generator is estimated with an impulse as the input and a noisy incident field pulse as output. The noise sources in the field generator include measurement

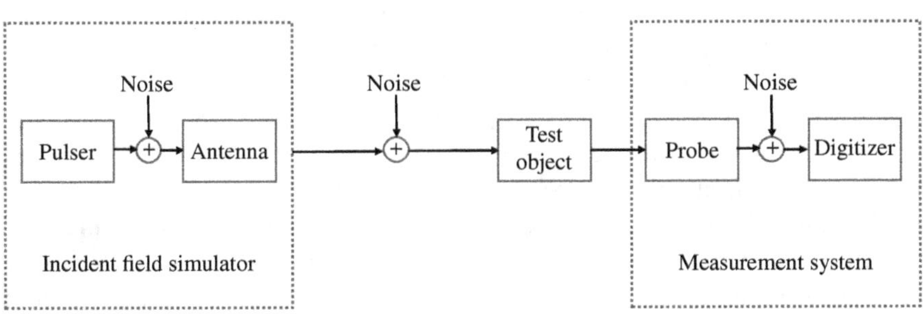

Figure 8.1 Electromagnetic test facility: signal processing model.

noise on the output of the pulse generator and environmental electromagnetic noise and reflections on the output of the antenna. The output of the generator is the input to the test object that is modeled as a linear system according to the Singularity Expansion Method [29]. The output of the object is sensed by a probe and digitized. This response is classically characterized by the spectral content of the signals using *FFT* transforms. Here, we apply model-based (parametric) techniques to estimate the variety of signals and models required to achieve this goal (see [26] for more details).

For our application of identifying a model for the pulser, we choose the *prediction error criterion* given by

$$\mathcal{J}(t) = E\{e^2(t)\}$$

where $e(t) = y(t) - \hat{s}(t)$ is the *prediction error* and $\hat{y}(t)$ is the best estimate of $y(t)$ given data up to $t - 1$.

An underlying autoregressive moving average with exogenous input (*ARMAX*) model can be used to characterize the *EMTF* facility as:

$$A(q^{-1})y(t) = B(q^{-1})u(t) + C(q^{-1})e(t)$$

where y, u, e are the process output, input, and noise sequences, and A, B, C are the respective polynomials in backward shift operator q^{-1} dimensioned N_a, N_b, N_c (see Section 4.2).

The crucial part of the identification process is estimating the *order* of the *ARMAX* model. Various order tests have been suggested to solve this problem (see Table 8.1) and a number of them are employed to initially screen models.

Once the order is determined, the parameters of the selected *ARMAX* (N_a, N_b, N_c) are estimated, then the adequacy of the fit is determined by examining the statistical properties of the corresponding prediction error or innovations sequence, $\{e(t)\}$ [26]. The prediction errors must be zero-mean and statistically white, that is, 95% of the sample correlation estimates must lie within the bounds established by the interval to ensure a viable fit.

For this example, we concentrate on identifying and evaluating the pulse generator unit. Here, the pulser is excited and an ensemble of measurements generated. The identification process is initiated by performing a series of order tests using the *FPE* and *AIC* criteria to select the model order, *ARMAX*(N_a, N_b, N_c), estimating the model parameters for the selected order, and finally performing signal and prediction error tests to assure a reasonable fit to the measured data. The results for the pulser identification are shown in Figure 8.2. Here, we see the *SErr-T* overlaying the estimated (solid) and measured (dashed) responses shown in Figure 8.2a. Note that the fit is quite reasonable and smooth compared to the original measurement. We also note the results of the *Order Test* displaying the *AIC* versus order in Figure 8.2b. Here we see that the minimum occurs for the *ARMAX*(5,4,1) model. These results are further validated from the prediction error whiteness test also

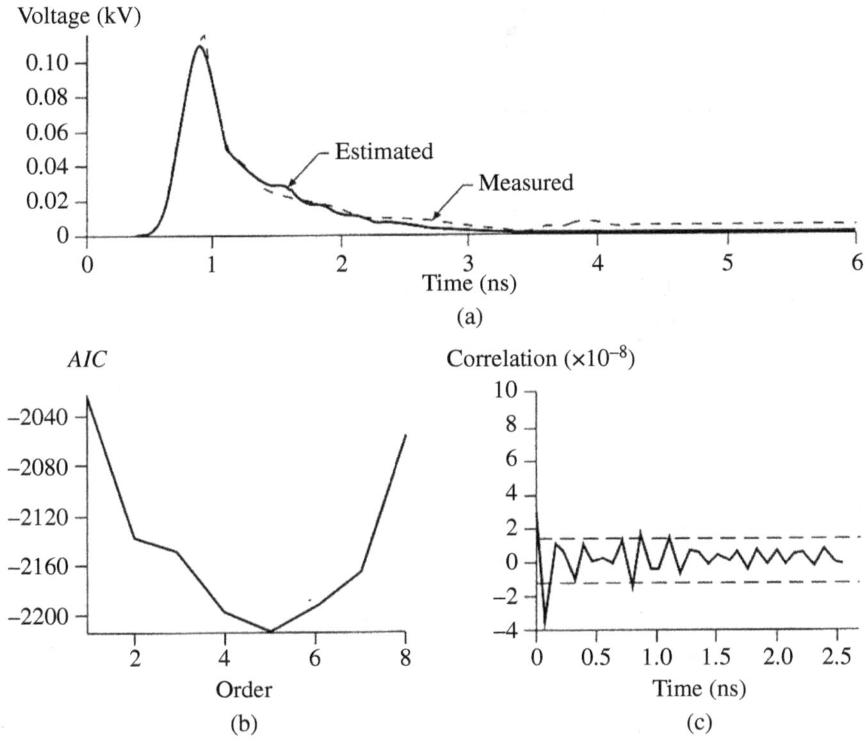

Figure 8.2 Pulser identification: (a) signal error, (b) order, and (c) whiteness tests.

shown in Figure 8.2c. Aside from an early lag correlation, the test indicates that the errors are statistically white.

The identified model is validated for prediction by calculating the residual error series over the ensemble of measurements, that is,

$$\epsilon_k(t) = y_k(t) - \hat{s}(t)$$

where k corresponds to the kth record in the ensemble and $\hat{s}$ is obtained from the estimated model parameters.

Once the parameters are determined, the corresponding standard errors can be calculated to ascertain the adequacy of the model, that is, the standard error, bias error, and RMS error are given by,

$$\sigma_\epsilon = \sqrt{E\{\epsilon^2(t)\} - E^2\{\epsilon(t)\}}$$
$$\beta_\epsilon = E\{\epsilon(t)\} - 0$$
$$RMS_\epsilon = \sqrt{\sigma_\epsilon^2 + \beta_\epsilon^2}$$

with the results summarized in Table 8.2.

Table 8.2 *EMTF* models and validation.

Type	Pulse generator	Probe
	Order	
	5	2
	Coefficient estimates	
a_1	−1.262	−0.904
a_2	1.101	0.231
a_3	−0.926	
a_4	0.445	
a_5	−0.169	
b_0	0.110	0.529
b_1	0.218	−0.530
b_2	0.324	
b_3	0.399	
b_4	0.243	
c_1	0.559	0.519
	Prediction error	
μ_e	0.244×10^{-5}	1.83×10^{-5}
σ_e	5.60×10^{-4}	5.38×10^{-4}
	Validation	
β_ε	1.14×10^{-2}	-49×10^{-5}
σ_ε	9.8×10^{-3}	1.008×10^{-3}
RMS_ε	1.5×10^{-2}	1.11×10^{-3}

The identified generator parameters along with the 2nd-order probe model are then transformed to the continuous domain using the impulse-invariant transformation completing the example [5].

△△△

Next, we investigate the development of physically-based metrics that can be applicable to a wide variety of problems especially those consisting of highly resonant processes.

8.2 Physical Performance Metrics

Physical performance methods have evolved from a variety of application areas with specific interpretations that are related to the corresponding phenomenology under analysis. In this section, we focus on those metrics that have evolved from the structural (vibration) dynamics area that can be applied to generic estimation problems. For instance, the modal analysis of vibrational responses is a common technique that can be applied directly to intrinsic material properties investigated in resonant ultrasonic spectroscopy [30–33]. Thus, although these techniques provide physical insight into the dynamics of structures, their corresponding modal *decomposition* metrics can easily be extrapolated to many other application areas. Any of those areas that require the decomposition of a signal into sinusoidal responses (Fourier transforms) or system responses that can be represented by eigen-decompositions containing resonant frequencies are viable candidates.

With this in mind, recall that the *modal decomposition* of a system characterized by its transfer function or frequency response functions can be characterized in terms of its underlying modes or modal frequencies [30, 31]. The multichannel *FRF*-matrix can be expressed in partial fraction form in terms of residues as well as modal parameters (frequencies, shapes, participation) as before in Eq. (7.40) as:

$$\mathbf{F}(\omega) = \sum_{k=1}^{K} \frac{\mathbf{R}_k}{j\omega - \lambda_k} + \frac{\mathbf{R}_k^*}{j\omega - \lambda_k^*} = \sum_{k=1}^{K} \frac{\mathbf{\Phi}_k \mathbf{L}_k}{j\omega - \lambda_k} + \frac{\mathbf{\Phi}_k^* \mathbf{L}_k^*}{j\omega - \lambda_k^*} \tag{8.51}$$

for the *residue matrix* at the kth-mode $\mathbf{R}_k \in C^{N_o \times N_i}$, the *rank one* mode shape and participation factors, $\mathbf{\Phi}_k \in R^{N_o \times 1}$, $\mathbf{L}_k \in R^{1 \times N_i}$. K the number of poles defined by $\{\lambda_k\}$ with $\lambda_k = \sigma_k \pm j\omega_k$ and $(\{\sigma_k\}, \{\omega_k\})$ the respective damping coefficients and natural frequencies. The mode shapes $\mathbf{\Phi}_k$ embedded in the residue matrices can be extracted from the *PSD*-matrix by performing an singular value decomposition (*SVD*) [31, 34].

In principle, any response can be expressed directly in "modal coordinates" $\{\mathbf{q}(t)\}$ as

$$\mathbf{y}(t) = \mathbf{\Phi}\mathbf{q}(t) = \sum_{k=1}^{K} \mathbf{\Phi}_k \mathbf{q}_k(t) \tag{8.52}$$

with *covariance matrix*

$$\mathbf{R}_{yy}(\tau) = E\{\mathbf{y}(t+\tau)\mathbf{y}(t)\} = E\left\{\mathbf{\Phi}\mathbf{q}(t)\mathbf{q}'(t)\mathbf{\Phi}'\right\} = \mathbf{\Phi}\mathbf{R}_{qq}(\tau)\mathbf{\Phi}' \tag{8.53}$$

and the power spectrum that follows from the Fourier relationship (Wiener–Khintchine Theorem) [5]

$$\mathbf{S}_{yy}(\omega) = \mathcal{F}[\mathbf{R}_{yy}(\tau)]$$

as

$$\mathbf{S}_{yy}(\omega) = \mathbf{\Phi}\mathbf{S}_{qq}(\omega)\mathbf{\Phi}' \tag{8.54}$$

Decomposition of the *PSD* by applying the *SVD* technique is

$$\mathbf{S}_{YY}(\omega) = \mathbf{U}(\omega)\mathbf{\Sigma}(\omega)\mathbf{V}^\dagger(\omega) = \mathbf{U}(\omega) \times \mathbf{\Sigma}(\omega) \times \mathbf{U}^\dagger(\omega) \tag{8.55}$$

since $\mathbf{S}_{YY}(\omega) \geq \mathbf{0}$ and $\mathbf{U}(\omega) = [\mathbf{u}_1(\omega) \;\cdots\; \mathbf{u}_K(\omega)]^\dagger$ with $\boldsymbol{\Sigma}(\omega) = \text{diag}[\sigma_1(\omega) \;\cdots\; \sigma_K(\omega)] \ni \sigma_1 > \sigma_2 > \cdots > \sigma_K$.

This decomposition can be used to relate the *SVD* to the modal representation of Eq. (8.54) [35, 36]. For instance, the singular vectors are directly related to the mode shapes and the singular values are related to the modal responses. Therefore, these parameters can be used to define equivalent single degree-of-freedom (*SDOF*) systems. Near a resonance (ω_o), the first singular value corresponds to the dominant mode at that frequency. The rank of the output *PSD*-matrix at a particular frequency indicates the number of modes (order) contributing to the structural response (see Eq. (8.52)).

Assuming that a single dominant mode exists at ω, then the *SDOF* system in terms of the *SVD* parameters is

$$\mathcal{F}(\omega) = \frac{\overbrace{R_o}^{\boldsymbol{\Phi}_1 \ell_1}}{\underbrace{j\omega + \sigma_o - j\omega_o}_{\sigma_1}} + \frac{R_o^*}{\underbrace{j\omega + \sigma_o + \omega_o}_{\sigma_1^*}} \tag{8.56}$$

and

$$\begin{aligned} \sigma_o &= \zeta \omega_n \\ \omega_o &= \omega_n \sqrt{1 - \zeta^2} \end{aligned} \tag{8.57}$$

where ω_n is the *undamped natural frequency* and ζ is the *damping ratio* of the system,

8.2.1 Spectral Peaks: Picking/Histogram

Once the *FRF* or *PSD* are available, then spectral peak-picking and histograms are simple techniques to locate resonant frequencies, when the *SNR* is high. In this case, even damping can be estimated from the 3 dB-width of the spectrum at a distinct resonance.

Spectral *peak-picking* is based on estimating the *average* spectrum of the processed data and locating the peaks at each resonant frequency as well as their half-power bandwidth (damping) [30]

$$\omega_i = \max \;\mathcal{F}(\omega)|_{\omega=\omega_i} \qquad \text{[Peak-Picking]} \tag{8.58}$$

where $\{\omega_i\}$; $i = 1, \ldots, N_\omega$ is the set of resonant peaks.

Spectral histograms are obtained by selecting resonant peaks over a set of multichannel measurements (sensors) or an ensemble of single-channel measurements. The *spectral histogram* is based on estimating the batch power spectrum of the processed data, locating the peaks at each resonant frequency, and creating a frequency histogram [2]

$$H_{\Omega}(\omega) := p_{\Omega}(\omega_i) = \frac{n_{\omega_i}}{N \times \triangle_{\omega}}, \quad |\omega - \omega_i| \leq \frac{\triangle_{\omega}}{2} \qquad \text{[Spectral Histogram]} \tag{8.59}$$

where p is the estimated probability at the ith-peak, $\triangle_{\omega}$ is the bin-width, N_{ω} is the total number of estimated peaks.

8.2.2 Modal Assurance Criterion (*MAC*)

Once the resonant frequencies and damping parameters are available providing estimates of the underlying system eigenvalues (modes), estimates of the corresponding eigenvectors (mode shapes) can be extracted in the usual manner [30, 31]. A related correlation-based metric that provides a direct comparison of mode shapes is the *Modal Assurance Criterion (MAC)*. It is a statistical indicator enabling a measure of correlation or equivalency consistency between mode shapes that is important especially when comparing pairs of shape vectors derived from analytical models to those obtained from experimental measurements [37]. Similar to the correlation coefficient, it is bounded between 1 and 0, indicating a fully consistent (correlated) mode shape or complete lack of consistency (uncorrelated). However, in practice *MAC*-values greater than 0.9 are considered fully consistent, while those in the range of 0.1 – 0.2 lack consistency. Not to be confused with a true orthogonality metric, the *MAC* is still considered an reasonable approximation in practice.

Mathematically, the *MAC* is defined in terms of mode shapes (eigenvectors) $\boldsymbol{\phi}_i$ by

$$\mathcal{MAC}(i,j) := \frac{|\boldsymbol{\phi}_i^{\dagger} \times \boldsymbol{\phi}_j|^2}{|\boldsymbol{\phi}_i^{\dagger}\boldsymbol{\phi}_i| \times |\boldsymbol{\phi}_j^{\dagger}\boldsymbol{\phi}_j|} \tag{8.60}$$

$$\mathcal{MAC}(i,j) \approx \begin{cases} 1 \text{ if } i = j & \text{[Consistent]} \\ 0 \text{ if } i \neq j & \text{[Not Consistent]} \end{cases}$$

In the structural analysis application, the *MAC*-comparison is employed to [37]:

- *Match* modes;
- *Measure* against analytical mode shapes;
- *Estimate* test modal vector from a variety of excitation locations;
- *Estimate* shapes from various parameter estimation techniques; and
- *Modified* shapes after structural changes.

Typically, a *MAC*-matrix is constructed by estimating the metric for each mode, row, and columns, resulting in a diagonal matrix of ones and zeros elsewhere in the ideal case [37].

Next, we investigate criteria based on the "balanced" identification of state-space modal models from time domain data developed in Chapter 6.

8.2.3 Hankel/SVD Criteria

Recall the development of a variety of criteria for the identification of state-space models $\left(\Sigma_{ABC} = \{A, B, C\}\right)$ based on the *SVD* decomposition of the underlying Hankel matrix in Section 6.3 and the concept of a balanced realization. From systems theory, the Hankel data (impulse response) matrix can be factored into the controllability and observability matrices and an *SVD* performed, that is,

$$\mathcal{H}_{K,K} = \underbrace{\mathcal{O}_K \times \mathcal{C}_K}_{\text{Observability/Controllability}} = \underbrace{\mathcal{U}_K \times \Sigma_K \times \mathcal{V}_K'}_{SVD} = \underbrace{\mathcal{U}_K \Sigma_K^{\frac{1}{2}}}_{\mathcal{O}_K} \times \underbrace{\Sigma_K^{(\frac{1}{2})'} \mathcal{V}_K'}_{\mathcal{C}_K}$$

where the positive definite, diagonal matrix Σ_K is defined by

$$\Sigma_K := \text{diag}\left[\sigma_1, \sigma_2, \ldots, \sigma_K\right] \text{ for } \sigma_1 \geq \sigma_2 \geq \cdots \sigma_K$$

Recall that a *balanced realization* is given by the equality of the diagonalized Gramians (Section 6.2):

$$\Sigma_{AB}(K) = \Sigma_{AC}(K) = \Sigma_K$$

with the corresponding *Hankel singular values* obtained as the roots of the observability and controllability Gramian product

$$\sigma_{ABC} = \sqrt{\Sigma_{AC} \times \Sigma_{AB}}$$

The *Hankel norm* is a measure of system input/output energy that is given by the *maximum* Hankel singular value:

$$\|\mathcal{H}_{K,K}\|_2 = \max\{\sigma_i\} \qquad \text{for} \qquad i = 1, \ldots, K$$

and the *balanced similarity transformation* is specified in terms of the product matrix and its decompositions (Cholesky and *SVD*) as:

$$T_{\text{bal}} = \Sigma_{AB}^{\frac{1}{2}} \times V_{\mathcal{P}} \times \Sigma_{\mathcal{P}}^{-\frac{1}{2}} \quad \text{and} \quad T_{\text{bal}}^{-1} = \Sigma_{\mathcal{P}}^{-\frac{1}{2}} \times U_{\mathcal{P}}' \times \Sigma_{AC}^{\frac{1}{2}}$$

where we define the product matrix, in terms of its Cholesky square-root decomposition as well as its corresponding *SVD* as

$$\mathcal{P} = \Sigma_{AC}^{\frac{1}{2}} \times \Sigma_{AB}^{\frac{1}{2}} = U_{\mathcal{P}} \times \Sigma_{\mathcal{P}} \times V_{\mathcal{P}}' \tag{8.61}$$

with $\Sigma_{\bullet}^{\frac{1}{2}}$ the respective square-root matrices of the observability and controllability of Gramians.

Summarizing these metrics developed from the SVD of the Hankel matrix (Section 6.6) are:

- *Hankel order estimation (SVD, Best Rank)* of Table 8.1 $\Rightarrow \hat{N}_x$
- *Hankel singular values.* $\sigma_{ABC} = \sqrt{\Sigma_{AC} \times \Sigma_{AB}}$
- *Hankel norm* (energy). $\|\mathcal{H}_{K,K'}\| = \max\{\sigma_i\};\ i = 1, \dots, N_x$
- *SVD model error.* $\mathcal{M}_\epsilon = 2\sum_{k=\hat{N}_x+1}^{K} \sigma_k$
- *SVD model order estimation.* $\hat{N}_x = \max\left\{N : \log\hat{\sigma}_N > \left(\log\hat{\sigma}_{\max} + \log\hat{\sigma}_{\min}\right)\right\}e$

Based on these metrics and the concept of a balanced realization, the following criteria have evolved [38, 39].

8.2.4 Modal Observability Correlation (*MOC*) Criterion

Assuming that the state-space model is based on the *SVD*-decomposition and transformed (similarity) to a balanced realization $\left(\Sigma_{ABC}\right)$, then an efficient observability-based metric that provides a methodology to compare mode shapes similar to the *MAC* has been developed. The *Modal Observability Correlation (MOC) Criterion* is a method of comparing mode shapes through a correlation coefficient using the columns of the modal observability matrix, $\overline{\mathcal{O}}_{\mathrm{i}}$ estimated using a *balanced realization* model [38]:

$$\mathcal{MOC}(i,j) := \frac{|\overline{\mathcal{O}}_{\mathrm{i}}^{\dagger} \times \mathcal{O}_{\mathrm{j}}|^2}{|\overline{\mathcal{O}}_{\mathrm{i}}^{\dagger}\overline{\mathcal{O}}_{\mathrm{i}}|^2 \times |\mathcal{O}_{\mathrm{j}}^{\dagger}\mathcal{O}_{\mathrm{j}}|^2} \tag{8.62}$$

$$\mathcal{MOC}(i,j) \approx \begin{cases} 1 \text{ if } i = j \\ \\ 0 \text{ if } i \neq j \end{cases}$$

In summary, introducing the contribution of each mode in the *MOC*-relation, then two modes are highly correlated to each other, if they have almost the same contributions, that is, similar eigenvalues, eigenvectors, and shapes (≈ 1); therefore, the *MOC*-matrix is almost diagonal with some very small off-diagonal terms. Thus, the effect of each mode's contribution yields a high degree of correlation validating the correspondence of the estimated model to the simulated "truth" model [38].

8.2.5 Modal Singular Value (*MSV*) Criterion

When a signal/system is decomposed into individual modal components, the contribution to its total response is determined by the strength (amplitude) of each particular mode. The *Modal Singular Value* (*MSV*) metric characterizes the contribution of each mode to the total response as [39], that is, for the ith-mode with data $\ell \leq N$, we have:

$$\mathcal{MSV}(i) = \sqrt{|\hat{C}_i| \times \sum_{\ell=1}^{K} \hat{\lambda}^{\ell-1} \times |\hat{B}_i|} \approx \sqrt{\frac{|\hat{C}_i| \times |\hat{B}_i|}{1 - |\hat{\lambda}_i|}} \quad \text{for } i = 1, \ldots, N_m$$

where $\hat{C}_i$ and $\hat{B}_i$ are the estimated measurement and input transmission matrices provided by the particular estimation method applied and λ_i is the ith-eigenvalue (pole) of the system. Employing this metric for modal order reduction can be facilitated by investigating the superposition of these individual responses (see [39, 40] for more details).

8.2.6 Stabilization Diagram (*SDIAG*)

A *Stabilization Diagram* (*SDIAG*) is the graphical comparison of *identified* modal frequencies as a function of model order where the true eigen frequencies will *repeat* as the order is increased while mathematical (computational) modes will not maintain their consistency (repeatability) from model to model as the order is increased [30, 31]. This characteristic feature forms the basis of the SDIAG that is employed to estimate the modes (see Section 7.2).

Thus, the *SDIAG* provides a methodology to extract viable physical modes from data by classifying modal frequency and damping estimates as stable or unstable based on the closeness or small errors that physical (repeatable) modes exhibit. The error constraints[1] applied to determine modal closeness and consistency are:

$$SD(\omega, \zeta) \Rightarrow \begin{cases} \dfrac{\omega_k - \omega_{k-1}}{\omega_k} \leq \varepsilon_\omega \\ \dfrac{\zeta_k - \zeta_{k-1}}{\zeta_k} \leq \varepsilon_\zeta \end{cases}$$

where $\{\varepsilon_\omega, \varepsilon_\zeta\}$ is the corresponding error tolerance limits usually specified at 1%. A particular mode is classified as *stable*, if both the frequency and damping satisfy these constraints or if not, the modes are classified as *unstable* [41–43].

Thus, *SDIAG* is simply a plot of model order versus the frequencies and damping estimated at each of these orders with the motivation that a "true" physical mode of the system should appear with *consistent* frequency, damping, and mode shape at various orders, while spurious modes exhibit a more erratic appearance.

---

1 Another constraint usually incorporated in *SDIAG* is the *modal assurance criterion* for the comparison ($MAC \leq \epsilon_{MAC}$) of mode shapes (model versus estimates).

8.2.7 Modal Frequency Tracker

Another, more elegant approach, to tracking modal frequencies utilizes estimation theory culminating in the design of an optimal, least-squares solution—the Kalman filter. The modal frequency tracker design is a model-based processor that has been applied successfully in a wide variety of applications [2]. The underlying frequency estimator/tracker is based on the assumption that the frequency change is constant over the sampling interval ($\dot{f}(t_{k+1}) \approx \dot{f}(t_k)$) and the model uncertainty can be characterized by Gaussian process noise leads to the following set of discrete-time, Gauss–Markov stochastic equations that

$$\begin{aligned} f(t_{k+1}) &= f(t_k) + \Delta t_k \dot{f}(t_k) + w_1(t_k) && \text{[Frequency]} \\ \dot{f}(t_{k+1}) &= \dot{f}(t_k) + w_2(t_k) && \text{[Rate]} \end{aligned} \tag{8.63}$$

where w is zero-mean, Gaussian with $\mathbf{w} \sim \mathcal{N}(0, R_{ww})$. The corresponding measurement is also contaminated with instrumentation noise represented by zero-mean, Gaussian uncertainties as

$$y(t_k) = f(t_k) + v(t_k) \qquad \text{[Measurement]}$$

such that $v \sim \mathcal{N}(0, R_{vv})$.

A combination of both process and measurement systems can be placed in a discrete-time ($t_k \to t$), state-space framework by defining the state vector as $\mathbf{f}(t) := [f(t)|\dot{f}(t)]'$ giving the corresponding *Modal-Frequency Gauss–Markov model* of Eq. (8.63).

With this underlying frequency model established, we know that the optimal solution to the state estimation or frequency tracking problem is provided by the Kalman filter [2], that is,

$$\hat{\mathbf{f}}(t+1|t+1) = \hat{\mathbf{f}}(t+1|t) + K(t+1)e(t+1)$$

or component-wise (states), we have

$$\begin{aligned} \hat{f}(t+1|t+1) &= \hat{f}(t+1|t) + K_1(t+1)e(t+1) \\ \hat{\dot{f}}(t+1|t+1) &= \hat{\dot{f}}(t+1|t) + K_2(t+1)e(t+1) \\ \hat{y}(t+1|t) &= \hat{f}(t+1|t) \\ e(t+1) &= y(t+1) - \hat{y}(t+1|t) \end{aligned} \tag{8.64}$$

where $e(t)$ is the innovations/residual sequence and $K(t)$ is the gains or weights. This notation is defined by the conditional mean, $\hat{f}(t+1|t) := E\{f(t+1)|y(t), \ldots, y(0)\}$, that is, the estimate of $f(t+1)$ based on all of the available data up to time t.

Since we are primarily interested in a real-time application, we restrict the processor to reach steady-state in Eq. (8.64). That is, the Kalman gain ($K(t) \to K_{ss}$)

becomes a constant (steady-state) that can be *pre-calculated* directly from the discrete Riccati equation to give the frequency tracker relations [2].

This completes the discussion of physical metrics that can be applied directly to the signal/system decomposition problem, next we investigate a high-order modal frequency problem and apply some of these metrics.

8.3 Case Study: Resonant Modal *MCK* System

In this case study, we simulate a set of noisy uncertain data from a highly resonant system and apply some of the physical performance metrics to the estimated model. The *truth model* consists of the following set of modal frequencies

$$\{\lambda_i\} = \{180, 188, 190, 204, 217, 219, 247, 249, 276, 277, 322,$$
$$\times\, 326, 367, 798, 959, 1134\} \text{ Hz}$$

We synthesized the set of noise-free *MIMO* (3-inputs), truth data at 2%-damping-ratio for 45-outputs (measurements) at a sampling frequency of 3.2 KHz and estimated the corresponding power spectrum. The results are shown in Figure 8.3a where we observe the eigen frequencies selected by a

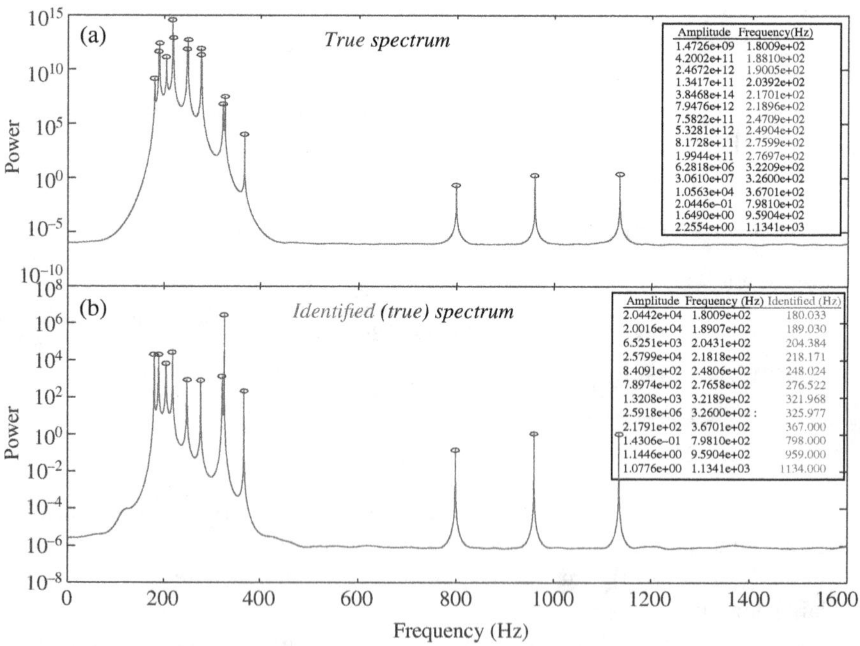

Figure 8.3 True/identified system: (a) True power spectrum (16 modes). (b) Identified power spectrum: peak-picking-parametric ID.

peak-picking algorithm shown in the inset. The "truth" frequencies are captured in the data. Next, we added zero-mean, uncorrelated (white) Gaussian noise at a standard deviation of 0.001 to the truth data and identified a 32-order, state-space model using the *N4SID* subspace algorithm of Section 6.5. The identification results are captured in the power spectrum of the estimated impulse response from the identified model and are shown in Figure 8.3b where we see the estimated eigen frequencies using both the peak-picking and the subspace identified model parameters. Note that the algorithms are not able to identify (resolve) the closely-spaced frequencies (e.g. 188 and 190 Hz).

Once the parametric model is identified, the spectral estimates of the multichannel temporal response data are performed channel-by-channel to validate the model performance. These results are shown in Figure 8.4 where we see that the identified batch spectrum basically overlays itself indicating consistency in Figure 8.4a along with the eigen-frequency histogram in Figure 8.4b. The histogram is a solid indicator of true rather than spurious frequencies available by the number of identified resonances at a given eigen frequency at each channel.

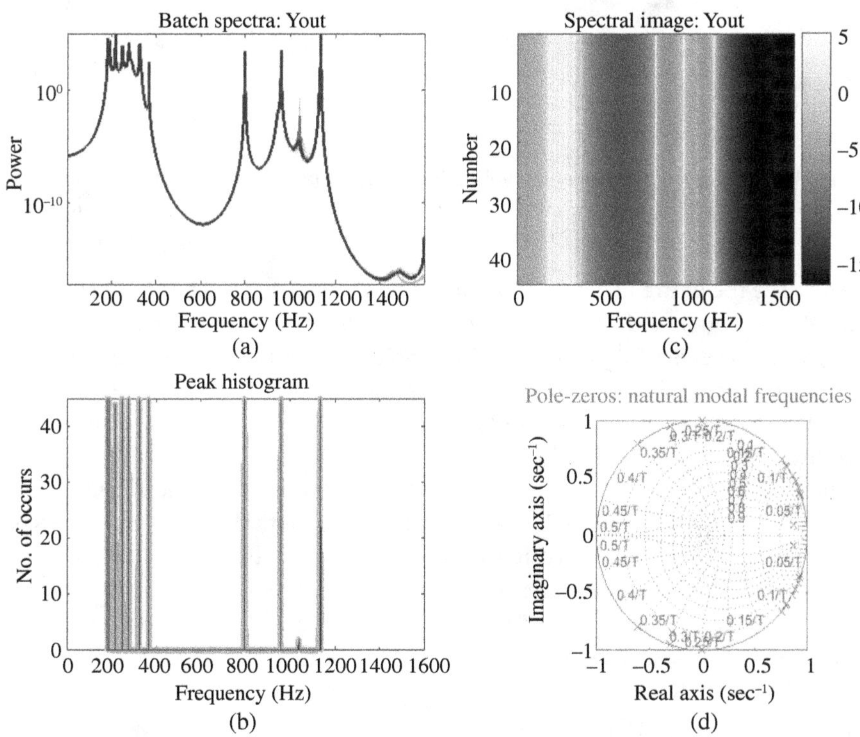

Figure 8.4 Identified system analysis (batch): (a) identified spectrum. (b) Peak-frequency histogram. (c) Batch spectrum image. (d) Pole-zero locations (unit circle).

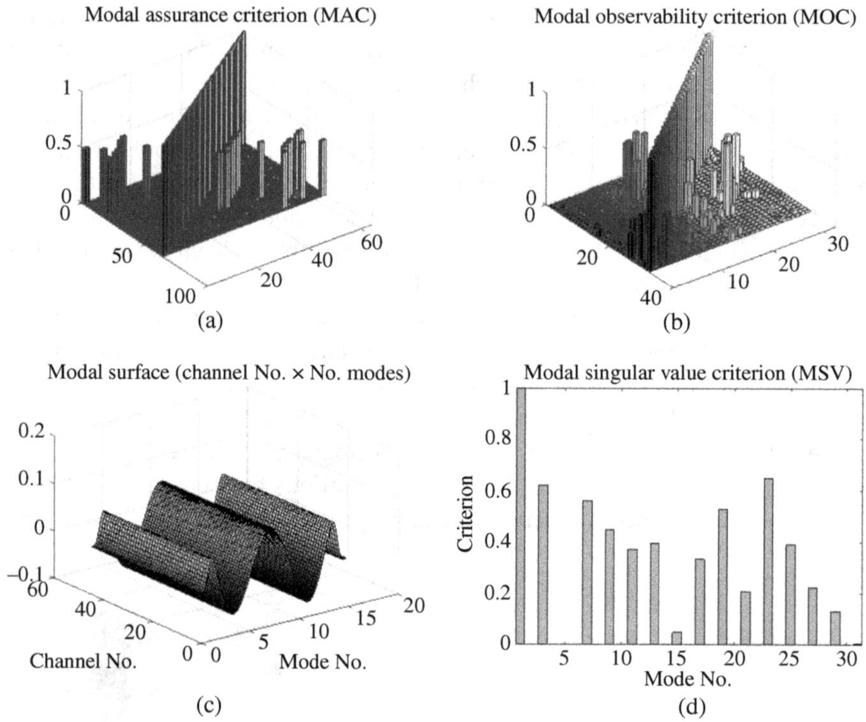

Figure 8.5 Identified system modal validation analysis: (a) modal assurance criterion. (b) Modal observability criterion. (c) Identified model surface (spatial). (d) Modal singular value criterion.

We observe the entire batch (45 channels) in an image defined by *channel (no.)* × *frequency* × *power* with the white vertical lines again indicating the consistency of a particular eigen frequency. Finally, a pole-zero plot is shown in Figure 8.4d with the corresponding continuous eigen-frequency annotated at a particular discrete pole position. All poles should be located on or within the unit circle for stability.

The modal analysis *validation* techniques are applied to the identified system with the results are shown in Figure 8.5 where we see both the *MAC* in Figure 8.5a and *MOC* in Figure 8.5b. Both criteria indicate a match between the "truth model" and identified modes as the strong *unit diagonal* with spurious modes of lower amplitude created by the uncertainty or slightly imprecise estimates. The estimated mode shape surface is shown Figure 8.5c along with the *MSV* criterion in Figure 8.5d completing the study.

8.4 Summary

A variety of performance metrics useful to test the qualitative/quantitative results of signal/system decompositions are developed in this chapter. Two metric approaches, one statistically motivated and the other physically motivated are discussed. The statistical-based approaches incorporate "sanity testing" metrics: the zero-mean (*Z-M*), whiteness (*W-T*), *WSSR*, and standard error tests. The identification order testing metrics consist of the *AIC*, *MDL* description, and the *FPE* along with the *SVD*-based methods. Physically-based metrics evolve primarily from the structural dynamics community with a suite of metrics involving modal decomposition and analysis. These approaches are applied and discussed in a case study based on highly resonant mechanical system model.

MATLAB® Notes

MATLAB® and its *Signal Processing Toolbox* (**signal**) can be used to perform whiteness testing using the Durbin–Watson statistical test (**dwtest**) along with the usual correlation command (**xcorr**). The *Statistics and Machine Learning Toolbox* provide a wealth of statistical commands ranging from basic calculations such as mean, median, standard deviation, variance, histogram, maximum likelihood estimator, and others (**mean**, **median**, **std**, **var**, **hist**, **mle** etc.) to the more sophisticated estimation methods like normalizing data evolving from the **zscore** to the Cholesky decomposition of a covariance matrix (**cholcov**) and robust estimation of the covariance (**robustcov**). More model-based metrics evolve from the *System Identification Toolbox* enabling the comparison of the identified model with measurement data (**compare**), error residual calculations (**residual**), prediction employing the identified model (**predict**) and (**forecast**) as well as goodness testing using the (**goodnessoffit**) command. The usual model order metrics are also available such as the final prediction error (**fpe**) and the Akaike information criterion (**aic**).

References

1 Anderson, B. and Moore, J. (1979). *Optimal Filtering*. Englewood Cliffs, NJ: Prentice-Hall.

2 Candy, J. (2006). *Model-Based Signal Processing*. New Jersey: Wiley/IEEE Press.

3 Wilsky, A.S. (1976). A survey of design methods for failure detection in dynamic systems. *Automatica* 12 (6): 601–611.

4 Mehra, R.K. and Peschon, J. (1971). An innovations approach to fault detection and diagnosis in dynamic systems. *Automatica* 7 (5): 637–640.

5 Candy, J. (2019). *Model-Based Processing: An Applied Subspace Identification Approach*. New Jersey: Wiley/IEEE Press.

6 Orfanidis, S. (1985). *Optimum Signal Processing*. New York: MacMillan.

7 Martin, W.C. and Stubberud, A.R. (1976). The innovations process with application to identification. In: *Control and Dynamic Systems* (ed. C.T. Leondes). New York: Academic Press.

8 Schweppe, F. (1973). *Uncertain Dynamic Systems*. Englewood Cliffs, NJ: Prentice-Hall.

9 Bendat, J. and Piersol, A. (1980). *Engineering Applications of Correlation and Spectral Analysis*. New York: Wiley.

10 Bendat, J. and Piersol, A. (2010). *Random Data: Analysis and Measurement Procedures*, 2e. New York: Wiley.

11 Candy, J. (2016). *Bayesian Signal Processing, Classical, Modern and Particle Filtering*, 2e. Hoboken, NJ: Wiley/IEEE Press.

12 Akaike, H. (1974). A new look at the statistical model identification. *IEEE Trans. Autom. Control* 19 (6): 716–723.

13 Sakamoto, Y., Ishiguro, M., and Kitagawa, G. (1986). *Akaike Information Criterion Statistics*. Boston, MA: D. Reidel/Kluwer Academic.

14 Ljung, L. (1987). *System Identification: Theory for the User*. Englewood Cliffs, NJ: Prentice-Hall.

15 Ljung, L. and Soderstrom, T. (1983). *Theory and Practice of Recursive Identification*. Cambridge, MA: MIT Press.

16 Soderstrom, T. and Stoica, P. (1989). *System Identification*. New York: Academic Press.

17 Kay, S. (1988). *Modern Spectral Estimation*. New Jersey: Prentice-Hall.

18 Hayes, M. (1996). *Statistical Digital Signal Processing and Modeling*. New York: Wiley.

19 Haykin, S. (1993). *Adaptive Filter Theory*. Englewood Cliffs, NJ: Prentice-Hall.

20 Jategaonkar, R., Raol, J., and Balakrishna, S. (1982). Determination of model order for dynamical systems. *IEEE Trans. Syst. Man Cybern.* 12 (1): 56–62.

21 van Overschee, P. and De Moor, B. (1996). *Subspace Identification for Linear Systems: Theory, Implementation, Applications*. Boston, MA: Kluwer Academic Publishers.

22 Verhaegen, M. and Verdult, V. (2007). *Filtering and System Identification: A Least-Squares Approach*. Cambridge: Cambridge University Press.

23 Katayama, T. (2005). *Subspace Methods for System Identification*. London: Springer.

24 Pham-Gia, T. and Hung, T. (2001). The mean and median absolute deviations. *Math. Comput. Modell.* 34: 921–936.

25 Zoubir, A. and Iskander, D. (2004). *Bootstrap Techniques for Signal Processing*. Cambridge University Press, Cambridge.

26 Candy, J. and Zicker, J. (1987). Electromagnetic signal processing: an estimation identification application. *Automatica* 23 (2): 175–187.

27 Bevensee, R., Deadrick, F., Miller, E., and Okada, J. (1977). Validation and calibration of the LLL transient electromagnetic measurement facility. *LLNL Report*, UCRL-52225.

28 Miller, E. (ed.) (1986). *Transient Electromagnetic Measurements*. New York: Van Nostrand.

29 Baum, C. (1971). On the singularity expansion method for the solution of EM interaction problems. *AFWL Interaction Note 88*.

30 Reyenders, E. (2012). System identification methods for modal analysis: review and comparison. *Arch. Comput. Meth. Eng.* 19 (1): 51–124.

31 Rainieri, C. and Fabbrocino, G. (2014). *Operational Modal Analysis of Civil Engineering Structures*. New York: Springer.

32 Gawronski, W. (2004). *Advanced Structural Dynamics and Active Control of Structures*. New York: Springer.

33 Migliori, A. and Sarrao, J. (1997). *Resonant Ultrasound Spectroscopy: Applications to Physics, Materials Measurements, Nondestructive Evaluation*. New York: Wiley.

34 Avitabile, P. (2017). *Modal Testing: A Practitioner's Guide*. New York: Wiley.

35 Brinker, R., Zhang, L., and Andersen, P. (2000). Modal identification from ambient responses using frequency domain decomposition. In: *Proceedings of the 18th International Modal Analysis Conference (IMAC)*, 625–630.

36 Brinker, R., Andersen, P., and Zhang, L. (2001). Modal identification of output-only systems using frequency domain decomposition. *Smart Mater. Struct.* 10: 441–445.

37 Pastor, M., Binda, M., and Harcarik, T. (2012). Modal assurance criterion. *Procedia Eng.* 48: 543–548.

38 Yaghoubi, V. and Abrahamson, T. (2013). The modal observability correlation as a modal correlation metric. Topics in Modal Analysis, Volume 7: Proceedings of the 31st IMAC, A Conference and Exposition on Structural Dynamics.

39 Juang, J. (1994). *Applied System Identification*. Upper Saddle River, NJ: Prentice-Hall PTR.

40 Juang, J. and Pappa, R. (1985). An eigensystem realization algorithm for modal parameter identification and model reduction. *J. Guidance* 8 (5): 620–627.

41 Van der Auweraer, H. and Peeters, B. (2004). Discriminating physical poles from mathematical poles in high order systems: use and automation of the stabilization diagram. In: *Proceedings of the Instrumentation and Measurement Technology Conference (IMTC) 2004*, 2193–2198.

42 Verboven, P., Guillaume, P., Cauberghe, B. et al. (2003). Stabilization charts and uncertainty bounds for frequency-domain linear least squares estimators. PhD thesis. Vrije Universiteit Brussel.

43 Wu, C., Liu, H., Qin, X., and Wang, J. (2017). Stabilization diagrams to distinguish physical modes and spurious modes for structural parameter identification. *J. Vibroeng.* 19 (4): 2777–2794.

9

Applications

9.1 Modal Decomposition: Sounding Rocket Flight

Sounding rocket testing is well-founded providing a reasonable approach to evaluate the performance of both sensor suites and processing techniques, especially in a noisy, complex, real-time environment [1–5]. A typical sounding rocket along with an accompanying experimental test bed is shown in Figure 9.1. The payload consists of a set of "deck plates" enabling the inclusion of a variety of self-contained experiments that communicate and extract power from the available rocket avionics as illustrated in Figure 9.1a. The corresponding experimental unit consists of the following instrumented components: deck plate, enclosure, circuit boards, and CREPE (thin plate) as shown in Figure 9.1b,c. It was designed to test the performance of miniature *MEM* accelerometers with accompanying analog-to-digital (A/D) converters along with gathering vibrational response data from a drum-like, *CREPE* attached below the deck plate housing the experiment. Accelerometer data were digitized, collected, and telemetered to the ground monitoring station for eventual processing.

9.1.1 Experimental Test Unit Design and Analysis

In this section, we discuss the development (design) of the experimental unit with the compartmental deck fixture including the thin plate (CREPE) mounted under the deck plate of the sounding rocket in order to provide a significant modal response to the various flight stages to follow. Simulation results of the "free" CREPE responses using COMSOL® are shown in Figure 9.2 where the structure is excited by a swept chirp input between 35 and 1250Hz [6, 7]. The structural design is shown in Figure 9.2a along with the predicted mode shapes in Figure 9.2b with light and sharp dark lines indicating maxima/minima amplitudes in Figure 9.2c with the asterisks ($*$) denoting the accelerometer positions. We observe the rich spectral response created by the drum-like resonances offering a well-designed

Signal Processing: An Applied Decomposition Approach, First Edition. James Vincent Candy.

Companion website: www.wiley.com/go/CandySignProc

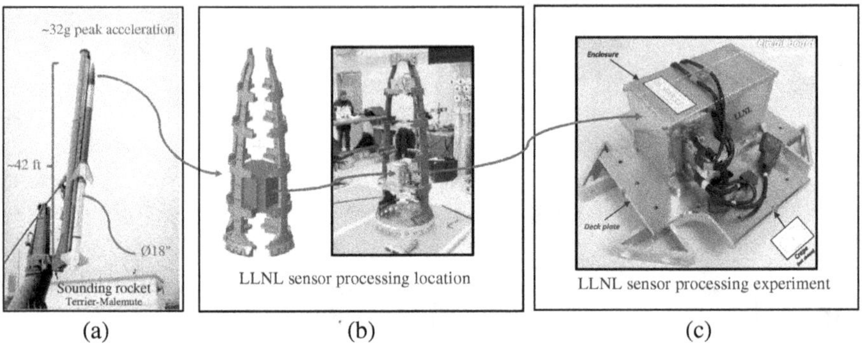

Figure 9.1 Sounding rocket experiment: (a) Sounding rocket (Terrier-Malemute) two-stage rocket assembly. (b) Payload with various decks for experiments. (c) Processing experimental unit with enclosed electronics and sensors (deck plate, enclosure, circuit boards, and CREPE).

test object to evaluate the sensor hardware acquisition system as well as the proposed signal processing to extract the modal frequencies. Besides predicting the modal frequencies, the simulation provides the response (time series) data for processing. The identified modal frequencies are shown in Figure 9.2d which are used as the "true" values in the figures and tables to follow.

In order to validate the COMSOL model response, a ground test unit is developed to excite the entire structure. Here, we measure CREPE accelerometer outputs along with their corresponding power spectra. Results from the ground test validate sensor performance along with the modal response (frequencies) of the CREPE and the signal processing techniques are utilized to extract them.

9.1.2 Sounding Rocket Flight Testing

The results of the sounding rocket flight are discussed in this section, where we consider a set of various processing options. First, we simply investigate the entire flight record and process it in its entirety to observe the overall spectral content. Next, we examine each of the individual component responses of the experimental unit. Finally, we investigate both multiple input/multiple output (*MIMO*) processing and compare it to output-only processing (ignoring the input measurements) along with both single/multichannel approaches [8–13].

The overall test flight data is shown in Figure 9.3 where we show the average (Z-direction) raw data in Figure 9.3a incorporating *all* of the various flight-stages. Here, the objective was to observe the frequency content and validate the operation of our *MEM* accelerometers as well as the pre-processing performed. In Figure 9.3b, we see the processed input/output responses from the deck plate/CREPE measurements with the corresponding power spectra in Figure 9.3c

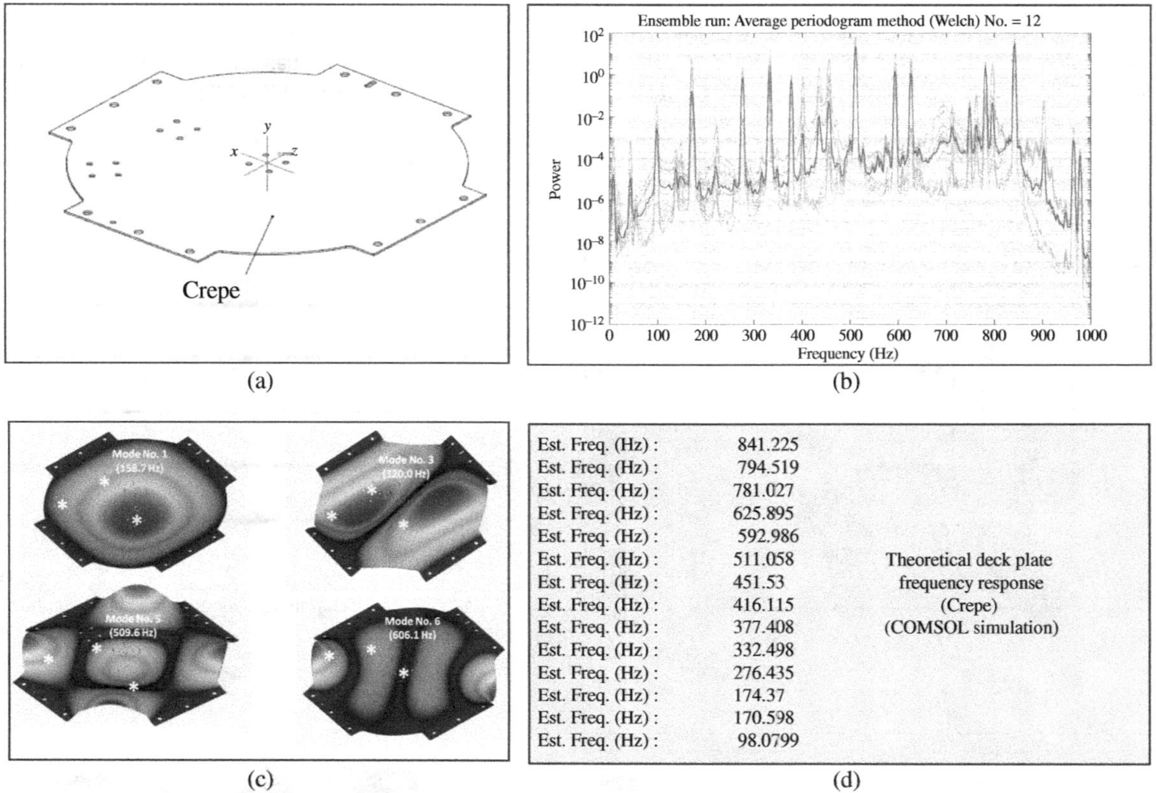

Figure 9.2 CREPE simulation (COMSOL) results for chirp (35-to-1250 Hz): (a) CREPE assembly. (b) Mode shapes/frequencies and sensor locations (*). (c) Ensemble modal frequency spectra (light gray) and average (dark gray). (d) CREPE (thin plate) modal frequencies.

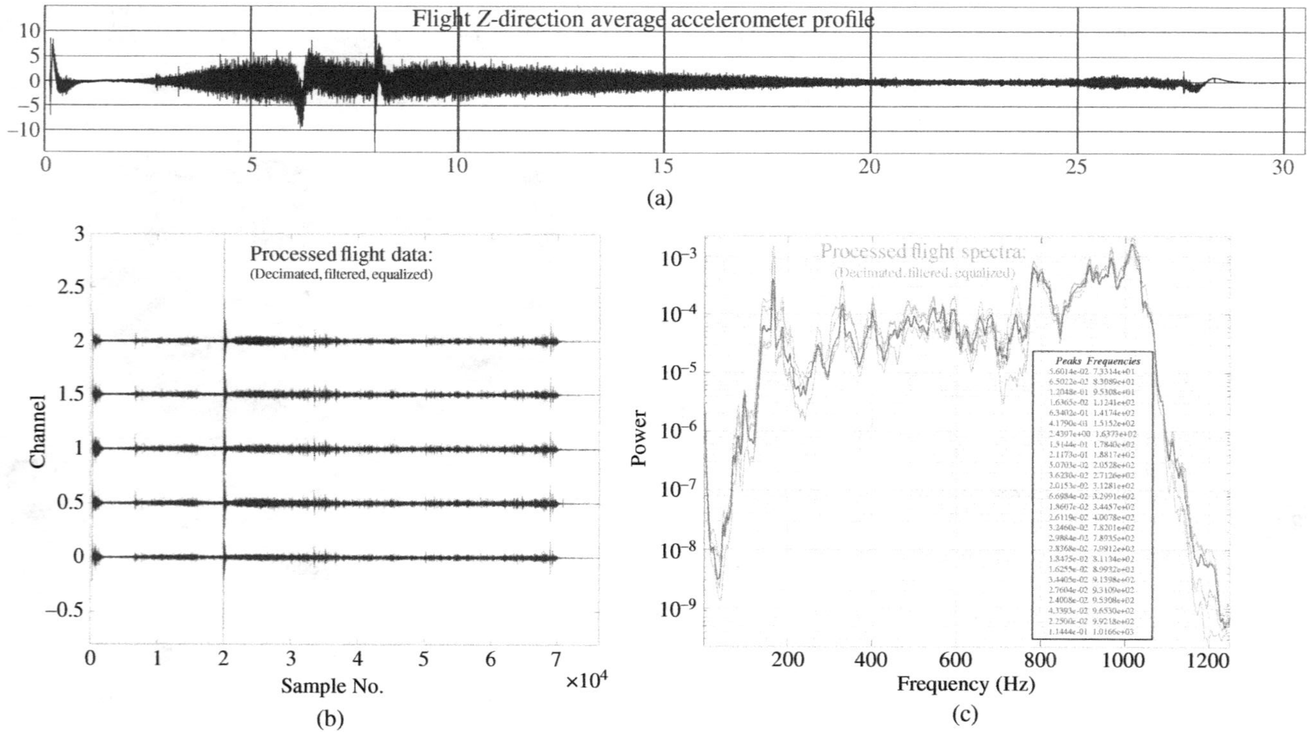

Figure 9.3 Sounding rocket flight test data: (a) Raw (average) flight data. (b) Processed input (deck plate). (c) Output (CREPE) power spectra: ensemble and average with peak frequencies (list).

along with the list of resonant peaks. The results show that the sensor measurements and processor designs are valid and capable of extracting the modal frequencies of interest. Next, we investigate the spectral response of the CREPE over various stages of the sounding rocket flight as spectra and raw tracks are illustrated in Figure 9.4. Again using a smoothed spectral estimator, we observe a very consistent response of the CREPE test object during each of the stages as can be observed by the overlapping resonant peaks. These results indicate that any of the stage data are viable for processing and extracting modal frequencies. Also, if any "changes" occur during flight, they can be detected by comparing the stage spectra.

An average spectral comparison of the raw accelerometer measurements for both deck plate and CREPE is displayed along with the CREPE COMSOL "truth" data for signal processing operations including *MIMO* and output-only techniques using both the subspace (see Section 6.4) as well as single input/single output (*SISO*) lattice (Burg) (see Section 5.4) techniques on each channel individually [9–13]. The CREPE spectral results for the sounding rocket flight are shown in Figure 9.5 with the deck plate included for comparison as well as the synthesized (COMSOL) truth spectra. Both deck plate (solid line) and crepe (dashed line) raw spectra are shown on the upper portion of the figure with the identified spectra consisting of large peaks below. The true (COMSOL) spectra are characterized by strong resonance peaks (dashed), while the weaker peak resonances of the output-only identifications using an unweighted principal component (*UPC*) state-space identifier are shown (*UPC* (thick dashed) and Burg lattice (thick solid)) as well [12, 14]. It is interesting to note that the identified modal frequencies (squares) from the *UPC* algorithm align quite well with the true (COMSOL) resonances as well as the corresponding spectral estimates [7]. It is also clear from the figure that the raw average spectra of the deck plate and CREPE data overlap somewhat in the low-frequency band (150–200 Hz) with higher frequency resonances occurring in the CREPE. Comparing the COMSOL® simulations with the identified spectra, especially the extracted modal frequencies ("squares"), it is noted that most of the frequencies are extracted after processing. Comparing the output-only *UPC* to the single channel lattice (dashed to solid) indicates that the lattice with high order (25 modes) is capable of extracting the dominant modal frequencies as illustrated in the figure.

We observe the quantitative results including the COMSOL truth data (peak detection) for the input/output, output-only and single-channel processors. The *input/output* processor uses the measured deck plate accelerometer data (rigid body vehicle modes) as input exciting the CREPE, while the *output-only* processors for both single/multichannel processors are also shown. What is somewhat surprising is the ability of the single channel processor to extract the modal frequencies almost as well as the multichannel processors. However, since we

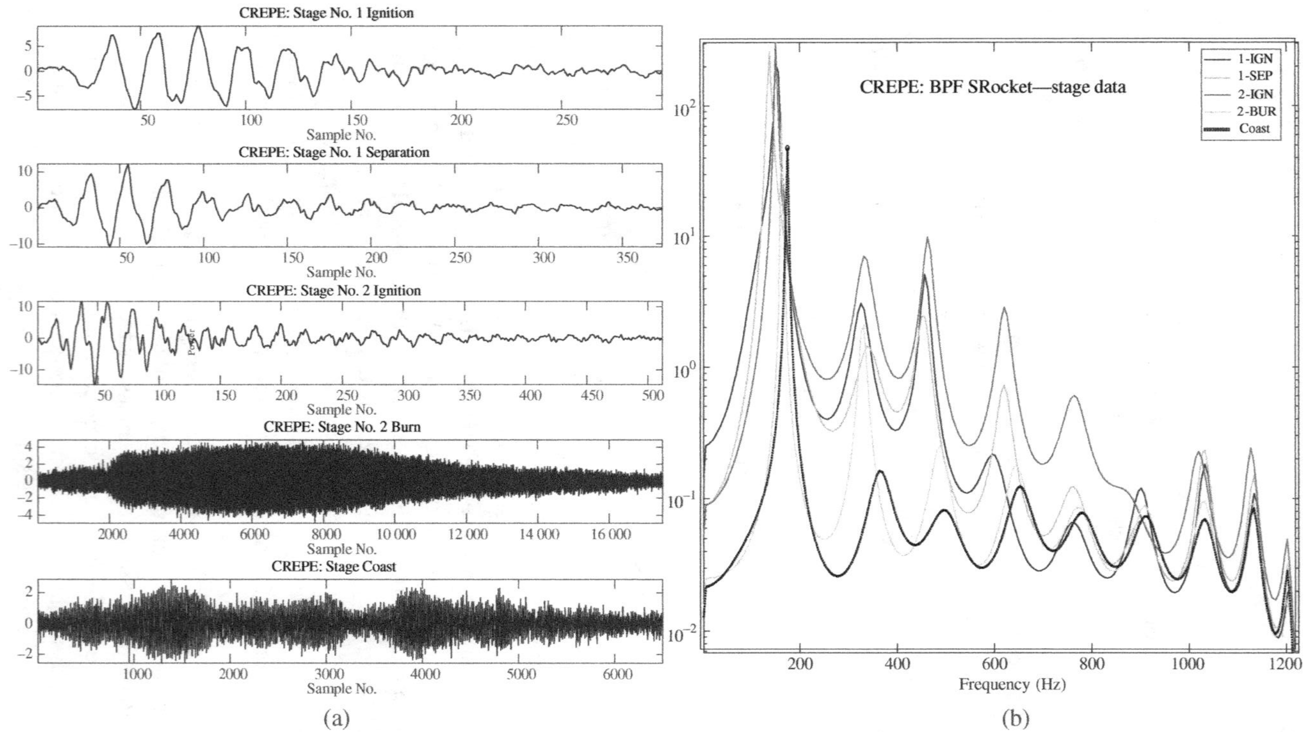

Figure 9.4 CREPE flight test stage spectra: (a) Extracted stage processed (decimated, band-pass filtered) vibrations for first-stage ignition, first-stage separation, second-stage ignition, second-stage burn-out, and terminal coast. (b) Average spectral estimates.

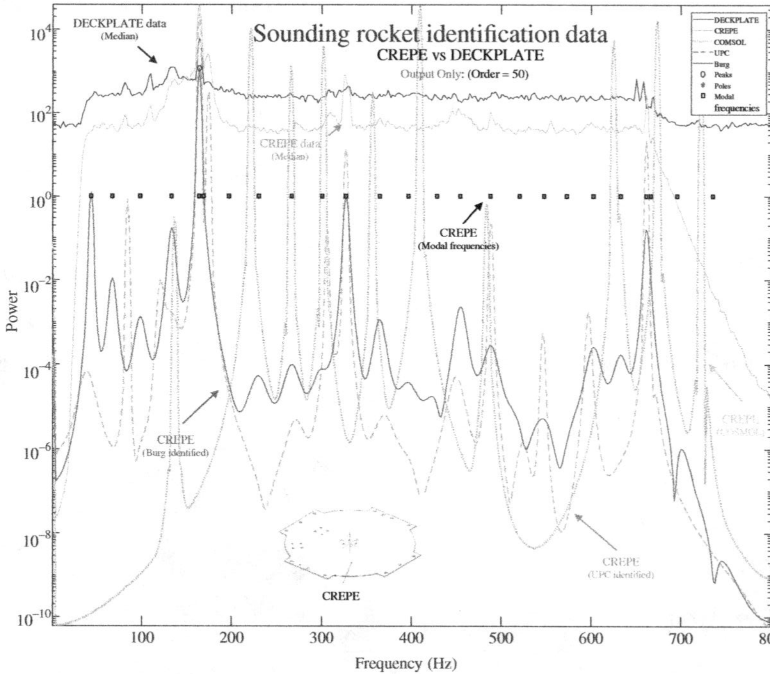

Modal frequencies

(I/O)	(TRUE)	(O/O-UPC)	(O/O-Burg)
Frequency (Hz) : 800	(747.90)	----------	735.552
Frequency (Hz) : 667.752		668.151	695.898
Frequency (Hz) : 661.135		661.323	666.444
Frequency (Hz) : 649.221		651.180	661.438
Frequency (Hz) : 627.376	(626.3)	627.919	633.207
Frequency (Hz) : 596.553	(593.2)	552.950	602.436
Frequency (Hz) : 546.069		544.306	547.867
Frequency (Hz) : 524.668	(511.3)	522.971	520.574
Frequency (Hz) : 488.265		488.062	488.237
Frequency (Hz) : 449.281	(454.9)	451.956	454.126
Frequency (Hz) : 400.455		----------	396.787
Frequency (Hz) : 370.205	(377.6)	372.472	364.830
Frequency (Hz) : 327.04	(332.6)	330.820	327.350
Frequency (Hz) : 308.955		307.784	----------
Frequency (Hz) : 305.884		304.533	300.931
Frequency (Hz) : 271.181	(276.6)	270.021	266.682
Frequency (Hz) : 243.53		----------	----------
Frequency (Hz) : 174.28	(174.4)	174.686	168.864
Frequency (Hz) : 169.323	(170.7)	163.819	164.018
Frequency (Hz) : 163.932		158.281	----------
Frequency (Hz) : 131.584		144.673	133.133
Frequency (Hz) : 120.334		133.932	----------
Frequency (Hz) : 84.2209	(98.0)	78.369	98.071
Frequency (Hz) : 38.9473	(43.4)	40.722	43.769
Frequency (Hz) : 38.4381		----------	----------

Figure 9.5 Sounding rocket flight test data results: Average power spectra for raw accelerometer measurements and *MIMO* and output-only data for (I/O) *N4SID*, *UPC*, and Lattice algorithms.

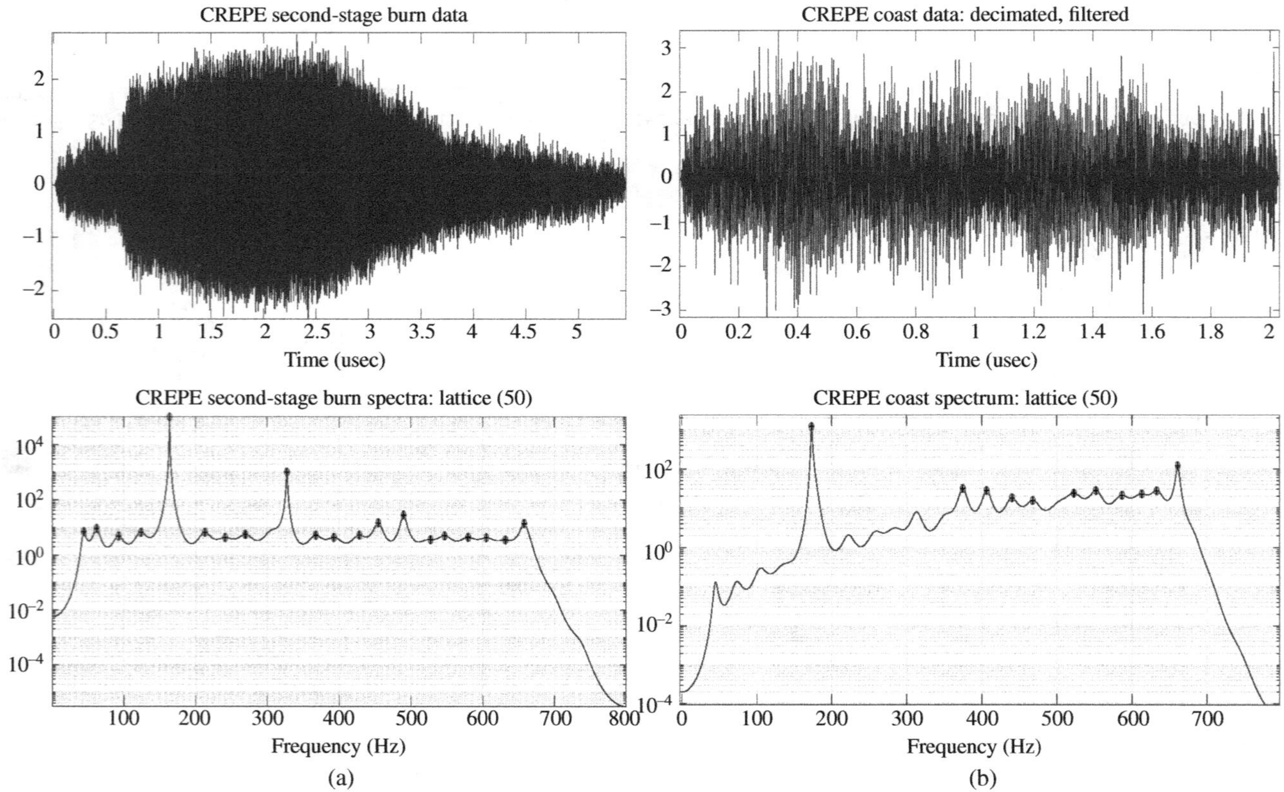

Figure 9.6 Sounding rocket flight test data for second-stage burn-out and terminal coast data: (a) Average burn-out data and spectra. (b) Average terminal coast data and spectra.

only have two viable channels available, then the modes appear on both as weak. Here, coupling is not an issue nullifying the strength of the *MIMO* processors.

We analyze the performance of the processors during two stages of the sounding rocket flight: (i) second-stage burn-out and (ii) terminal coast prior to impact for measuring the CREPE accelerometer data. We chose to average the multichannel data to compare the performance of the *SISO* techniques with those of the *MIMO* (I/O processor). These data are shown in Figure 9.6 along with their corresponding power spectra for the burn-out in Figure 9.6a and coast in Figure 9.6b. Here we used a large order ($N = 50$) model corresponding to 25-modes in extracting the CREPE modal frequencies. We applied "output-only" processors: (i) I/O (*N4SID*) processor; (ii) *UPC* processor; and (iii) lattice (Burg) processor to the average CREPE data. A summary of the spectral estimation results for all three processors investigated is shown in Figure 9.7 for both datasets: second-stage burn-out, terminal coast. The accompanying list is from the I/O (*N4SID*) processor estimates with the modal frequencies corresponding to those closest to the true (COMSOL) values predicted. The lattice and *UPC* methods perform similarly, while the I/O processor provides much better signal-to-noise ratio (*SNR*) capability as indicated by spectrum and the detected modal frequencies extracted.

Quantitatively, the modal frequency estimates for the second-stage burn-out data are given in Table 9.1 with those for the terminal coast in Table 9.2. The modal frequency estimates are as expected with those using the "full" I/O processor resulting in an average 1.8% deviation with an average relative error (modal frequency) of 2.4%. Both the *UPC* and lattice algorithms also performed reasonably well with relative errors below 5%. However, in some cases, the modal frequency standard deviations were high, especially for the lattice. The terminal coast data results are somewhat deceiving being skewed by some large frequency errors. In this case, the relative errors of the I/O and *UPC* techniques are high, but all lie below the 5% threshold. In this case, the frequency deviations are quite reasonable with the lattice estimates not quite as good. Part of this problem is that the number of samples available during this part of the flight is limited.

9.1.3 Summary

In this application, we have investigated the performance of robust single/multichannel spectral estimation techniques for single input/single output (*SISO*) as well as *MIMO* channel data using both input/output *N4SID* and output-only *UPC* processors. This choice for extracting the expected modal frequencies from noisy measurement data has the potential for real-time application and also provides a "data compression" feature that is desirable for bandlimited telemetry systems. The *MIMO*-data is processed using the identified state-space model of the mechanical system by both input/output as well as output-only

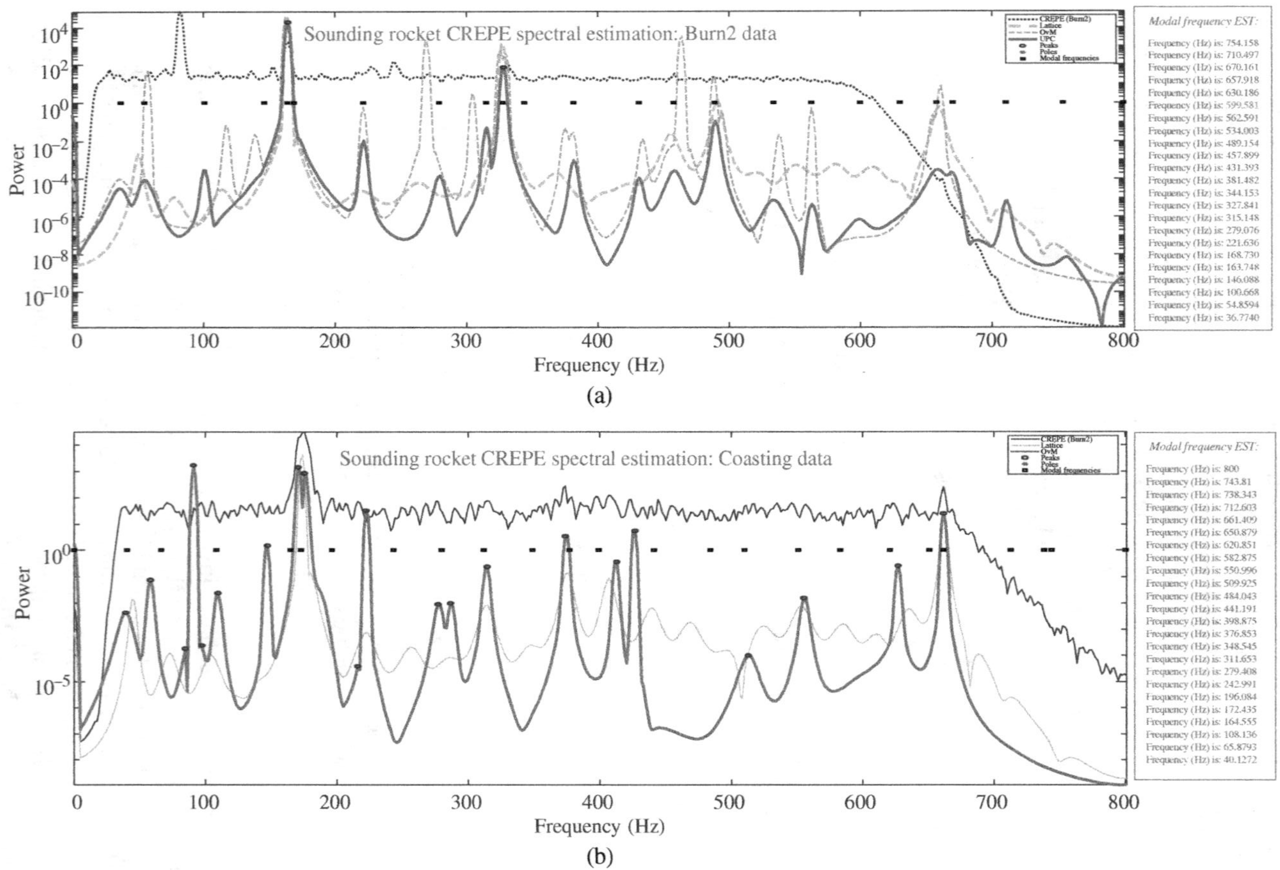

Figure 9.7 Sounding rocket flight test spectra for second-stage burn-out and terminal coast data: (a) Burn-out power spectrum and list of I/O (*N4SID*) modal frequencies. (b) Terminal coast power spectrum and list of I/O (*N4SID*) modal frequencies.

Table 9.1 MODAL identification/timing statistics: sounding rocket flight second-stage burn-out data.

MODAL frequency estimates (error deviation/relative error ($\pm\sigma$ (ϵ))			
Frequency (Hz)	**I/O (*N4SID*)**	**UPC**	**Lattice**
747.9	769.6±3.2 (2.9)	789.7±0.5 (5.6)	—
626.3	625.8±2.0 (0.1)	622.8±0.7 (0.5)	628.0±7.1 (0.3)
593.2	595.2±0.8 (0.3)	592.3±2.6 (0.2)	590.0±11.4 (0.5)
511.3	497.6±0.7 (2.7)	496.8±0.8 (2.8)	496.9±6.4 (2.8)
377.6	370.4±0.7 (0.9)	367.1±0.9 (2.8)	383.9±7.1 (0.5)
332.6	330.9±0.1 (0.5)	329.7±0.2 (0.9)	314.1±6.4 (5.6)
276.6	270.4±2.6 (2.2)	266.9±1.4 (3.5)	288.9±9.9 (4.4)
174.4	—	—	—
170.7	165.6±0.0 (3.0)	165.5±0.0 (3.0)	166.3±5.4 (2.6)
98.0	87.4±1.2 (10.8)	105.6±4.2 (7.8)	108.8±1.2 (11.0)
43.4	44.8±1.0 (3.2)	50.8±1.3 (17.1)	49.6±0.8 (14.2)
Avg $\sigma(\%\epsilon)$	±1.8 (2.4%)	±1.5 (3.9%)	±6.2 (4.0%)

Table 9.2 MODAL identification/timing statistics: sounding rocket flight terminal coast data.

MODAL frequency estimates (error deviation/relative error ($\pm\sigma$ (ϵ))			
Frequency (Hz)	**I/O (*N4SID*)**	**UPC**	**Lattice**
747.9	785.0±1.1 (5.0)	785.6±5.1 (5.0)	—
626.3	615.3±1.0 (1.8)	626.8±0.7 (0.0)	618.7±17.0 (1.2)
593.2	589.8±0.5 (0.6)	598.7±0.0 (0.9)	597.1±11.1 (0.7)
511.3	519.2±0.2 (1.6)	522.1±4.2 (2.1)	498.2±13.0 (2.6)
454.9	444.1±0.1 (2.4)	449.1±1.1 (1.8)	456.3±10.7 (0.3)
377.6	372.1±0.2 (1.5)	371.3±0.2 (1.7)	393.9±9.9 (4.3)
332.6	329.7±4.1 (0.9)	334.7±2.2 (2.6)	353.9±11.7 (6.4)
276.6	259.9±0.6 (6.0)	277.6±3.2 (0.4)	260.7±3.2 (5.7)
174.4	175.2±0.8 (0.5)	173.8±0.1 (0.3)	174.3±0.6 (0.1)
170.7	164.2±3.7 (3.8)	162.4±1.8 (8.3)	—
98.0	93.6±2.6 (4.5)	115.4±2.7 (17.8)	105.8±0.5 (8.0)
43.4	54.2±0.5 (24.9)	46.8±0.8 (7.3)	43.9±0.3 (1.1)
Avg $\sigma(\%\epsilon)$	±1.3 (4.5%)	±1.6 (4.0%)	±7.8 (3.0%)

processors. Each of these approaches has a variety of trade-offs, especially in the development of a real-time system [14–17].

We started by describing the sounding rocket experiment and objectives of (i) sensor test/evaluation and (ii) the development of processing techniques for this application. After performing theoretical studies of the experimental test unit including the outputs from a sophisticated finite element (COMSOL®) from synthesized time series, a ground test unit was developed to validate the theoretical models and test the processing techniques prior to flight testing. The majority of the structural modal frequencies were extracted from the noisy measurement data for both *SISO*/*MIMO* processors yielding excellent results.

9.2 Vibrational Response of a Cylindrical Structure: Identification and Modal Tracking

In this section, we study the application of the subspace identification of a vibrating cylindrical structure excited by a random input, specifically an impact device (instrumented hammer). The investigation of the response of this cylindrical object in a noisy environment (a pipe in air), to impact excitations provided by a cam-driven, hammer-like surface, randomly excited by an electric motor is of interest in order to develop a real-time processing system.

For the cylinder, we have the following "theoretical" (analytical) modal frequencies:

$$f_i = \{51.7, 143.5, 281.3, 465.0, 694.7, 970.2, 1291.7, 1435.3, 2813.1 \text{ Hz}\}$$

A simple free–free experiment for this application is shown in Figure 9.8 where we observe the 60 in. (1.125 in. OD) copper pipe along with the excitation system (cam-driven hammer) signal conditioner (10 KHz filter), accelerometer, and digitizing oscilloscope. Data were sampled at 10 KHz and decimated to 5 KHz yielding a 2.5 KHz Nyquist frequency for spectral analysis.

The basic approach is to excite the pipe with the repetitive hammer blows at a high rate and measure its response. Once the noisy data are digitized, they are pre-processed by bandpass filtering (eighth order Butterworth) between 40 and 1.25 KHz followed by a third order equalization (whitening) filter to enhance the low-band modal frequencies. The spectrum is estimated along with the corresponding spectral peaks to extract the modal frequencies. We investigated a variety of estimators to perform spectral decompositions ranging from the simple discrete Fourier transform (*FFT*) to the sophisticated MUltiple SIgnal Classification (*MUSIC*) approach requiring a singular value decomposition (*SVD*) of a Toeplitz covariance matrix [13] (see Section 3.1). Once the spectral peaks are extracted *locating* the corresponding modal frequencies, they are either provided

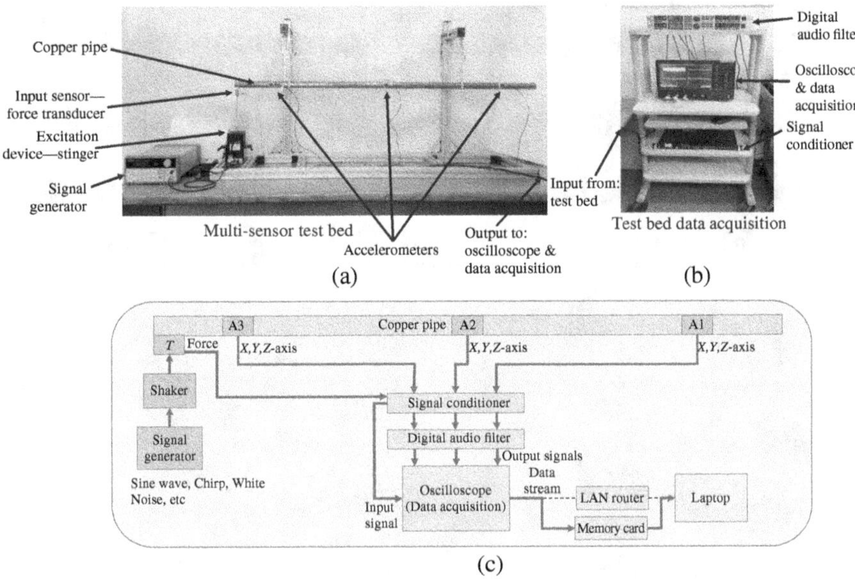

Figure 9.8 Cylindrical object (copper pipe) vibrational response experiment: (a) pipe, cam-driven hammer excitation, tri-axial accelerometer, signal conditioner (filter), data acquisition system (digitizing oscilloscope). (b) data acquisition system (oscilloscope). (c) experimental system diagram.

as input to a frequency tracker for enhancement and/or simply stored for eventual probability density function (*PDF*) estimation using a histogram or smoothed kernel density estimator for monitoring and eventual detection.

A *MIMO* experiment was performed as before demonstrating the capability of extracting the modal frequencies of high interest. Typical pre-processed response data are shown in Figure 9.9 with the acquired signals (*XYZ*-channels) in Figure 9.9a and the corresponding power spectra in Figure 9.9b. Here, we observe the modal frequencies shown at the major spectral peaks capturing the dominant resonances. We are primarily focused on the first six (6) dominant modal frequencies to characterize the cylinder.

The *MIMO* subspace identification algorithm (*N4SID*) is applied to the response data and the results are summarized in Figure 9.10. In Figure 9.10a, we see the estimated impulse response (average) and its corresponding power spectrum. We observe the "average" (identified) power spectrum in Figure 9.10b as well as the extracted modal (poles) frequencies (squares) from the identified model system matrix ($\hat{A}$). The *SVD*-order estimation test is shown in Figure 9.10c. Based on our prior knowledge of the cylinder and its modes, we decided to select a 28th order (14 mode) representation. The estimated modal frequencies obtained from the

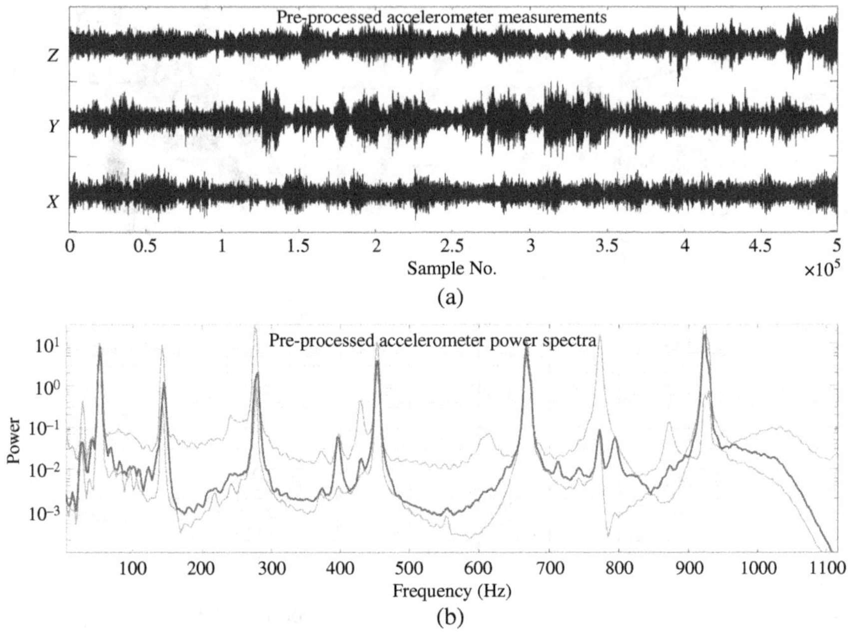

Figure 9.9 Multichannel cylindrical object data: (a) pre-processed tri-axial channel data. (b) tri-axial power spectra.

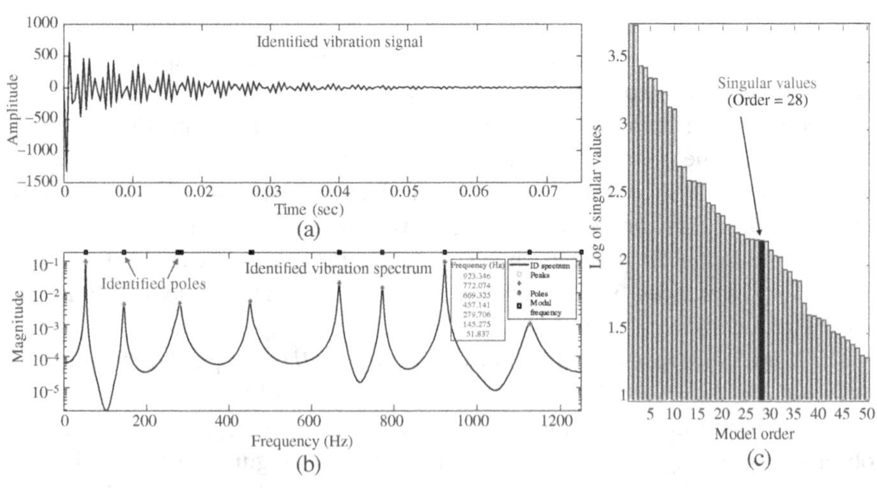

Figure 9.10 Cylindrical object multichannel identification results: (a) identified average impulse response. (b) identified average power spectrum and modal frequencies (poles). (c) order estimation (singular values) with $N_x = 28$ corresponds to 14-modal frequencies.

Table 9.3 Subspace identification: Six (6) modal frequencies.

			Model error/validation results			
Channel No.	**Residual Error**	**Average Model error**	**Deviation Model error**	**MAD Model error**	**Z-Mean 3.1×10^{-2}**	**WSSR Below τ**
1	1.3×10^{-11}	3.3×10^{-3}	4.2×10^{-3}	3.3×10^{-3}	1.6×10^{-8}	Yes
2	1.3×10^{-11}	3.3×10^{-3}	4.1×10^{-3}	3.3×10^{-3}	4.5×10^{-9}	Yes
3	1.3×10^{-11}	3.1×10^{-3}	3.8×10^{-3}	3.1×10^{-3}	1.1×10^{-8}	Yes
AVG	1.3×10^{-11}	3.2×10^{-3}	4.0×10^{-3}	3.2×10^{-3}	1.2×10^{-8}	Yes

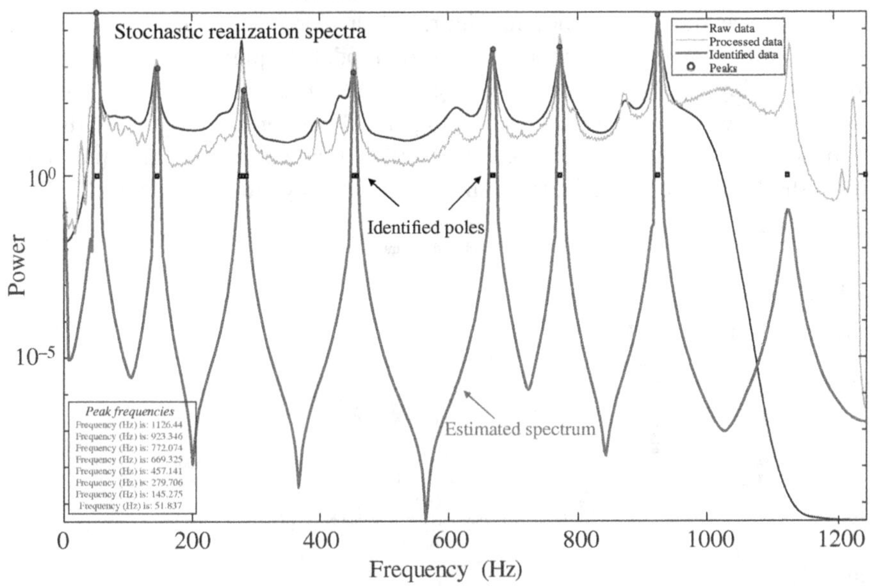

Figure 9.11 Identification of the cylindrical object (pipe) power spectral: raw, processed and identified spectra with poles and modal frequencies.

discrete-to-continuous transform ($\mathcal{Z} \rightarrow \mathcal{S}$) are also listed in the figure. We note that they are quite reasonable and are given in Table 9.3 with their corresponding relative errors. Finally, the identification is validated by comparing the raw, pre-processed and identified spectra along with the corresponding poles shown in Figure 9.11. Here it is clear from the figure and the table that the subspace identifier has extracted the primary modal frequencies from the noisy *MIMO* vibrational data.

In order to investigate the potential real-time performance, the pre-processed data are buffered into 20*K*-sample windows creating an ensemble. The *N4SID*-method is applied using a 28th order model to fit each member of the *MIMO* data ensemble. Using the identified 14-mode model, the power spectrum is estimated along with the poles ("+") obtained from the transformed eigenvalues of the A-matrix as shown in Figure 9.12a At each step (window buffer), the estimated poles are displayed in Figure 9.12b as circles with embedded "+" symbols along with the theoretical poles shown as thick bars in the figure. Tracking is achieved (visually) for a given modal frequency, if the circles lie within or near the thick bars. Even with the limited buffer data, the *N4SID*-approach appears to reliably "track" the poles mimicking the expected performance of this approach for a real-time system.

The ensemble statistics for the 14-mode identification are shown in Figure 9.13 where we see the scatter plot in Figure 9.13a along with the corresponding 99.9% confidence interval about the mean modal frequencies. The estimates are reasonably precise as shown in Figure 9.13a and in Table 9.3 as well. In Figure 9.13b, we observe the corresponding modal frequency histogram with most of the identified frequency bins heavily populated indicating very high probabilities of the "theoretical" modal frequencies estimated by the subspace tracker [16].

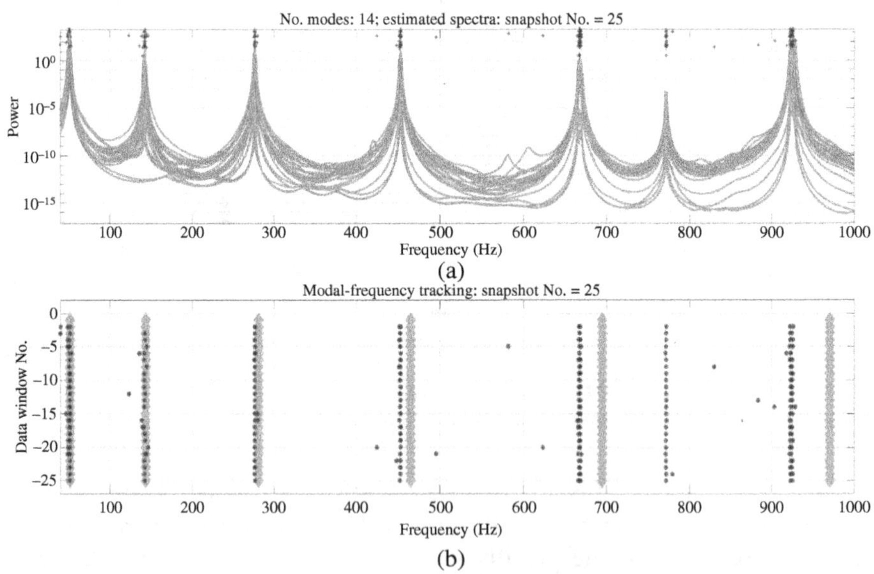

Figure 9.12 Subspace identification of cylindrical object, 14-mode model: (a) estimated transfer function spectra of identified modal models and raw modal frequency estimates (+). (b) modal frequency tracking (raw) results (circles with embedded +) and "truth" bands (thick bars).

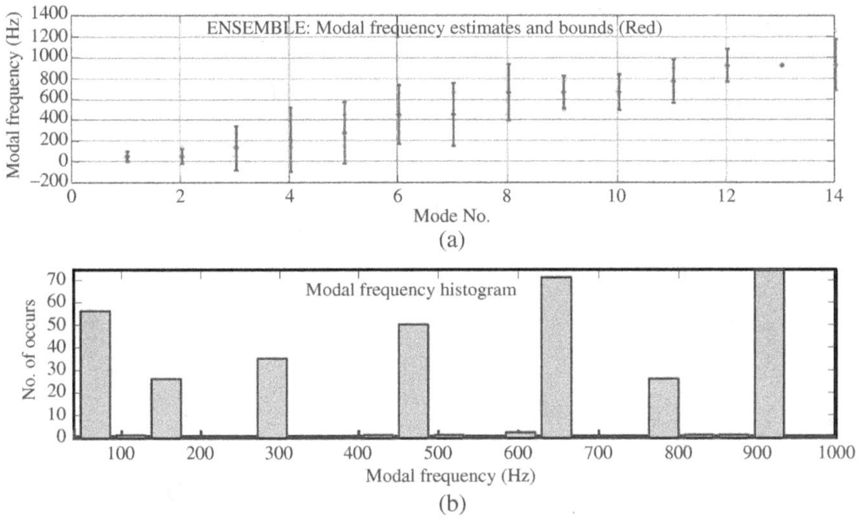

Figure 9.13 Post-processing of modal frequency tracking (ensemble) statistics: (a) modal frequency scatter plot with 99.9% confidence limits about the mean. (b) modal frequency histogram indicating high probabilities in identified modal frequency bins.

Recall that the optimal solution to the state estimation or frequency tracking problem is provided by the Kalman filter (see Section 8.2 for details), that is, [12],

$$\hat{\mathbf{f}}(t+1|t+1) = \hat{\mathbf{f}}(t+1|t) + K(t+1)e(t+1)$$

Post-processing of the tracking frequencies using a Kalman filter is shown in Figure 9.14 for the 14-mode tracks [16]. It is clear from the figures that by post-processing (off-line), the modal frequency tracks are improved from those provided online primarily because of the availability of the "batch" statistics. That is, the *batch* post-processor performing both outlier correction and tracker smoothing is superior to the on-line tracks of Figure 9.14 primarily because the ensemble statistics of 25 data windows are now available to improve the estimates as shown in the figure. The light diamonds in the figure are the theoretical or considered the "true" values of the poles—those circles on or close to these represent excellent estimates of the modal frequencies. The darker diamonds are the batch median estimates giving a visual precision estimate of the modal frequencies.

Next, we calculate the various validation metrics to observe how well they predict/validate a reasonable realization of this data. The results are shown in Table 9.3 below. From the table, we examine the average results of each of the statistics. Over the ensemble of 25-realizations and subspace identifications, the average residual error is 1.3×10^{-11} as expected since the synthesized signal is essentially deterministic (100 dB *SNR*). The calculated *model errors* are reasonable

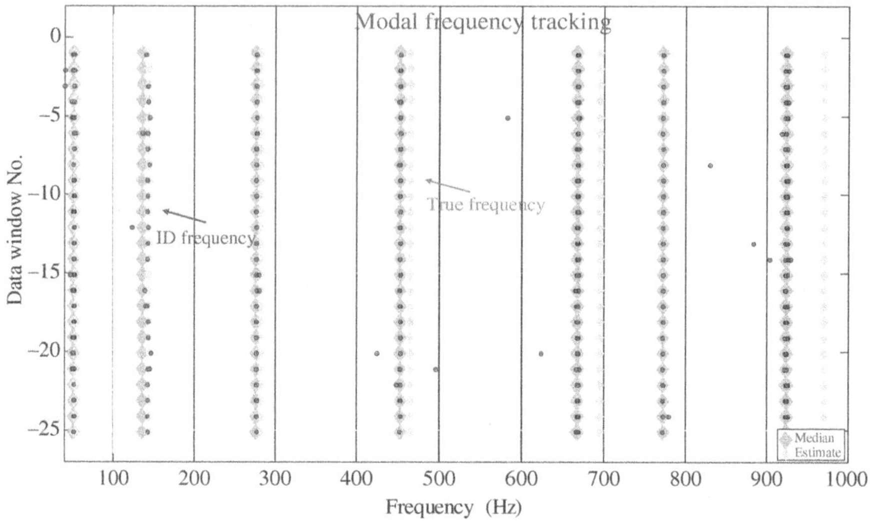

Figure 9.14 Post-processed modal frequency tracking of cylindrical object data for an identified 14-mode model: diamonds represent theoretical (light) and median (dark) values.

for the corresponding average, standard deviation and *MAD* statistics on the order of 10^{-3} corresponding to less than 1% error. We also perform a zero-mean test along with the *WSSR* test. The results are quite reasonable substantiating the "matches" shown in the figures.

9.2.1 Summary

In this application, we discuss the development of techniques to monitor the modal frequency performance of a cylindrical object subjected to random vibrations created by an impact hammer-like surface. It is demonstrated that modal frequencies can be extracted using spectral estimators coupled to a simple peak detection scheme leading to the development of a sophisticated model-based (Kalman-filter) tracker. This is achieved in both single and multi-channel accelerometer measurements. Spectral estimators in the single channel cases range from a simple windowed-FFT to the sophisticated *MUSIC* algorithm (see Section 5.3) requiring a *SVD* of a data matrix. The multichannel (tri-axial accelerometer) is also investigated. A sophisticated multichannel subspace processor is applied directly to this noisy data set with good results. Here, because of its ability to capture all of the modal frequencies available, the subspace processor extracted and tracked each of the underlying modal frequencies quite well.

This completes the application of performance metric techniques to a vibrating cylindrical (pipe) structure.

9.3 Resonant Ultrasound Spectroscopy

Properties of materials are an important aspect in a variety of disciplines ranging from the exploration of the universe by analyzing lunar rocks to investigating the worthiness of substrates that might be incorporated in acoustic transducers. Geophysical investigations explore materials throughout the Earth's crust, while the elastic properties of materials are critical in structural acoustics [4, 18–21]. Such disciplines must rely on investigations that incorporate metallurgy, crystallography, or even biological membranes that are all concerned with the elastic properties of materials [18, 22]. When subjected to stress forces that stretch or compress, materials can expand or contract. Whether or not they return to their original state is fundamentally dependent on their intrinsic properties—their elastic coefficients that govern their response to stress [18, 22]. Propagating stress forces in solids are essentially "elastic" waves that are longitudinal, shear, surface, or other types [22].

Resonant ultrasonic spectroscopy (*RUS*) is a methodology capable of measuring a change in modal resonant frequencies of material structures. It is an ultrasonic-based elastic modulus measurement technique that employs mechanical resonance estimates or equivalently resonant vibrational modal estimation techniques to extract material elastic properties [22]. These resonances are determined uniquely by the object's shape, elastic coefficients, symmetry, and material density [21]. *RUS* encompasses an inversion technique based on the eigen-frequencies of a simple regular sample geometry to estimate the elastic tensor of a solid. Obtaining an accurate and complete set of resonant frequencies is a critical first step in the *RUS* process. In this application, a model-based decomposition technique to extract and validate material resonances from complicated measurement spectra is discussed.

The response of solids to stress is completely captured by their intrinsic elastic coefficients [23]. Typically, under specific constraints, the resultant strain produced is proportional to the applied stress satisfying Hooke's law [24]. Perhaps, the most important characteristic of these coefficients is that they essentially determine the "stiffness" of the solid to forces that do not alter its volume [25–28]. With this in mind, one methodology that has evolved to extract these critical parameters is based on the solution of the elastic wave equation given by

$$\rho\omega^2 u_i + C_{ijk\ell}\frac{\partial^2 u_i}{\partial x_{j\ell}^2} = 0 \qquad \text{[Elastic Wave Equation]} \tag{9.1}$$

where u_i is the displacement field, $C_{ijk\ell}$ is the elastic tensor, ρ is the material density, and ω is the resonant (material) frequency that reflects the size, shape, and elastic composition with free boundary conditions on the specimen under investigation [24]. Here the usual approach utilizing Hamilton's principle involves minimizing the action defined as the difference between kinetic and potential energy such that over any time interval and sample volume, this variation is zero [23–25].

$$\delta \int_t \int_V (KE - PE)\, dV dt = 0 \tag{9.2}$$

with the kinetic and potential energies defined by

$$\underbrace{KE := \frac{1}{2}\rho\omega^2 u_i(\mathbf{x}) u_i(\mathbf{x});}_{\text{Kinetic Energy}} \quad \underbrace{PE := \frac{1}{2} C_{ijk\ell} \frac{\partial u_i(\mathbf{x})}{\partial x_j} \frac{\partial u_k(\mathbf{x})}{\partial x_\ell}}_{\text{Potential Energy}}$$

The condition established by Eq. (9.2), is satisfied if and only if u_i and ω are at a *resonance* of the sample [22, 23]. Therefore, exciting or vibrating a specimen sinusoidally by sweeping over a band of frequencies that encompass the resonant frequencies of the sample enables the extraction of its intrinsic elastic coefficients. It is this property that provides the coupling of the set of resonant frequencies $\{\omega_m\}$ to the elastic coefficients $\{C_{ijk\ell}\}$ leading to the elastic coefficient estimation problem that is typically solved using numerical optimization techniques.

The solution of this minimization problem begins with determining a suitable approximation for the displacement field, u_i, such that Eq. (9.2) is satisfied. Following the Rayleigh–Ritz approach, a set of basis functions $\{\Phi_n\}$; $n = 1, \ldots, N_\Phi$ over the corresponding spatial directions are defined such that

$$u_i = \sum_{n=1}^{N_\Phi} a_{i,n}\Phi_n \quad \text{for} \quad i = 1, 2, 3 \qquad \text{[Displacement Basis Expansion]} \tag{9.3}$$

where

$$\Phi_n(x, y, z) = (x/L_x)^\ell (y/L_y)^m (z/L_z)^p$$

for

$$L_x, L_y, L_z$$

the specimen dimensions and $n = (\ell, m, p)$ a function label of a set of nonnegative integers (ℓ, m, p) [22, 27].

Performing this approximate minimization of the Lagrangian ($\mathcal{L}$) in matrix form of Eq. (9.2)

$$\min_\alpha \mathcal{L} = \alpha^T \left(\omega^2 \mathbf{M}\alpha - \mathbf{K}\alpha\right) \tag{9.4}$$

where $\mathbf{M}$ is the underlying normalized mass and $\mathbf{K}$ the stiffness matrices [22, 23, 27, 28] enables the development of an eigen-problem. That is, applying

Hamilton's principle constraining $\mathcal{L}$ to be stationary to small (eigenfunction) perturbations evolve directly to the condition $\frac{\partial \mathcal{L}}{\partial \alpha} \approx 0$ enabling the problem to be expressed in a form as an eigen-problem.

Therefore, small, first-order perturbations in a consistent stress-free boundary condition leads to the following result that $\{\omega, \alpha_j\}$ following generalized eigen-solution

$$[\lambda\ \mathbf{M}\ -\ \mathbf{K}]\ \alpha = \mathbf{0} \quad \text{for} \quad \lambda = (2\pi\rho\mathbf{f})^2 \qquad \text{[Eigen-Model]} \tag{9.5}$$

The choice of various basis functions leads to a variety of solutions that can be employed in the inversion process to extract the elastic coefficients. Here we use the $\{\Phi_n(x, y, z)\}$ of Eq. (9.3) as our basis set.

9.3.1 *RUS* Methodology

Material resonances can be obtained by driving the sample through a set of frequencies and recording its dynamic response. This approach is based on the principle that elastic bodies isolated in free space will resonate at discrete frequencies corresponding to the modes of the system. The specimen's intrinsic elastic tensor is coupled to the eigen-frequencies of the system modes. *RUS* incorporates these characteristics to infer elastic parameters from a suite of the measured resonance frequencies.

A typical *RUS* methodology is shown in Figure 9.15 where piezoelectric crystals are excited by a swept harmonic oscillator over a frequency range that brackets the fundamental resonance of the sample on the low side, and as many higher order eigen-mode frequencies on the high side as necessary to provide sufficient data to perform an accurate fit. These are the measured frequencies $\{\mathbf{f}_m\}$. A sophisticated forward model that can predict resonant frequencies based on geometrical and an a priori estimate of the elastic coefficients provides the set of estimated frequencies $\{\mathbf{f}_o\}$ [27, 28]). These frequencies are incorporated into a resonant frequency error criterion used to estimate $\{\hat{\mathbf{C}}_{ijk\ell}\}$ over the elastic coefficient space converging to the "true" solution. Here, the required mass and stiffness matrices are updated by the current coefficient estimates and used in a forward eigen-model to predict the resulting resonant frequencies $\{\mathbf{f}\}$. This solution to the inverse problem is accomplished using *forward modeling* that eventually converges to the required solution. The complete process is illustrated in this figure showing the interaction of a eigen-model and a generic optimization technique [22, 25, 29].

A wide variety of optimization techniques can be applied to solve the inversion problem of minimizing the frequency error function [25, 29].

$$\varepsilon_{\mathbf{f}} = \mathbf{f}_{\mathbf{m}} - \mathbf{f}(\mathbf{C}) \qquad \text{[Frequency Error]} \tag{9.6}$$

Depending on the particular optimization technique, the elastic coefficients are updated, and the iteration is repeated until the error criterion is minimized

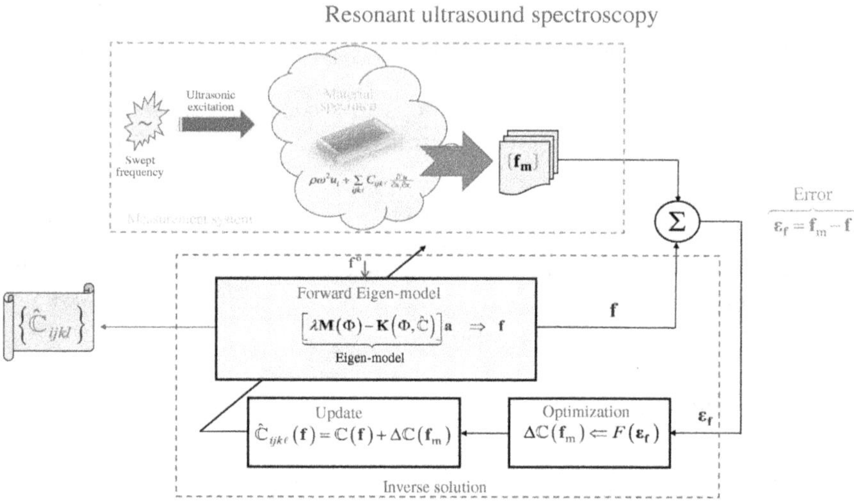

Figure 9.15 Resonant ultrasound spectroscopy (RUS): Ultrasonic frequency measurements are obtained and provided to an optimization algorithm to estimate the elastic coefficients through an iterative inversion process using forward (eigen-model) modeling techniques.

converging to the true elastic coefficients. This is known as a method of solving the inverse problem through forward modeling.

We summarize the generic approach as: (i) obtain the resonant frequency measurements; (ii) provide current elastic coefficients to the forward model; (iii) calculate resonant frequency estimates using these coefficients and forward eigen-model; (iv) minimize the frequency error (optimization method) [25]; and (v) update the elastic coefficients and iterate [22, 27, 28].

This completes the discussion of the *RUS* approach, next we consider the underlying basic dynamic representations applied to the *RUS* problem.

9.3.2 Modal Analysis: FRF and Frequency Histogram

The primary data required are the multichannel, frequency response function matrix (*FRF* matrix) that is a collection of *FRF*s generated at each reference input, that is, $N_O \times N_I$ *frequency response matrix* is defined by:

$$\mathbf{F}(\omega) := \begin{bmatrix} F_{11}(\omega) & F_{12}(\omega) & \cdots & F_{1N_I}(\omega) \\ F_{21}(\omega) & F_{22}(\omega) & \cdots & F_{2N_I}(\omega) \\ \vdots & \vdots & & \vdots \\ F_{N_O1}(\omega) & F_{N_O2}(\omega) & \cdots & F_{N_ON_I}(\omega) \end{bmatrix} \quad \text{[Frequency Response Matrix]} \tag{9.7}$$

where $F_{ij} \in C^{1\times N_f}$ is the *i*th output (*FRF* function) and *j*th input (reference).

The *FRF* matrix can be *decomposed* directly using a *SVD* technique at *each frequency* to provide a more numerically robust, economical and practical technique [30]. In any case, the underlying matrix is obtained by performing a *SVD* (svd ($\mathbf{F}(\omega)$)),

$$\mathbf{F}(\omega) = \mathbf{U}(\omega) \times \mathbf{\Sigma}(\omega) \times \mathbf{V}^{\dagger}(\omega) \quad \text{[FRF-Matrix (SVD)]} \tag{9.8}$$

for $\mathbf{U}$, $\mathbf{V}$, $\mathbf{\Sigma}$, the respective left and right singular (eigenvector) matrices with the corresponding diagonal matrix of ordered singular values (largest-to-smallest)—all functions of radian frequency, ω, that is,

$$\mathbf{\Sigma}(\omega) = \text{diag}\left[\sigma_{11}(\omega)\cdots\sigma_{N_\sigma N_\sigma}(\omega)\right]$$

with the set of singular values contained in Σ, $\mathbf{U}$ contains mode shape information, while $\mathbf{V}$ contains the modal participation vectors.

Another modal analysis tool is the indicator function that incorporates much of the resonant (modal) frequency information. The complex mode indicator function (*CMIF*) is a set of frequency-dependent singular value functions of the decomposed *FRF*-matrix. Mathematically, we define the *CMIF*, $\mathbf{C}(\omega) \in C^{N_\sigma \times N_f}$, for N_σ the number of dominant singular values and n_i, the number of reference inputs or equivalently the set of spectral (frequency) lines with $\omega = \{\omega_i\}; i = 1, \dots, N_i$ as (see Section 7.2 for more details) [30–33]

$$\text{CMIF} := \mathbf{C}(\omega) = \mathbf{\Sigma}(\omega) \qquad \text{[Complex Mode Indicator Function]} \tag{9.9}$$

The *CMIF* is based on the so-called Expansion Theorem such that at every frequency the column vector $\mathbf{f}_j(\omega)$ of the *FRF*-matrix is the summation of modal vectors [31–33]. Since $\mathbf{C}(\omega)$ is based on the diagonals of $\mathbf{\Sigma}(\omega)$; therefore, the *SVD* generates this set of curves, say $\{\mathbf{C}_k(\omega\})$, such that

$$\{\mathbf{C}_k(\omega)\} = \text{diag}[\Sigma(\omega) := \boldsymbol{\sigma}_k(\omega)]; \quad k = 1, \dots, N_\sigma \qquad \text{[CMIF]} \tag{9.10}$$

is ordered from largest (magnitudes) to smallest with the primary indicator curve (strongest) indicating a local minimum or maximum at each of the natural frequencies (ω_n) of the structure. This curve is followed by secondary curves indicating existing repeated roots with subsequent curves indicating high-order repeated roots—these repeated poles exist for axisymmetric (e.g. cylindrical) structures [31, 32]. Thus, the *CMIF* provides a metric that identifies the dominant modal vectors participating at each frequency along with a measure of the magnitude or strength (dominance) of all modes at a particular frequency with the corresponding singular vectors ($\mathbf{u}_m(\omega)$) providing estimates of the contributing modal vectors associated with each singular values (function). The peaks of $\mathbf{C}(\omega)$ provide estimates of damped natural (modal) frequencies of each identified mode with the total number of peaks yielding an estimate of the number of existing modes characterizing the structural response.

9.3.3 Model-Based Decomposition Approach

The model-based decomposition approach to processing *FRF* data consists of estimating state-space models from the noisy experimental data by applying subspace identification techniques. The fundamental idea is to "fit" a multichannel (*MCK*) model in state-space form defined by the set of model matrices, Σ_M and employ it to "parametrically" estimate the resonant frequencies directly as well as enhance the *FRF*s and impulse response functions (*IRF*) for especially noisy, uncertain data providing improved estimates of the resonant frequencies. The approach is to apply *MIMO* system identification techniques to fit the raw multichannel *IRF*/*FRF* data using the *N4SID* algorithm. Thus, this technique is to estimate the number of modes and apply subspace identification techniques to extract a structural (state-space) model directly from each *IRF* or *FRF* of the output measurement datasets. Therefore, starting with the *IRF* matrix, subspace identification proceeds.

Here, the *SVD* of a data matrix derived from each individual response provides a means of obtaining the order (2 $\times$ modes) of the underlying *RUS* system—that is, an estimate of the number of resonant frequencies [13]. Once the order is obtained, then the corresponding state-space model, $\Sigma_M = \{A_M, B_M, C_M, D_M\}$, is estimated and used to obtain a parametric modal frequency estimate by performing an eigen-decomposition of the system matrix, $\lambda(A_M)$ as well as providing enhanced *IRF*/*FRF*s.

We summarize the model-based approach to *RUS* estimation as:

Model-Based Decomposition

- *extract* the multichannel IRF from the FRF data
- *approximate* the number of modes for the MCK model from the SVD data tests
- *identify* the modal state-space model, $\Sigma_M = \{A_M, B_M, C_M, D_M\}$
- *perform* eigen-decomposition of $\lambda(A_M)$
- *estimate* the corresponding resonant frequencies the identified system, $\hat{\Sigma}_M$.

Next, we apply this approach to a well-known structure—a parallel piped.

9.3.4 Application: Parallel Piped Structure

The model-based approach to estimating the modal frequencies from the *RUS* measurements is founded on fitting or identifying a structural model that captures the vibrational response of the material under investigation. The advantage is that the underlying model representing a structural system is capable of processing multichannel measurements after first extracting the impulse responses from the frequency domain (sweeps) data generating a multiple-output system.

For this investigation, we choose a solid isotropic 304*L* stainless-steel sample, with a regular parallelepiped (RPP) geometry. Here, both simulated and

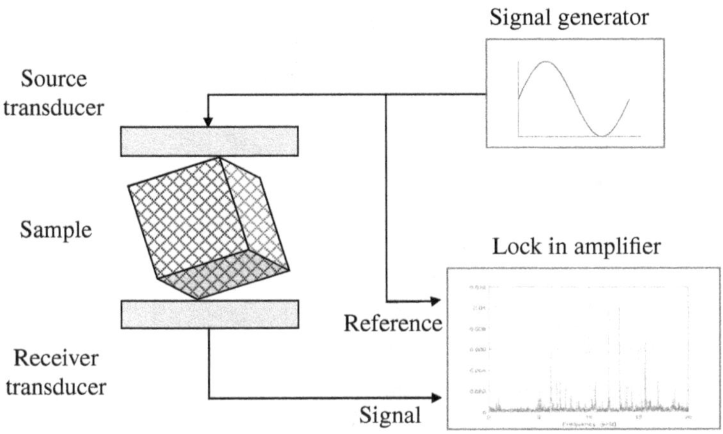

Figure 9.16 Resonant ultrasound spectroscopy measurement system and acquired spectrum.

experimental data are generated to investigate the performance of a state-space, model-based decomposition approach. A sophisticated *RUS* simulation is performed to generate a set of tri-axial (*XYZ*) outputs to synthesize a typical data. The *RUS* application, shown in Figure 9.16, illustrates the experimental set-up for gathering the frequency domain datasets. A piezoelectric transducer is excited by sinusoids producing frequency domain sweeps across bands of high interest for the material specimen. By repositioning this specimen, an ensemble of swept frequency output data is generated for the multichannel physics-based processor. The solid RPP structure is excited by the transducer using the frequency domain sinusoidal sweeps for each position to create the multiple output ensemble. Using this approach enables a set of multiple measurements to be generated with the expectation of exciting weak modes in some locations.

9.3.4.1 Synthesized Data: RPP Structure

Initially, the structural vibrations are simulated using a sophisticated finite element model (FEM) implemented using the COMSOL® commercial tool to synthesize *RUS* frequency spectra [7]. The data are simulated in 20 Hz sampling frequency intervals up to a 35 KHz to match the subsequent experiment. The tri-axial (*XYZ*) response is illustrated in Figure 9.17a for each individual direction along with the composite ensemble average. Here, we observe the peak modal frequencies at various magnitudes along with a corresponding histogram ensuring valid resonant frequencies in Figure 9.17b expected from a set of experimental measurements. The peak resonance estimates are annotated in the list inset. The *CMIF* is also estimated over the set of data confirming the corresponding measured resonances as shown in Figure 9.17c.

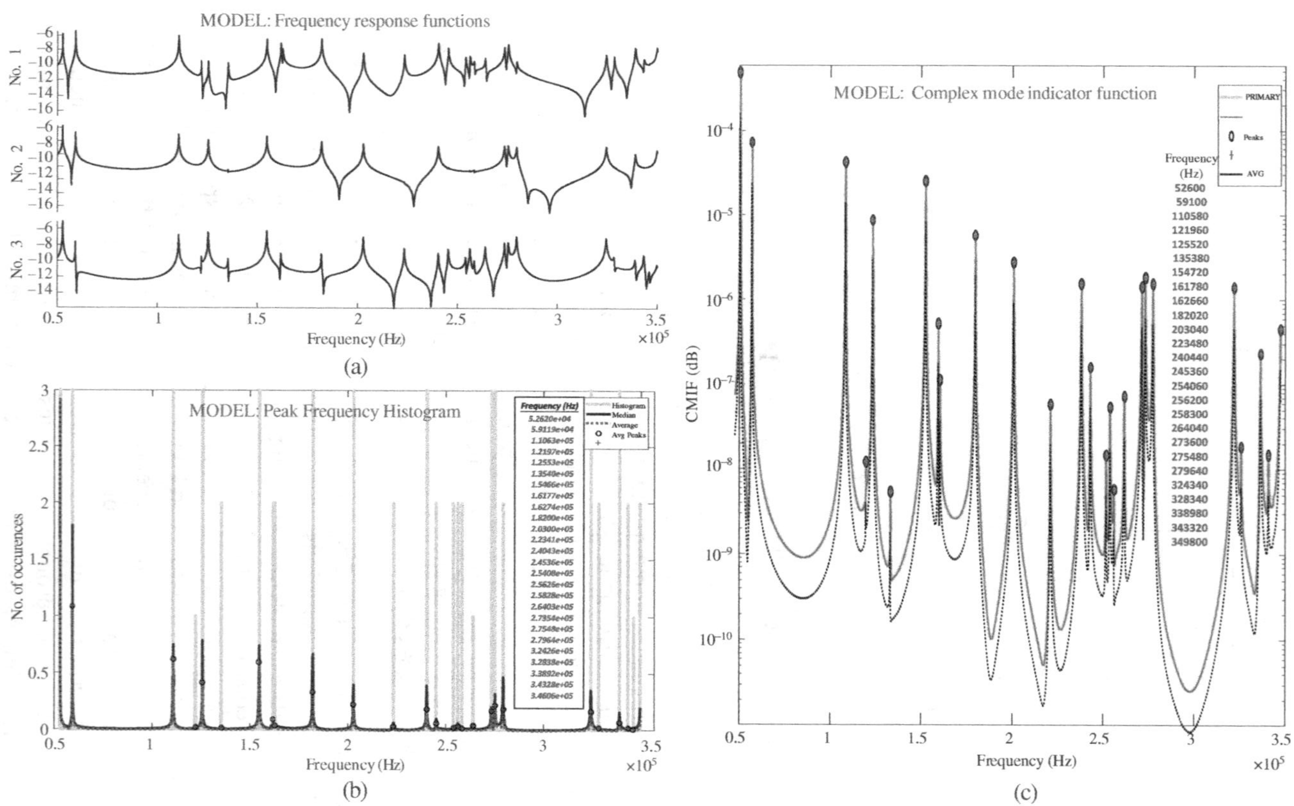

Figure 9.17 Simulated (MODEL) solid parallelepiped (RPP) structure frequency (tri-axial) measurements: (a) resonant frequency (log-magnitude) three-channel responses (*FRFs*). (b) peak frequency histogram, overlay average *FRF* and peak frequency list. (c) *CMIF* with average (dotted), frequency peaks, list, and estimates.

The solid parallelepiped frequency swept *FRF* data are converted to the impulse response time series through the inverse Fourier transform providing the impulse response function (*IRF*) data for input to the multichannel state-space processor.

For the synthesized model data, a 64th-order (32 modes), multichannel physics-based processor was applied to this multichannel dataset with the results shown in Figure 9.18 where we observe the singular values of the data matrix used to estimate the number of states (modes) ranging from 30 to 100 each filled and annotated on the order diagram [13]. We selected the model order (dark) and used the performance metrics of percent model fit (% fit). Each of the tri-axial channel output measurement estimates are shown in the figure with percentage fits: 84%, 97%, and 98% providing outstanding results with the synthesized data (light) overlayed by its predicted estimate (solid dark line) for each channel. The average measurement spectrum (light) and its corresponding composite *MIMO* and average impulse response (dark) spectra for the 64th-order model are also shown along with the modal frequency peak list. It is clear from these results that the multichannel physics-based approach provides a reasonable methodology to extract even weakly excited modes.

9.3.5 Experimental Data: RPP Structure

It was found that the "best" performance could be achieved by acquiring multiple measurements generating an ensemble obtained by repositioning the RPP vertices and repeating the swept-frequency excitation. The primary motivation for this approach is based on the premise that some material resonant frequencies are weakly excited in a particular excitation position but can be strongly excited in another position and therefore, extracted by summing (averaging) over the multiple positions.

As an application consider the following set of test data extracted from this solid parallelepiped positioned and held by its vertices in a test stand consisting of two PZT transducers: one for excitation at its apex and one diagonally at its bottom vertex for reception (see Figure 9.19). The *RUS* experiment for the RPP is performed by generating five (5) individual frequency sweeps at different positions (vertices) in order to generate the ensemble. The results are shown in Figure 9.19, where we observe the occurrence of spectral peaks in a variety of positions as well as the major resonances aligning quite well. The frequency histogram shown in Figure 9.19b illustrates the strongest (and weakest) resonances recorded in three or more locations. The accompanying list estimated using simple peak detection provides the subsequent frequencies identified. To confirm these peaks, the raw *CMIF* is illustrated along with a spectrally smoothed (Gaussian window) estimate. The specific peaks are determined by peak detection and listed in the figure. The data are then transformed to the time domain using the inverse Fourier transform.

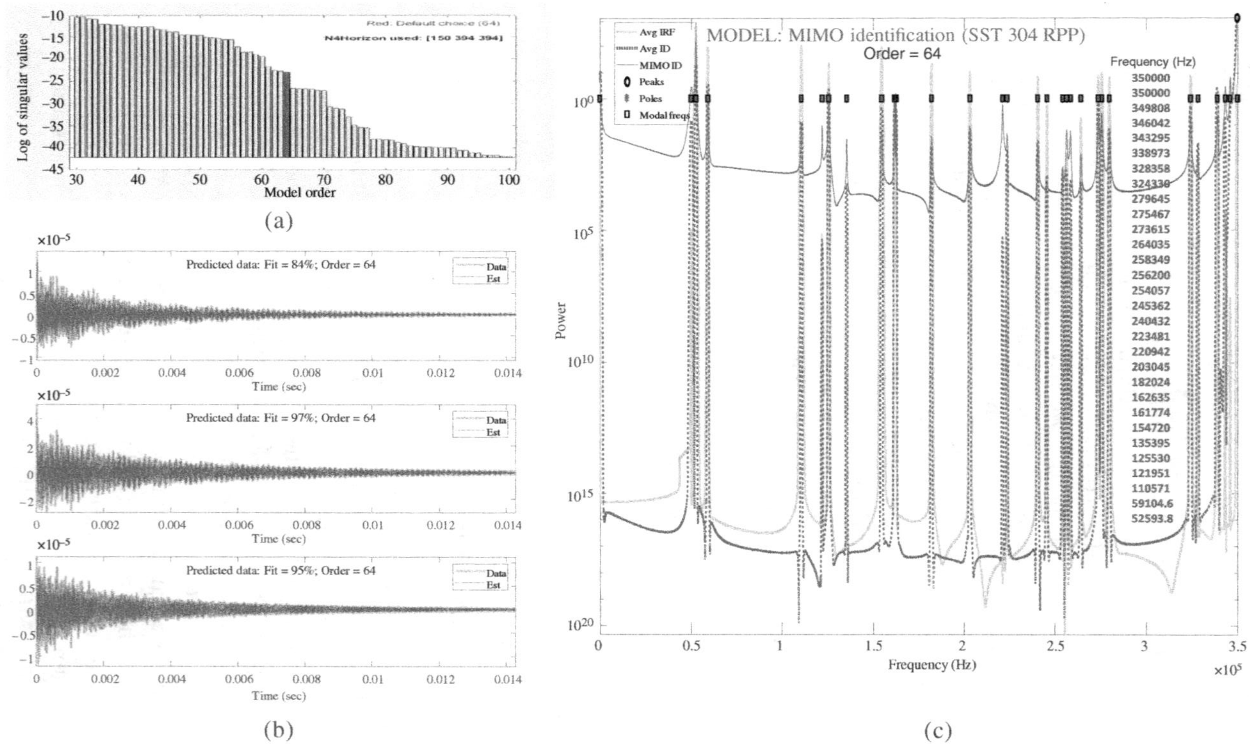

Figure 9.18 Estimated *MIMO* state-space model for synthesized (MODEL) solid parallelepiped (RPP) structure measurements: (a) SVD order estimation (64th-order). (b) predicted output measurements (thick line) and estimates (thin line) with model fits of 84%, 97%, and 95%. (c) MIMO composite and average (dotted line) resonant frequency estimates with data parametric modal frequency estimates (squares) along with model-based resonant frequency list (inset).

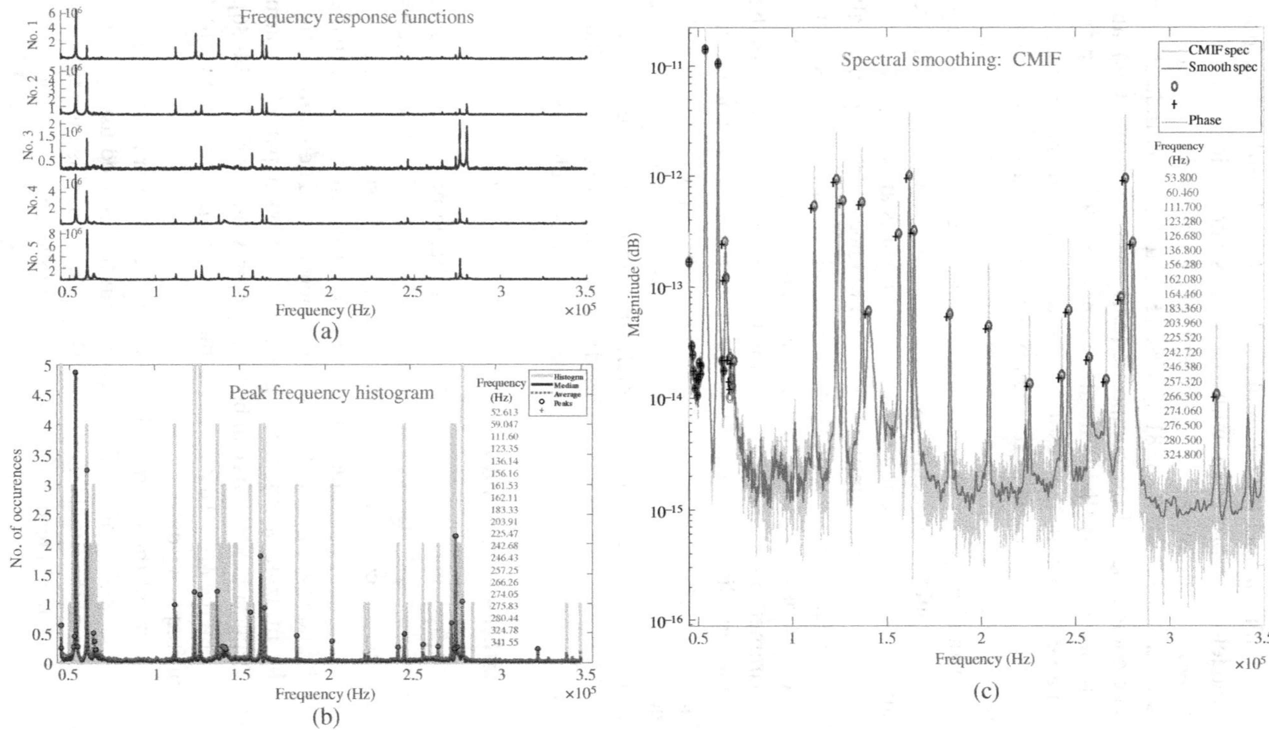

Figure 9.19 RUS experimental (DATA) solid parallelepiped (RPP) structure frequency measurements: (a) resonant frequency (log-magnitude) 5-channel output response (*FRF*). (b) peak frequency histogram, overlay average *FRF* and peak frequency list. (c) *CMIF* data (turquoise) with smoothed, Gaussian-window, estimate along with frequency peaks (circles) and detailed list.

9.3.6 Model-Based Decomposition Processor

The experimental multichannel *RUS* data of Figure 9.17 was provided as input to the *MIMO* model-based processor with the results shown in Figure 9.20. A 100-mode model was fit to the noisy experimental ensemble dataset. The predicted fits are reasonable at: 81%, 81%, 62%, 75%, and 78% as indicated in Figure 9.20a. The comparison of the composite (dark), average (dark dots), and noisy average experimental data (light) are shown in Figure 9.20b. The estimates are quite reasonable extracting the resonant frequencies with the parametric estimates indicated by the squares in the figure. The estimated frequencies are shown in the list (inset). Again, the weak resonances are enhanced and extracted illustrating the outstanding performance of the model-based approach. The relative errors using the simulated model (COMSOL) eigen-frequencies as "truth" are used as an accuracy criterion. The multichannel processor is almost "perfect" in the resonant frequency estimates captured, 0.0%, but not necessarily capturing all (32 modes) of the available eigen-frequencies. This condition is probably a result of weak modal excitations of particular frequencies in the synthesized datasets. It is clear that the parametric and peak-based estimates are very consistent with the model-based results enabling more of the resonant frequencies to be identified. The results of processing the experimental data are not as impressive with fewer of the eigen-frequencies extracted from the noisy experimental datasets. Here, the resonant frequency estimates indicated extremely small average relative errors and 0.527% for the *MIMO* (26 modes), but still missed some of the frequencies available in the *MIMO* case again due to weak excitations even with such high-order models.

9.3.7 Elastic Coefficient Estimation

The estimated eigen-frequencies are utilized in an inversion process to estimate the elastic tensor of the 304*L* stainless steel sample. The *RUS* inversion process is based on applying the cost function defined by Eq. (9.6) such that a set of elastic coefficients $\{Ci\}$ are determined that minimize the difference between the experimental, f_n^{meas} and the calculated eigen-frequencies f_n^{calc}. The minimization is accomplished by utilizing a stochastic process based on a particle swarm optimization (PSO) algorithm, details can be found in previous work [25, 34–36]. A comparison between the *MIMO* data is performed with respect to experimental and modeled eigen-frequency data.

The stochastic nature of the optimization process cannot guarantee an exact solution; however, the quality of the fitted results can be assessed using a measure of the total error, of the fitted eigenvalues (Eq. (9.6)). For this investigation, the total error is calculated in terms of a percentage as follows:

$$\%\ error = 100 \times \sum_{i=1}^{M} \frac{\left(f_i^{Meas} - f_i^{Fit}(C_{ij})\right)^2}{\left(f_i^{Meas}\right)^2} \tag{9.11}$$

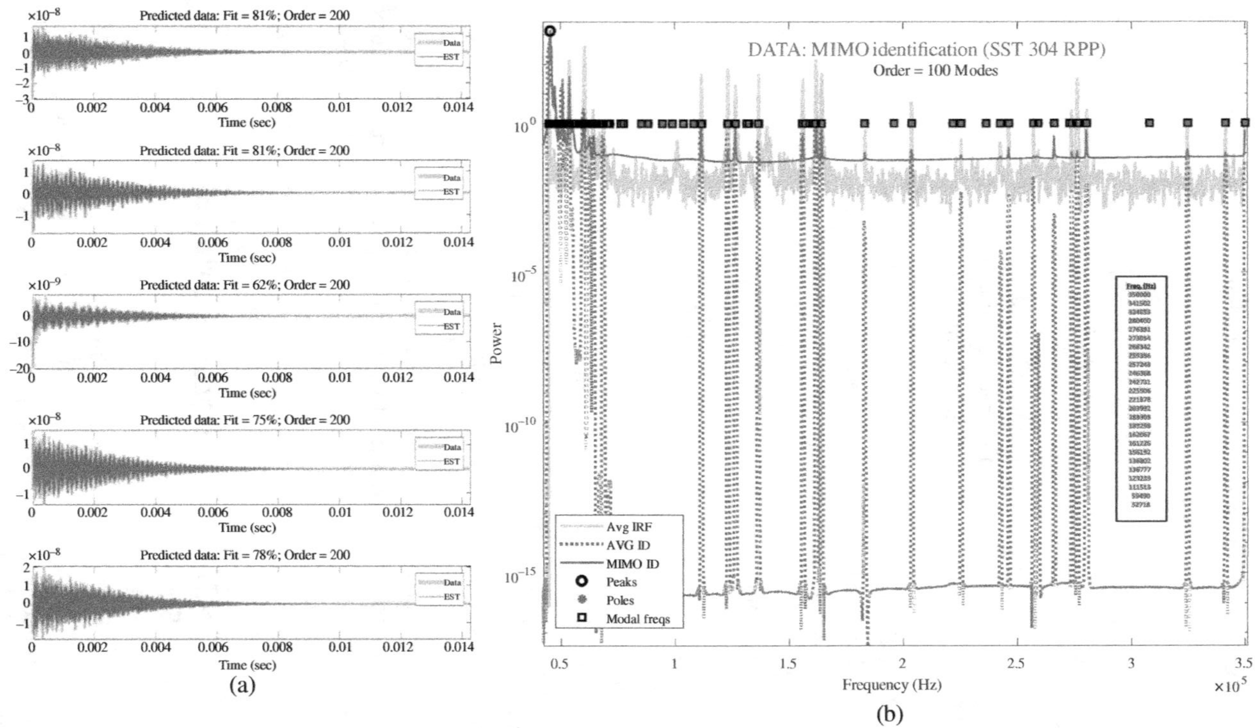

Figure 9.20 Estimated *MIMO* state-space model (100 modes) of the *RUS* experiment (DATA) for the noisy solid parallelepiped (RPP) structure measurements: (a) predicted output measurements and estimates with model fits of 81%, 81%, 62%, 75%, and 78%. (b) average *MIMO* resonant frequency estimates (dots) with data with composite spectrum along with parametric modal frequency estimates (squares) and the corresponding model-based resonant frequency list shown.

The PSO fitting process was repeated five times on all the datasets. The average for the elastic parameters C_{11} and C_{44} along with derived values for the Young's modulus, E, Poisson ratio, v, and shear modulus, G. For isotropic materials,

$$\begin{aligned} G &= C_{44} \\ E &= C_{44}\frac{(3C_{11}-4C_{44})}{(C_{11}-C_{44})} \\ v &= \frac{(C_{11}-2C_{44})}{2(C_{11}-C_{44})} \end{aligned} \tag{9.12}$$

The calculated isotropic elastic coefficients C_{11} and C_{44} along with the sample dimensions and measured density are incorporated into a 3*D* FEM of the 304 SST RPP using COMSOL® [7]. The modeled spectra provide a controlled set of data for analyzing the *FRF* spectra.

The estimated elastic coefficients from the modeled (COMSOL®) eigenvalues finite element solution and the measured eigenvalues are comparable. The fitting error is less than 0.05%, for the first 18 eigenvalues and less than 0.07% for the first 28, which demonstrates the consistency between the experimental and modeled results completing the application.

9.3.8 Summary

This application is focused on the analysis of material (structural) vibration data to extract the elastic coefficients of the specimen under investigation using *RUS* methods. The approach to extract resonant frequencies from spectral data ranging from simple peak picking to sophisticated model-based decomposition is discussed. The emphasis is primarily on a physics-based approach as a technique that can be used to estimate and validate the extracted resonance frequencies leading to more accurate elastic coefficient estimates—the underlying goal of this effort. With reliable resonant frequencies as input to numerical optimization techniques, the underlying material elastic coefficient estimates are good for both simulated and experimental data cases.

9.4 Model-Based Subsystem Decomposition of an 8-Story (8-Mass) Structure

In structural analysis, it is often the requirement that the response of a critical subsystem must be extracted from the response of the overall structure even when not directly excited. Prior knowledge of the targeted subsystem bandwidth may be enough to design a simple filter, if there is minimal overlap with other component systems. If this is not satisfactory and a specific spectral response is required, then a model-based decomposition approach can be applied using an available free-free response of the subsystem from calibration or acceptance tests. Response data are available from a simulation model or calibration/acceptance tests can also be processed to extract a subsystem response that can then be used to identify a modal model enabling subsequent analysis—the approach in this application.

There exist a variety of signal processing approaches that may be applied to attack this problem depending on the a priori information available. One approach utilizes Wiener filtering theory based on measurement data available such as model reference estimation (using free–free or calibration data) and developing a "shaping" filter capable of extracting or matching an available estimated (known) response [12]. This is the primary method developed for model-based decomposition using modal identification coupled with the filter. We first apply a subspace identification technique to perform the decomposition in modal coordinates (model-based decomposition) and then employ the shaping filter concept to extract various subsystems. We start with the structural model.

9.4.1 Subspace Structural Identification

The application of a subspace identification technique to a structure represented by a *LTI*, *MIMO*, mass-spring-damper mechanical system consisting of 8-modes, or 16-states (see [37] for details) measured by three-output accelerometers excited by a random input is developed. Structurally, the system mass ($\mathcal{M}$) is characterized by an identity matrix, while the coupled spring constants in (N/m) are given by the tri-diagonal matrix

$$\mathcal{K} = \begin{bmatrix} 2400 & -1600 & 0 & 0 & 0 & 0 & 0 & 0 \\ -1600 & 4000 & -2400 & 0 & 0 & 0 & 0 & 0 \\ 0 & -2400 & 5600 & -3200 & 0 & 0 & 0 & 0 \\ 0 & 0 & -3200 & 7200 & -4000 & 0 & 0 & 0 \\ 0 & 0 & 0 & -4000 & 8800 & -4800 & 0 & 0 \\ 0 & 0 & 0 & 0 & -4800 & 10400 & -5600 & 0 \\ 0 & 0 & 0 & 0 & 0 & -5600 & 12000 & -6400 \\ 0 & 0 & 0 & 0 & 0 & 0 & -6400 & 13600 \end{bmatrix}$$

the damping matrix is constructed using the relation (Raleigh damping)

$$C_d = 0.680\mathcal{M} + 1.743 \times 10^{-4}\mathcal{K} \quad \left(\frac{N\,s}{m}\right)$$

The measurement system consists of three (3) accelerometers placed to measure the modes at the 1, 4, and 8 locations on the structure. The accelerometer data are acquired and digitized at a sampling frequency of 50 Hz ($\triangle t = 0.02$ sec). The input signal from a randomly excited stinger rod is applied at a specified spatial location with governing MCK relations given by

$$\mathcal{M}\ddot{\mathbf{d}}(t) + C_d\dot{\mathbf{d}}(t) + \mathcal{K}\mathbf{d}(t) = \mathcal{B}_p\mathbf{p}(t)$$

The structural system model is transformed to state-space such that

$$\dot{\mathbf{x}}(t) = \left[\begin{array}{c|c} 0 & I \\ \hline -\mathcal{M}^{-1}\mathcal{K} & -\mathcal{M}^{-1}C_d \end{array}\right]\mathbf{x}(t) + \left[\begin{array}{c} 0 \\ \hline \mathcal{M}^{-1}\mathcal{B}_p \end{array}\right]\mathbf{p}(t)$$

with accelerometer measurements

$$\mathbf{y}(t) = C\mathbf{x}(t) + \mathbf{v}(t) \tag{9.13}$$

where $\mathbf{v}$ is a Gaussian noise vector and the input and measurement matrices are given by

$$\mathcal{B}_p = \mathbf{e}_{N_x}; C = \begin{bmatrix} \mathbf{e}_1' \\ \mathbf{e}_4' \\ \mathbf{e}_8' \end{bmatrix}$$

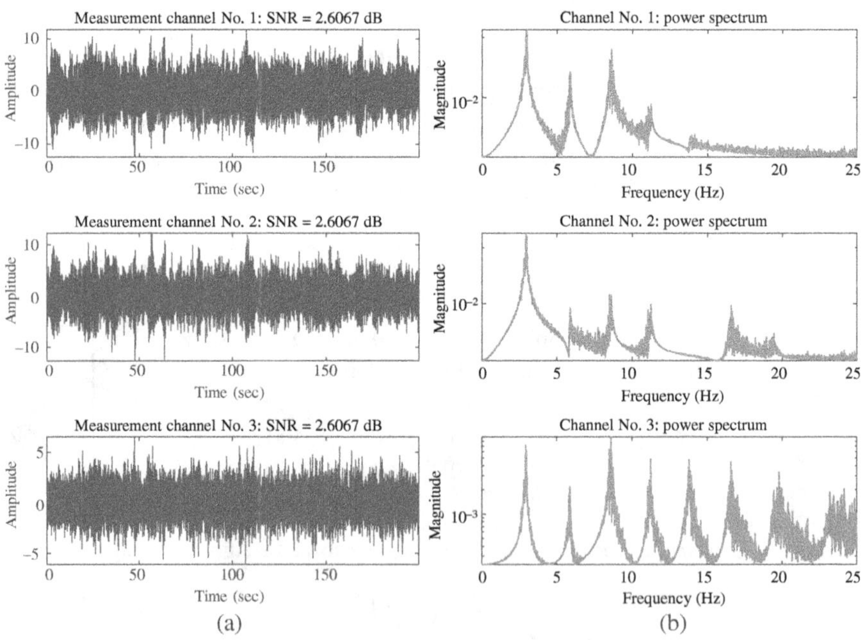

Figure 9.21 Structural vibrations of 8-mode mechanical system (SNR=0 dB): (a) accelerometer responses of 3-output system. (b) Fourier power spectra of channel (No. 1, 4, 8) responses.

for $\mathbf{e}_{N_x}$ an N_x-column vector and $\mathbf{e}_i'$ is a unit row vector with a 1 in the ith column and the excitation $\mathbf{p}$ is a unit variance, pseudo-random sequence. Three accelerometer outputs of the synthesized vibrating structure are recorded for 200 sec with the vibrational responses shown in Figure 9.21 along with their corresponding power spectra where we see a persistently excited system ideal for identification.

The subspace identification algorithm (*N4SID*) is applied to the response data and the results are summarized in Figure 9.22. We observe the estimated power spectra (thin lines) of each channel measurement along with the "average" identified spectra (thick line) in Figure 9.22a as well as the extracted modal frequencies (squares) from the identified model system matrix (A). The *SVD*-order test is shown in Figure 9.22b. This test reveals a 16th-order (8-mode) representation as adequate with the large gap shown to indicate the proper order. The poles/zeros (discrete) populating the unit circle are shown in Figure 9.22c with the estimated modal frequencies obtained from the discrete-to-continuous transform ($\mathcal{Z} \rightarrow \mathcal{S}$) listed in Figure 9.22d corresponding to those of the simulation.

In order to validate the identified modal structure, we investigate two complimentary criteria: the stabilization diagram (*SDIAG*) and the modal assurance criterion (*MAC*). The results of these validation tests are shown in Figures 9.23 and 9.24. The *SDIAG* confirms the stability of the physical poles at the 8-mass structural system modal frequencies except for the weakest resonance at 11.8 Hz. On the other hand, the estimated shapes and *MAC* confirm, at least qualitatively, the true mode shapes compared to those identified by the strong anti-diagonal values close to unity in Figure 9.23b with the accompanying shapes in Figure 9.23a. Both providing a reasonable set of validation metrics.

The corresponding Hankel metrics are:

$$\begin{aligned} \text{Hankel order estimation}: &\quad \hat{N}_x = 16 \\ \text{Hankel singular values}: &\quad \sigma_i = \{18.8,\ 17.9,\ 10.4,\ 10.6,\ 6.4,\ \ldots, 2.3\} \\ \text{Hankel norm (energy)}: &\quad \|\mathcal{H}_{K,K'}\| = 1.2 \times 10^{-5} \\ \text{SVD model error}: &\quad \mathcal{M}_\epsilon = 3.0 \times 10^{-5} \end{aligned}$$

Next, we calculate the various validation (physical-based) metrics to observe how well they predict/validate a reasonable realization of this data. The results are shown in Table 9.4. From the table, we examine the average results of each of the statistics. Over an ensemble of 100-realizations and subspace identifications, the average residual error is 1.3×10^{-11} as expected, since the synthesized signal is essentially deterministic (100 dB *SNR*). The calculated *model errors* are reasonable for the corresponding average, standard deviation, and *MAD* statistics on the order of 10^{-3} corresponding to less than 1% error in completing the validation.

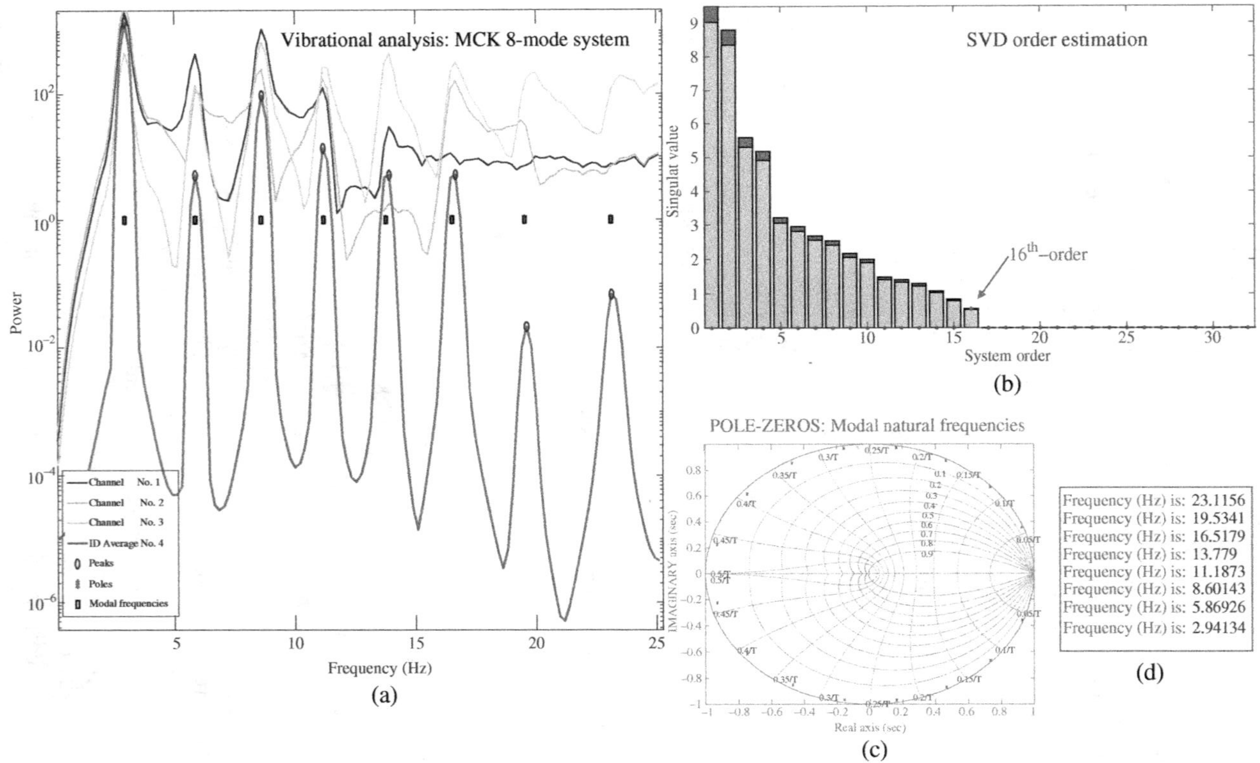

Figure 9.22 Structural vibration subspace identification 8-mode mechanical system. (a) power spectra: outputs/average with poles (boxes). (b) singular values of the Hankel matrix. (c) poles/zeros. (d) Estimated modal frequencies.

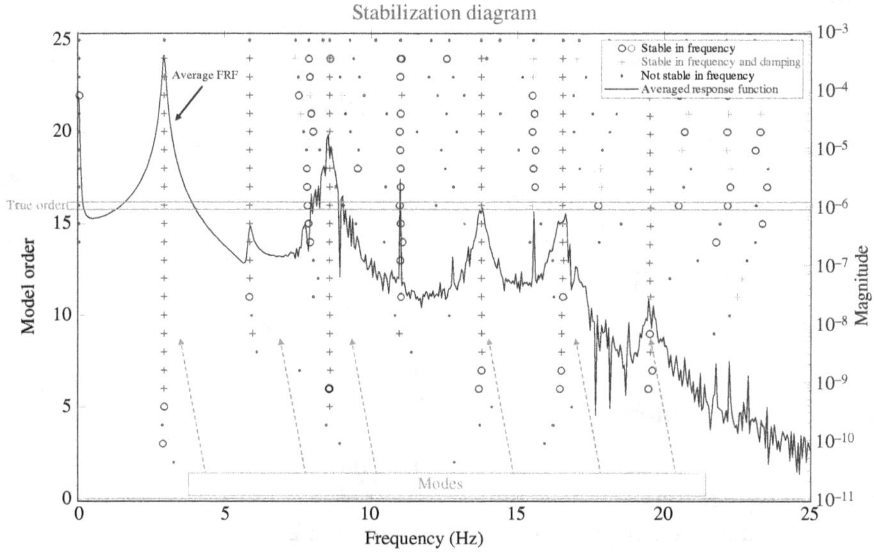

Figure 9.23 8-Mass structural stabilization diagram metric: modal identification (true order = 8 modes—16 poles).

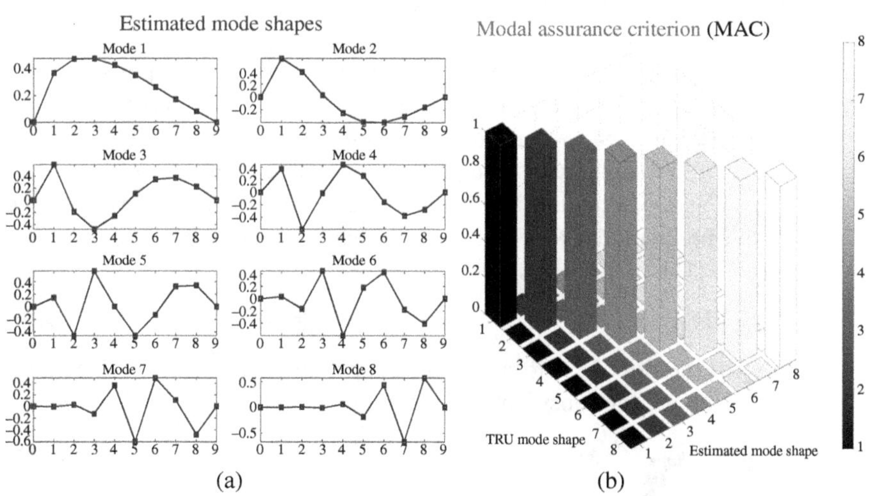

Figure 9.24 8-Mass structure mode shapes: (a) shape estimates. (b) modal assurance criterion metric.

Table 9.4 Statistical metrics for identification of 8-mass structural system.

Model Errror/validation results				
Channel No.	**Residual Error**	**Average Model error**	**Deviation Model error**	**MAD Model error**
1	1.3×10^{-11}	3.3×10^{-3}	4.2×10^{-3}	3.3×10^{-3}
2	1.3×10^{-11}	3.3×10^{-3}	4.1×10^{-3}	3.3×10^{-3}
3	1.3×10^{-11}	3.1×10^{-3}	3.8×10^{-3}	3.1×10^{-3}
AVG	1.3×10^{-11}	3.2×10^{-3}	4.0×10^{-3}	3.2×10^{-3}

9.4.2 Shaping Filters

Consider the design of a shaping filter based on prior knowledge to enhance/ extract subsystem (system) data. When a priori information about a system under investigation is available, then an approach to estimate a subsystem is to perform optimal shape filtering. A shaping filter is based on having an available response of the "desired" filtered output through simulation or testing (calibration, etc.) and using it to constrain the designed filter response. This approach is the classical mean-squared error solution to optimal filtering (Wiener) design [12]. For this application, the desired signal is a subsystem response (noise-free or filtered) obtained during calibration.

The filter design is performed using the system response as input using the a priori subsystem response to constrain the design. The basic structure of the shaping filter design is characterized by the composite signal as input $y(t)$. We assume that a desired subsystem signal, $y_{sub}(t)$ is available from free–free tests during assembly. Therefore, by filtering the measured composite signal, we have the solution to this problem that is well-known and leads to the optimal (Wiener) filter design given by [12]

$$\mathbf{F}_{shape} = \mathbf{R}_{yy}^{-1} \times \mathbf{r}_{yy_{sub}} \tag{9.14}$$

The purpose of the filter in this case is to "shape" the composite measurement signal to approximate the subsystem response, that is,

$$\hat{y}_{sub}(t) = \mathbf{F}_{shape}(t) * y(t) \tag{9.15}$$

completing the process.

Next, we design the processor and apply it to a subsystem extraction for the 8-mass structural problem. Thus, we investigate the application of the shaping filter subsystem estimator to extract modal frequencies to the structural responses illustrated in Figure 9.23. The structure is excited by both a deterministic impulse and a random input at a 0 dB *SNR* with the "true" modal frequencies at:

$$f_{modal} = \{2.94, 5.87, 8.60, 22.29, 13.78, 16.52, 19.54, 23.12\}\text{Hz}$$

A modal decomposition of the system indicates more of the structural details as shown in Figure 9.25 along with the corresponding power spectra of the composite signal (dark) as well as each of the individual modal signals and their corresponding spectra. It is this decomposition that eventually leads to the subsystem extraction techniques discussed earlier.

The shaping filter provides a viable approach to solving the subsystem extraction problem. The first step is to design a filter using a "desired" response or equivalently impulse response of the desired signal. For instance, if we construct a reference *Sub*56 response for these modal outputs, then they can be used as our specified response. The shaping filter is then designed using these data as input and desired response as output to construct the filter impulse response using the mean-squared error (Wiener) solution (Section 7.2) and then apply it (convolution) to any of the data sets available.

The design of the shaping filter is illustrated in Figure 9.26 where the output of the filter is shown in Figure 9.26a and b with the corresponding time series and power spectra (light) and median (dark) along with the peaks at *Sub*56 as desired. The actual desired filter design spectra are shown in Figure 9.26c. The results of applying the various design spectra to the subsystems *Sub56*, *Sub567*, and *Sub34* are shown in Figure 9.27a–c, respectively for both deterministic and random excitations.

9.4.3 Subsystem Modal Extraction

The final step in subsystem extraction is to recover the individual modal responses from the outputs of the multichannel shaping filter outputs. As shown in Figure 9.27, modal subsystem extraction can be accomplished using a black-box ($\hat{\Sigma}_{ABCD}$) subspace identification (*N4SID*) technique and a modal transformation applying similarity transform of Section 6.1 (T_M) to obtain $\Sigma_{A_M B_M C_M D_M}$ [13]:

$$\Lambda = T_M A T_M^{-1} = \text{diag}\left(\underbrace{\begin{bmatrix} \sigma_1 & \omega_1 \\ -\omega_1 & \sigma_1 \end{bmatrix}}_{\Lambda_1}, \ldots, \underbrace{\begin{bmatrix} \sigma_{N_x} & \omega_{N_x} \\ -\omega_{N_x} & \sigma_{N_x} \end{bmatrix}}_{\Lambda_{N_x}} \right) \tag{9.16}$$

with

$$B_M = T_M \times B; C_M = C \times T_M^{-1} \quad \text{and} \quad D_M = D$$

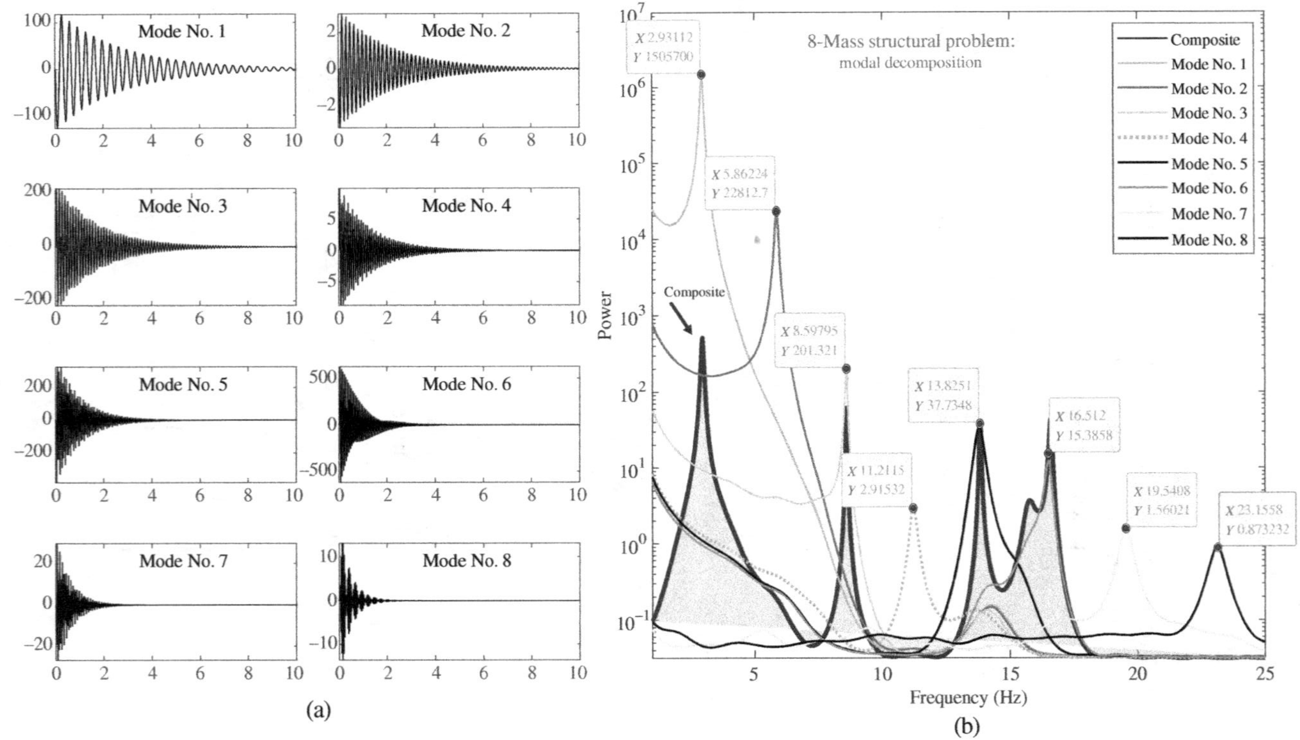

Figure 9.25 Structural 8-mass modal response simulation: (a) impulse responses. (b) composite and modal power spectral responses.

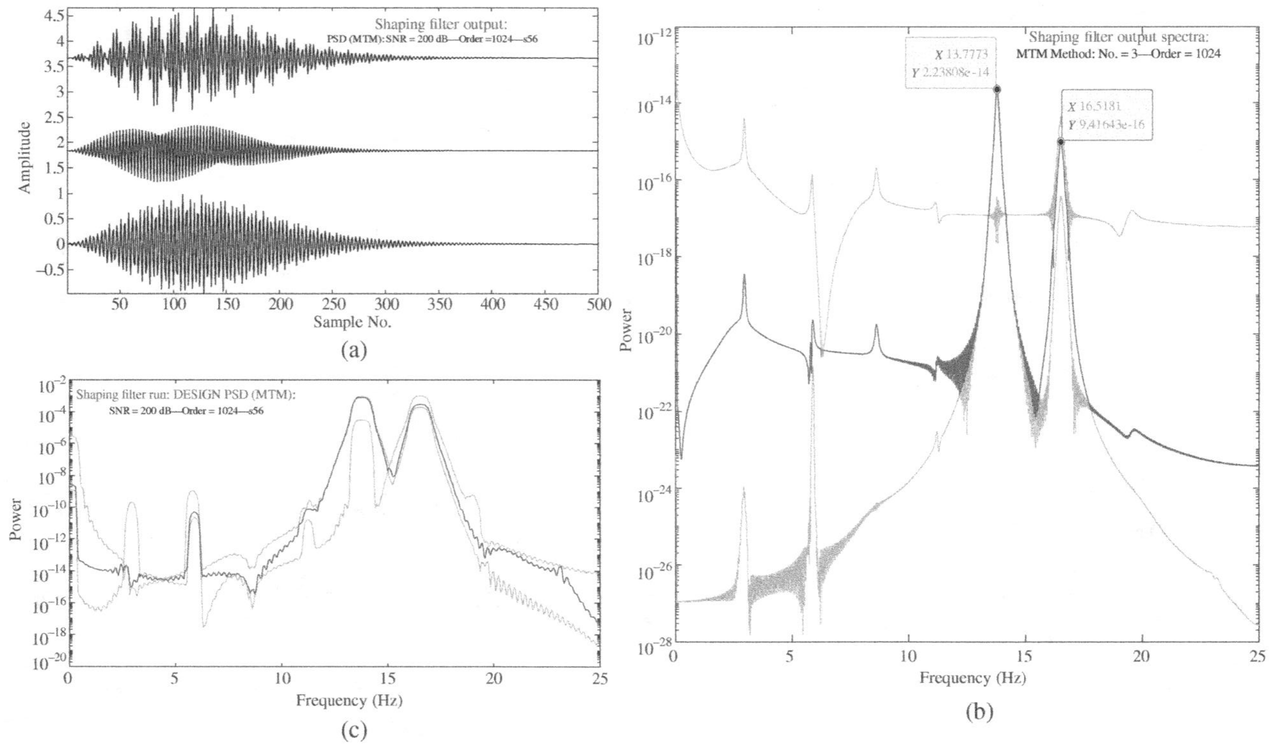

Figure 9.26 8-Mass structure shaping filter subsystem design (impulse response data): (a) subsystem 5–6: filter output. (b) subsystem 5–6: filter output spectra: 13.78, 16.52 Hz. (c) subsystem 5–6 design spectrum: 13.78, 16.52 Hz.

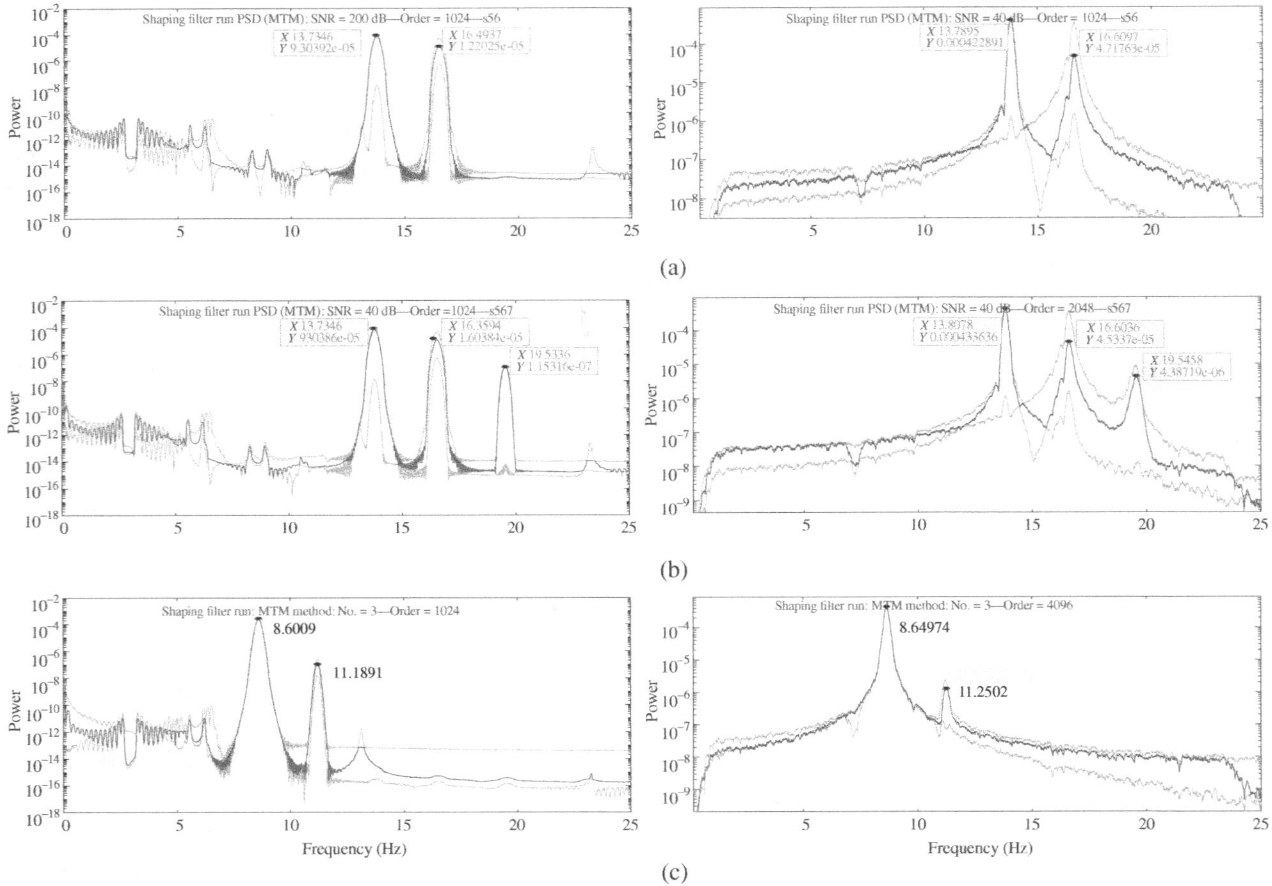

Figure 9.27 8-Mass structure shaping filter subsystem estimation (impulse response/noise excitations): (a) subsystem 5-6: 13.78 Hz, 16.52 Hz. (b) subsystem 5–6–7: 13.78, 16.52, 19.54 Hz. (c) subsystem 3–4: 8.60, 11.19 Hz.

which leads to the individual ith-modal responses (discrete) as:

$$\mathbf{x}_i(t) = \Lambda_i \mathbf{x}_i(t-1) + \mathbf{B}_{\Lambda_i} \mathbf{u}_{\Lambda_i}(t-1)$$
$$y_i(t) = \mathbf{C}_{\Lambda_i} \mathbf{x}_i(t) + \mathbf{D}_{\Lambda_i} \mathbf{u}_{\Lambda_i}(t)$$

Consider a subsystem defined as subsystem 5 – 6 – 7 with resonances at 13.78, 16.52, and 19.54 Hz to be extracted using the shaping filter design resulting in the spectra in Figure 9.28b. The multichannel filter outputs are provided to the identifier assuming that they represent the subsystem impulse responses. For the 3-mode subsystem a sixth-order model was identified from the filter outputs with the results shown in Figure 9.28.

Here, we observe that the identified subsystem model has accurately captured the subsystem as indicated by the model fit metric (%Fit) based on the data ($y(t)$) and its identified model estimate ($\hat{y}(t)$) as

$$\%Fit = 100 \times \left(1 - \frac{y(t) - \hat{y}(t)}{y(t) - \mathrm{E}\{y(t)\}}\right) \tag{9.17}$$

with statistics of 67%, 87%, 88% for each individual measurement channel shown in Figure 9.28a. The corresponding average power spectra are shown in Figure 9.28 with the resonant peaks (squares) parametrically estimated from the eigen-decomposition of the identified system matrix **A**, that is,

$$\lambda\,(\mathbf{A}) = \{13.7788, 16.5263, 19.5402\,\mathrm{Hz}\}$$

matching those of the synthesized 8-mass structural system.

The modal decomposition into each of the individual modal subsystems is shown in Figure 9.29 where the average composite subsystem estimated by the shaping filter is shown as *Mode No. O* as a time series in the left column and a spectrum on the right. Each of the subsystem modal frequencies is indicated by the spectral peaks. Using the identified model, each modal subsystem response and spectrum is also illustrated in the figure with the corresponding modal frequency listed.

Applying the *N4SID*-identifier and transforming to modal coordinates provides the following decomposition of the *discrete* modal subsystems along with the corresponding modal frequency as:

$$\Lambda_1 = \begin{bmatrix} -0.4714 & 0.8531 \\ -0.8531 & -0.4715 \end{bmatrix};\ B_1 = \begin{bmatrix} -3.713 \\ 2.186 \end{bmatrix};\ f_1 = 16.518\,\mathrm{Hz}\ \text{[Subsystem No. 1]}$$

$$\Lambda_2 = \begin{bmatrix} -0.7457 & 0.6127 \\ -0.6127 & -0.7457 \end{bmatrix};\ B_2 = \begin{bmatrix} -2.045 \\ 2.718 \end{bmatrix};\ f_2 = 19.527\,\mathrm{Hz}\ \text{[Subsystem No. 2]}$$

$$\Lambda_3 = \begin{bmatrix} -0.1569 & 0.9677 \\ -0.9677 & -0.1569 \end{bmatrix};\ B_3 = \begin{bmatrix} 1.0170 \\ 2.5950 \end{bmatrix};\ f_3 = 13.779\,\mathrm{Hz}\ \text{[Subsystem No. 3]}$$

This completes the application to the 8-mass structural system.

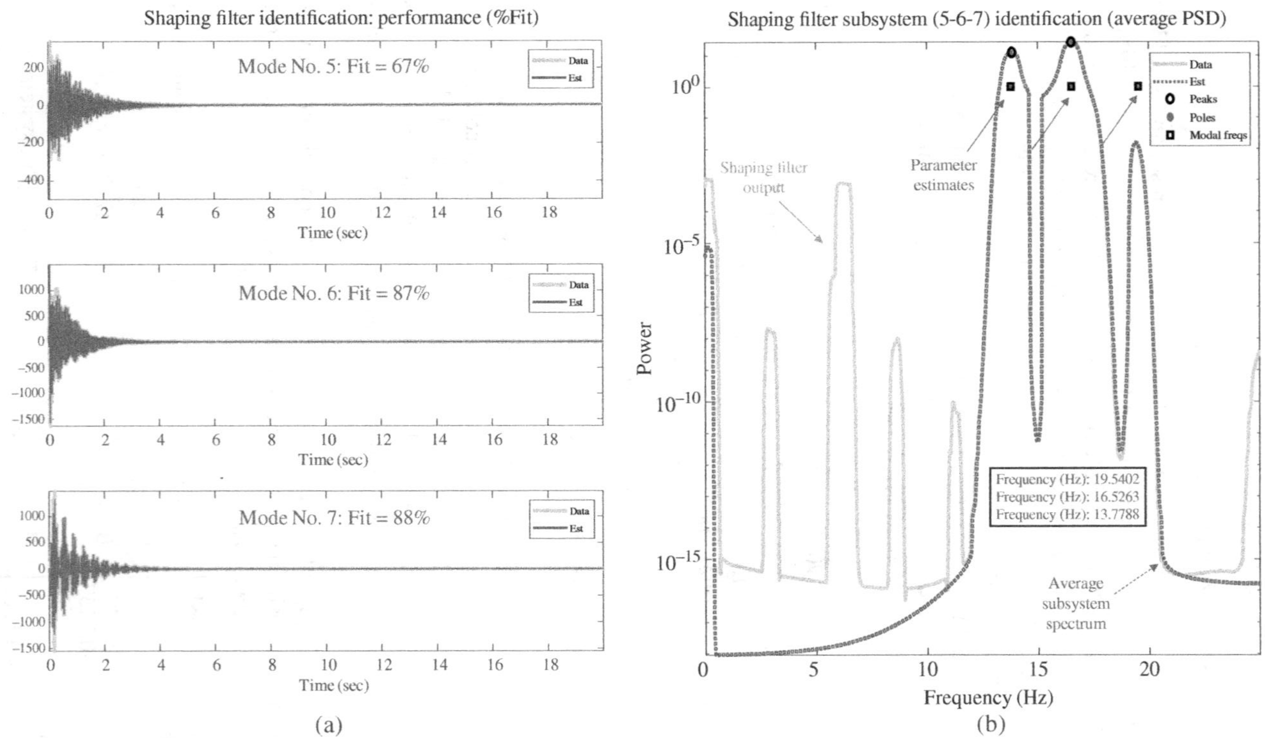

Figure 9.28 8-Mass structure shaping filter multichannel subsystem identification: (a) identified subsystem 5 – 6 – 7, fit (67%, 87%, 88%). (b) average subsystem spectra (light) identified spectrum (dark dots) with parametric frequencies at 13.78, 16.53, 19.54 Hz.

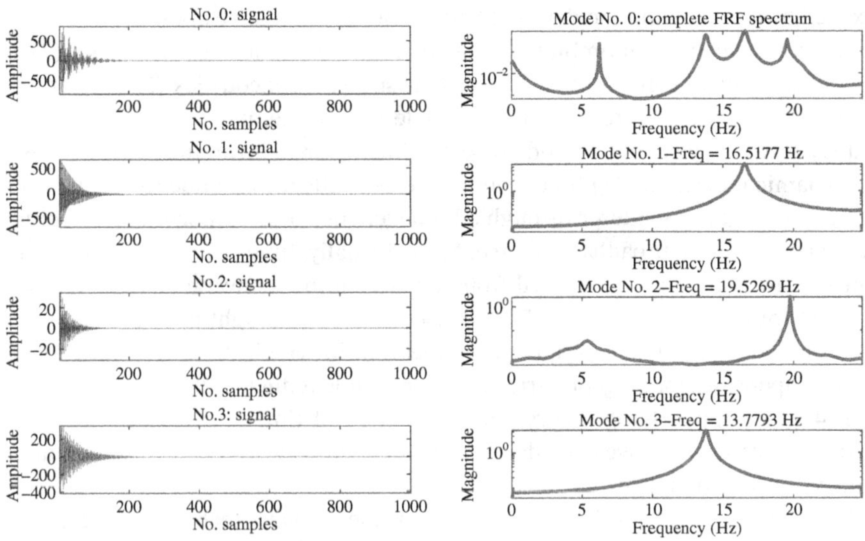

Figure 9.29 8-Mass structure multichannel subsystem decomposition: extracted signals (composite No. 0 and No. 1 – 3 modal subsystems) and spectra with indicated modal frequencies.

9.4.4 Summary

The development of processing techniques to extract the response of an embedded subsystem from a complex system under investigation is discussed. It is demonstrated that a targeted subsystem can be characterized effectively by its modal decomposition. Using a priori subsystem information, a shaping filter method is developed and applied to an 8-mass structure to evaluate and analyze the performance of the approach.

9.5 Data-Based Decomposition: Time-Reversal Processing

When a source propagates through a spatiotemporal medium, a propagating wave is created with its resulting wavefront distorted, much like dropping a pebble into a puddle. If the medium is homogeneous and the source resides in the near field, then a spherical-type wavefront evolves, but if the medium is inhomogeneous, then a distorted wavefront results. In the first case, simple time-delay processing is sufficient to enhance the field at a given point; however, for inhomogeneous media, the required time delays and amplitude are more difficult to estimate [38]. The use of delay estimation becomes limited in an inhomogeneous medium excited by a broadband incident field requiring an alternative approach to solve a focusing or localization problem. A viable alternative called "time-reversal (*T/R*) processing" has been proposed with great success in acoustics [39–41]. It has been shown that time reversal is applicable to spatiotemporal phenomena that satisfy a wave-type equation and possess a time-reversal invariance property [39].

Dynamic focusing using time reversal is essentially a technique to "focus" on a reflective target or scatterer through a homogeneous or inhomogeneous medium that is excited by a broadband source. More formally, time-reversal focusing converts a divergent wave generated from a source into a convergent wave focused on that source as illustrated in Figure 9.30. It can be thought of as an "optimal" spatiotemporal filter that adapts to the medium in which the wavefront evolves and compensates for all geometric distortions while reducing the associated noise [41, 42]. The underlying theory and application of time-reversal techniques to acoustical problems have been developed along with a wide range of applications and proof-in-principle experiments [43].

Space-time decompositions are based on these focusing techniques that are of interest, especially in the biomedical area where ultrasonic waves propagating in tissue media offer the potential to perform noninvasive treatments. For instance, this approach has been demonstrated for the focusing and shattering of painful kidney stones in lithotripsy [40]. Focusing in a homogeneous medium is a matter

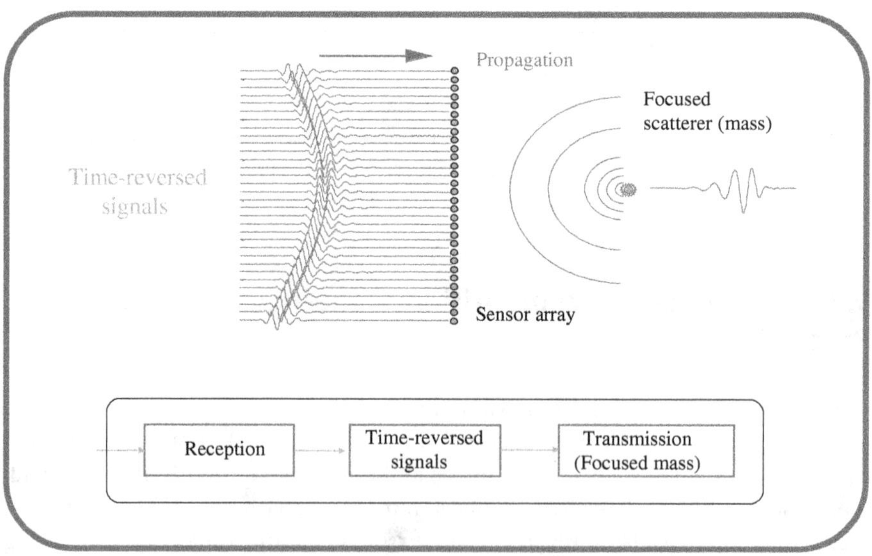

Figure 9.30 Time reversal focusing: After reception of scattered field, the temporal signals are reversed and retransmitted into the medium where the acoustic energy is focused on the scatterer.

of selecting a focal point, calculating the appropriate delays, and corresponding gains for application to an L-element beamforming array [39, 44]. The signal received at the focal point is amplified L times the original transmitted waveform due to the overall gain of an array [44]. Spatially, these signals propagate through the medium and coherently (phase) add at the prescribed focal point to achieve the gain. Clearly, as in lithotripsy, the transmitted energy can be increased at the source to noninvasively to shatter a kidney stone. Phased arrays in radar systems are common and use this basic focusing principle to actively locate and track targets [44]. Synthetic aperture focusing techniques in nondestructive evaluation of materials is another popular focusing technique using a single transducer to synthesize an array for imaging purposes [39, 44].

When the medium is inhomogeneous, then all bets are off, unless the medium is known (model) a priori or "blind" time-reversal processing is used. In the former case, focusing can be achieved by selecting the focal point and using a propagation model of the medium to "synthesize" the correct temporal data that can be time reversed and retransmitted through the inhomogeneous media. If the model is reasonable, then we should expect the propagated ultrasonic waves to coherently superpose at the focal point producing the desired effect. Time reversal, on the other hand, does not require a model to focus, it merely insonifies the media and time reverses the received signal a number of times until the largest scatterer is detected—iterative time reversal. Reversing these signals will focus on the target. In this application, we start with the simple time-reversal focusing principle and then develop an eigen-decomposition approach developed in [43, 45].

Thus, time-reversal processing is a focusing technique that can be used to eliminate the aberrations created by an inhomogeneous or random medium illuminated by propagating waves. This technique can be used to "focus" on the principal scatterer dominating a pulse-echo response. The applicability of time-reversal processing to focus energy without the need to model the medium is a tantalizingly important property, since most media are unknown and random (in the worst case) and frankly temporal coherence (time delay) processing no longer is applicable. A T/R technique simply processes the multichannel time series radiated from the region under investigation, collects the array data, digitizes, time-reverses the temporal array signals, and retransmits them back through the medium to focus on each scatterer.

The basic principle of time-reversal processing, in its simplest form, can succinctly be characterized by the following. Consider the spatiotemporal propagation of a source, $s(r_o; t)$ located at r_o and time t through a medium characterized by the Green's function (impulse response) $G(r, r_o; t)$ from the source at r_o to location r. This operation is given by convolution to yield the received signal, that is,

$$R(r; t) = G(r, r_o; t) * s(r_o; t) \quad \overset{F}{\Leftrightarrow} \quad R(r; \omega) = G(r, r_o; \omega) S(r_o; \omega) \tag{9.18}$$

where for simplicity, we assume a unity scattering coefficient. We have also included the equivalent Fourier transform representation. Based on the underlying theory, we "retransmit" or "back-propagate" from r, through the medium, back to the original source position at r_o, and we choose to transmit the time-reversed signal, $R(r; -t)$, then we have that

$$\hat{s}(r_o; t) = G(r_o, r; t) * R(r; -t) \quad \overset{F}{\Leftrightarrow} \quad \hat{S}(r_o; \omega) = G(r_o, r; \omega)\, R^*(r; \omega) \tag{9.19}$$

utilizing the Fourier transform conjugation property. But substituting the reversed signal into Eq. (9.19) and invoking the Reciprocity Theorem ($G(r_o, r; t) \equiv G(r, r_o; t)$) interchanging source and receiver positions, we obtain

$$\hat{s}(r_o, t) = G(r_o, r; t) * G(r_o, r; -t) * s(r_o; -t) \overset{F}{\Leftrightarrow} \hat{S}(r_o; \omega) = \underbrace{|G(r, r_o; \omega)|^2}_{K} S^*(r_o; \omega) \tag{9.20}$$

which implies that the reversed signals retransmitted through the medium will "focus" the enhanced energy (with gain K) back to the original source position with no change in phase precisely demonstrating the broadband version of phase conjugation. The *T/R*-technique simply processes the multichannel data radiated from the region under investigation, collects the array data, digitizes, time-reverses these temporal array signals, and retransmits them back through the medium to focus on each scatterer (see Figure 9.30). This completes the background on time-reversal processing, next we discuss the to solve this application problem.

9.5.1 Iterative Time-Reversal Decomposition

Suppose we have an N_L-element array of sensors sampling the scattered field of a medium characterized by its underlying Green's function, $g(r_\ell, r_k; t)$, then the field received at the ℓth-sensor induced by the N_S-scatterers at the ith-iteration is given by

$$R_i(\mathbf{r}_\ell; t) = \sum_{k=0}^{N_s-1} g(\mathbf{r}_\ell, \mathbf{r}_k; t) * q_i(\mathbf{r}_k; t), \quad \ell = 0, \dots, N_L - 1 \quad \text{[CATCH]} \tag{9.21}$$

where the scattered return function is defined by

$$q_i(\mathbf{r}_k; t) = c(\mathbf{r}_k; t) * s_i(\mathbf{r}_k; t) \quad \text{for} \quad k = 0, \dots, N_s - 1 \tag{9.22}$$

with $c(r_k; t)$ the corresponding reflectivity or scattering coefficient. The field induced at the kth-scatterer by the incident field, $e_i(r_k; t)$ is given by

$$s_i(\mathbf{r}_k; t) = \sum_{\ell=0}^{N_L-1} g(\mathbf{r}_k, \mathbf{r}_\ell; t) * e_i(\mathbf{r}_\ell; t), \quad k = 0, \dots, N_s - 1 \quad \text{[PITCH]} \tag{9.23}$$

In time-reversal processing during each phase of the "transmit-receive" or "pitch-catch" cycle, the medium is insonified with the time-reversed field transmitted from the array. Initially, we can transmit (pitch) a planar wave field or a pulse, $e(\mathbf{r};t)$, for instance, to excite all of the inherent scatterers as in Eq. (9.23). The scatterers, in turn, retransmit or scatter the wave energy back to the array (catch) such that we receive the initial scattered field at the ℓth-sensor as in Eq. (9.21) initiating the pitch-catch cycle $R_0(\mathbf{r}_\ell;t)$. For the next iteration, the transmitted excitation is the "time-reversed" field received at the previous iteration (reverse), that is,

$$e_i(\mathbf{r};t) = R_{i-1}(\mathbf{r};-t) \qquad \text{[REVERSE]} \tag{9.24}$$

inducing the next field "pitched" to the kth-scatterer from the reversed field as $s_i(\mathbf{r}_k;t)$ completing the ith-iteration cycle. We depict the pitch-catch-reverse cycle of time-reversal focusing in Figure 9.31. Here, we observe the basic sequence of events.

We can use the iteration cycle to perform the required analysis. The focusing capability of the T/R-processor was proven mathematically through an *eigen-decomposition* of the T/R-operator, when the scatterer can be characterized by a simple scattering matrix, the T/R-processor extracts the largest eigenvalue. That is, the scattering matrix can be *decomposed* into its constituent eigenvalues/eigenfunctions as $\lambda_1(\omega) > \cdots \lambda_{N_S}(\omega)$ with the T/R selecting, $\lambda_1(\omega)$, the largest [43, 45]. This property can be demonstrated through a simple simulation of targets embedded in a homogeneous medium. If we were able to put a receiver

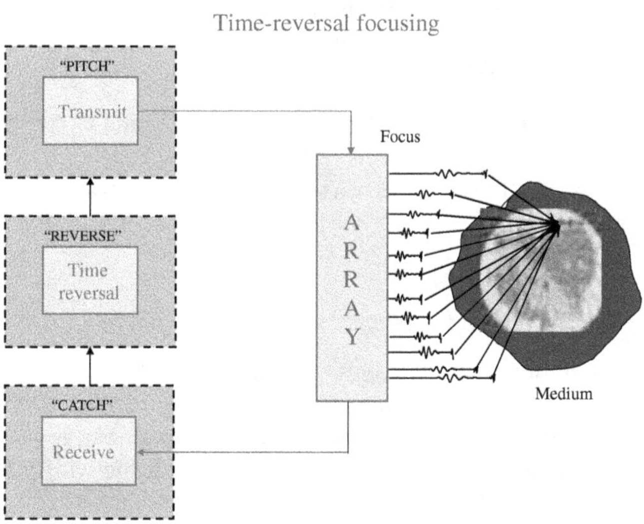

Figure 9.31 *T/R* focusing PITCH-CATCH-REVERSE sequence.

at each targeted scatterer and observe the acoustic energy, we see that after seven iterations the largest scatterer survives and still has a return as observed in Figure 9.32. We discuss eigen focusing next.

It is possible to perform this cycle "iteratively" by focusing on the strongest target, receiving its scattered field, and removing it from the time series data as depicted in Figure 9.32. Thus, the scattered field would be "decomposed" with each individual scatterer extracted. Here, the strongest scatterer signature is removed from the data sequentially until no other scatterers exist [46, 47]. This completes the section on iterative time-reversal processing, next we consider the eigen-decomposition approach to the extraction problem.

9.5.2 Eigen-decomposition Time-reversal Extraction

In this section, we discuss the eigen-decomposition approach to time-reversal scattering extraction by focusing. We briefly outline the approach discussed in [43] and [45].

The signals received at the array are given by

$$R(\mathbf{r};t) = G(\mathbf{r};t) * q(\mathbf{r};t) \qquad \text{[CATCH]} \tag{9.25}$$

and at the scatterers (targets) are

$$s(\mathbf{r};t) = G^T(\mathbf{r};t) * e(\mathbf{r};t) \qquad \text{[PITCH]} \tag{9.26}$$

with the reversal operation performed, after reception, as

$$e(\mathbf{r};t) = R(\mathbf{r};-t) \qquad \text{[REVERSE]} \tag{9.27}$$

where we have that $s(\mathbf{r};t), q(\mathbf{r};t) \in \mathbf{R}^{N_s \times 1}, R(\mathbf{r};t), e(\mathbf{r};t) \in \mathbf{R}^{N_L \times 1}$ and $G(\mathbf{r};t) \in \mathbf{R}^{N_L \times N_s}$.

Next, let us assume that the medium consists of a set of point scatterers allowing only single scattering, then at the kth-scatterer, the received field is retransmitted or reflected by the scatterer as

$$q(\mathbf{r}_k;t) = c(\mathbf{r}_k;t) * s(\mathbf{r}_k;t) \quad \text{for} \quad k = 0,\ldots,N_s - 1 \tag{9.28}$$

where $c(\mathbf{r}_k;t)$ is the corresponding reflectivity or scattering coefficient.

With this relation in mind, we can define the overall multichannel wave system transfer (function) matrix in the temporal frequency domain by taking the appropriate Fourier transforms and substituting the results. That is, starting with the CATCH and substituting the Fourier transform of Eq. (9.28), we obtain

$$R(\mathbf{r};\omega) = G(\mathbf{r};\omega)Q(\mathbf{r};\omega) = G(\mathbf{r};\omega)C(\mathbf{r};\omega)S(\mathbf{r};\omega) \tag{9.29}$$

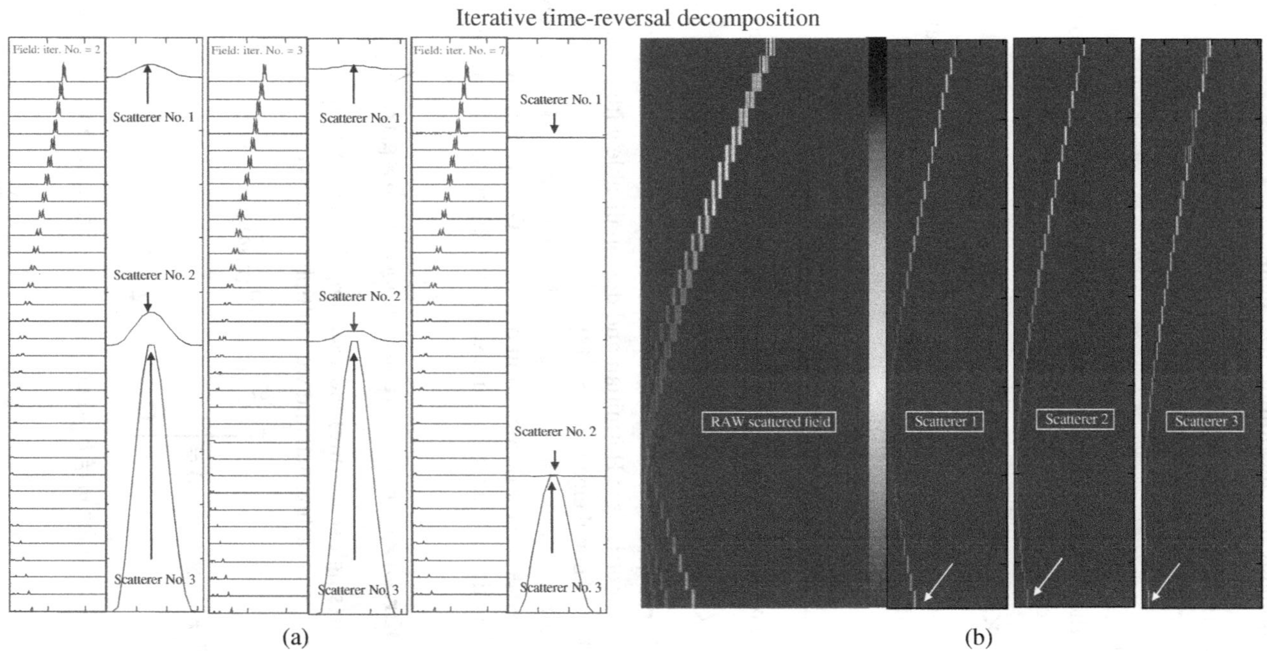

Figure 9.32 *T/R* focusing pitch-catch-reverse iteration sequences: (a) iterations: receiver field and source amplitudes. (b) scatterer decomposition.

where $C(\mathbf{r};\omega)$ is diagonal due to the single scattering assumption. Using the PITCH in the frequency domain and substituting, we obtain

$$R(\mathbf{r};\omega) = \underbrace{G(\mathbf{r};\omega)C(\mathbf{r};\omega)G^T(\mathbf{r};\omega)}_{\mathbf{K}(\mathbf{r};\omega)} E(\mathbf{r};\omega) \tag{9.30}$$

with the $N_L \times N_L$ transfer matrix defined by $\mathbf{K}(\mathbf{r};\omega)$

If we perform the "iterative time-reversal" process described in Section 9.5.1, then Eq. (9.30) becomes

$$R_i(\mathbf{r};\omega) = \mathbf{K}(\mathbf{r};\omega) \times E_i(\mathbf{r};\omega) \quad \text{for} \quad i = 0, 1, 2, \ldots \tag{9.31}$$

in terms of the ith-iteration (pitch-catch-reverse). Performing the reversal operation after the initial excitation vector, E_o and received vector, R_o are available, we have that

$$E_1(\mathbf{r};\omega) = R_o^*(\mathbf{r};\omega) = \mathbf{K}^*(\mathbf{r};\omega) \times E_o^*(\mathbf{r};\omega) \tag{9.32}$$

and therefore, the received signal for the first iteration (Eq. (9.31)) becomes

$$R_1(\mathbf{r};\omega) = \mathbf{K}(\mathbf{r};\omega) \times E_1(\mathbf{r};\omega) = \mathbf{K}(\mathbf{r};\omega) \times \left[\mathbf{K}^*(\mathbf{r};\omega) \times E_o^*(\mathbf{r};\omega)\right] \tag{9.33}$$

where $\left[\mathbf{K}(\mathbf{r};\omega)\mathbf{K}^*(\mathbf{r};\omega)\right]$ is defined as the *time-reversal operator*. Since this operator is complex and symmetric, it is Hermitian and diagonalizable with positive, real eigenvalues, and orthogonal eigenvectors [48].

It is possible to focus transmitted energy exclusively on the kth-scatterer by using the time-reversed vector (kth-column of $G(r;t)$) Green's function or equivalently transfer function, for as the input excitation signal, then from Eq. (9.30), we have [45]

$$R(\mathbf{r};\omega) = \mathbf{K}(\mathbf{r};\omega) \times \mathbf{g}_k^*(\mathbf{r};\omega) = G(\mathbf{r};\omega)C(\mathbf{r};\omega)G^T(\mathbf{r};\omega) \times \mathbf{g}_k^*(\mathbf{r};\omega) \tag{9.34}$$

Since C is diagonal (single scattering) and assuming that the scatterers are "ideally resolved," then the g_k^* is orthogonal and we multiply out this expression yielding,

$$C(\mathbf{r};\omega)G^T(\mathbf{r};\omega) \times \mathbf{g}_\mathbf{k}^*(\mathbf{r};\omega) = \begin{bmatrix} C_{11}(\mathbf{r};\omega)\underline{G}_1^T(\mathbf{r};\omega)\mathbf{g}_k^*(\mathbf{r};\omega) \\ \vdots \\ C_{N_sN_s}(\mathbf{r};\omega)\underline{G}_{N_s}^T(\mathbf{r};\omega)\mathbf{g}_k^*(\mathbf{r};\omega) \end{bmatrix} \tag{9.35}$$

but $\underline{G}_m^T(\mathbf{r};\omega)\mathbf{g}_k^*(\mathbf{r};\omega) = 0$ for $m \neq k$, then only the kth-column survives with element

$$\lambda_k(\omega) \equiv \left(C(\mathbf{r};\omega)G^T(\mathbf{r};\omega) \times \mathbf{g}_k^*(\mathbf{r};\omega)\right) \underline{e}_k = C_{kk}(\mathbf{r};\omega)\sum_{\ell=1}^{N_L} |g_{k\ell}(\mathbf{r};\omega)|^2 \tag{9.36}$$

where $\underline{e}_k$ is an N_L-unit vector with unity at the kth-row. Multiplying Eq. (9.34) using the orthogonality property and the unit vector, we obtain

$$\mathbf{K}(\mathbf{r};\omega) \times \mathbf{g}_k^*(\mathbf{r};\omega) = R(\mathbf{r};\omega) = \lambda_k(\omega) G(\mathbf{r};\omega)\underline{e}_k = \lambda_k(\omega)\mathbf{g}_k(\mathbf{r};\omega) \tag{9.37}$$

and therefore

$$\mathbf{K}(\mathbf{r};\omega)\mathbf{g}_k^*(\mathbf{r};\omega) = \lambda_k(\omega)\mathbf{g}_k(\mathbf{r};\omega) \tag{9.38}$$

Pre-multiplying both sides by $\mathbf{K}^\dagger$ and using the fact that the conjugate of Eq. (9.38) is

$$\mathbf{K}^\dagger(\mathbf{r};\omega) \times \mathbf{g}_k(\mathbf{r};\omega) = \lambda_k(\omega) \times \mathbf{g}_k^*(\mathbf{r};\omega) \tag{9.39}$$

yields the eigen-equation

$$\left[\mathbf{K}^\dagger(\mathbf{r};\omega)\mathbf{K}(\mathbf{r};\omega)\right] \mathbf{g}_k^*(\mathbf{r};\omega) = \lambda_k^2(\omega)\mathbf{g}_k^*(\mathbf{r};\omega) \tag{9.40}$$

with eigenvalue, $\lambda_k^2(\omega)$, and eigenvector, $\mathbf{g}_k^*(\mathbf{r};\omega)$, also an eigenvector of $\mathbf{K}$.

What this result implies is that for ideally separated scatterers, it is possible to focus on each individually. It is also interesting to note that to solve this problem we can employ the *SVD* of the Hermitian matrix, $\mathbf{K}^\dagger\mathbf{K}$, that is

$$\left[\mathbf{K}^\dagger(\mathbf{r};\omega)\mathbf{K}(\mathbf{r};\omega)\right] = \mathbf{V}(\mathbf{r};\omega) \times \mathbf{\Lambda}(\omega) \times \mathbf{V}^\dagger(\mathbf{r};\omega) \tag{9.41}$$

where the $N_L \times N_L$-matrix, $\mathbf{\Lambda} = \text{diag}\left[\lambda_0^2(\omega) \cdots \lambda_{N_s-1}^2(\omega) \quad 0 \cdots 0\right]$ of rank N_s contains the singular values (squared-eigenvalues). The singular values are ordered as $\left(\lambda_0^2(\omega) > \lambda_2^2(\omega) \cdots > \lambda_{N_s-1}^2(\omega)\right)$ corresponding to the strongest scatterer progressing down to the weakest.

We summarize a few of the interesting properties of the time-reversal decomposition:

- individual scatterers and eigenvectors are functions of temporal frequency, ω, that is, $\{\lambda_k^2(\omega), \mathbf{g}_k^*(\mathbf{r};\omega)\}$;
- individual scatterers are uniquely identified by their eigenvectors at a given temporal frequency;
- each scatterer (kth-scatterer) has its own unique propagation path which is completely captured by its corresponding eigenvector, $\mathbf{g}_k^*(\mathbf{r};\omega)$;
- the kth-singular value of $\mathbf{K}^\dagger\mathbf{K}$ is related to the kth-eigenvector or kth-column vector of V, that is, $\mathbf{V}(\mathbf{r};\omega)\,\underline{e}_k = \mathbf{g}_k(\mathbf{r};t)$.

With this information in mind, we summarize the eigen-decomposition approach to focusing in Figure 9.33. The transfer function matrix is created by exciting each array element with a broadband pulse (impulse like), the time-series data at the receive array is acquired and then Fourier transformed. Next, a *SVD* of the Hermitian product of the transfer function matrix is performed at the

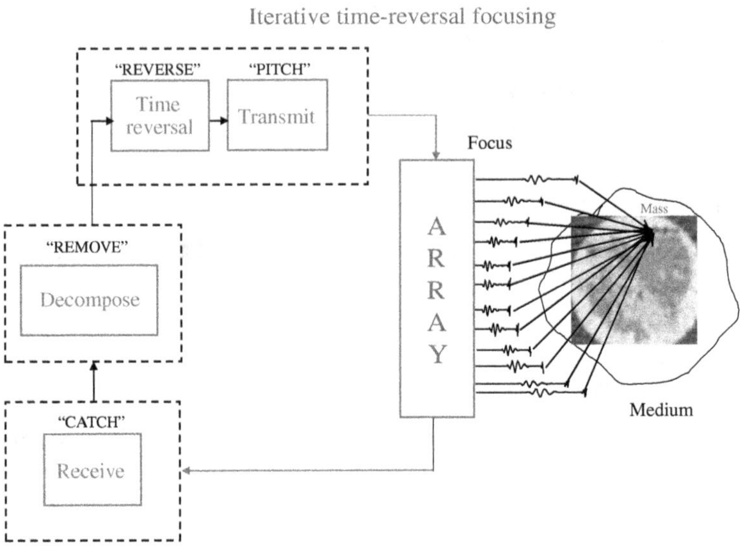

Figure 9.33 Eigen-decomposition selective focusing using the singular value decomposition of the rime reversal operator.

given temporal frequency selected to find the singular values and corresponding eigenvectors used as the eigen-weights (delays and gains). Once the weights are calculated, a narrowband planar pulse signal at the temporal frequency is transmitted into the medium. The waves created coherently superpose at the scatterer location corresponding to the selected singular value.

Thus, we see how the transfer matrix is used to perform the required eigen-decomposition and focus energy at the desired target. As an application of eigen-focusing, we show the results of extracting flaws in glass, which is a homogeneous medium (aluminum) using *T/R*. As shown in Figure 9.34, we see the image of the glass with the two flaws displayed. Here, we merely performed the processing with a spherical wave model (beamforming) to obtain the image. Next, we focused on the strongest flaw using this eigen-decomposition procedure and chose the eigenvectors corresponding to the two strongest singular values (eigenvalues). Using a narrowband pulsed signal weighted by the eigenvectors (delay and attenuation), we transmitted the wave back to the medium and imaged the received energy as before (spherical wave model) using the processor. The results shown in Figures 9.34b and 9.34c, clearly demonstrate that using the transmitted eigenvector as the incident field the propagating waves superpose at the correct flaw position for each flaw illustrating the capability of this decomposition completely the application.

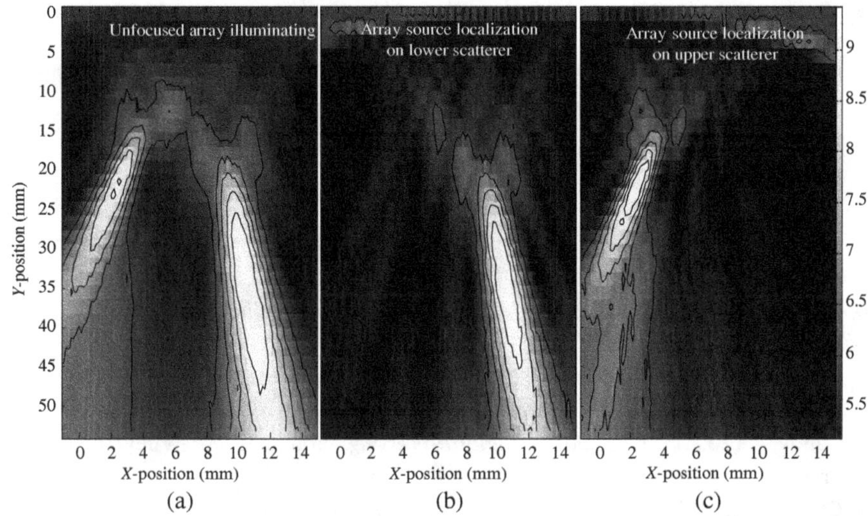

Figure 9.34 Eigen-decomposition selective focusing in aluminum: (a) composite image of material. (b) decomposed strongest target (flaw) extraction. (c) decomposed weaker target (flaw) extraction).

9.5.3 Summary

In the spatiotemporal domain that supports a wide variety of applications, decomposition methods are highly desirable. Here, a powerful methodology that simply reverses a temporal signal is discussed and shown to separate and extract individual targets (sources) from noisy environments.

References

1. Junger, M. and Feit, D. (1986). *Sound: Structures and Their Interaction*. Cambridge, MA: MIT Press.
2. Hatch, M.R. (2001). *Vibration Simulation Using MATLAB and ANSYS*. Boca Raton, FL: Chapman and Hall/CRC Press.
3. Gawronski, W.K. (2004). *Advanced Structural Dynamics and Active Control of Structures*. London: Springer.
4. Rainieri, C. and Fabbrocino, G. (2014). *Operational Modal Analysis of Civil Engineering Structures*. London: Springer.
5. Candy, J., Franco, S., Ruggiero, E. et al. (2017). Anomaly detection for a vibrating structure: a subspace identification/tracking approach. *J. Acoust. Soc. Am.* 142 (2): 680–696.

6 Merovitch, L. (1996). *Analytic Methods in Vibrations*. New York: MacMillan Publishers.

7 COMSOL (2023). Multiphysics® Version 5.1a. http://www.comsol.com (14 June 2024).

8 Juang, J. (1994). *Applied System Identification*. Upper Saddle River, NJ: Prentice-Hall PTR.

9 van Overschee, P. and De Moor, B. (1996). *Subspace Identification for Linear Systems: Theory, Implementation, Applications*. Boston, MA: Kluwer Academic Publishers.

10 Katayama, T. (2005). *Subspace Methods for System Identification*. London: Springer.

11 Verhaegen, M. and Verdult, V. (2007). *Filtering and System Identification: A least-Squares Approach*. Cambridge: Cambridge University Press.

12 Candy, J. (2006). *Model-Based Signal Processing*. Hoboken, NJ: Wiley/IEEE Press.

13 Candy, J. (2019). *Model-Based Processing: An Applied Subspace Identification Approach*, 511 pages. Hoboken, NJ: Wiley.

14 Candy, J., Case, J., Fisher, K. et al. (2019). Signal Processing for Sounding Rocket Flights: Campaign No. 2 Results. LLNL Report, *LLNL-TR-786174*.

15 Candy, J., Case, J., Fisher, K. et al. (2020). Vibrational processing of a dynamic structural flight system: a multichannel spectral estimation approach. *J. Acoust. Soc. Am.* 147 (4): 2694–2712.

16 Candy, J.V., Case, J.E., Fisher, K.A. et al. (2019). Multichannel processing of vibrational measurements: a constrained subspace application. *J. Acoust. Soc. Am.* 146 (4): 2350–2372.

17 Irvine, T. (1999). *Application of the Newton-Raphson Method to Vibration Problems*. Vibrationdata Publications.

18 Maynard, J. (1996). Resonant ultrasound spectroscopy. *Phys. Today* 49: 26–31.

19 Balakirev, F., Ennaceur, S., and Migliori, R. (2019). Resonant ultrasound spectroscopy: the essential toolbox. *Rev. Sci. Instrum.* 90: 121401, 1–14.

20 Ulrich, T. and McCall, K. (2002). Determination of elastic moduli of rock samples using resonant ultrasound spectroscopy. *J. Acoust. Soc. Am.* 111 (4): 1667–1674.

21 Remillieux, M., Ulrich, T., Payan, C. et al. (2015). Resonant ultrasound spectroscopy for materials with high damping and samples of arbitrary geometry. *J. Geophys. Res.: Solid Earth* 1250: 4898–4916.

22 Migliori, A. and Sarrao, J. (1997). *Resonant Ultrasound Spectroscopy: Applications to Physics, Materials Measurements, Nondestructive Evaluation*. New York: Wiley.

23 Zadler, B., Le Rousseau, J., Scales, J., and Smith, M. (2004). Resonant ultrasound spectroscopy: theory and application. *Geophys. J. Int.* 156: 154–169.

24 Kinsler, L., Frey, A., Coppens, A., and Sanders, J. (1982). *Fundamentals of Acoustics*. New York: Wiley.

25 Fisher, K. (2020). Estimation of elastic properties of an additively manufactured lattice using resonant ultrasound spectroscopy. *J. Acoust. Soc. Am.* 148 (6): 4025–4036.

26 Green, R. (1973). *Ultrasonic Investigation of Mechanical Properties*, Treatise on Materials Science and Technology, vol. 3. New York: Academic Press.

27 Visscher, W., Migliori, A., Bell, T., and Reinert, R. (1991). On the normal modes of free vibration of inhomogeneous and anisotropic elastic objects. *J. Acoust. Soc. Am.* 90 (4): 2154–2162.

28 Plesek, J., Kolman, R., and Landa, M. (2004). Using finite element method for the determination of elastic moduli by resonant ultrasound spectroscopy. *J. Acoust. Soc. Am.* 116 (1): 282–287.

29 Shen, F., Fan, F., Wang, R. et al. (2020). Particle swarm optimization-based identification of the elastic properties in resonant ultrasound spectroscopy. *IEEE Trans. Ultrason. Ferroelectr. Freq. Control* 67 (7): 1412–1423.

30 Reyenders, E. (2012). System identification methods for modal analysis: review and comparison. *Arch. Comput. Meth. Eng.* 19 (1): 51–124.

31 Shih, C., Tsuei, T., Allemang, R., and Brown, D. (1988). Complex mode indicator function and its applications to spatial domain parameter estimation. *Mech. Syst. Signal Process.* 2 (4): 367–377.

32 Allemang, R. and Brown, D. (2006). A complete review of the complex mode indicator function (CMIF) with applications. Proceedings of ISMA Conference, Leuven, Belgium.

33 Papai, F. and Szucs, I. (2015). The role of singular values of measured frequency response function matrix in modal damping estimation. *Tehnicki Vjesnik* 22 (3): 729–734.

34 J. Elmer, "Wire-base additive manufacturing of stainless-steel components," *Welding J.*, 99, S8-S24, doi:https://doi.org/10.29391/2020.99.002, 2020.

35 Candy, J., Fisher, K., and Elmer, J. (2022). Processing of resonant frequency ultrasonic measurements for an additive manufacturing application. LLNL Report, *LLNL-TR-833002*.

36 Elmer, J.W., Fisher, K., Gibbs, G. et al. (2022). Post-build thermomechanical processing of wire arc additively manufactured stainless steel for improved mechanical properties and reduction of crystallographic texture. *Addit. Manuf.* 50: 102573.

37 Cara, F., Juan, J., Alarcon, E. et al. (2013). Modal contribution and state space order selection in operational modal analysis. *Mech. Syst. Signal Process.* 38: 276–298.

38 Dorme, C. and Fink, M. (1995). Focusing in transmit-receive mode through and inhomogeneous media: the time-reversal matched filter approach. *J. Acoust. Soc. Am.* 98 (2): 62–71.
39 Fink, M. (1992). Time reversal of ultrasonic fields—Part I: Basic principles. *IEEE Trans. Ultrason. Ferroelect. Freq. Control* 39: 555–567.
40 Fink, M. (1993). Time reversal mirrors. *J. Phys. D: Appl. Phys.* 26: 1333–1350.
41 Fink, M. (1996). Time reversal in acoustics. *Contemp. Phys.* 37 (2): 95–109.
42 Chambers, D., Lehman, S., and Candy, J. (2004). Time reversal and the spatio-temporal matched filter. *J. Acoust. Soc. Am.* 115 (6): 3095–3110.
43 Prada, C., Wu, F., and Fink, M. (1991). The iterative time-reversal mirror: a solution to self-focusing in pulse-echo mode. *J. Acoust. Soc. Am.* 90 (1): 1119–1129.
44 Dudgeon, D.E. and Mersereau, R.M. (1984). *Multidimensional Digital Signal Processing*, Prentice-Hall Signal Processing Series. Prentice-Hall.
45 Prada, C., Thomas, J.L., and Fink, M. (1995). The iterative time-reversal process: analysis of convergence. *J. Acoust. Soc. Am.* 97 (1): 62–71.
46 Candy, J. (2003). Reversing time: a way to unravel distorted communications. *J. Acoust. Soc. Am.* 113 (4 Pt. 2): 2212.
47 Candy, J. (2002). Method and apparatus for dynamic focusing of ultrasound energy. Patent, (US 6,490,469 B2) and International Patent (WO 01/69283 A2).
48 Strang, G. (1976). *Linear Algebra*. New York: Academic Press.

A

Probability and Statistics Overview

A.1 Probability Theory

Defining a sample space (outcomes), Ω, a field (events), B, and a probability function (on a class of events), Pr, we can construct an *experiment* as the triple, $\{\Omega, B, \Pr\}$.

Example A.1 Consider the experiment, $\{\Omega, B, \Pr\}$ of tossing a fair coin, then we see that

$$\begin{array}{lll} \text{Sample space:} & \Omega & = \{H, T\} \\ \text{Events:} & B & = \{0, \{H\}, \{T\}\} \\ \text{Probability:} & \Pr(H) & = p \\ & \Pr(T) & = 1 - p \end{array}$$

△△△

With the idea of a sample space, probability function, and experiment in mind, we can now start to define the concept of a discrete random signal more precisely. We define a discrete *random variable* as a real function whose value is determined by the outcome of an experiment. It assigns a real number to each point of a sample space Ω, which consists of all the possible outcomes of the experiment. A random variable X and its realization x are written as

$$X(\omega) = x \qquad \text{for} \qquad \omega \epsilon \Omega \tag{A.1}$$

Consider the following example of a simple experiment.

Example A.2 We are asked to analyze the experiment of flipping a fair coin, then the sample space consists of a head or tail as possible outcomes, that is,

$$\Omega = \left\{ 0 \quad H \quad T \right\} \Longrightarrow X(\omega) = x$$

$$\omega = \{H, T\}$$

Signal Processing: An Applied Decomposition Approach, First Edition. James Vincent Candy.

Companion website: www.wiley.com/go/CandySignProc

If we assign a 1 for a head and 0 for a tail, then the random variable X performs the mapping of

$$X(\omega = H) = x(H) = 1$$
$$X(\omega = T) = x(T) = 0$$

where $x(.)$ is called the sample value or realization of the random variable X.

△△△

A *probability mass function* is defined in terms of the random variable, that is,

$$P_X(x_i) = \Pr\left(X(\omega_i) = x_i\right) \tag{A.2}$$

and the *probability distribution function* is defined by

$$F_X(x_i) = \Pr\left(X(\omega_i) \le x_i\right) \tag{A.3}$$

These are related by

$$P_X(x_i) = \textstyle\sum_i F_X(x_i)\delta\left(x - x_i\right) \tag{A.4}$$
$$F_X(x_i) = \textstyle\sum_i P_X(x_i)\mu\left(x - x_i\right) \tag{A.5}$$

where δ and μ are the unit impulse and step functions, respectively.

It is easy to show that the distribution function is a monotonically increasing function (see Papoulis [1] for details) satisfying the following properties,

$$\lim_{x_i \to -\infty} F_X(x_i) = 0$$

and

$$\lim_{x_i \to \infty} F_X(x_i) = 1$$

These properties can be used to show that the mass function satisfies

$$\sum_i P_X(x_i) = 1$$

Either the distribution or probability mass function completely describes the properties of a random variable. Given either of these functions, we can calculate probabilities that the random variable takes on values in any set of events on the real line. To complete our coin-tossing example, if we define the probability of a head occurring as p, then we can calculate the distribution and mass functions as shown in the following example.

Example A.3 Consider the coin-tossing experiment and calculate the corresponding mass and distribution functions. From the previous example, we have

Sample space:	Ω	=	$\{H, T\}$
Events:	B	=	$\{0, \{H\}, \{T\}\}$

$$\begin{array}{llcl}
\text{Probability:} & P_X(x_1 = H) & = & p \\
 & P_X(x_0 = T) & = & 1 - p \\
\text{Random variable:} & X(\omega_1 = H) & = & x_1 = 1 \\
 & X(\omega_2 = T) & = & x_2 = 0 \\
\text{Distribution:} & F_X(x_i) & = & \begin{cases} 1 & x_i \geq 1 \\ 1 - p & 0 \leq x_i \leq 1 \\ 0 & x_i < 0 \end{cases}
\end{array}$$

the mass and distribution functions for this example are shown in Figure A.1. Note that the sum of the mass function value must be 1 and that the maximum value of the distribution function is 1 satisfying the properties mentioned previously.

△△△

If we extend the idea that a random variable is now a function of time as well, then we can define a stochastic process as discussed in Chapter 2. More formerly, a random or *stochastic process* is a two-dimensional function of t and ω:

$$X(t,\omega) \qquad \omega\epsilon\Omega, \quad t\epsilon T \tag{A.6}$$

where T is a set of index parameters (continuous or discrete) and Ω is the sample space.

We list some of the major theorems in probability theory and refer the reader to more detailed texts [1, 2].

$$\begin{array}{llcl}
\text{Univariate:} & \Pr(X) & = & P_X(x) \\
\text{Bivariate:} & \Pr(X,Y) & = & P_{XY}(x,y) \\
\text{Marginal:} & \Pr(X) & = & \sum_y P_{XY}(x,y) \\
\text{Independent:} & \Pr(X,Y) & = & P_X(x) \times P_Y(y) \\
\text{Conditional:} & \Pr(X|Y) & = & P_{XY}(x,y)/P_Y(y) \\
\text{Chain Rule:} & \Pr(X,Y,Z) & = & \Pr(X|Y,Z) \times \Pr(Y|Z) \times \Pr(Z)
\end{array}$$

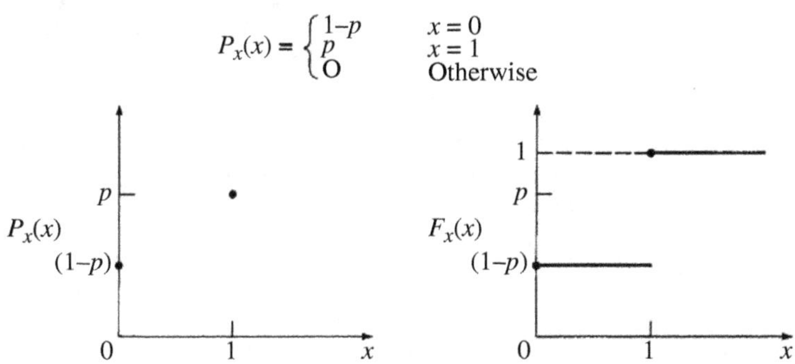

Figure A.1 Probability mass and distribution functions for coin tossing experiment.

For a random variable, we can define basic statistics in terms of the probability mass function. The *expected value* or *mean* of a random variable X, is given by

$$m_x = E\{X\}$$

and is considered the typical or representative value of a given set of data. For this reason, the mean is called a measure of central tendency. The degree to which numerical data tend to spread about the expected value is usually measured by the *variance* or equivalently, *auto-covariance* given by

$$R_{xx} = E\{(X - m_x)^2\}$$

The basic statistical measures are called *ensemble statistics* because they are measured across the ensemble ($i = 1, 2, \ldots$) of data values, that is, the expectation is always assumed over an ensemble of realizations. We summarize these statistics in terms of their mass function as:

$$\begin{array}{lll}
\text{Expected Value: } m_x = & E\{X\} & = \sum_i X_i\, P_X(x_i) \\
N\text{th-moment:} & E\{X^n\} & = \sum_i X_i^n P_X(x_i) \\
N\text{th-moment about mean:} & E\{(X - m_x)^n\} & = \sum_i (X_i - m_x)^n P_X(x_i) \\
\text{Mean Squared } (N = 2)\text{:} & E\{X^2\} & = \sum_i X_i^2 P_X(x_i) \\
\text{Variance: } R_{xx} = & E\left\{(X_i - m_x)^2\right\} & = \sum_i (X_i - m_x)^2 P_X(x_i) \\
\text{Covariance:} & R_{xy} & = E\left\{(X_i - m_x)(Y_j - m_y)\right\} \\
\text{Standard Deviation:} & \sigma_{xx} & = \sqrt{R_{xx}} \\
\text{Conditional Mean:} & E\{X|Y\} & = \sum_i X_i\, P(X_i|Y) \\
\text{Joint Conditional Mean:} & E\{X|Y, Z\} & = \sum_i X_i P(X_i|Y, Z) \\
\text{Conditional Variance:} & R_{x|y} & = E\{(X - E\{X|Y\})^2|Y\}
\end{array}$$

These basic statistics possess various properties that enable them to be useful for analyzing operations on random variables, some of the more important[1] are:

$$\begin{array}{lll}
\text{Linearity:} & E\{ax + b\} & = aE\{x\} + b = am_x + b \\
\text{Independence:} & E\{xy\} & = E\{x\}E\{y\} \\
\text{Variance:} & R_{xx}(ax + b) & = a^2 R_{xx} \\
\text{Covariance:} & & \\
\quad \text{Uncorrelated:} & E\{xy\} & = E\{x\}E\{y\} \quad \{R_{xy} = 0\} \\
\quad \text{Orthogonal:} & E\{xy\} & = 0
\end{array}$$

Note that the expected value operation implies that for stochastic processes these basic statistics are calculated *across* the ensemble. For example, if we want to calculate the mean of a process, that is,

$$m_x(t) = E\{X(t, \omega_i) = x_i(t)\}$$

1 Recall that independence states that the joint mass function can be factored, $\Pr(x, y) = \Pr(x) \times \Pr(y)$ which leads to these properties.

we simply take the values of $t = 0, 1, \ldots$ and calculate the mean for each value of time across ($i = 1, 2, \ldots$) the ensemble. Dealing with stochastic processes is similar to dealing with random variables except that we must account for the time indices (see Chapter 2 for more details).

Next let us define some concepts about the probabilistic information contained in a random variable. We define the (self) *information* contained in the occurrence of the random variable $X(\omega_i) = x_i$, as

$$I(x_i) = -\log_b P_X(x_i) \tag{A.7}$$

where b is the base of the logarithm which results in different units for information measures (base = 2 → bits) and the *entropy* or *average information* of $X(\omega_i)$ as

$$H(x_i) = -E\{I(x_i)\} = \sum_i P_X(x_i) \log_b P_X(x_i) \tag{A.8}$$

Consider the case where there is more than one random variable. Then we define the *joint* mass and distribution functions of an N-dimensional random variable as

$$P_X(x_1, \ldots, x_N), \qquad F_X(x_1, \ldots, x_N)$$

All of the basic statistical definitions remain as before, except that we replace the scalar with the joint functions. Clearly, if we think of a stochastic process as a sequence of ordered random variables, then we are dealing with joint probability functions, that is, a collection of time-indexed random variables. Suppose we have two random variables, x_1, and x_2 and we know that the latter has already assumed a particular value, then we can define the *conditional* probability mass function of x_1 given that $X(\omega_2) = x_2$ has occurred by:

$$\Pr(x_1 \mid x_2) := P_X(X(\omega_1) \mid X(\omega_2) = x_2) \tag{A.9}$$

and it can be shown from basic probabilistic axioms (see Papoulis [1]) that

$$\Pr(x_1 \mid x_2) = \frac{\Pr(x_1, x_2)}{\Pr(x_2)} \tag{A.10}$$

Note also that this expression can also be written as

$$\Pr(x_1, x_2) = \Pr(x_2 \mid x_1) \Pr(x_1) \tag{A.11}$$

Substituting this equation into Eq. (A.10) gives *Bayes' rule*, that is,

$$\Pr(x_1 \mid x_2) = \Pr(x_2 \mid x_1) \frac{\Pr(x_1)}{\Pr(x_2)} \tag{A.12}$$

If we use the definition of joint mass function and substitute it into the previous definitions, then we can obtain the *probabilistic chain rule* [1–3],

$$\Pr(x_1, \ldots, x_N) = \Pr(x_1 \mid x_2, \ldots, x_N) \Pr(x_2 \mid x_3, \ldots, x_N) \cdots \Pr(x_{N-1} \mid x_N) \Pr(x_N) \tag{A.13}$$

Along with these definitions follows the idea of conditional expectation, that is,

$$E\{x_i \mid x_j\} = \sum_i X_i \ \Pr(x_i \mid x_j) \tag{A.14}$$

With the conditional expectation defined, we list some of their basic properties:

1. $E_x\{X|Y\} = E\{X\}$, if X and Y are independent
2. $E\{X\} = E_y\{E\{X|Y\}\}$
3. $E_x\{g(y)X|Y\} = g(y)E\{X|Y\}$
4. $E_{xy}\{g(Y)X\} = E_y\{g(Y)E\{X|Y\}\}$
5. $E_x\{c|Y\} = c$
6. $E_x\{g(Y)|Y\} = g(Y)$
7. $E_{xy}\{cX + dY|Z\} = cE\{X|Z\} + dE\{Y|Z\}$

The concepts of information and entropy can also be extended to the case of more than one random variable. We define the *mutual information* between two random variables, x_i and x_j as

$$I(x_i; x_j) = \log_b \frac{P_X(x_i \mid x_j)}{P_X(x_i)} \tag{A.15}$$

and the *average mutual information* between $X(\omega_i)$ and $X(\omega_j)$ as

$$I(X_i; X_j) = E_{x_i x_j}\{I(x_i, x_j)\} = \sum_i \sum_j P_X(x_i, x_j) I(x_i, x_j) \tag{A.16}$$

which leads to the definition of *joint entropy* as

$$H(X_i; X_j) = -\sum_i \sum_j P_X(x_i, x_j) \log_b P_X(x_i, x_j) \tag{A.17}$$

This completes the section on probability theory, next let us consider an important multivariable distribution and its properties.

A.2 Gaussian Random Vectors

In this section, we consider the multivariable Gaussian distribution used heavily in this text to characterize Gaussian random vectors, that is, $\mathbf{z} \sim \mathcal{N}(\mathbf{m}_z, \mathbf{R}_{zz})$ where $\mathbf{z} \in \mathcal{R}^{N_z \times 1}$ and defined by

$$\Pr(\mathbf{z}) = (2\pi)^{-N_z/2} |\mathbf{R}_{zz}|^{-1/2} \exp\left(-\frac{1}{2}(\mathbf{z} - \mathbf{m}_z)' \mathbf{R}_{zz}^{-1} (\mathbf{z} - \mathbf{m}_z)\right) \tag{A.18}$$

where the vector mean and covariance are defined by

$$\mathbf{m}_z := E\{\mathbf{z}\} \quad \text{and} \quad \mathbf{R}_{zz} = \mathrm{Cov}(\mathbf{z}) := E\{(\mathbf{z} - \mathbf{m}_z)(\mathbf{z} - \mathbf{m}_z)'\}$$

Certain properties of the Gaussian vectors are useful such as:

- *Linear transformation.* Linear transformations of Gaussian variables are Gaussian, that is, if $\mathbf{z} \sim \mathcal{N}(\mathbf{m}_z, \mathbf{R}_{zz})$ and $\mathbf{y} = \mathbf{Az} + \mathbf{b}$, then

$$Y \sim \mathcal{N}(\mathbf{Am}_z + \mathbf{b}, \mathbf{AR}_{zz}\mathbf{A}') \tag{A.19}$$

- *Uncorrelated Gaussian vectors.* Uncorrelated Gaussian vectors are independent.
- *Sums of Gaussian variables.* Sums of independent Gaussian vectors yield Gaussian distributed vectors with mean and variance equal to the sums of the respective means and variances.
- *Conditional Gaussian vectors.* Conditional Gaussian vectors are Gaussian distributed, that is, if $\mathbf{x}$ and $\mathbf{y}$ are jointly Gaussian, with

$$\mathbf{m}_z = E\left\{\begin{matrix}\mathbf{x}\\ \mathbf{y}\end{matrix}\right\} = \begin{bmatrix}\mathbf{m}_x\\ \mathbf{m}_y\end{bmatrix}; \text{and } \mathbf{R}_{zz} = \text{Cov}(\mathbf{z}) = \begin{bmatrix}\mathbf{R}_{xx} & \mathbf{R}_{xy}\\ \mathbf{R}_{yx} & \mathbf{R}_{yy}\end{bmatrix}$$

  then the conditional distribution for $\mathbf{x}$ and $\mathbf{y}$ is also Gaussian with conditional mean and covariance given by

$$\mathbf{m}_{\mathbf{x}|\mathbf{y}} = \mathbf{m}_x + \mathbf{R}_{xy}\mathbf{R}_{yy}^{-1}(\mathbf{y} - \mathbf{m}_y)$$

$$\mathbf{R}_{\mathbf{x}|\mathbf{y}} = \mathbf{R}_{xx} - \mathbf{R}_{xy}\mathbf{R}_{yy}^{-1}\mathbf{R}_{yx}$$

  and the vectors $\mathbf{x} - E\{\mathbf{x}|\mathbf{y}\}$ and $\mathbf{y}$ are independent.
- *Gaussian conditional means.* Let $\mathbf{x}$, $\mathbf{y}$, and $\mathbf{z}$ be jointly distributed Gaussian random vectors and let $\mathbf{y}$ and $\mathbf{z}$ be *independent*, then

$$E\{\mathbf{x}|\mathbf{y}, \mathbf{z}\} = E\{\mathbf{x}|\mathbf{y}\} + E\{\mathbf{y}|\mathbf{z}\} - \mathbf{m}_x$$

A.3 Uncorrelated Transformation: Gaussian Random Vectors

Suppose, we have a Gaussian random vector, $\mathbf{x} \sim \mathcal{N}(\mathbf{m}_x, \mathbf{R}_{xx})$, and we would like to transform it to a normalized Gaussian random vector with the mean, $\mathbf{m}_x$, removed so that, $\mathbf{z} \sim \mathcal{N}(0, \mathbf{I})$. Assume that the mean has been removed ($\mathbf{z} \rightarrow \mathbf{z} - \mathbf{m}_x$), then there exists a nonsingular transformation, $\mathbf{T}$, such that $\mathbf{z} = \mathbf{Tx}$ and therefore

$$\mathbf{R}_{zz} = \text{Cov}(\mathbf{z}) = \text{Cov}\left((\mathbf{Tx})(\mathbf{Tx})'\right) = \mathbf{TR}_{xx}\mathbf{T}' = \mathbf{I} \tag{A.20}$$

Thus, we must find a transformation that satisfies the relation

$$\mathbf{R}_{zz} = \mathbf{I} = \mathbf{TR}_{xx}\mathbf{T}' \tag{A.21}$$

Since $\mathbf{R}_{xx}$ is a positive semi-definite, symmetric matrix it can always be factored into matrix square roots ($\mathbf{R} = \mathbf{UU}' = \mathbf{R}_{xx}^{1/2}\mathbf{R}_{xx}^{T/2}$) using a Cholesky decomposition [4]; therefore, Eq. (A.21) implies

$$\mathbf{R}_{zz} = \mathbf{I} = (\mathbf{TU})\,\mathbf{U}'\mathbf{T}' \tag{A.22}$$

or simply that

$$\mathbf{T} = \mathbf{U}^{-1} = \mathbf{R}_{xx}^{-1/2} \qquad \text{(inverse matrix square root)} \tag{A.23}$$

and therefore

$$\mathbf{R}_{zz} = \left(\mathbf{R}_{xx}^{-1/2}\mathbf{U}\right)\mathbf{U}'\mathbf{R}_{xx}^{-T/2} = \mathbf{R}_{xx}^{-1/2}\mathbf{R}_{xx}\mathbf{R}_{xx}^{-T/2} = \mathbf{I} \tag{A.24}$$

the desired result.

A.4 Toeplitz Correlation Matrices

An $sN \times N$ correlation matrix, $\mathbf{R}$, is a Hermitian Toeplitz matrix ($R^H = (R^*)'$) which has the following general properties [4]:

1. The set of distinct eigenvalues, $\{\lambda_i\}$ of a Hermitian matrix are *real*.
2. The set of eigenvectors, $\{\mathbf{e}_i\}$ of a Hermitian matrix corresponding to the distinct eigenvalues, $\{\lambda_i\}$, are *orthogonal* $\longrightarrow$ $\mathbf{e}_i^H\mathbf{e}_i = 0 \text{if}\, \lambda_i \neq \lambda_j$.
3. Any Hermitian matrix admits an eigen-decomposition such that:

$$R = E\Lambda E^H \text{ with } E \text{ unitary}(E^{-1} = E^H)$$

4. The inverse of a Hermitian matrix can be obtained for the eigen-decomposition as:

$$R^{-1} = \left(E\Lambda E^H\right)^{-1} = E^{-H}\Lambda^{-1}E^{-1} = E\Lambda^{-1}E^H = \sum_{i=1}^{N}\frac{1}{\lambda_i}\mathbf{e}_i\mathbf{e}_i^H$$

A.5 Important Processes

We define two important processes which will be used extensively to model random signals. The first is the *uniformly* distributed process which is specified by the mass function

$$p_X(x_i) = \frac{1}{b-a} \quad a < x_i < b \tag{A.25}$$

or simply

$$x \sim U(a, b)$$

where the random variable can assume any value within the specified interval. The corresponding mean and variance of the uniform random variable is given by

$$m_x = \frac{b+a}{2}, \qquad R_{xx} = \frac{(b-a)^2}{12}$$

Secondly, the *Gaussian* or *normal process* is defined by its probability mass function

$$p(X(t)) = \frac{1}{\sqrt{2\pi R_{xx}}} \exp\left\{ -\frac{1}{2}\frac{(X(t)-m_x)^2}{R_{xx}} \right\} \tag{4.2–26}$$

or simply

$$x \sim N(m_x, R_{xx})$$

where m_x and R_{xx} are the respective mean and variance completely characterizing the Gaussian process, $X(t)$. The *central limit theorem* makes Gaussian processes very important, since it states that the sum of a large number of independent random variables tend to be Gaussian, that is, for $\{x_i(t)\}$ independent then $y(t) \sim N(0,1)$ where

$$y(t) = \sum_{t=1}^{N} \frac{(x_i(t) - m_x(t))}{\sqrt{NR_{xx}(t)}}$$

Both of these distributions will become very important when we attempt to simulate stochastic processes on the computer. Other properties of the Gaussian process that are useful are:

1. *Linear transformation*. Linear transformations of Gaussian variables are Gaussian, that is, if $x \sim N(m_x, R_{xx})$ and $y = ax + b$, then

$$y \sim N(am_x + b, a^2R_{xx})$$

2. *Uncorrelated Gaussian variables*. Uncorrelated Gaussian variables are independent.
3. *Sums of Gaussian variables*. Sums of independent Gaussian variables yield a Gaussian distributed variable with mean and variance equal to the sums of the respective means and variances, that is,

$$x_i \sim N(m_x(i), R_{xx}(i)) \quad \text{and} \quad y = \sum_i k_i x(i)$$

Then

$$y \sim N\left(\sum_i k_i m_x(i), \sum_i k_i^2 R_{xx}(i)\right)$$

4. *Conditional Gaussian variables.* Conditional Gaussian variables are Gaussian distributed, that is, if x_i and x_j are jointly Gaussian, then

$$E\{x_i \mid x_j\} = m_{x_i} + R_{x_i x_j} R_{x_j x_j}^{-1} (x_j - m_{x_j})$$

and

$$R_{x_i|x_j} = R_{x_i x_i} - R_{x_i x_j} R_{x_j x_j}^{-1} R_{x_j x_i}$$

5. *Orthogonal errors.* The estimation error $\tilde{x}$ is orthogonal to the conditioning variable, that is, for $\tilde{x} = x - E\{x \mid y\}$, then

$$E\{y\tilde{x}\} = 0$$

6. *Fourth Gaussian moments.* The fourth moment of zero-mean Gaussian variables is given by

$$E\{x_1x_2x_3x_4\} = E\{x_1x_2\}E\{x_3x_4\} + E\{x_1x_3\}E\{x_2x_4\} + E\{x_1x_4\}E\{x_2x_3\}$$

This discussion completes the introductory concepts of probability and random variables which is extended to include stochastic processes in Chapter 2 and throughout the text.

References

1 Papoulis, A. (1965). *Probability, Random Variables and Stochastic Processes.* New York: McGraw-Hill.

2 Hogg, R. and Craig, A. (1970). *Introduction to Mathematical Statistics.* New York: MacMillan.

3 Jazwinski, A. (1970). *Stochastic Processes and Filtering Theory.* New York: Academic Press.

4 Bierman, G. (1977). *Factorization Methods for Discrete Sequential Estimation.* New York: Academic Press.

B

Projection Theory

B.1 Projections: Deterministic Spaces

Projection theory plays an important role in subspace identification primarily because subspaces are created by transforming or "projecting" a vector into a lower dimensional space—the subspace [1–3]. We are primarily interested in two projection operators: (i) orthogonal and (ii) oblique or parallel. The *orthogonal projection operator* "projects" • onto $\mathcal{Y}$ as $\mathcal{P}_{\bullet|\mathcal{Y}}$ or its complement $\mathcal{P}^{\perp}_{\bullet|\mathcal{Y}}$, while the *oblique projection operator* "projects" • onto $\mathcal{Y}$ "along" $\mathcal{Z}$ as $\mathcal{P}^{\|}_{\bullet|\mathcal{Y}\circ\mathcal{Z}}$. For instance, orthogonally projecting the vector $\mathbf{x}$, we have $\mathcal{P}_{\mathbf{x}|\mathcal{Y}}$ or $\mathcal{P}^{\perp}_{\mathbf{x}|\mathcal{Y}}$, while obliquely projecting $\mathbf{x}$ along $\mathcal{Z}$ is $\mathcal{P}_{\mathbf{x}|\mathcal{Y}\circ\mathcal{Z}}$.

Mathematically, the n-dimensional vector space $\mathcal{R}$ can be decomposed into a direct sum of subspaces such that

$$\mathcal{R} = \mathcal{R}_1 + \mathcal{R}_2 \tag{B.1}$$

then for the vector $\mathbf{r} \in \mathcal{R}^n$, we have

$$\mathbf{r} = \mathbf{r}_1 + \mathbf{r}_2 \quad \text{for} \quad \mathbf{r}_1 \in \mathcal{R}_1 \quad \text{and} \quad \mathbf{r}_2 \in \mathcal{R}_2 \tag{B.2}$$

with $\mathbf{r}_1$ defined as the projection of $\mathbf{r}$ onto $\mathcal{R}_1$, while $\mathbf{r}_2$ defined as the projection of $\mathbf{r}$ onto $\mathcal{R}_2$.

When a particular projection is termed parallel (||) or "oblique," the corresponding *projection operator* that transforms $\mathbf{r}$ onto $\mathcal{R}_1$ along $\mathcal{R}_2$ is defined as $\mathcal{P}^{\|}_{\mathbf{r}|\mathcal{R}_1\circ\mathcal{R}_2}$, then

$$\mathbf{r}_1 = \mathcal{P}^{\|}_{\mathbf{r}|\mathcal{R}_1\circ\mathcal{R}_2} \quad \text{or} \quad \mathbf{r}_2 = \mathcal{P}^{\|}_{\mathbf{r}|\mathcal{R}_2\circ\mathcal{R}_1} \tag{B.3}$$

with $\mathbf{r}_1$ defined as the oblique projection of $\mathbf{r}$ onto $\mathcal{R}_1$ along $\mathcal{R}_2$, while $\mathbf{r}_2$ defined as the projection of $\mathbf{r}$ onto $\mathcal{R}_2$ along $\mathcal{R}_1$r.

When the intersection $\mathcal{R}_1 \cap \mathcal{R}_2 = 0$, then the direct sum follows as

$$\mathcal{P}_{\mathbf{r}|\mathcal{R}_1\cup\mathcal{R}_1} = \mathcal{P}^{\|}_{\mathbf{r}|\mathcal{R}_1\circ\mathcal{R}_2} + \mathcal{P}^{\|}_{\mathbf{r}|\mathcal{R}_2\circ\mathcal{R}_1} \tag{B.4}$$

Signal Processing: An Applied Decomposition Approach, First Edition. James Vincent Candy.

Companion website: www.wiley.com/go/CandySignProc

The *matrix projection operation* (Q) is $P \times P'$ $(Q = PP')$ a linear operator in $\mathcal{R}^n$ satisfying the properties of linearity and homogeneity. It can be expressed as an *idempotent* matrix such that $P^2 = P$ with the null space of $\mathcal{R}^n$ given by $\mathcal{N}(P) = \mathcal{R}(I_n - P)$, Here $P \in \mathcal{R}^{nxn}$ is the projection matrix onto $\mathcal{R}(P)$ along $\mathcal{N}(P)$ iff P is idempotent. That is, the matrix P is an orthogonal projection *if an only if* it is idempotent $(P^2 = P)$ and symmetric $(P^T = P)$.

Suppose that $\mathbf{r}_1^T\mathbf{r}_2 = 0$, then $\mathbf{r}_1 \perp \mathbf{r}_2$ with the vectors defined as "mutually orthogonal." If $\mathbf{r}_1^T\mathbf{r}_2 = 0 \forall \mathbf{r}_1 \in \mathcal{R}_1$, then $\mathbf{r}_2$ is orthogonal to $\mathcal{R}_1$ expressed by $\mathbf{r}_2 \perp \mathcal{R}_1$ spanning $\mathcal{R}_2$ and called the *orthogonal complement* of R^n. If the subspaces R_1 and R_2 are orthogonal $R_1 \perp R_2$, then their sum is called the *direct sum* $R_1 \oplus R_2$ such that

$$R^n = R_1 \oplus R_2 = R_1 \oplus R_1^{\perp} \text{ with } \mathbf{r}_1 \in \mathbf{R}_1 \text{ and } \mathbf{r}_2 \in \mathbf{R}_1^{\perp} \tag{B.5}$$

When $\mathbf{r}_1$ and $\mathbf{r}_2$ are orthogonal satisfying the direct sum decomposition (above), then $\mathbf{r}_1$ is the *orthogonal projection* of $\mathbf{r}$ onto R_1

B.2 Projections: Random Spaces

If we define the projection operator over a random space, then the random vector $(\mathbf{w})$ resides in a *Hilbert space* (Ψ) with finite second-order moments, that is [2],

$$\Psi = E\{\mathbf{w}|E\{\|\mathbf{w}\|^2\} < \infty\} \qquad \text{where} \qquad \mathbf{E}\{\|\mathbf{w}\|^2\} = \sum_{\mathbf{i}} \mathbf{E}\{\mathbf{w}_\mathbf{i}^2\}$$

The projection operator $\mathcal{P}$ is now defined in terms of the expectation such that the projection of $\mathbf{w} \in \mathcal{W}$ onto $\mathbf{v} \in \mathcal{V}$ is defined by

$$\mathcal{P}_{\mathbf{w}|\mathcal{V}} := \mathbf{E}\{\mathbf{w}|\mathcal{V}\} = \mathbf{E}\{\mathbf{w}\mathbf{v}'\}\mathbf{E}\{\mathbf{v}\mathbf{v}'\}^{\#}\mathbf{v} = \text{cov}(\mathbf{w}, \mathbf{v})\text{cov}\,(\mathbf{v}, \mathbf{v})^{\#}\mathbf{v}$$

for # the pseudo-inverse operator given by: $V'\left(VV'\right)^{-1}$.

The corresponding projection onto the orthogonal complement $(\mathcal{V}^{\perp})$ is defined by

$$\mathcal{P}_{\mathbf{w}|\mathcal{V}}^{\perp} := \mathbf{w} - \mathbf{E}\{\mathbf{w}|\mathcal{V}\} = I - \mathbf{E}\{\mathbf{w}\mathbf{v}'\}\mathbf{E}\{\mathbf{v}\mathbf{v}'\}^{\#}\mathbf{v} = \mathbf{I} - \text{cov}(\mathbf{w}, \mathbf{v})\text{cov}\,(\mathbf{v}, \mathbf{v})^{\#}\mathbf{v}$$

Therefore, the space can be decomposed into the projections as

$$\Psi = \mathcal{P}_{\mathbf{w}|\mathcal{V}} + \mathcal{P}_{\mathbf{w}|\mathcal{V}}^{\perp}$$

The oblique (parallel) projection operators follow as well for a random space, that is, the *oblique projection* operator of $\mathbf{w}$ onto $\mathcal{V}$ along $\mathcal{Z}$

$$\mathcal{P}_{\mathbf{w}|\mathcal{V}\circ\mathcal{Z}}^{\|} := E\{\mathcal{W}|\mathcal{V}\}$$

and therefore, we have the operator decomposition for [2]

$$E\{\mathbf{w}|\mathcal{W} \cup \mathcal{V}\} = \mathbf{E}\{\mathbf{w}|\mathcal{W}\circ\mathcal{V}\} + \mathbf{E}\{\mathbf{w}|\mathcal{V}\circ\mathcal{W}\}$$

for $E\{\mathbf{w}|\mathcal{W}\circ\mathcal{V}\}$ is the oblique (parallel) projection of $\mathbf{w}$ onto $\mathcal{W}$ along $\mathcal{V}$ and $E\{\mathbf{w}|\mathcal{V}\circ\mathcal{W}\}$ is the oblique projection of $\mathbf{w}$ onto $\mathcal{V}$ along $\mathcal{W}$.

B.3 Projection: Operators

Let $A \in \mathcal{R}^{m\times n}$ then the *row space* of A is spanned by the *row vectors* $\mathbf{a}_i' \in \mathcal{R}^{1\times n}$, while the *column space* of A is spanned by the *column vectors* $\mathbf{a}_i \in \mathcal{R}^{m\times 1}$. That is,

$$A = \begin{bmatrix} \mathbf{a}_1' \\ \mathbf{a}_2' \\ \vdots \\ \mathbf{a}_m' \end{bmatrix}; \mathbf{a}_i' \in \mathcal{R}^{1\times n};\ i = 1, \ldots, m \qquad \text{[Row Vectors]}$$

and

$$A = \left[\mathbf{a}_1|\mathbf{a}_2|\cdots|\mathbf{a}_n\right]; \mathbf{a}_j \in \mathcal{R}^{m\times 1};\ j = 1, \ldots, n \qquad \text{[Column Vectors]}$$

Any vector $\mathbf{y}$ in the row space of $A \in \mathcal{R}^{m\times n}$ is defined by

$$\text{Row}(A) := \left\{ \mathbf{y} \in \mathcal{R}^{n\times 1} | \mathbf{y} = A'\mathbf{x} = \sum_{i=1}^{m} \mathbf{a}_i'\mathbf{x}_i;\ \mathbf{x} \in \mathcal{R}^{m\times 1} \right\}$$

and similarly for the column space

$$\text{Column}(A) := \left\{ \mathbf{y} \in \mathcal{R}^{m\times 1} | \mathbf{y} = A\mathbf{x} = \sum_{j=1}^{n} \mathbf{a}_j\mathbf{x}_j;\ \mathbf{x} \in \mathcal{R}^{n\times 1} \right\}$$

Therefore, the set of vectors $\{\mathbf{a}_i'\}$ and $\{\mathbf{a}_i\}$ provide a set of basis vectors spanning the respective row or column spaces of A and of course, Row(A) = Column(A').

B.3.1 Orthogonal (Perpendicular) Projections

Projections or more precisely, projection operators are well-known from operations on vectors (e.g. the Gram–Schmidt orthogonalization procedure). These operations, evolving from solutions to least-squares error minimization problems, can be interpreted in terms of matrix operations of their row and column spaces defined above. The orthogonal *projection operator* ($\mathcal{P}_\bullet$) is defined in terms of the row space of a matrix B by

$$\mathcal{P}_B = B'\left(BB'\right)^{-1}B \qquad \text{[Orthogonal Projection Operator]} \quad \text{(B.6)}$$

with its corresponding orthogonal complement operator as

$$\mathcal{P}_B^{\perp} = \left(I - \mathcal{P}_B\right) = I - B'\left(BB'\right)^{-1}B \quad \text{[Orthogonal Complement Operator]} \quad \text{(B.7)}$$

These operators when applied to a matrix "projects" the *row space* of a matrix A onto (|) the *row space* of B such that $\mathcal{P}_{A|B}$ given by

$$\mathcal{P}_{A|B} = A \times \mathcal{P}_B = A \times \underbrace{B'\left(BB'\right)^{-1}B}_{\mathcal{P}_B} \qquad \text{[Orthogonal Projection]} \quad \text{(B.8)}$$

and similarly, onto the orthogonal complement of B such that

$$\mathcal{P}^{\perp}_{A|B} = A \times \mathcal{P}^{\perp}_{B} = A \times \underbrace{\left(I - B'\left(BB'\right)^{-1}B\right)}_{\mathcal{P}^{\perp}_{B}} \quad \text{[Orthogonal Complement Projection]} \tag{B.9}$$

Thus, a matrix can be decomposed in terms of its inherent orthogonal projections as a direct sum as shown in Figure B.1a, that is,

$$A = \mathcal{P}_{A|B} + \mathcal{P}^{\perp}_{A|B} = A \times \mathcal{P}_{B} + A \times \mathcal{P}^{\perp}_{B} \quad \text{[Orthogonal Decomposition]} \tag{B.10}$$

Similar relations exist if the *row space* of A is projected onto the *column space* of B, then it follows that the projection operator, in this case, is defined by

$$\mathcal{P}_{B} = B\left(B'B\right)^{-1}B' \quad \text{[Orthogonal Projection Operator]} \tag{B.11}$$

along with the corresponding projection given by

$$\mathcal{P}_{A|B} = A \times \mathcal{P}_{B} = A \times \underbrace{B\left(B'B\right)^{-1}B'}_{\mathcal{P}_{B}} \quad \text{[Orthogonal Projection]} \tag{B.12}$$

Numerically, these matrix operations are facilitated by the LQ-decomposition (see Appendix C) as

$$\begin{bmatrix} B \\ -- \\ A \end{bmatrix} = \begin{bmatrix} L_{11} & | & O \\ -- & | & -- \\ L_{21} & | & L_{22} \end{bmatrix} \begin{bmatrix} Q_1' \\ -- \\ Q_2' \end{bmatrix} \tag{B.13}$$

where the *projections* are given by

$$\mathcal{P}_{A|B} = A \times \mathcal{P}_{B} = L_{21} \times Q_1'$$

$$\mathcal{P}^{\perp}_{A|B} = A \times \mathcal{P}^{\perp}_{B} = L_{22} \times Q_2' \tag{B.14}$$

B.3.2 Oblique (Parallel) Projections

The *oblique projection* (|) of the Row(A) onto Row(C) along ($\circ$) the Row(B) is defined by $\mathcal{P}_{A|C\circ B}$. This projection operation can be expressed as the *rows* of two nonorthogonal matrices (B and C) and their corresponding orthogonal complement. Symbolically, the projection of the rows of A onto the *joint row space* of B and C ($B \vee C$) can be expressed in terms of two oblique projections and one of an orthogonal complement. This projection is illustrated in Figure B.1b, where we see that the Row(A) is first orthogonally projected onto the joint row space of B and C (orthogonal complement) and then decomposed along B and C

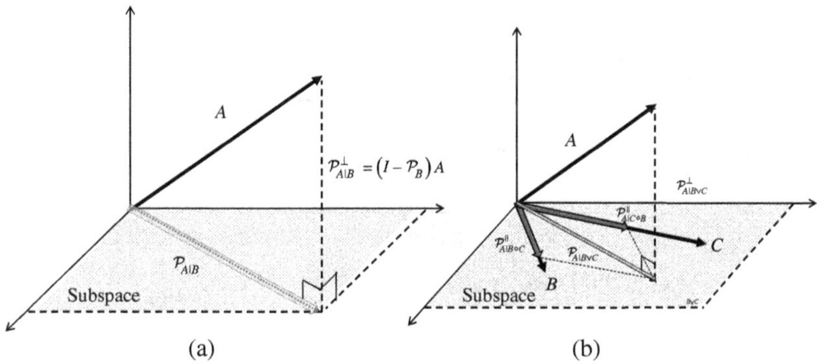

Figure B.1 Orthogonal and oblique projections: (a) orthogonal projection of A onto B ($\mathcal{P}_{A|B}$) and its complement ($\mathcal{P}^{\perp}_{A|B}$). (b) Oblique projection of A onto C along B ($\mathcal{P}_{A|C\circ B}$).

individually enabling the extraction of the oblique projection of the Row(A) onto Row(C) along ($\circ$) the Row(B).

$$\mathcal{P}_{A|B\vee C} = \mathcal{P}^{\|}_{A|B\circ C} + \mathcal{P}^{\|}_{A|C\circ B} + \mathcal{P}^{\perp}_{A|B\vee C} \tag{B.15}$$

Pragmatically, this relation can be expressed into a more "operational" form that is applied in Section 6.5 to derive the *N4SID* algorithm [1, 4], that is,

$$\mathcal{P}^{\|}_{A|C\circ B} = \left[A \times \mathcal{P}^{\perp}_{B}\right]\left[C \times \mathcal{P}^{\perp}_{B}\right]^{\#} \times C \qquad \text{[Oblique Projection]} \tag{B.16}$$

for # the *pseudo-inverse* operation (see Appendix C). Oblique projections also have the following properties that are useful in derivations:

$$\begin{aligned} \mathcal{P}^{\|}_{B|C\circ B} &= 0 \\ \mathcal{P}^{\|}_{C|C\circ B} &= C \end{aligned} \tag{B.17}$$

Analogously, the orthogonal complement projection can be expressed in terms of the oblique projection operator as

$$\mathcal{P}^{\perp}_{A|B} = \mathcal{P}^{\|}_{A|C\circ B} \times \mathcal{P}^{\perp}_{B} \tag{B.18}$$

Numerically, the oblique projection can be computed again using the LQ decomposition (see Appendix C)

$$\begin{bmatrix} B \\ -- \\ C \\ -- \\ A \end{bmatrix} = \begin{bmatrix} L_{11} & | & 0 & | & 0 \\ -- & | & -- & | & -- \\ L_{21} & | & L_{22} & | & 0 \\ -- & | & -- & | & -- \\ L_{31} & | & L_{32} & | & L_{33} \end{bmatrix} \begin{bmatrix} Q'_1 \\ -- \\ Q'_2 \\ -- \\ Q'_3 \end{bmatrix} \tag{B.19}$$

where the *oblique projection* of the Row(A) onto Row(C) along the Row(B) is given by

$$\mathcal{P}^{\|}_{A|C\circ B} = L_{32} \times L_{22}^{-1} \times C = L_{32} \times L_{22}^{-1} \times \left(L_{21}Q'_1 + L_{22}Q'_2\right) \tag{B.20}$$

Table B.1 Matrix projections.

Projection: operators, projections, numerics			
Operation	**Operator**	**Projection**	**Numerical (LQ decomposition)**
Orthogonal	**–**	**–**	**–**
$\mathcal{P}_{A\|B}$	$\mathcal{P}_{\bullet\|B} = B'(BB')^{-1}B$	$A \times B'(BB')^{-1}B$	$A \times (L_{21} \times Q_1')$
$\mathcal{P}_{A\|B}^{\perp}$	$\mathcal{P}_{\bullet\|B}^{\perp} = (I - B'(BB')^{-1}B)$	$A \times (I - B'(BB')^{-1}B)$	$A \times (L_{22} \times Q_2')$
Oblique	—	—	—
$\mathcal{P}_{A\|C\circ B}$	$\mathcal{P}_{\bullet\|C\circ B}$	$L_{32}L_{22}^{-1} \times C$	$L_{32} \times L_{22}^{-1}(L_{21}Q_1' + L_{22}Q_2')$
Relations	—	—	—
$\mathcal{P}_{\bullet\|B}$	—	—	Q_1Q_1'
$\mathcal{P}_{\bullet\|B}^{\perp}$	—	—	Q_2Q_2'
$\mathcal{P}_{A\|C\circ B}$	$\mathcal{P}_{\bullet\|C\circ B}$	$[A \times \mathcal{P}_B^{\perp}]\,[C \times \mathcal{P}_B^{\perp}]^{\#} \times C$	—
$\mathcal{P}_{A\|B}^{\perp}$	$\mathcal{P}_{\bullet\|B}^{\perp}$	$\mathcal{P}_{A\|C\circ B}^{\|\|} \times \mathcal{P}_B^{\perp}$	—

Some handy relationships between oblique and orthogonal projection matrices are given in Table B.1: (see [1–3, 5] for more details)[1]:

References

1 van Overschee, P. and De Moor, B. (1996). *Subspace Identification for Linear Systems: Theory, Implementation, Applications*. Boston, MA: Kluwer Academic Publishers.

2 Katayama, T. (2005). *Subspace Methods for System Identification*. London: Springer.

3 Verhaegen, M. and Verdult, V. (2007). *Filtering and System Identification: A Least-Squares Approach*. Cambridge: Cambridge University Press.

4 Behrens, R. and Scharf, L. (1994). Signal processing applications of oblique projection operators. *IEEE Trans. Signal Process.* 42 (6): 1413–1424.

5 Tangirala, A. (2015). *Principles of System Identification: Theory and Practice*. Boca Raton, FL: CRC Press.

1 Note the difference between projection operators and actual projections.

C

Matrix Decompositions

C.1 Singular Value Decomposition

The singular value decomposition (*SVD*) is an integral part of subspace identification methods [1–4]. It is characterized by an orthogonal decomposition of a rectangular $n \times m$ matrix $Z \in \mathcal{R}^{n\times m}$

$$Z = U \times \Sigma \times V' \tag{C.1}$$

where the matrices $U \in \mathcal{R}^{n\times n}$ and $V \in \mathcal{R}^{m\times m}$ are *orthogonal* such that $U'U = I_n$, $V'V = I_m$ along with the rectangular diagonal matrix, $\Sigma = \text{diag}[\sigma_1 \cdots \sigma_N]$ with $N < \min(n, m)$. The set of ordered *singular values* is given by $\{\sigma_i\};\ i = 1, \dots, N$ with the property that $\sigma_1 > \sigma_2 > \cdots \sigma_N$. The singular values of Z and the respective column and row vectors of U and V, u_i and v_i are the ith *left/right singular vectors* such that [1]

$$\begin{aligned} Zv_i &= \sigma_i u_i \\ A'u_i &= \sigma_i v_i \end{aligned}$$

which follows directly from the *SVD*, since $ZV = \Sigma U$ and $Z'U = \Sigma' V$.

Also pre-multiplying Eq. (C.1) by U' and post-multiplying by V and applying the orthogonality properties gives the relation

$$U' \times Z \times V = \Sigma = \text{diag}[\sigma_1 \cdots \sigma_N] \in \mathcal{R}^{n\times m} \quad \text{for} \quad N = \min(n, m) \tag{C.2}$$

It also follows that for a rank r matrix, the *SVD* can be expressed as

$$Z = U\Sigma V' = \left[U_r \mid U_{n-r}\right] \begin{bmatrix} \Sigma_r & \mid & 0 \\ - & - & - \\ 0 & \mid & 0 \end{bmatrix} \begin{bmatrix} V_r' \\ -- \\ V_{m-r}' \end{bmatrix} = U_r \Sigma_r V_r' \tag{C.3}$$

Signal Processing: An Applied Decomposition Approach, First Edition. James Vincent Candy.

Companion website: www.wiley.com/go/CandySignProc

Numerical errors occurring in this approximation are quantified in terms of the singular values as:

$$E_{abs} = \sqrt{\sum_{i=r+1}^{N} \sigma_i^2} \qquad \text{[Absolute Error]}$$

$$E_{rel} = \sqrt{\frac{\sum_{i=r+1}^{N} \sigma_i^2}{\sum_{i=1}^{N} \sigma_i^2}} \qquad \text{[Relative Error]} \tag{C.4}$$

There are various properties that the *SVD* of a matrix provides:

- The set of singular values of Z is equal to the positive square roots of the eigenvalues of $Z'Z$: $\sigma_i = +\sqrt{\lambda_i}$.
- For Z of rank r, then the singular values satisfy

$$\sigma_1 > \cdots \geq \sigma_r > \sigma_{r+1} = \cdots = \sigma_N = 0$$

- The rank condition gives

$$\begin{aligned} \rho(Z) &= r \\ null(Z) &= \text{span}\{v_{r+1}, \ldots, v_m\} \\ range(Z) &= \text{span}\{u_1, \ldots, u_r\} \end{aligned} \tag{C.5}$$

- The matrix decomposition can be expressed directly in terms of the singular values as

$$Z = U_r\, \Sigma_r\, V_r' = \sum_{i=1}^{r} \sigma_i\, u_i\, v_i' \tag{C.6}$$

- The *norms* are also expressed in terms of the singular values as:

$$\begin{aligned} \| Z\|_{\mathcal{F}} &= \sum_{i=1}^{N} \sigma_i \quad N < \min(n, m) \qquad \text{[Frobenious Norm]} \\ \| Z\|_2 &= \sigma_1 \qquad \text{[2-Norm]} \end{aligned} \tag{C.7}$$

- The *pseudo-inverse* of a matrix $\left(Z^{\#} = (Z'Z)^{-1}Z' \text{ (row space) or } Z^{\#} = Z'(ZZ')^{-1} \text{ (column space)}\right)$ is easily determined by

$$Z^{\#} = \left(U\, \Sigma\, V'\right)^{\#} = V\, \Sigma^{-1}\, U' \tag{C.8}$$

Associated with the *SVD* of Eq. (C.3) is the idea of *projections*, that is, projecting a vector onto a line or plane [4]. In terms of the *SVD*, we have the following *orthogonal projection matrices* [5]:

$$\begin{aligned} \mathcal{P}_{Z|U} &= U_1 \times U_1' = Z \times Z^{\#} \\ \mathcal{P}_{Z|U}^{\perp} &= U_2 \times U_2' = I - Z \times Z^{\#} \end{aligned}$$

$$\mathcal{P}_{Z|V} = V_1 \times V_1' = Z^{\#} \times Z$$
$$\mathcal{P}_{Z|V}^{\perp} = V_2 \times V_2' = I - Z^{\#} \times Z \tag{C.9}$$

Finally, the *SVD* simultaneously diagonalizes the projection, Gramian matrix, and pseudo-inverse [2]

$$\begin{aligned} \mathcal{P}_{Z|U} &= U \times \begin{bmatrix} I & 0 \\ 0 & 0 \end{bmatrix} \times U' && \text{[Projection]} \\ \mathcal{G}_Z &:= Z' \times Z = V \times \Sigma^2 \times V' && \text{[Gramian]} \\ Z^{\#} &= V \times \Sigma^{-1} \times U' && \text{[Pseudo-Inverse]} \end{aligned} \tag{C.10}$$

C.2 QR Decomposition

The *QR* decomposition of a matrix is the matrix implementation of the well-known Gram–Schmidt vector orthogonalization procedure [4]. It is defined by the product of an orthogonal matrix $Q \in \mathcal{R}^{n\times m}$ with $m \geq n$ and an upper triangular matrix $R_q \in \mathcal{R}^{m\times m}$, that is,

$$R = Q \times R_q \tag{C.11}$$

where $Q^T \times Q = I_n$ such that $Q^T R = R_q$ with Q^T is the orthogonal matrix transforming $R \to R_q$. It is numerically implemented using the *Householder transform*, $T = I - 2\mathbf{u}\mathbf{u}^T$ with the vector $\mathbf{u}$ defined by

$$\mathbf{u} = \pm \frac{v_1 - v_2}{\| v_1 - v_2 \|} \qquad \| \mathbf{u} \| = 1 \tag{C.12}$$

C.3 LQ Decomposition

The LQ decomposition is the (*dual* of the *QR* decomposition: $L = R'$, $Q = Q'$) where

$$Z = L \times Q \quad \text{with } L \text{ lower block triangular; and } Q \quad \text{orthogonal} \tag{C.13}$$

$$L = \begin{bmatrix} L_{11} & | & 0 \\ - & - & - \\ L_{21} & | & L_{22} \end{bmatrix} \quad \text{and} \quad Q = \begin{bmatrix} Q_1' \\ -- \\ Q_2' \end{bmatrix} \quad \text{for} \quad Q'Q = I$$

The LQ decomposition is a numerical (matrix) version of the *Gram–Schmidt orthogonalization* procedure [1, 4]. It contains the set of orthonormal basis vectors in Q ($Q'Q = I$) with $\mathbf{q}_i$ and L contains the coefficients of linear dependence or equivalently the relationship between the basis vectors of Z and those of Q.

In fact from this viewpoint, we can think of partitioning $Q = \left[Q_Z \mid Q_Z^\perp\right]$ such that a rank-deficient data matrix Z can (hypothetically) be decomposed as

$$Z = LQ = \left[L_Z \mid 0\right]\begin{bmatrix} Q_Z \\ -- \\ Q_Z^\perp \end{bmatrix} = L_Z \times Q_Z$$

The *projection matrix* associated with the decomposition is specified by [4]

$$\mathcal{P} = Q \times Q' \qquad \text{[Projection Matrix]}$$

References

1 Golub, G. and Van Loan, C. (1989). *Matrix Computations*, 2e. Baltimore, MD: Johns Hopkins University Press.
2 Scharf, L. (1990). *Statistical Signal Processing: Detection, Estimation, and Time Series Analysis*. Reading, MA: Addison-Wesley.
3 Katayama, T. (2005). *Subspace Methods for System Identification*. London: Springer.
4 Bretscher, O. (2009). *Linear Algebra with Applications*. Upper Saddle River, NJ: Prentice-Hall International.
5 Klema, V. and Laub, A. (1980). The singular value decomposition: its computation and some applications. *IEEE Trans. Autom. Control* AC-25 (2): 164–176.

Index

Signal Processing: An Applied Decomposition Approach, First Edition. James Vincent Candy.

Companion website: www.wiley.com/go/CandySignProc

e

p

z

www.ingramcontent.com/pod-product-compliance
Lightning Source LLC
Chambersburg PA
CBHW051609300825
31485CB00008B/15
9781394207442